Biophysical Chemistry of Proteins

Engelbert Buxbaum

Biophysical Chemistry of Proteins

An Introduction to Laboratory Methods

Engelbert Buxbaum
Ross University School of Medicine
P.O. Box 266
Biochemistry
Portsmouth Campus
266 Roseau
Dominica
engelbert_buxbaum@web.de

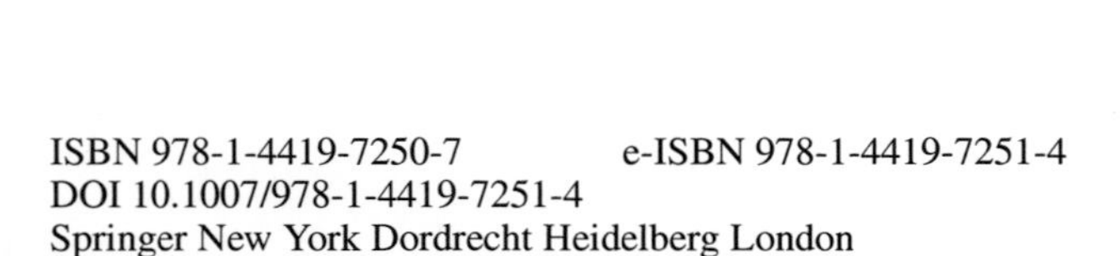

ISBN 978-1-4419-7250-7 e-ISBN 978-1-4419-7251-4
DOI 10.1007/978-1-4419-7251-4
Springer New York Dordrecht Heidelberg London

Printed on acid-free paper

Springer is part of Springer Science+Business Media (www.springer.com)

Preface

During undergraduate courses in biochemistry you learned *what* proteins do as enzymes, receptors, hormones, motors or structural components. The more interesting question, *how* proteins can achieve all these functions, is usually asked only in graduate courses, and in many cases it is a topic of ongoing research.

Here I want to present an overview of the methods used in such research projects, their possible applications, and their limitations. I have limited the presentation to a level where a general background in chemistry, physics, and mathematics is sufficient to follow the discussion. Quantum mechanics, where required, is treated in a purely qualitative manner. A good understanding of protein structure and enzymology is required, but these topics I have covered in a separate volume [44].

Apart from graduate training in protein science this book should also be useful as a reference for people who work with proteins.

After studying this book you should be able to collaborate with workers who have the required instruments and use these methods routinely. You should also be able to understand papers which make use of such methods. However, before embarking on independent research using these methods you are directed to the literature cited for a more in-depth, more quantitative coverage.

This book focuses on the biophysical chemistry of proteins. The use of nucleic acid-based methods [360], although in many cases very relevant and informative, is outside the scope of this text. Also only hinted at are modern approaches to computational biochemistry [20, 180, 231]. In the end, the models derived from such techniques have to be verified by experiments. If this book stimulates such studies, it has served its purpose.

Acknowledgements

I wish to thank all my students, friends, and colleagues who have given me their support and suggestions for this text, and who have gone through the arduous task of proof-reading. All remaining errors are, of course, mine. Please report any errors found and any suggestions for improvement to me (mailto://engelbert_buxbaum@web.de).

A big "thank you" goes to all those who have made software freely available, or who maintain repositories of information on the internet. Without your dedication, this book would not have been possible.

Portsmouth, Dominica *Engelbert Buxbaum*

Contents

Part VI Enzyme Kinetics

Part VII Protein–Ligand Interactions

Part I
Analytical Techniques

Chapter 1
Microscopy

The microscope is without question the most important instrument available to the biologist. The physiological function of proteins cannot be addressed without taking their localisation in a living cell and their interaction with other proteins into account. To answer these questions the microscope is an invaluable tool. http://micro.magnet.fsu.edu/primer/ covers microscopic techniques in much more detail than possible here.

1.1 Optical Foundations of Microscopy

Objects in cell biology range from an ostrich egg (20–30 cm) down to subcellular organelles. An impression of this range is given by Fig. 1.1. Although some objects can be seen with the naked eye, magnification is required for others.

1.1.1 KÖHLER-*Illumination*

The microscope is a system of lenses which create images of the illumination and object planes (see Fig. 1.2). In the illumination planes an image of the light source is created; in the object planes an image of the object. As you can see from the figure, light from each point of the light source is spread evenly across the entire object planes, passing the object in parallel beams from every azimuth. This creates a homogeneous, bright illumination; thus only a low power light source (low voltage halogen lamp of about 20 W) is required. The apertures have to be adjusted in position and diameter to avoid contrast reduction by stray light and to achieve maximum resolution.

1.1.1.1 Critical Illumination

Simpler microscopes with **critical** (NELSON) **illumination** are used in lab classes or clinical laboratories. In this case the image of the light source (the frosted glass

E. Buxbaum, *Biophysical Chemistry of Proteins: An Introduction to Laboratory Methods*, DOI 10.1007/978-1-4419-7251-4_1,

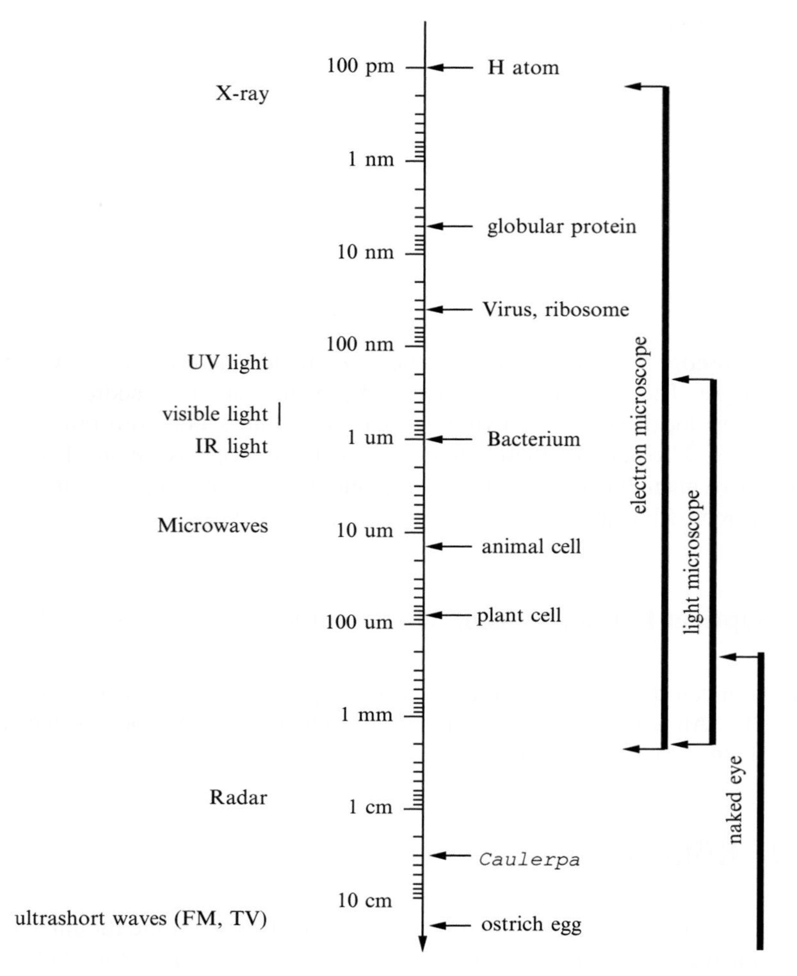

Fig. 1.1 Size of objects in cell biology, compared to the wavelengths of different electromagnetic waves

of a light bulb) is projected into the object plane by the condenser. Such microscopes are easier to use since condenser height and field diaphragm need not be readjusted each time the objective is changed; they are also considerably cheaper. With modern, multilayer-coated lens systems the increase in stray light and reduction in resolution is not relevant for $N_a \leq 0.5$. For micro-photography, however, this system is not suitable, since "hard" (high contrast) films amplify any inhomogeneity in the illumination.

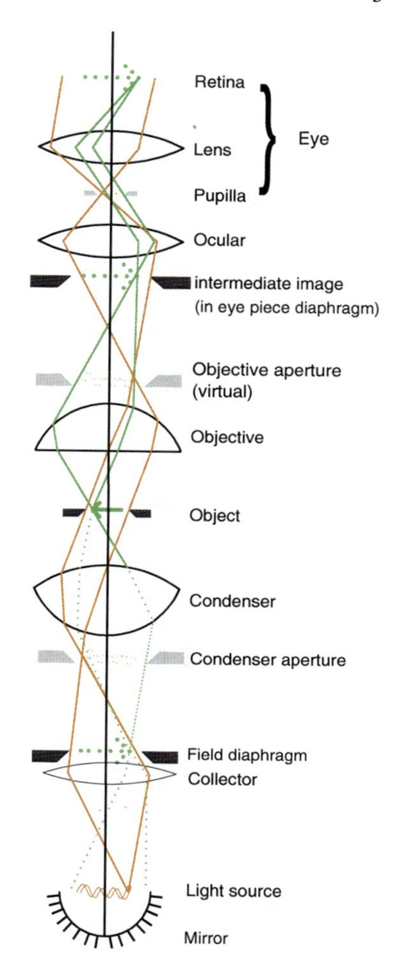

Fig. 1.2 Schematic diagram of a microscope with the illumination system originally introduced by AUGUST KÖHLER at Zeiss. The lens systems create images of both the object and the light source. The illumination apertures adapt the numerical aperture of the illumination system to that of the objective. The apertures in the object plane adjust the field of view. Thus the diameter of the light beam and its opening angle can be adjusted independently. This reduces the stray light inside the microscope. Careful adjustment of the size and position of the apertures is required to take full benefit of the microscope

1.1.2 The Role of Diffraction

Two factors influence the power of a microscope: resolution and contrast. **Contrast** can nearly always be increased by staining or by optical methods (see later), the microscope only needs to keep the level of stray light down. The **resolution** (the minimal distance between two objects that still allows them to be seen as separate), however, is subject to tight physical limitations.

Responsible for the image formation is the process of scattering the light waves on the object. The scattered light creates a primary interference pattern for each point of the condenser aperture in the objective aperture of the microscope (see Fig. 1.3 on p. 6 for a brief discussion of diffraction). The distance between the diffraction maxima depends on the **grid constant** (distance between structures) of the object.

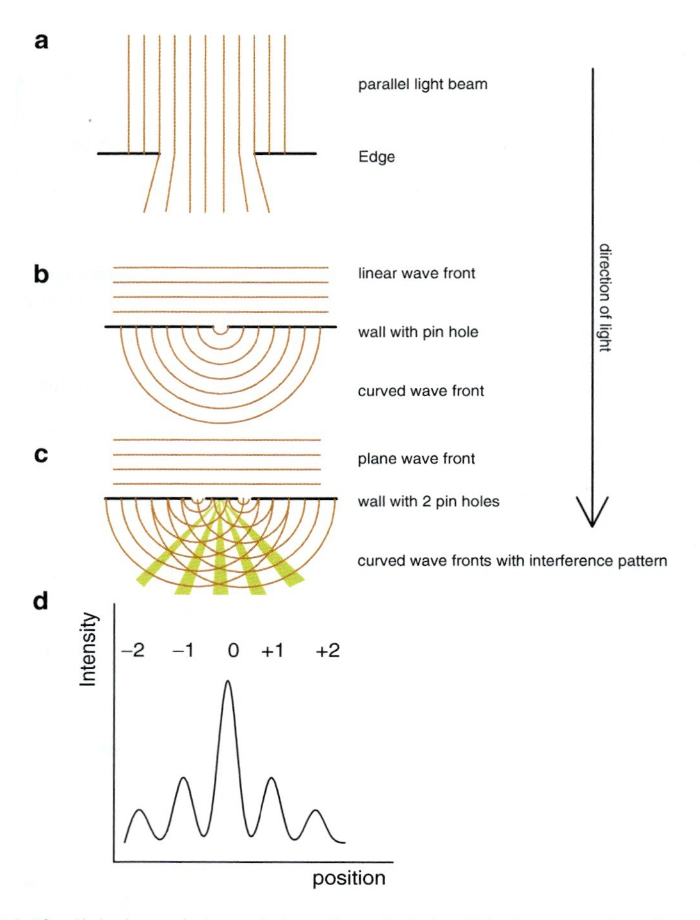

Fig. 1.3 (**a**) If a light beam is passed through a pin hole, light beams at the edge of the hole depart from their path. Simple beam optics can not explain such behaviour. (**b**) The same situation viewed by wave optics. A linear wave front (equivalent to parallel light beams) reaches a wall with a pin hole. Scattering results in a curved wave front. (**c**) If there are two pin holes, the resulting wave fronts interfere with each other. Out-of-phase waves cancel, in-phase waves amplify each other. Thus a pattern of bright and dark rings becomes visible. The central bright disk is called the maximum of zeroth order, the surrounding rings are numbered first, second... order. (**d**) The light intensity plotted as function of the position in the interference pattern. The maximum of zeroth order is much brighter than the first, higher order maxima are even weaker. Note that the interference rings are symmetrical, thus each ring results in two peaks, one to the right (positive numbers) and one to the left (negative numbers) of the centre

Since the light from these diffraction maxima continues to travel upwards, and since it comes from a single point, the beams are capable of **interference**. This creates a secondary interference pattern, the intermediate image, which is then magnified by the ocular. An animated discussion of interference may be found at http://www.doitpoms.ac.uk/tlplib/diffraction/index.php.

The simplest object is a tiny hole. Its image is a pattern of bright and dark rings, called AIRY-pattern.

1.1.3 *The Importance of the Numerical Aperture N_a*

As originally found by ERNST ABBE [1] the intermediate image is the more similar to the object, the more diffraction maxima are collected for it. The diffraction maximum of zeroth order contains most of the light, but no information, since it has not interacted with the object. This means that to reconstruct the object in the intermediate image, at least the diffraction maximum of first order must be in the objective aperture. Since finer object details give a higher distance of the refraction maxima from each other, the **resolution** capability of a microscope depends on the objective aperture.

The radius of this aperture is calculated to

$$r = \sin(\alpha)/n_m \tag{1.1}$$

with α being the opening angle of the light cone and n_m the **refractive index** of the medium. Optically important media and their refractive indexes can be found in Table 1.1.

On each objective you should find three numbers engraved: The magnification, the mechanical tubus length for which the objective was calculated (usually either 160 mm or infinity) and the **numerical aperture** N_a, with

$$N_a = \sin\left(\frac{1}{2} * \alpha\right) * n_m \tag{1.2}$$

The point-spread function of an ideal (abberation-free) lens is an AIRY-pattern (see Fig. 1.4):

$$h(r) = \left(\frac{2J_1(\frac{2\pi r N_a}{\lambda})}{\frac{2\pi r N_a}{\lambda}}\right)^2 \tag{1.3}$$

Table 1.1 Refractive index n of important media

Medium	n
Air	1.000
Water	1.330
Glycerol	1.460
Fused silica	1.462
Toluene	1.489
Glass	1.520
Immersion oil	1.520

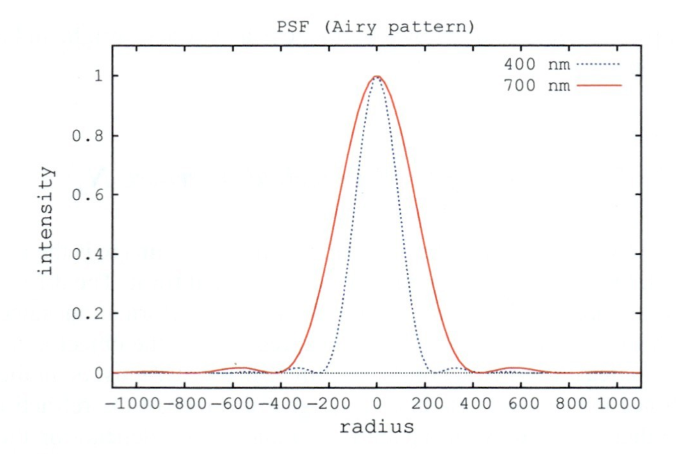

Fig. 1.4 Point-spread function of an ideal (abberation-free) lens. The diameter of the zeroth order maximum (AIRY-disk) increases with λ, hence the possible resolution decreases

with r the radius and $J_1(\zeta)$ the BESSEL-function of first kind and first order:

$$J_a(\zeta) = \sum_{m=0}^{\infty} \frac{(-1)^m}{m!\Gamma(m+a+1)} \left(\frac{\zeta}{2}\right)^{2m+a} \tag{1.4}$$

The Γ-function is a generalisation of the faculty-function for real arguments.

Since $J_1(\zeta)$ has its first minimum at $\zeta = 3.83$, the radius of the first minimum of the AIRY-pattern is

$$r = \frac{0.61\lambda}{N_\mathrm{a}} \tag{1.5}$$

The N_a of both the condenser and the objective determine the **obtainable resolution** d. The radius of the AIRY-disk of a point is

$$r_\mathrm{Airy} = \frac{0.61\lambda}{N_\mathrm{a}(\text{objective}) + N_\mathrm{a}(\text{condenser})} \tag{1.6}$$

We can imagine that each point of an object results in its own AIRY-pattern, resulting in a mosaic image. Experience shows that two points are seen as separate if the distance between their AIRY-disks is at least equal to r_Airy, then the minimum between the overlapping disks has about 80 % of the intensity of the maxima. However, due to optical imperfections the radius of the AIRY-disk is somewhat larger than that calculated above, empirically a factor of 112 % is assumed.

N_a also determines the **light flux** Φ, the axial resolution r_a (which is considerably worse than the lateral resolution), and the **depth of focus** T:

$$\Phi = \Phi_0 * \pi * N_\mathrm{a} \tag{1.7}$$

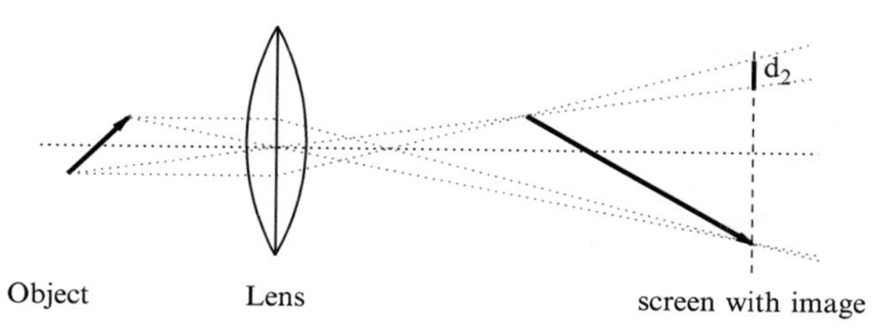

Fig. 1.5 Lenses have a limited depth of focus. While the image of the tip of the arrow is focused on the screen, the image of the bottom is not. Instead, it is drawn out into a dispersion spot with diameter d_2. This effect is stronger in stronger lenses with shorter working distance

$$r_a = \frac{0.88 n\lambda}{n - \sqrt{n^2 - N_\mathrm{a}^2}} \tag{1.8}$$

$$T = \pm \frac{n\lambda}{2N_\mathrm{a}^2} \tag{1.9}$$

with d_2 being the diameter of the acceptable dispersion spot (see Fig. 1.5).

As a rule of thumb the total magnification (magnification of ocular × magnification of objective × photographic magnification) should be in the range of 500–1 000 × N_a. Higher magnifications do not reveal further details, so one speaks of "empty magnification". Magnification compares the angular size of images at a distance of 25 cm and is hence a rather arbitrary parameter.

Remember. It is the numerical aperture, not the magnification of an objective that is really important.

The numerical aperture of the condenser should be equal to that of the objective, thus condensers have a diaphragm. This allows adjustment of the condenser aperture to the aperture of the objective. Although this setting would give the highest resolution, in practice the condenser aperture is set slightly lower than that of the objective. This increases the depth of focus. Since most objects are more or less spherical this means that their borders in the image become thicker and more visible (see Fig. 1.6). This is often, but incorrectly, described as improved contrast.

1.1.4 Homogeneous Immersion

Since the maximum angle of the opening cone of an objective is 180° and the refractive index of air is approximately 1, the maximum N_a possible in air is theoretically 1.0, in practice 0.95. For this reason some systems are corrected for the use of immersion oil between object and objective (in some cases also between condenser and object). Immersion oil has the same refractive index as glass (1.5), resulting in $N_\mathrm{a} \leq 1.4$ (see Fig. 1.7). Such systems are marked "Oil" and should not be used

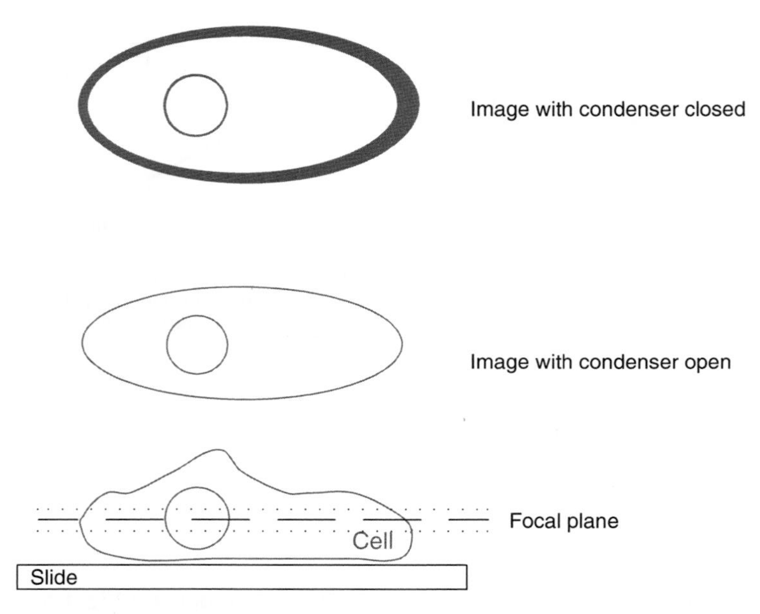

Fig. 1.6 Closing the condenser aperture a little more than required for KÖHLER-illumination increases the depth of focus of the image (*dotted lines* above and below the focal plane), resulting in thicker, darker outlines of structures. For an optimum balance between apparent contrast and resolution, condenser aperture should be 20–30 % smaller than that of the objective

without immersion oil as the resulting image would be very poor. For observation under UV-light, fused silica is used for the lens and the coverglass, in this case glycerol is used for homogeneous immersion.

1.1.5 Lens Aberrations

Lenses suffer from chromatic and spherical aberrations (see Fig. 1.8).

Chromatic aberration is caused by **dispersion**, that is, different refractive index n_m for different wavelengths. This results in separate images for light of various wavelengths, and in addition these images have different magnification. A lens made from normal glass refracts blue light stronger than red. The error is corrected by using lens systems rather than single lenses, and the various lenses are made from different materials. **Achromatic** lenses are corrected for red and blue light. Since green light can easily be removed by a strong red filter, such relatively cheap lenses are often used in black and white photography. **Apochromatic** lenses are corrected for four different wavelengths: deep blue, blue, green, and red. Such systems are used for colour photography. However, they are not only very expensive, the high

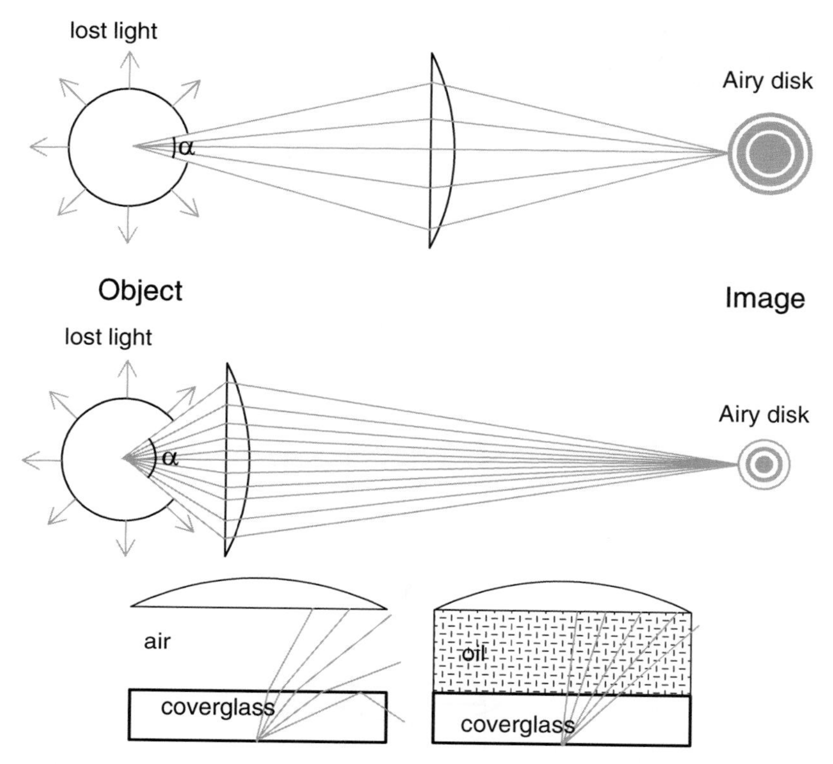

Fig. 1.7 *Top*: Relationship between working distance, numerical aperture, brightness, and resolution of an objective, the object is a single bright point. Systems with high numerical aperture collect more light and give smaller AIRY-disks (= higher resolution). *Bottom*: If the air between the coverglass and lens is replaced by immersion oil with the same refractive index as glass, refraction of the light beams is avoided and a higher N_a becomes possible (**homogeneous immersion**)

number of lenses inside also causes increased stray light, thus reducing contrast. Fluorite systems are in between achromats and apochromats. The use of new glass materials in the last 30 years has improved the quality of achromats so much that those made by premium manufacturers are sufficient for all but the most demanding applications.

Spherical aberration is caused by lenses having different focal lengths for light beams entering at the centre and in the periphery, even for monochromatic light. Again, this is disturbing mostly in micro-photography, during observation we can simply play with the focus. Lenses corrected for spherical aberration are called **plan**. Modern lens systems from reputable manufacturers are so good that an investment into plan lenses is not usually required.

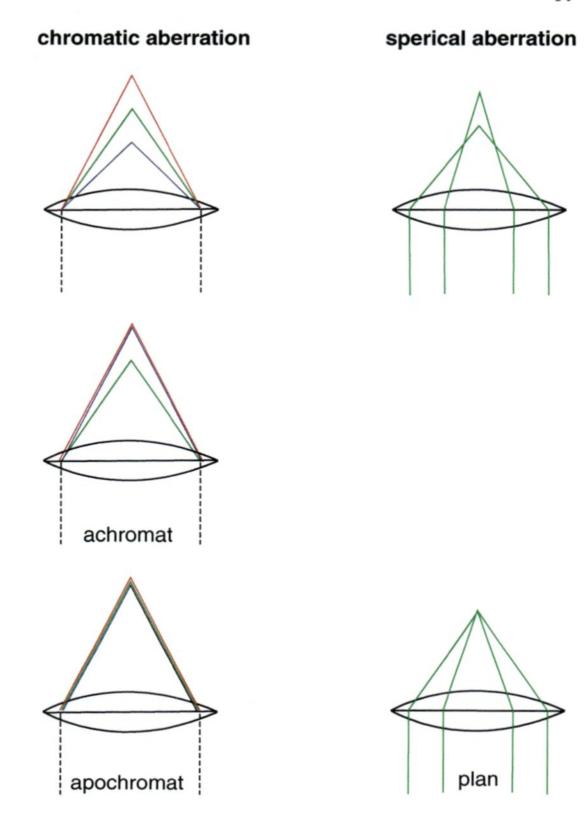

Fig. 1.8 Correction of lens aberrations in microscopic lens systems. For details see text

1.1.6 Special Methods in Light Microscopy

An object can be thought of as a simple grid that creates an interference pattern in the objective aperture like in Fig. 1.9 in the objective aperture. The maximum of zeroth order is brightest, but its light has not interacted with the object and hence contains no information. The final image produced by the microscope depends on the treatment of the various interference maxima (see also Fig. 1.10):

Bright field microscopy This is the normal mode. All diffraction maxima, including zeroth order, are allowed to form the image. Resolution of bright-field microscopy is limited by the RAYLEIGH-condition: objects are visible in a microscope only if they have at least twice the size of the wavelength of the light.

Dark field microscopy The zeroth order diffraction maximum is removed by a stop in the condenser, which is projected into the objective aperture. Objects appear bright on a dark background. Note that in dark field (and fluorescence)

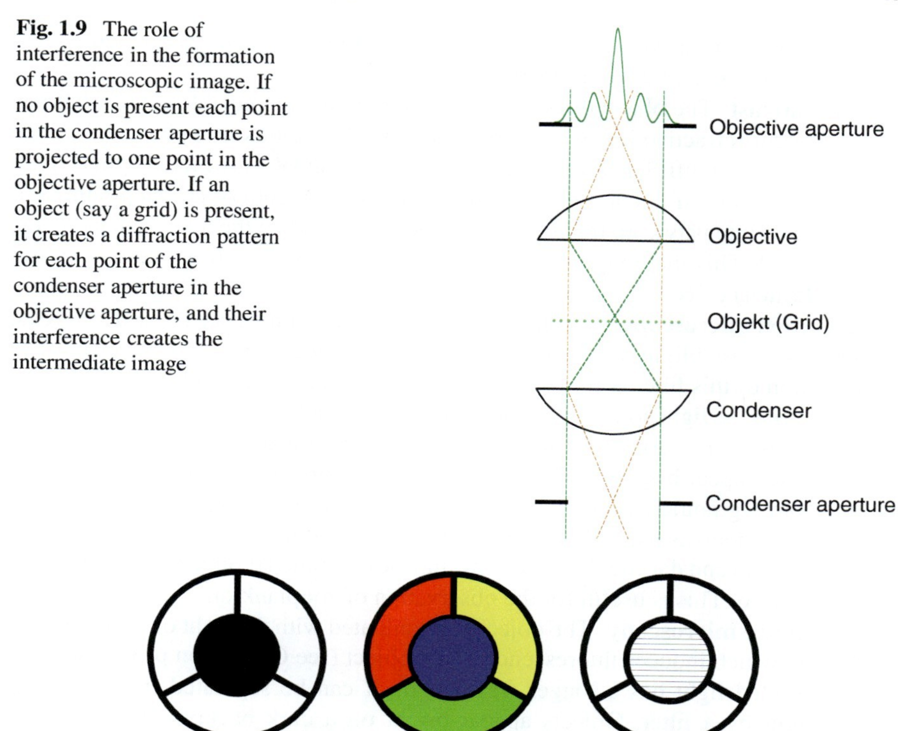

Fig. 1.9 The role of interference in the formation of the microscopic image. If no object is present each point in the condenser aperture is projected to one point in the objective aperture. If an object (say a grid) is present, it creates a diffraction pattern for each point of the condenser aperture in the objective aperture, and their interference creates the intermediate image

Fig. 1.10 Condenser filter for special effects. *Top*: A darkfield filter stops light of zeroth order, whilst a RHEINBERG-filter colours zeroth order light differently from that of higher orders. In darkfield polarisation light of zeroth order is polarised. *Bottom*: Filters for oblique and oblique darkfield illumination

microscopy the RAYLEIGH-condition does not apply, hence the term **ultramicroscopy** is sometimes used.

Oblique bright field Only diffraction maxima of one side (including zeroth order) form the picture. In theory this doubles resolution, but contrast improvement by shadowing is more important. In science this method is rarely used since we can not distinguish between shadowing by different refractive index and different thickness.

Oblique dark field like oblique bright field, but zeroth order maximum is removed.

Rheinberg contrast Different colour filters for light from the zeroth order and higher order maxima. Of aesthetic value only.

Phase contrast The object does not absorb light to an appreciable extend, but has a different refractive index from the medium. This results in a phase difference between the undiffracted (zeroth order maximum) and the diffracted light (higher order maxima). If a phase plate is used to delay the zeroth order maximum (and reduce its intensity), subtractive interference occurs when the intermediate image is formed. This method is of extreme importance for the observation of **living (unstained) cells**.

Polarisation microscopy A polarisation filter in the condenser is used to create linear polarised light. A second polariser filter is mounted on the eye piece. The direction of this filter is placed at right angle to that in the condenser, so that all light from the light source is absorbed by either one filter or the other, creating a dark background. Any birefringent object will turn the direction of the polarised light and appear bright. Biological examples for birefringent objects are muscle and starch granules. In the lab protein crystals are birefringent. Geologists and material scientists use polarisation microscopes routinely.

Polarisation and darkfield microscopy may be combined if only zeroth order light is polarised. This is useful for the observation of zooplankton.

Fluorescence microscopy The object is illuminated with the light of a short wavelength, which induces fluorescence in the object (see Chap. 6 on p. 39). Because the emitted light has a longer wavelength, it can be separated from the exiting light by a filter. Objects appear bright on a dark background, with very high contrast. Techniques like fluorescence lifetime measurement, FÖRSTER resonance energy transfer (FRET) and fluorescence spectroscopy can all be performed on specialised microscopes [141, 142]. Because of the sensitivity of this method, small amounts of fluorescent compounds can be detected (vital stains, immunofluorescence microscopy). Illumination is usually through the objective (see Fig. 1.11), using a dichroic (from Gr. "two colour") mirror which has several layers of coating of different refractive index. At the interface between these layers part of the light is reflected, the rest is refracted. Depending on the thickness of the layers the reflected and refracted light of some wavelengths undergo constructive, of other wavelengths destructive, interference. Therefore, the mirror reflects some wavelengths, but transmits others.

The high light intensity required for the exciting beam results in phototoxicity, which can be reduced by stroboscopic illumination, or by adding protective substances into the mounting medium (*e.g.*, ascorbic acid).

Two-photon fluorescence More recently dyes were developed which can absorb 2 **photons** and hence excite fluorophores with half the exciting wavelength. Infrared lasers (titanium sapphire) are used for excitation. Because absorbtion of two photons by the same dye molecule within the lifetime of the excited state is improbable, the required flux-density is reached only in the aperture of a high N_A condenser. Thus the emission comes from one point of the sample only, similar to confocal microscopy. This point is scanned in x-, y- and z-directions through the sample to get a 3-D image.

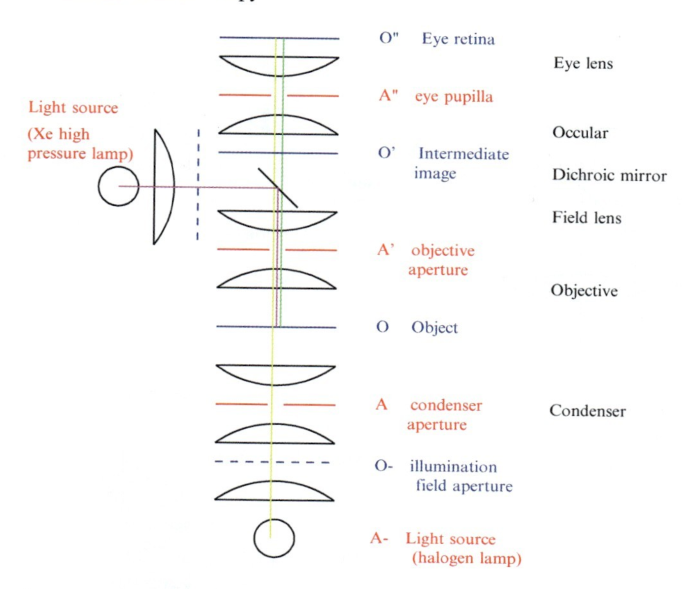

Fig. 1.11 A fluorescence microscope is built like a normal microscope (see Fig. 1.2), but has an additional light source in a side arm. This is usually a high pressure Xe or Hg lamp (**danger**: it can explode if incorrectly handled), which generates UV light with good efficiency. Recently, LEDs as excitation sources have also become popular. The light is reflected by a dichroic (purists say: dichromatic) mirror through the objective (which acts as a condenser) onto the object (*purple* beam). The induced fluorescent light (*green* beam) passes back through the objective, through the dichroic mirror (which is transparent to long wavelengths), into the ocular and the eye. Light from the halogen lamp in the foot of the microscope (*yellow* beam) can be used to compare the fluorescent with the transmission image

Fluorescence activated cell sorter The fluorescent light is no longer observed, but measured and used to sort cells (or other objects) into groups. The number of objects in each group is counted; additionally it is possible to purify cells with selected properties (see Fig. 1.8).

Confocal microscopy The depth of focus in a microscope is limited, in particular at high magnification. Light from out of focus planes creates a haze in the microscopic image, which reduces contrast. From Fig. 1.12 you can see that light from a focused object passes through a certain point along the optical axis. In 1955 MARVIN MINSKY introduced a pinhole at this point to reduce stray light. In addition, he used a small light point to excite fluorescence and moved the sample under that point in *x*- and *y*-direction. Thus weakly fluorescing objects could be seen next to very bright ones. The idea was ahead of its time: it simply took too long to scan the sample. The invention of the laser changed this; scanning the laser across the sample can be done relatively quickly; lasers are also very bright and can be focused onto a small spot. The fluorescent light is detected by a photomultiplier and fed into a computer. Once a X/Y-scan is completed, the

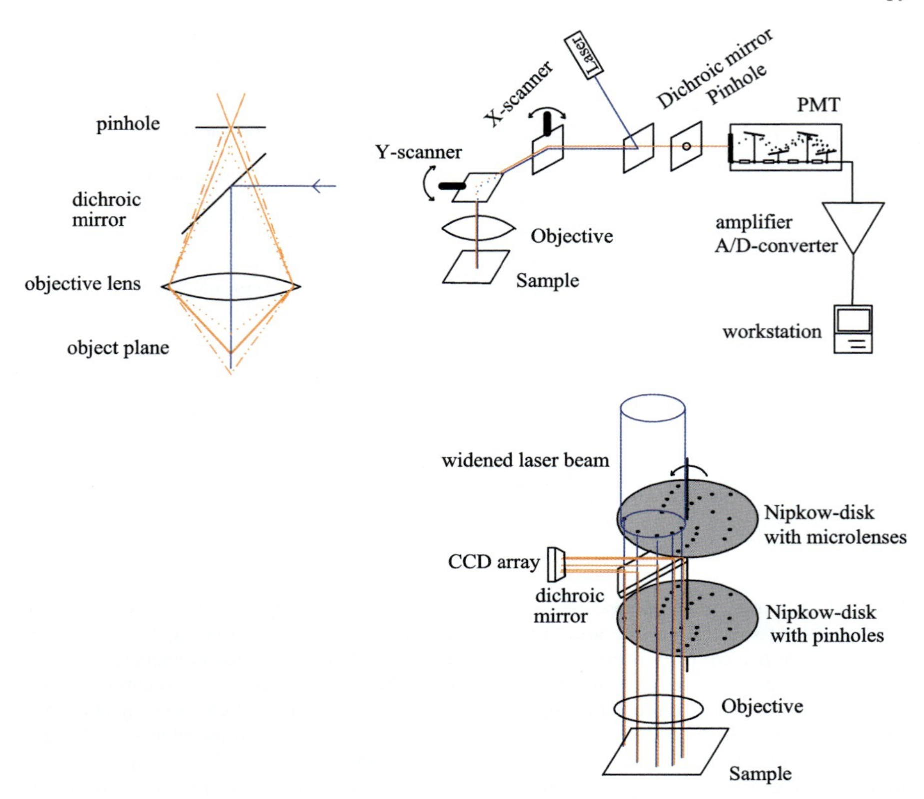

Fig. 1.12 The confocal microscope. *Top left*: Light coming from an in-focus object passes though a single point. Only a small fraction of the light from objects higher and lower than the in-focus object also pass through that point. A pin-hole at the focus thus removes most of the light coming from out-of-focus objects. The diameter of the pin hole should be the diameter of the AIRY-disk. *Top right*: Schematic diagram of a laser scanning confocal microscope. The light from a monochromatic source (usually a laser) is brought into the optical path via a dichroic mirror. Two oscillating mirrors move the beam across the sample, and align the resulting fluorescent light with the optical system. After passing the dichroic mirror the light goes through the pin-hole (which removes out-of-focus light) to a photomultiplier tube. The resulting electrical signal feeds through an analog/digital converter (ADC) into a computer system. Scanning of the laser beam prevents fluorescent light from nearby objects in the focus-plane from entering the optics, increasing lateral resolution. *Bottom* NIPKOW confocal microscope: The laser light is spread to illuminate a section of a rotating NIPKOW-disk. Light passing the holes is focused onto the sample by the objective. Fluorescent light emitted by the sample is focused by the same objective lens onto pinholes in a second NIPKOW-disk. Thus several sample points are imaged onto a CCD-array at the same time, increasing the frame rate

object is moved up or down a little and the scan is repeated. After several such scans a three-dimensional representation of the object is created in the computer, which contains quantitative information on the light intensity in each point of the object.

To further increase the rate at which pictures can be taken, several points of the object may be scanned at the same time using a rotating NIPKOW-disk. Such systems can either use lasers or white light for illumination.

Stimulated emission depletion fluorescence Since the lens projecting the laser light onto the specimen shows diffraction, spot size is limited to about 200 nm, this limits the resolution of the confocal microscope. In newer instruments two laser beams are coupled into the microscope, one to excite fluorescence and one to stimulate the emission of fluorescent light (note: do not mix up stimulated emission, which is reversible, with irreversible bleaching). The emission-stimulating beam has zero intensity in the centre but high intensity in the periphery [154]. Thus fluorescence is observed only in a spot much smaller than the diffraction spot. The new resolution becomes:

$$\Delta d = \frac{\lambda}{2n \sin\alpha \sqrt{1 + I/I_{\text{satt}}}} \tag{1.10}$$

with I the intensity of the emission stimulating beam and I_{satt} the saturation intensity (intensity that reduces fluorescence probability by 63 %) for the emission stimulating beam. This is a constant which depends on the fluorescent dye. With this technique, a resolution down to 20 nm is possible.

Computer de-convolution Advances in digital image processing and ever increasing computer performance have now made it possible to remove the haze of normal microphotographs. This is called de-convolution of the image. Light from different depths is transferred through a lens system in different ways. This can be described mathematically using **point spread functions**, applying them in reverse de-convolutes the image.

Selective plane illumination microscopy is another way to limit the excitation of fluorescence to a single plane. The light is focused not into a point, but a sheet by a cylindrical lens. The light sheet enters the sample from the side. Since only one plane is illuminated, out of plane light can not lower contrast. Contrary to confocal microscopy an entire section is captured at once, and scanning is required only in z-direction. This high scanning velocity allows real time videos to be produced. In addition, penetration depth is higher in SPIM, allowing larger objects to be studied (*e.g.*, entire fruit fly or fish embryos in developmental biology). Also, phototoxicity and fluorescence bleaching are reduced because the exciting light is distributed over a much larger area.

1.2 The Electron Microscope

1.2.1 Transmission Electron Microscopy

Visible light has wavelengths between about 380 (violet) and 780 nm (dark red). As can be seen from (1.6), the resolution of a microscope is inversely proportional to the wavelength of the light used. In normal bright field microscopy, the separation distance is about twice the wavelength (*i.e.*, 0.8 μm min.), in dark field and fluorescence microscopy slightly lower separation distances are possible.

Because electrons can behave like electromagnetic waves with very short wavelength, it is possible to construct microscopes that use an electron beam rather than a light beam. This idea was first developed by ERNST RUSKA in the early 1930s (NP 1986). Electrons are emitted from a glowing cathode (or a tungsten tip → **field emission electron microscopy**), focused by a WEHNELT-cylinder and accelerated by an electric field of 10^4 to 10^6 V to high energies (= short wavelengths). For biological samples, 80–120 keV are most often used. Magnetic fields are used instead of glass lenses. These fields are created by coils, the refracting power is determined by the current flowing through these coils.

Electrons can travel only in high vacuum, thus all objects observed in an EM must be completely dry. This limits EM to fixed, dehydrated samples. Some attempts have been made to circumvent this limitation ("wet EM"). In that case the objects are mounted onto a thin, electron-transparent plastic foil and illuminated by the electrons through the foil. Back-scattered electrons also pass through the foil before being collected to form the image.

Thermal stress for the sample in the EM is very high, and this can damage the sample during observation and cause artifacts. Special stabilisation procedures (which are expensive and time consuming) can be used to limit this problem. In special cases (EM crystallography, see Sect. 34.4 on p. 341) the sample stage is cooled with liquid nitrogen.

Because electrons do not travel far in solid materials, samples need to be very thin. Biological materials are usually embedded in plastic, cut on an ultramicrotome with knifes made from glass or diamond, and placed on small copper grids. Contrast is increased by staining with salts of heavy metals like lead, osmium, or uranium.

1.2.1.1 Cryo-Electron Tomography

If unfixed cells are snap-frozen to −190 °C so that water vitrifies (becomes glass-like) rather than crystallises, cell structures down to protein complexes are preserved. The object is then scanned in the electron microscope under low-dose condition and under constant cooling. It is rotated, so that pictures are taken from several angles. From these tilt-series, three-dimensional reconstructions of the object are made in a computer using similar techniques as in X-ray tomography in medicine. The structures visible then need to be identified–that is, they are automatically compared with known structures (template matching)—and known proteins are docked into the complex. Identified components can be marked, for example, by colour coding. With this technique a resolution of about 5 nm is achievable. Noise reduction in the cameras used will probably push that limit in the coming years. For a review see [87].

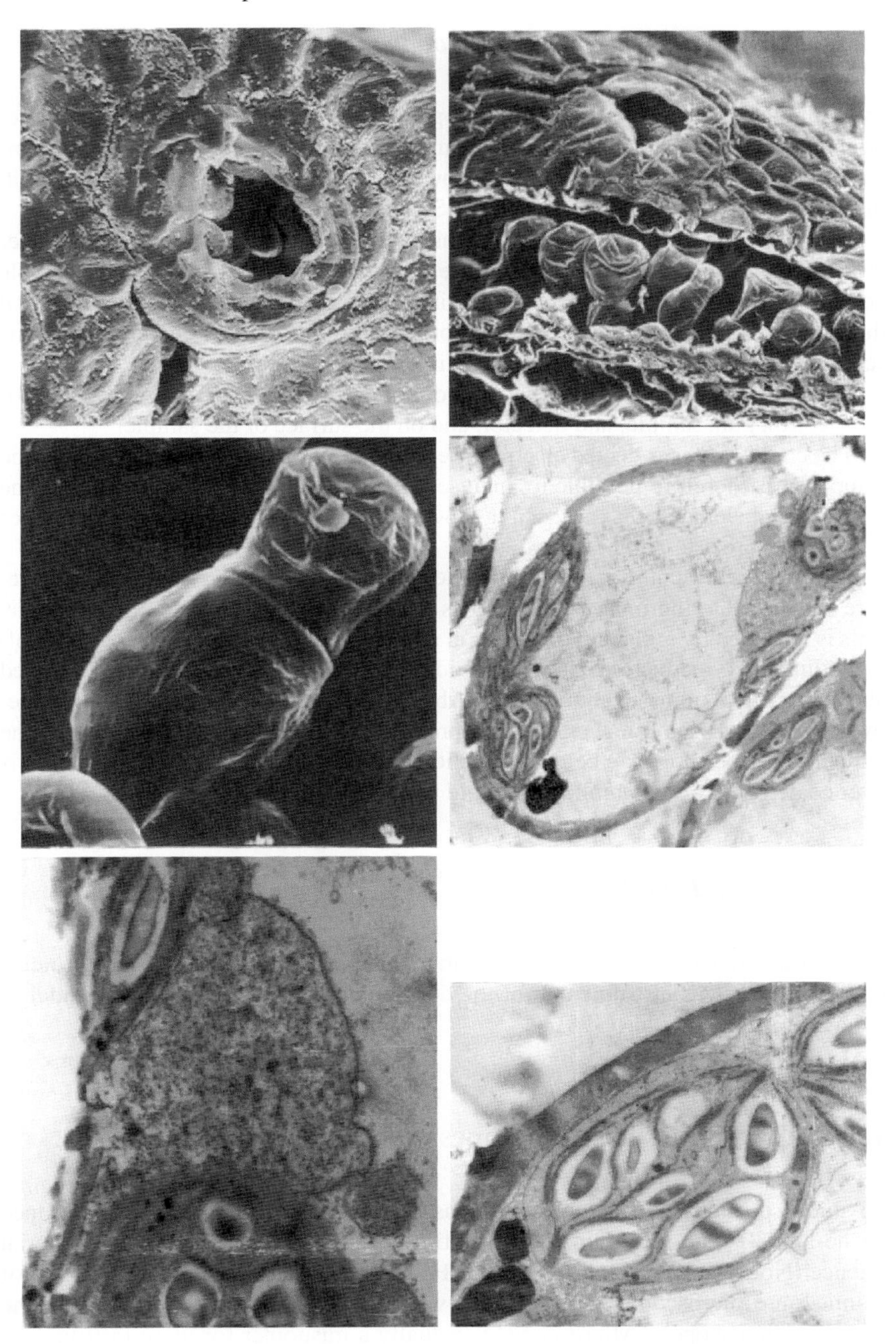

Fig. 1.13 Raster scanning electron microscopy (first three pictures) and transmission electron microscopy (last three pictures) of the liver wort *Marchantia polymorpha*. The pictures show a leaf with vegetative buds and a single cell, with nucleus and chloroplasts, the latter containing starch granules

1.2.2 Scanning Electron Microscopy

In scanning electron microscopy, developed by M. V. ARDENNE in 1940, three-dimensional information is obtained from the sample. This is an advantage compared to TEM, which is paid for by lower resolution. Note that confocal light microscopy was developed in analogy to SEM.

A SEM works with two electron beams, one in the microscope proper, the other in a BRAUNIAN tube. The two beams are moved synchronously across the sample and the screen of the monitor tube. The intensity of the beam in the monitor (and therefore the brightness of a particular spot) is controlled by the number of electrons scattered from the sample into a detector in the microscope.

Thus the brightness of a spot on the monitor depends on the relative positions of the electron gun, sample, and detector, and on the surface of the sample.

The energy of the reflected electrons depends on the chemical composition of the sample; it is therefore possible to analyse the sample and quantitatively determine the chemical elements that occur in it. However, this method is neither sensitive nor particularly precise.

Again, a high vacuum is required inside the microscope, and samples must therefore be dried. Drying must be done so as not to change the surface of the sample. This is achieved by replacing the water in a biological sample first with organic solvent (dimethoxymethane), then with liquid carbon dioxide. The CO_2 is then warmed above its critical point (to about 45 °C). Above the critical point the phase border between liquid and gas vanishes. The gas pressure is then slowly reduced (so that the temperature does not drop below the critical point!) through a valve. The sample is later coated with a very thin layer of gold to make it conducting.

1.2.3 Freeze Fracture

Freeze fracture (see Fig. 1.14) is a valuable technique to study membrane proteins in the TEM. It was instrumental for the development of the fluid-mosaic model of the biological membrane [378].

1.3 Other Types of Microscopes

To study specimens of molecular dimensions special techniques were developed in the last 30 years, leading to a revolution in biology and (even more) material science (Nobel price to G. BINNIG and H. ROHRER 1986 for the development of atomic force microscopy (AFM) together with E. RUSKA for the development of the electron microscope). "Microscope" is actually a misnomer for these instruments, since they do not look at the sample, but "feel" its surface.

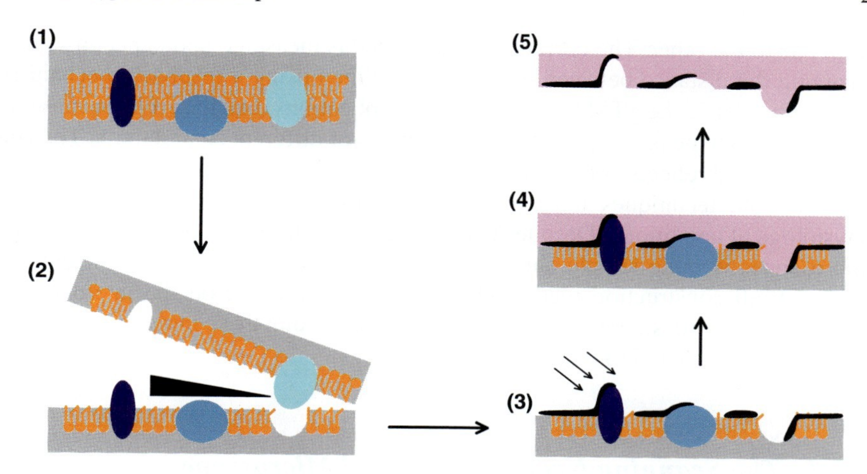

Fig. 1.14 Freeze-fracture. **(1)** The sample is vitrified in water. **(2)** The block is fractured so that the break occurs between the leaflets of a membrane. The resulting block is then etched under a vacuum, so that some of the water on the surface sublimates away. **(3)** The surface is sputtered with platinum from an angle (usually 45°), so that elevations on the sample cause shadowing. **(4)** The platinum layer is reinforced with carbon, which is translucent to electrons. **(5)** The block is immersed in water, the original sample washed away, and the platinum/carbon film fished from the surface with a sample net for TEM

1.3.1 The Atomic Force Microscope

In AFM a thin probe (ideally with a tip of a single atom) is scanned in x,y-direction over the specimen by a mechanism driven by piezoelectric crystals [106]. The force generated by the specimen on the tip is measured and minimised by a feedback circuit, keeping the tip just above the specimen. Thus the elevation profile (z-direction) of the specimen is recorded. By increasing the force generated by the tip it is possible to move atoms or molecules around (nano-technology). About 1–2 min are required for a complete scan, resolution is about 1 nm.

The AFM may be used to measure not only the shape of an object, but also surface properties like elasticity, charge or the presence of binding sites for ligands (**lab on a tip**). By measuring the force required to dissociate the interaction between a ligand fixed on the tip and a protein the affinity can be determined [106, 195]. Protein unfolding can be followed by binding the protein to both the support and the tip and then slowly increasing the distance between tip and support. The force required to break bonds is recorded.

1.3.2 The Scanning Tunnelling Microscope

The STM is similar to the AFM discussed above, but the tip is moved a small (sub-atomic) distance above the specimen. Between the tip and specimen a voltage is applied, and because of the small distance between them electrons can tunnel from one to the other. The current thus generated is measured and kept constant by changing the distance between probe and specimen using a feedback mechanism. In protein science the STM has found fewer applications than the AFM because the current through the protein can take several paths of different conductivity. Thus the signal obtained reflects not only the height of the protein at a particular position.

With both techniques it is possible to investigate unfixed specimens under physiological conditions. Despite their atomic resolution the construction of such microscopes is quite simple and can be done by any reasonably competent mechanic. Both construction plans and software are available on the internet (*e.g.*, http://www.geocities.com/spm_stm/Project.html), putting these instruments into the reach of ambitioned high schools or amateurs.

1.3.3 The Scanning Near-Field Optical Microscope

The SNOM is a zwitter between the scanning probe and the optical microscope. It is a form of fluorescence microscope where the excitation beam is sent through a glass fibre. The fibre is tapered at the end to a diameter of only a few nanometers. This end is scanned over the sample, illuminating only a tiny spot. The fluorescent light emitted by the sample is collected by a photomultiplier. The resolution of the instrument is therefore limited only by the diameter of the probe tip.

In such a thin fibre the light no longer undergoes total reflection, therefore the outside of the fibre must be coated with aluminium. Once the fibre diameter becomes smaller than the wavelength of the light, light propagation is by tunnelling. This part of the probe therefore needs to be very short.

Chapter 2
Single Molecule Techniques

Assume a sample of a fairly dilute protein solution, 1 nM (=50 ng/ml if M_r = 50 kDa). Then 1 μl will contain 1 fmol of protein. Multiplied with AVOGADRO's number (6.022×10^{23} mol^{-1}) this still corresponds to 6×10^8 protein molecules.

Any experiment performed on this sample of protein, say, the determination of its affinity for a ligand, will return a value averaged over a huge number of molecules. That is fine as far as it goes, but what is the distribution? Is it normal or skewed? What is the standard deviation? To answer these questions, one has to perform experiments on single molecules and repeat those on many different specimens. One has to determine what percentage of time a protein molecule has or has not ligand bound at a given concentration to determine the affinity. Or one has to measure the force required to pull a ligand from a protein molecule to determine the binding energy. Such measurements are of particular interest for proteins that move. What is the torque generated by rotation in ATP-synthase? Can it explain the synthesis of ATP? How much force does a single myosin head generate when moving along an actin fibre? Only in the last 20 years have we started to answer such questions.

2.1 Laser Tweezers and Optical Trapping

It is immediately apparent that–in a homogeneous sample in dynamic equilibrium–if a certain fraction of molecules has ligand bound, then each molecule has ligand bound for the same fraction of time. Thus the probability P_{PL} of finding the protein molecule with its ligand becomes:

$$P_{\mathrm{PL}} = \frac{[\mathrm{PL}]}{[P] + [\mathrm{PL}]} = \frac{\frac{[\mathrm{PL}]}{[P]}}{\frac{[P]}{[P]} + \frac{[\mathrm{PL}]}{[P]}} = \frac{K_{\mathrm{eq}}}{1 + K_{\mathrm{eq}}}. \tag{2.1}$$

Light can be seen as particles (photons) moving at high speed (c), giving them a momentum $\mathfrak{p} = h/\lambda$ and an energy $E = h\nu = hc/\lambda$. If the direction of travel is changed, a force is generated which is equal to $\vec{\mathfrak{F}} = -d\vec{\mathfrak{p}}/dt$. This force has to be vectorially added for all photons acting at a given moment.

E. Buxbaum, *Biophysical Chemistry of Proteins: An Introduction to Laboratory Methods*, DOI 10.1007/978-1-4419-7251-4_2,

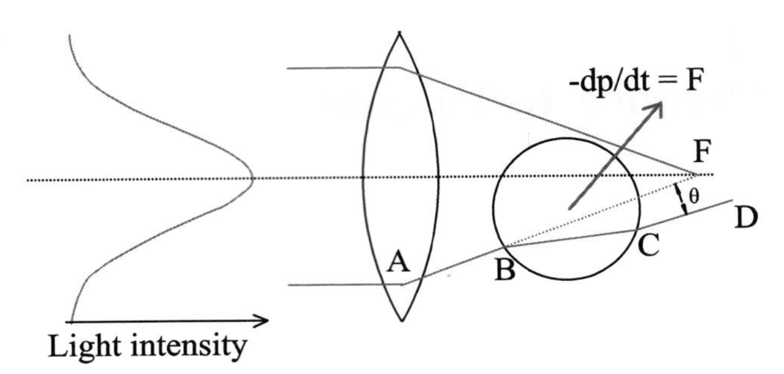

Fig. 2.1 Principle of the optical tweezers. As light enters the bead (which has a much higher refractive index than the medium) it is refracted along the path ABCD, resulting in a change of momentum of the photons. This pushes the bead into a position on the optical axis and just below the focus of the lens. Note that the light intensity in a laser beam has GAUSSIAN distribution

If a laser beam (near infrared CW laser with 700–1 300 nm, *e.g.*, a Nd-YAG laser at 1 064 nm, with 10^6–10^8 W/cm^2) is focused through a high N_A lens (1.25–1.40, magnification 40–100 times, oil immersion) onto a bead of high refractive index material the bead is held on the optical axis just below the focus of the lens (see Fig. 2.1). The laser beam is steerable, thus the bead can be moved and a precisely defined force applied to a molecule linked with one end to the bead and with the other to the substrate [12, 299].

In practice, two counter-propagating laser beams are often used with high N_a lenses but low N_a beams. Each objective not only focuses its beam onto the sample, but also catches the light coming from the other beam for analysis. A setup for an optical tweezers experiment for student labs is described in [21].

Since its invention this technique has been used to study:

- The binding energy between two molecules, which can be calculated from the force required to pull them apart.
- The forces generated by molecular motors (*e.g.*, actin/myosin, ATP synthase, RNA polymerase) and their step width.
- The heterogeneity of the dissociation constants in a population of molecules.
- Free energy of folding and unfolding of molecules. Pulling the ends of a protein apart with increasing force results first in a stretching of the coils, then breakage of tertiary and secondary structure and finally in the rupture of the primary structure.

Besides laser tweezers, magnetic tweezers may also be used, where a magnetic bead is held in a magnetic field. The main advantage is that they allow twisting of the molecule attached by rotation of the bead.

Chapter 3
Preparation of Cells and Tissues for Microscopy

Literally thousands of detailed recipes and procedures are given in the handbook started by ROMEIS [286]—this should be consulted before embarking on such studies.

Phase contrast microscopy can observe living cells. Fluorescent microscopy and so-called **vital stains** are used to make cellular structures fluorescent without harming them. Genetic engineering allows proteins of interest to be coupled to naturally fluorescent proteins (**green fluorescent protein** (**GFP**) and its descendants with other emission spectra), making it possible to localise a protein in the living cell and measure interaction with other proteins.

3.1 Fixing

Apart from those exceptions, microscopy is usually performed on dead cells. Thus it is important to prevent post-mortem changes (drying out, decomposition, protease digestion) in the cell structure as much as possible. This is achieved by **fixing** the samples with appropriate chemicals. Such chemicals include protein-precipitating agents (heavy metals like Hg or Cr, alcohols, acids) or protein cross-linkers (methanal, glutardialdehyde) and lipid-stabilisers (osmium tetroxide) in various proportions. Development of fixatives suitable for a particular investigation is more art than science, and hundreds of such mixtures have been described. For most purposes 4 % methanal and 2 % glutardialdehyde in buffered saline are adequate, good preservation of anatomical detail is also achieved by AFA (25 % ethanol, 4 % methanal and 10 % acetic acid). The volume of the fixative should be at least 10 times that of the sample, samples should be as small as possible so that the fixative reaches all cells quickly. Particularly rapid is perfusion fixing where the fixative is injected through the aorta of a freshly killed animal and the vena cava is opened to allow fluid to drain. Invertebrates sometimes change their shape during killing, so to prevent this they need to be anaethesised. Suitable recipes for various taxa are again found in [286].

E. Buxbaum, *Biophysical Chemistry of Proteins: An Introduction to Laboratory Methods*, DOI 10.1007/978-1-4419-7251-4_3,

3.2 Embedding and Cutting

Samples should spend no more than 24 h in fixative to avoid excessive hardening. After that they are dehydrated by incubation in an alcohol series (30, 50, 70, 95, 100, 100 % for 12 h each), transferred into an inter-medium which is miscible with both ethanol and xylene (*t*-butanol, methylsalicilate) and finally in xylene. This solvent is then replaced by paraffine for light microscopy or methylmetacrylate for both light and electron microscopy. Tissues embedded in this way can be stored indefinitely.

From these blocks very thin sections (1–20 μm for light and 0.1 μm for electron microscopy) are cut on a **microtome**. The sections are floated onto a drop of water on a slide, for paraffine sections the water must be warm enough to stretch the sections, but not so hot that the paraffin melts. As the water evaporates, the sections firmly attach to the slides.

3.3 Staining

Electron microscopic preparations can be observed at this stage. Preparations for light microscopy have to be stained first. This is achieved by dissolving the paraffine in xylene, then taking the slide back to water by graded ethanol series. Since the sections are so thin each bath is applied only for a few minutes. Most slides I have seen, even those prepared by professional pathologists, were heavily overstained, masking the finer details. There really should be only a hint of colour. Low contrast can always be improved with a phase-contrast microscope, details lost in a sea of colour can not be brought back.

After staining, either with antibodies and/or with chemical dyes, the section is embedded in mounting medium and covered with a cover glass. Mounting media can be water miscible (glycerol gelatine, polyvinylalcohol) or xylene-miscible (usually mixtures of unpublished composition sold under brand names like Entellan, Mallinol or Permount). For unstained specimens the mounting medium must have a refractive index as different from the specimen as possible to get good contrast, for stained specimens a refractive index near that of the specimen is preferred.

3.4 Laser Precision Catapulting

If a UV-laser (*e.g.*, N_2-laser at 337 nm) is used instead of IR, biological material can be cut on an inverted microscope with minimal thermal damage to the tissue. Focusing the laser just below the sample will then catapult it into a collection vessel. Thus homogeneous tissue samples (*e.g.*, only cancer or only normal cells) collected can then be used for proteomics or genomics applications. This can also be done with tissue archives that have been formalin-fixed and paraffin embedded (see [292] for a review on the use of such material in proteomics and Fig. 19.2 on p. 202 for a discussion of formaldehyde action).

Chapter 4
Principles of Optical Spectroscopy

4.1 Resonant Interaction of Molecules and Light

Many analytical techniques involve the interaction of a sample with light (IR, visible and UV). While microwave radiation excites **rotation** of a molecule ($\Delta E \approx 4\,\text{kJ/mol}$) and IR-light oscillatory movement of atoms around their position in a molecule (**vibration**, $\Delta E \approx 40\,\text{kJ/mol}$), visible and UV light move **electrons** to higher energy levels ($\Delta E \approx 400\,\text{kJ/mol}$). X-rays excite nucleons ($\Delta E \approx 4000\,\text{kJ/mol}$).

There are three kinds of orbitals: σ (single bonds), π (double- and triple bonds), and *n* (lone electron pairs in hetero-atoms like N and O). They can occur in the ground state (binding orbitals, σ, π and *n*) and in an excited state (anti-binding orbitals, σ^*, π^*). In optical spectroscopy, $\pi \rightarrow \pi^*$ and $n \rightarrow \pi^*$ transitions are the most important, transitions to the high energy σ^*-orbitals are not used in biochemistry.

For transition metals, $d \rightarrow d^*$-transitions are also important, however, absorption is generally weak because these transitions are spin-forbidden. Field splitting of d-orbitals by ligands of transition metal complexes leads to blue-shift and intensification of the bands. Ligands can be sorted by the strength of this effect: $I^- < Br^- < Cl^- < SCN^-$ (S-bonded) $< F^- < OH^- <$ oxalate $< H_2O < NCS^-$ (N-bonded) $< NH_3 \simeq$ pyridine < ethylenediamine < dipyridyl < o-phenantroline $< NO_2^- < CN^- \simeq CO$.

Strong absorbance is caused by **charge–transfer-complexes** (electron donor–acceptor complexes) between central atom and ligands, where an electron is passed from one to the other upon excitation by UV or visible light. Transfer can be from the p-orbitals of the ligand to d- or s-orbitals of the metal (metal ions of high valency), or it can be from the d-orbitals of the (low valency) metal to a π^*-orbital of the ligand. If one molecule contains several transition metal ions of different valency, electrons may also be passed between them, *e.g.*, in Prussian blue (Fe^{2+} and Fe^{3+}). Unlike $d \rightarrow d^*$-transitions charge transfer is allowed by selection rules, they have a high absorbtion coefficient ($\approx 50\,000\,\text{l mol}^{-1}\,\text{cm}^{-1}$). Charge-transfer-complexes are often found in oxidoreductases.

According to the PAULI-principle each orbital can have a maximum of two electrons, which need to have different spin (angular momentum). If an electron is

E. Buxbaum, *Biophysical Chemistry of Proteins: An Introduction to Laboratory Methods*, DOI 10.1007/978-1-4419-7251-4_4,

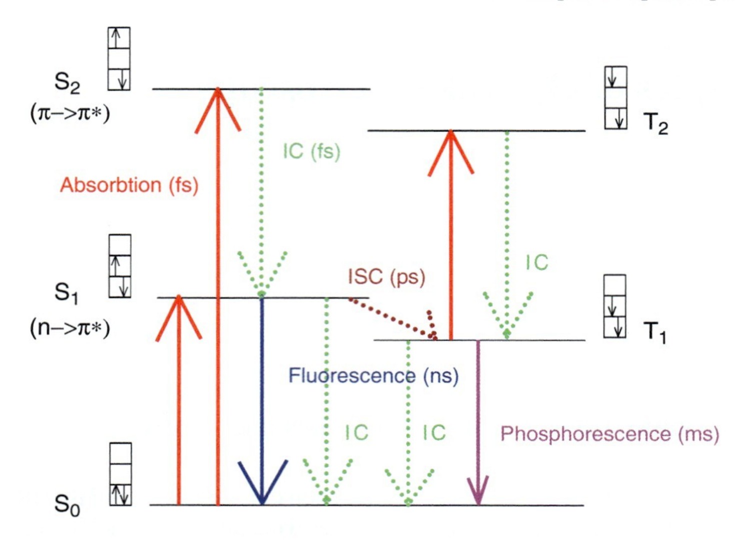

Fig. 4.1 JABLONSKI-diagram of the energy levels in optical spectroscopy

moved to a higher orbital, it can either maintain its spin or change it. If the spin is maintained, we speak of a **singlet state (S)**, if it is changed, of a **triplet state (T)** (see Fig. 4.1). Return from a singlet state to the ground state is fast, while return from a triplet state involves another spin reversal and is less likely, and hence slower.

Figure 4.1 is oversimplified, as each of the states in fact has an entire set of sub-states, caused by the vibration and rotation of the molecule (at a time scale of 10^{-13} s). This causes the rather broad peaks observed in optical spectroscopy. If the sub-states of the ground-state overlap with those of the excited state, direct return of an electron to the ground state is possible by **internal conversion (IC)** (FRANCK–CONDON-principle). The energy-difference is converted to molecular vibration, that is, heat.

If the sub-states of a ground- and an excited state do not overlap, direct return to the ground state is not possible. Instead, the electron first falls to S_1 by internal conversion, and returns from there to S_0 by emission of light. The energy difference between S_2 and S_1 is lost as molecular vibration (heat). The light emitted thus has less energy, and hence a *longer* wavelength, than the light absorbed (STOKES law). This phenomenon is called **fluorescence**.

Note, however, that in rare cases several light quanta (two or three) are absorbed successively by an electron, pushing it to higher and higher energy levels. Under these circumstances, the emitted light can have a *shorter* wavelength than the absorbed. Because such two- or three-photon events are improbable, they have become of practical significance only with the advent of laser light, where high photon flux densities occur. Fluorescence microscopy with such "anti-STOKES photons" results in a particularly low background.

Instead of returning to S_0, an electron at S_1 can undergo **intersystem crossing (ISC)** and arrive at the first triplet state T_1 with reversed spin. Return from there to the ground state is delayed, as it involves another spin reversal. Thus emission of light in **phosphorescence** is slow compared to fluorescence, the lifetime of the exited state may be up to 2 s, compared to less than 1ns for fluorescence.

Thermal motion can cause some of the molecules to be in an excited state. The probability of this depends on the energy gap between the ground and excited state and the available thermal energy, which is kT, with k = BOLTZMANN constant (1.3806×10^{-23} J K^{-1}) and T = absolute temperature (in K). The ratio of molecules in the ground and excited state is given by the BOLTZMANN distribution:

$$\frac{\eta_1}{\eta_2} = e^{\Delta E/kT} \tag{4.1}$$

For UVIS-spectroscopy ΔE is quite high, and virtually all molecules are in the ground state at room temperature. However, for IR spectroscopy a sizable fraction of molecules are in the excited state, and in NMR the ratio η_1 / η_2 is almost (but not quite) 1. A ratio of 1 can only be achieved if either $\Delta E = 0$ or $T = \infty$. The sensitivity of a spectroscopic method depends on the ratio η_1/η_2, as only molecules in E_1 can be pushed to E_2. Hence much larger sample sizes are required in NMR- than in UVIS-spectroscopy.

4.2 The Evanescent Wave

If light is passed from a medium of high- refractive index (n) into one with lower, some of the light is reflected at the interface between the media (see Fig. 4.2). The amount of light reflected depends on the angle θ of the light beam with the interface. Beyond a certain angle total reflection occurs. This behaviour is described by SNELL's law:

$$n_1 \sin(\theta_1) = n_2 \sin(\theta_2) \tag{4.2}$$

For a certain critical angle θ_C the light beam will be refracted in such a way that it passes along the interface between both media, *i.e.*, $\theta_2 = 90°$. This is true for

$$\theta_C = \arcsin(n_2/n_1) \tag{4.3}$$

However, because of quantum mechanical effects, some light will penetrate a small distance into the low refractive index medium. The intensity of this light decreases sharply with distance from the interface according to:

$$I_z = I_0 e^{-z/d} \tag{4.4}$$

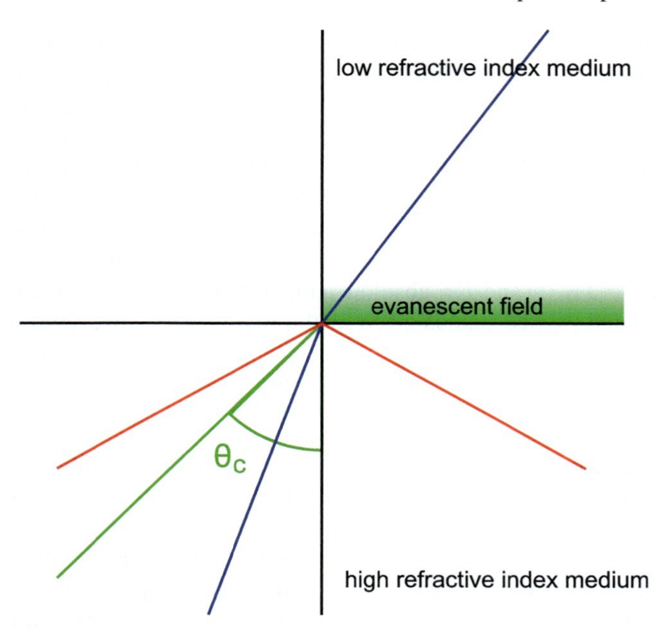

Fig. 4.2 Generation of an evanescent wave. Light is passed from a high-into a low-refractive index medium. If the incident angle is below a critical threshold θ_C, the light is refracted towards the high-refractive index medium (*blue line*). Above the critical angle, light is totally reflected back into the high refractive index material (*red line*). If the incident angle is θ_C, the light is refracted along the interface between both materials. Only the evanescent field penetrates into the low-refractive index material, the intensity decreases exponentially with distance from the interface

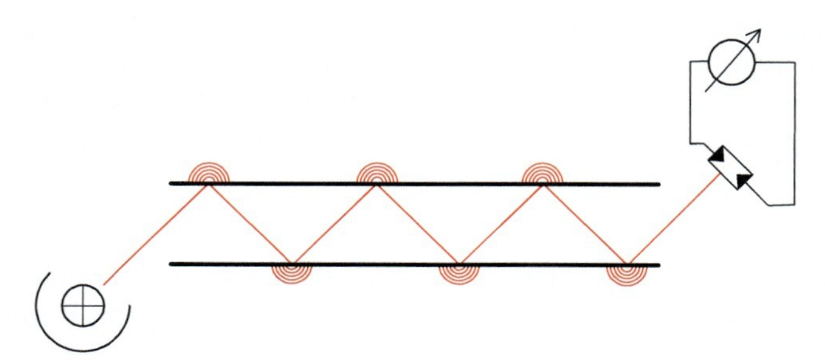

Fig. 4.3 A light beam is coupled into a piece of glass, so that repeated total internal reflection occurs. At each reflection point the light can interact with molecules on the surface of the glass

$$d = \frac{\lambda}{4\pi\sqrt{n_1^2 \sin^2\theta_1 - n_2^2}} \tag{4.5}$$

where d is called **penetration depth** of the light beam and z is the distance to the interface. Since d is only in the order of nm (which is about the size of proteins), methods based on evanescent waves can measure the events that happen about 1 molecule diameter away from a surface, with very little background caused by molecules further away. Many optical methods can be performed also with evanescent waves.

Even in those cases where total reflection occurs, light will penetrate a short distance into the low-refractive index medium, where it can interact with sample molecules. Depending on the angle of entry a light beam may be refracted several times before reaching the detector, increasing sensitivity (see Fig. 4.3). This is called **total internal reflection spectroscopy**.

Chapter 5
Photometry

5.1 Instrumentation

The simplest device to measure interaction of a sample with light is the absorbance spectrophotometer. Monochromatic light is passed through the sample (or, in some cases, reflected from it). If absorption occurs, the light intensity that arrives at a detector is lower than what would arrive in the absence of the sample.

Better instruments use a beam splitter after the monochromator to create two light beams: one passes through the sample and one through a reference cuvette filled with pure solvent or reagent blank (see Fig. 5.1). Thus any artifact caused by variations in light intensity or by absorbtion by the solvent are automatically eliminated (**double beam spectrophotometer**).

In modern instruments, the signal from the detector is usually digitalised, which allows convenient data handling by a computer.

A key part of a photometer is the **monochromator**, because measurements need to be taken with monochromatic light. This can be either a filter, a prism, or a grating, the latter being most common.

Some light will be **scattered** on the grating, depending on the shape of the groves. The groves need to be as flat as possible, they are optimised for low scattering at a particular wavelength and angle of incident (**blaze angle**). Modern gratings are made by laser holography, their groves have a concave cross-section to reduce stray light further.

Cuvettes should have high optical quality. For visible light they can be made of normal laboratory glass, for UV-light cuvettes need to be made of fused silica, usable down to about 180 nm. To reduce cross-contamination disposable cuvettes made of plastic may be used. Polystyrene is cheapest, but absorbs UV-light strongly below 320 nm. Polymetacrylate can be used down to about 280 nm, certain proprietary plastics to about 230 nm (all wavelengths for 50 % transmission).

5.2 LAMBERT–BEER's Law

What influence will the sample have on the intensity of light reaching the detector?

E. Buxbaum, *Biophysical Chemistry of Proteins: An Introduction to Laboratory Methods*, DOI 10.1007/978-1-4419-7251-4_5,

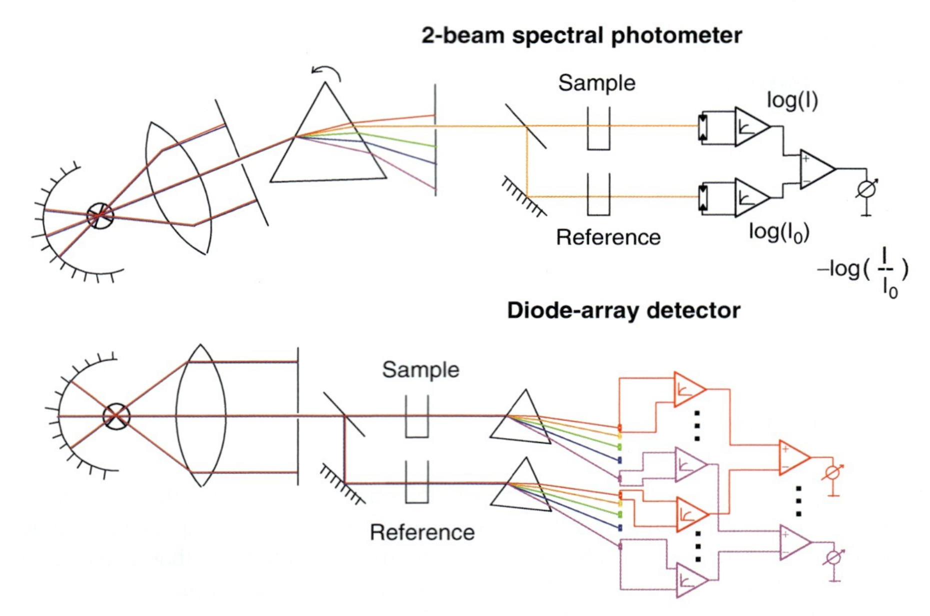

Fig. 5.1 *Top*: Principle of an absorbtion spectrophotometer. Light is passed through a monochromator (prism, grid, or filter) and a beam splitter. The resulting two monochromatic light beams pass through a sample or reference cuvette, respectively, onto detectors. Their relative intensity is converted into an electrical signal. Light source can be a tungsten filament lamp for visible and near-IR light, UV light is usually produced by a deuterium-lamp. Most research photometers have both, you need to select the one appropriate for your measurements. *Bottom*: The light may be passed through the sample directly, and then through a grid or prism which projects the spectrum onto an array of (usually 256 or 512) light sensitive diodes. This way an entire absorbtion spectrum can be determined in one go (about 0.1 s per spectrum, but averaging over several spectra is often required for better S/N ratio), albeit with lower sensitivity and resolution. Such instruments tend to be used as chromatographic detectors, where different peaks may absorb at different wavelengths

Assume that a sample absorbs just half of the light that passes through it. Then a second, identical sample will absorb half of the light that passed through the first, and the detector will see only 1/4 of the light generated by the source. Thus doubling the thickness of the sample (d) results in doubling of **absorbance** ($\mathcal{E} = -\log(I/I_0)$). Thickness is usually measured in cm, because standard cuvettes used in spectroscopy have $d = 1$ cm.

By the same reasoning, if the amount of absorbing substance in the second sample were dissolved in the first sample, doubling its concentration (c), we would also expect doubling of the absorbance.

We also know that there are substances that cause strong absorbtion at a certain wavelength even in minute concentration, while other substances do not absorb the same wavelength at all (but may cause absorption at a different wavelength). This property can be expressed mathematically as a wavelength dependent constant, the **molar absorbtion coefficient** $\epsilon(\lambda)$, which has the unit $\mathrm{l\,mol^{-1}\,cm^{-1}}$.

These considerations lead directly to LAMBERT–BEER's law:

$$\mathcal{E} = -\log\left(\frac{I}{I_0}\right) = \epsilon(\lambda) * c * d \tag{5.1}$$

the absorbance $\mathcal{E}$ caused by the sample, defined as the negative logarithm of the quotient of outgoing (I) and incoming (I_0) light intensity, is proportional to its concentration (c, measured in mol/l) and its thickness (d, measured in cm). The proportionality constant is the wavelength dependent molar absorption coefficient, ϵ.

A more formal way to derive this equation starts with the assumption that a given volume of solution V contains a certain number of molecules n with a certain probability P to absorb a light quantum of a given wavelength. Then the cross-section for absorption $k = Pn/V$ and $I = I_0 e^{knd} = I_0 e^{-\epsilon c d}$. The absorption probability is $k = 3.28 \times 10^{-21}\epsilon$.

LAMBERT–BEER's law is, however, valid only under certain conditions:

- The light must be (as nearly as practical) monochromatic. Since absorbtion of light is wavelength dependent, polychromatic light results in a non-linear relationship between concentration and absorbance.
- The sample must be free of dust particles, gas bubbles, or other objects that cause light scattering. Vacuum filtration of all samples is a good way to achieve this.
- The sample must be so dilute that interactions between molecules do not influence the absorbance. Additionally, as absorbance increases the light sensor sees less and less light, until measurements become influenced by noise. The noise level depends on the quality of the photometer. As a rule of thumb, $\mathcal{E} > 1$ (90 % light absorbed) will cause non-linear $\mathcal{E}$ vs c plots in most cases but problems can occur earlier with some substances (see Fig. 5.2). Before establishing a new

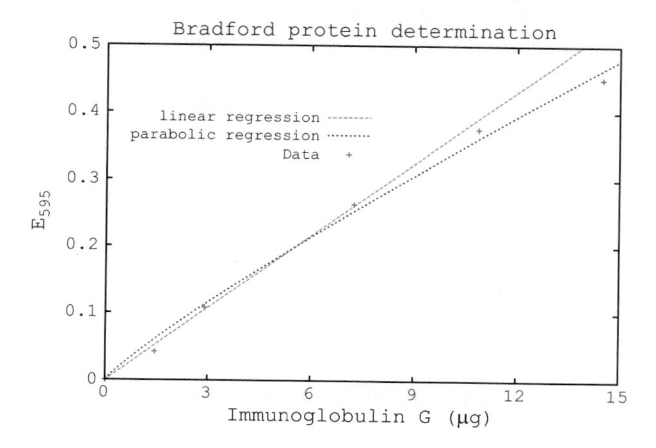

Fig. 5.2 LAMBERT–BEER's law is valid only for small concentrations. Here, different amounts of Immunoglobulin G were used to establish a standard curve for protein determination according to BRADFORD [36]. The last point clearly deviates from the linear regression line. For such high concentrations, either the sample would have to be diluted, or a parabolic regression curve used

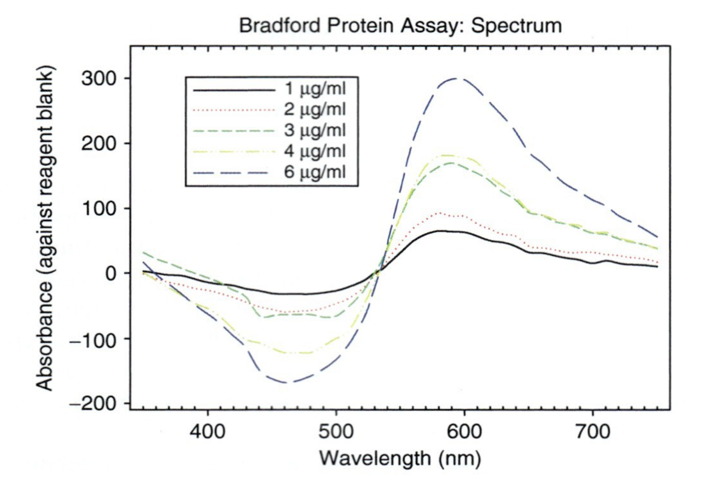

Fig. 5.3 Spectrum of the BRADFORD protein assay [36]. Maximum sensitivity is obtained at 590 nm, the isosbestic point is at about 535 nm

assay, the concentration range where LAMBERT-BEER's law is valid must be checked. Higher concentrations may however be used if a parabolic standard curve is established.

5.2.1 The Isosbestic Point

Figure 5.3 shows the spectrum of BRADFORD's protein assay for different protein concentrations. Maximum difference between the samples is at 590 nm, this is the wavelength one would select to measure protein concentrations. There is an isosbestic point at 535 nm, where the absorbance is independent of the protein concentration. Differences between samples at this wavelength are caused by experimental artifacts, for example inhomogeneities in the bottom of 96-well plates. Plate readers therefore measure the ratio of absorbance at two wavelengths, the second wavelength should be at the isosbestic point; in the case of the BRADFORD assay, one would measure the ratio of A_{590}/A_{535}. If the spectrum has several isosbestic points, select the one closest to the measurement wavelength so that scattering is similar at both wavelengths.

5.3 Environmental Effects on a Spectrum

The absorbtion spectrum of a substance depends on the polarity of its solvent. If a change from a less to a more polar solvent leads to a shift of the absorbance maximum toward red (longer wavelengths, lower energy), we speak of a **bathochromic** effect; a shift towards blue (shorter wavelengths, higher energy) is called a **hypsochromic** effect. Bathochromic effects are usually seen in $\pi \rightarrow \pi^*$-transitions, hypsochromic in n $\rightarrow \pi^*$-transitions. Higher solvent polarity stabilises the more polar, that is excited, state of an orbital, which explains the bathochromic effect of high polarity solvents on the $\pi \rightarrow \pi^*$-transition. At the same time, polar solvents will form hydrogen bonds with *n*-orbitals, which need to be broken for a $n \rightarrow \pi^*$-transition. This requires energy, explaining the hypsochromic effect.

Interactions between chromophores in highly organised macromolecules can reduce (**hypochromic effect**) or increase (**hyperchromic effect**) the molar absorbtion coefficient at a particular wavelength. Thus the absorbtion of a DNA-solution at 260 nm will increase sharply with temperature around the melting point of DNA. This has to do with interactions between dipole moments of the chromophores (attraction and repulsion of partial charges, which stabilise or destabilise a particular state). According to the KUHN–THOMAS-rule a hyperchromic effect needs to be compensated by a hypochromic effect at a different wavelength, that is, the integral under the absorbtion curve is constant. In case of DNA, hypochromic effects are observed in the far UV region, which is not accessible with normal laboratory equipment.

Chapter 6
Fluorimetry

Fluorescence can be used to determine the concentration of a substance quantitatively (see Fig. 6.1). A more thorough introduction into the various fluorescent techniques is given in [219]. Even very specialised techniques, discussed below, can also be performed on microscopes [141, 142].

Because in photometry the difference between the incoming and outgoing light intensities is measured, this method requires fairly high concentration to be precise (small differences between large values are notoriously difficult to determine). By comparison, in fluorescence spectroscopy the signal is measured quasi as a difference from zero. Thus fluorescence is more sensitive than absorbance spectroscopy by 3–4 orders of magnitude. Fluorescence intensity is a linear function of the concentration of the sample, at least until the concentration becomes large enough that internal filtering reduces the measured intensity. As a general rule, sensitivities of different techniques are absorbance $<$ fluorescence $\leq$ radioactivity. Assay costs increase in the same order.

Internal filtering can be caused by reabsorption of the emitted light by other molecules, this process will be more important if the absorbance and emission spectra of a fluorophore partially overlap (small STOKES-shift). A further possible cause for loss of fluorescence is oligomerisation of the fluorophore, in most cases such dimers or higher oligomers do not fluoresce. However, for a few compounds excited dimers (**excimers**) show fluorescence at a longer wavelength than the monomers.

One cause of error in fluorescence measurements to be aware of is higher order stray light. Most fluorimeters use grid monochromators instead of prisms. As you may recall from your physics lessons, these project several spectra which partly overlap. If you want monochromatic light of $\lambda = 900$ nm, you also get 1/2 $\lambda = 450$ nm, 1/3 $\lambda = 300$ nm and so on. Such higher order light must be filtered out with a low pass filter (which allow light of low energy = long wavelengths to pass), as it can cause false peaks in the spectrum, especially with turbid samples like membrane suspensions.

Fluorophores (see Fig. 6.2 and Table 6.1 for common examples) may show hypsochromic or bathochromic effects, which can be used to probe their environment (at the surface or the interior of a protein, within a membrane or in an aqueous solution). Tryptophane fluorescence is often used for this purpose.

E. Buxbaum, *Biophysical Chemistry of Proteins: An Introduction to Laboratory Methods*, DOI 10.1007/978-1-4419-7251-4_6,

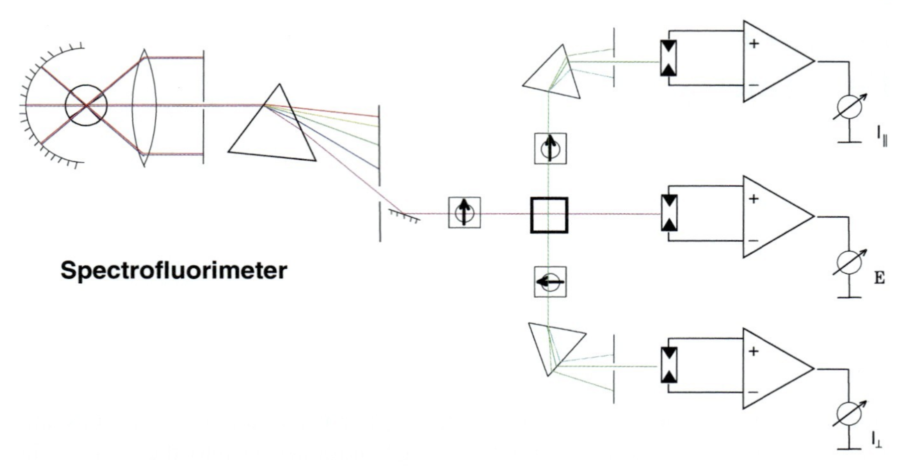

Fig. 6.1 Schematic diagram of a complete fluorimeter. The light (usually from a xenon or mercury high pressure arc lamp: **danger**, may explode if incorrectly handled) is passed first through a monochromator, then through a polariser, and finally through the sample. Fluorescent light is collected at an angle of 90°, and passed through another polariser and a monochromator. Its intensity is then determined by a light sensitive device, usually a photomultiplier. The polariser can have parallel or vertical orientation with respect to the polariser of the exciting beam, thus two detection systems are required. A third one measures the absorbance of the sample, and, if that is considered constant, can be used to correct for variations in the intensity of the light source. For reasons of cost, most fluorimeters have only one detector. Excitation and/or emission monochromator can be replaced by filters, if a fixed wavelength is to be used. This increases light throughput and hence sensitivity of the instrument

6.1 Fluorescent Proteins

Phycobiliproteins are used as light antenna by algae, capturing light and donating it to chlorophylls. Purified phycobiliproteins show intense fluorescence with a quantum yield of 0.98, together with a large STOKES-shift. In addition, because they contain many bilin fluorophores they have absorption coefficients of several 10^6 l mol cm^{-1}. Thus they can be used for extremely sensitive assays, detecting a few hundred target molecules. Their disadvantages include price and high molecular mass (several 100 kDa), which makes separation of conjugated and non-conjugated probes difficult.

Green fluorescent protein (GFP) was isolated from the jelly fish *Aequorea victoria* [415]. It accepts energy from a chemiluminescent protein (which produces blue light) and translates it into green light by **bioluminescence resonance energy transfer (BRET)**. The function of bioluminescence in these animals is unknown. Formation of the fluorophore proceeds spontaneously in the absence of other proteins and cofactors except oxygen (see Fig. 6.3). Because of its intensive green fluorescence, GFP has become a favourite marker in molecular biology. The genetic information for GFP is attached to the gene for the protein under investigation, so that a fusion product is generated. Both the amount of target protein produced and its

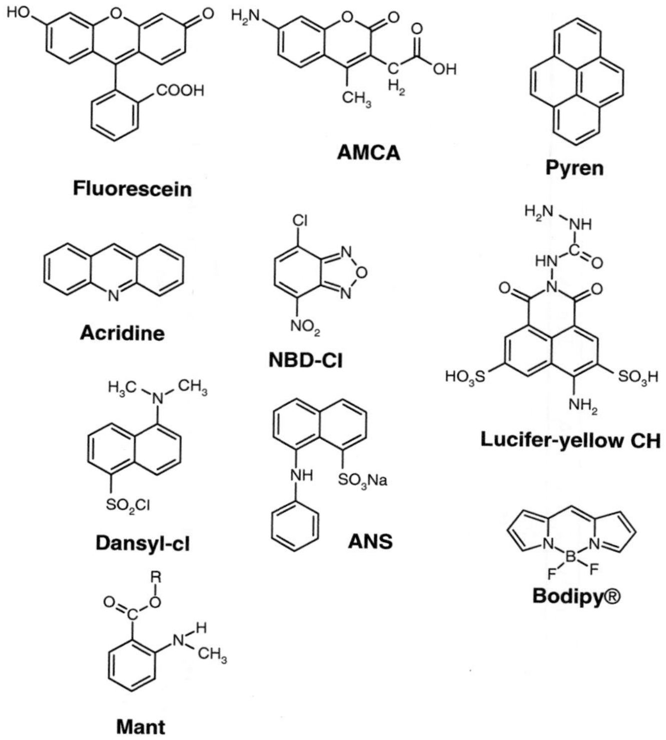

Fluorescein AMCA Pyren

Acridine NBD-Cl Lucifer-yellow CH

Dansyl-cl ANS Bodipy®

Mant

Fig. 6.2 Some common fluorescent probes. The abbreviations are expanded in Table 6.1

subcellular localization can then be studied by fluorescence video microscopy. Several mutant forms (blue, cyan, yellow, red... fluorescent protein) have been designed by changing amino acids in the vicinity of the fluorophore, allowing multi-label studies.

6.2 Lanthanoid Chelates

Lanthanoids (older literature: lanthanides, from Greek "to lie hidden") or rare earths are the elements between Lanthanum (57) and Hafnium (72) in the periodic system. In this cluster of elements the 4f-orbital is filled successively. Since, however, the f-orbitals are deep inside the electron cloud (hidden by 5s and 5p), this has little effect on their chemical properties, which resemble those of Sc and Y. All these elements occur in the oxidation state +III, some also in +II (Sm, Eu, Tm, Yb) or +IV (Ce, Pr, Nd, Tb, Dy). The f-electrons can be pushed to higher energy levels by UV-light, resulting in luminescence. Because the f-electrons are hidden, their energy levels are independent of the ligand environment, resulting in narrow emission lines

Table 6.1 Common fluorescent probes and their properties

Name	Abbreviation	λ_{ex} (nm)	$\epsilon(1\ mol^{-1}\ cm^{-1})$	λ_{ex} (nm)	Φ_F	Remark
Acridine	–	362	11 000	462	0.01	
7-Amino-4-methylcoumarin	AMCA	345	19 000	445	0.64	
1-Anilinonaphtalene-6-sulphonate	1,6-ANS	372	7 800	480	0.37	Fluorescence stronger in hydrophobic environment
4-Bora-3a,4a-diaza-s-indacene	Bodipy					Fluorescent properties depend on substituent
4-Chloro-7-nitrobenz-2-oxa-1,3-diazole	NBD-Cl	465	8 000	535	0.3	Fluorescence stronger in hydrophobic environment
5-Dimethylaminonaphtalene-1-sulphonyl chloride	Dansyl-Cl	372	3 900		0.7	Fluorescent only after reaction with amines
Fluorescein	–	394	88 000	518	0.93	pH-dependent ($pK_a = 6.4$)
Lucifer yellow	–	428	11 000	536	0.2	Aldehyde fixable cell tracer
Methylanthroyl-	Mant-	368	6 500	437	0.4–0.5	Small residue used to label substrates
Pyrene	–	340	37 000	376	0.06	Excimer-formation ($\lambda_{ex} = 470$ nm)
Tryptophane	Trp	280	5 600	355	0.13	
Tyrosine	Tyr	275	1 400	304	0.14	
Phenylalanine	Phe	258	200	282	0.02	

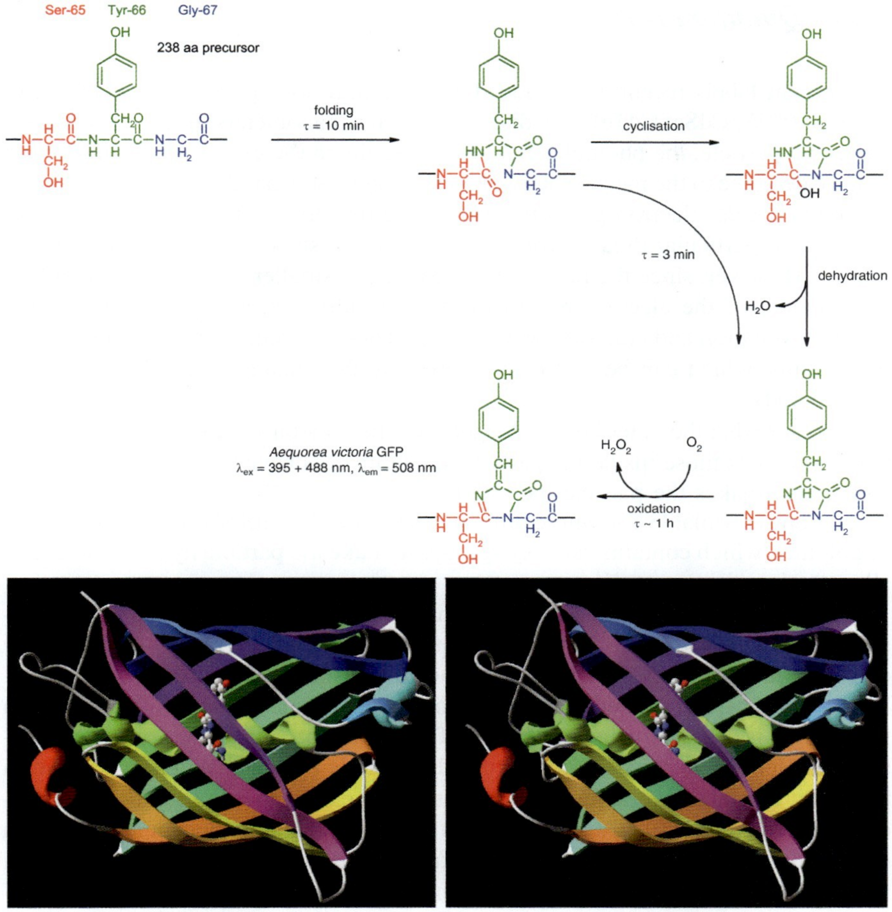

Fig. 6.3 *Top*: Formation of the fluorophore of GFP. The only extrinsic factor required is oxygen. *Bottom*: Stereo-diagram of the crystal structure of GFP (PDB-code 1ema). The fluorophore is located in an α-helix (green), which is protected from environmental influences by a surrounding β-barrel

(half width 10 nm) that are characteristic for the respective element. However, the excitation transitions are forbidden, therefore luminescence depends on sensitisation by ligands that absorb light energy to achieve a triplet state, which can transfer energy to the lanthanoid ion 4f-electrons. Lanthanoids form nine coordination bonds. Such complexes have large STOKES-shifts with no overlap between excitation and emission lines, and hence no self-quenching. Fluorescent lifetimes are extremely long (μs), making these complexes ideal for time-resolved fluorescence and FRET studies. The brightest complexes are those of Eu^{3+} (615 nm), followed by Tb^{3+} (545 nm) and Sm^{3+} (645 nm) [56, 264].

6.2.1 *Quantum Dots*

Fluorescent labels recently introduced are quantum dots, particles of photoactive material (CdS, CdSe, CdTe) with diameter of a few nanometers. If light is absorbed by such a particle, the photoelectrical effect results in the excitation of an electron of the calcogene to the metal (valence band to conduction band), resulting in the formation of a hole–electron-pair. This pair can be treated quantum-mechanically like an atom. In particular the electron has a preferred distance from the hole (BOHR-radius). However, since the radius of the particle is smaller than the BOHR-radius, confinement of the electron requires energy, resulting in a larger band gap and smaller excitation and emission wavelengths. These become dependent on the particle radius, which can be tightly controlled by the manufacturer (2 nm → blue, 6 nm → red).

To ensure that the outer layer calcogene and metal atoms see the same electronic environment as those inside the particle, quantum dots have an outer shell of ZnS, which also makes the particle inert and non-toxic to cells ([76], no toxicity data are currently available for whole animals). Outside the ZnS-shell is an outer coat of polymer, which contains carboxyl-groups to make the particle hydrophilic and to allow conjugation to proteins.

Advantages of quantum dots over conventional fluorophores include high absorbtion (up to $10^7\,\mathrm{l\,mol^{-1}\,cm^{-1}}$) and quantum efficiency (up to 0.95), narrow emission lines, long-term photostability, lack of self-quenching, inertness under cellular conditions, and a moderately long fluorescent lifetime (5–40 ns). The excitation wavelength is characteristic for a family of quantum dots, this makes multi-label assays simpler. Their most significant disadvantage is blinking: After emission of a photon, it can take several ms before a dot can be exited again. Blinking can be reduced by 100 mM DTT. Currently, almost half of all quantum dots in a preparation may be dark, *i.e.*, do not fluoresce. Hopefully, better manufacturing processes will reduce this problem. For recent reviews on quantum dots see [56, 159, 274].

6.3 Fluorescence Quenching

6.3.1 *Dynamic Quenching*

An excited electronic state can decay by several processes, fluorescence being only one of them. Each of these processes is first order, hence

$$v = \frac{\mathrm{d}n}{\mathrm{d}t} = k_i n \qquad (6.1)$$

with k_i being the rate constant for process i. Then the **quantum efficiency** Φ for fluorescence (photons of fluorescent light emitted divided by photons of exciting light absorbed) is given by:

$$\begin{aligned}\Phi &= \frac{n_F}{n_a} \\ &= \frac{k_F}{k_F + k_{isc} + k_{ic} + k_Q} \\ &= \frac{k_F}{\sum_{i=1}^{n} k_i} \\ &= k_F \tau \end{aligned} \tag{6.2}$$

with fluorescence, intersystem crossing, internal conversion, and quenching competing for excited molecules. τ is the lifetime of the excited state. Quenching of fluorescence occurs when the excited molecule collides with some paramagnetic molecule that can accept its energy. I^-, O_2 and acrylamide are commonly used quenchers. The STERN–VOLMER-equation describes how the fluorescence intensity changes with quencher-concentration $[Q]$:

$$\frac{I_0}{I} - 1 = K_q[Q]\tau_0 \tag{6.3}$$

with K_q = quenching constant of the quencher Q. If $I_0/I - 1$ is plotted as a function of quencher concentration (STERN–VOLMER-plot), K_q can be determined from the slope if the fluorescence lifetime of the fluorophore is known.

The efficiency of quenching depends on how much access the quencher has to the fluorophore. If the fluorophore is buried deep inside a protein molecule, then water soluble quenchers have limited or no access to it and quenching will not occur. If the protein is unfolded, for example by addition of urea, quenching will increase and fluorescence quantum yield will decrease.

Such measurements can determine:

- If a fluorophore bound to a membrane protein is inside the membrane, cytoplasmic or extracellular. If the fluorophore is buried inside the membrane, lipophilic but not hydrophilic quenchers can reach it. If it is extracellular, it can be reached by hydrophilic quenchers without permeabilisation of the plasma membrane. If it is intracellular it can be reached by hydrophilic quenchers, but only after the membrane has been permeabilised (for example with digitonin).
- Structural changes during the catalytic cycle of an enzyme where fluorophores buried, say, in ES may become accessible, say, in EP.
- The sensitivity of a protein against unfolding by heat, denaturant, pressure or other parameters.

A special case of fluorescence quenching is **excimer** (excited dimer) formation, which occurs in certain fluorophores when the exited fluorophore collides with a second, non-exited fluorophore-molecule. The fluorescence of that excimer is shifted to longer wavelengths. Excimer-formation may be used to measure complex-formation between proteins: If Ca-ATPase (a P-type transmembrane ATPase) molecules are labelled with pyrene (a molecule that shows excimer formation), association of ATPase-molecules into a catalytically active dimer brings the pyrene-molecules close enough together for excimer-formation. Additional lipid will not reduce the excimer-fluorescence, but unlabelled Ca-ATPase will [227]. Similar results can also be obtained by fluorescence energy transfer studies on Ca-ATPase.

6.3.2 *Static Quenching*

While in dynamic quenching the quencher accepts the energy of the excited fluorophore during a collision, static quenching is caused by complex formation between the quencher and the fluorophore in its ground state. If the complex does not show fluorescence, complex formation results in a reduction of the effective concentration of the fluorophore. Since this reduction can be described by the law of mass action, static quenching can be described by

$$\frac{I_0}{I} - 1 = K_a[Q] \tag{6.4}$$

which is similar to the STERN–VOLMER-equation except that the fluorescence lifetime does not occur.

It may be necessary to experimentally distinguish static from dynamic quenching. Dynamic quenching depends on the *collision* of molecules and is diffusion controlled. It increases with temperature and decreases as the viscosity of the medium is increased. Static quenching is caused by *binding* of quencher to fluorophore, and the association constant depends on the temperature according to the VAN'T HOFF-equation (see p. 410). Viscosity of the medium will have little effect on static quenching. However, these are only limiting cases, in many experiments you will see a combination of static and dynamic quenching.

6.4 FÖRSTER Resonance Energy Transfer

FRET is similar to fluorescence quenching, except that energy transfer requires the fluorophore ("donor") and the "quencher" ("acceptor") to be in resonance. This happens if the acceptor is itself a fluorophore with an absorbance spectrum that overlaps (at least in part) with the emission spectrum of the donor (see Fig. 6.4). Under these conditions the acceptor can accept the energy from the donor over distances of up to 10 nm (100 Å), while dynamic quenching requires actual collision between fluorophore and quencher. Additionally, the transition dipole moments of acceptor and donor must be oriented properly. Note that FRET occurs directly between the electron clouds of the two molecules, rather than by fluorescent emission by the donor and re-absorbtion by the acceptor. Experimentally, FRET leads to the appearance of fluorescence at the emission maximum of the acceptor, whilst the fluorescence

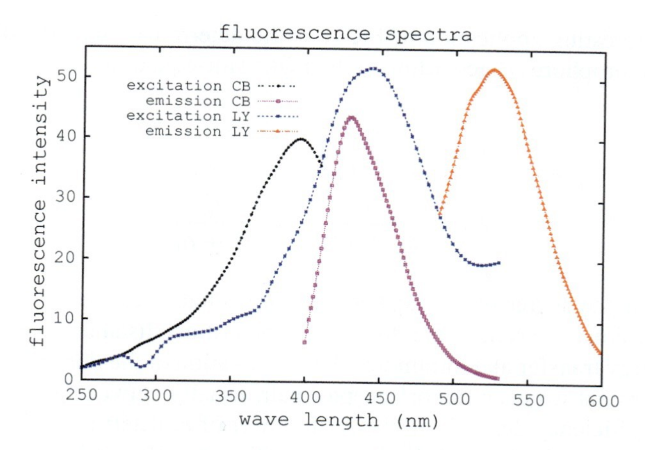

Fig. 6.4 Fluorescence spectrum of Cascade blue (a pyrene derivative) and Lucifer yellow. The emission spectrum of Cascade blue and the excitation spectrum of Lucifer yellow overlap (*i.e.*, the dyes are in resonance), hence fluorescence energy transfer is possible between them

intensity, and the fluorescence lifetime, at the emission wavelength of the donor is reduced. Because the excited state has an additional pathway for decay, photobleaching of the donor is also reduced, which is particularly useful when FRET is measured under a microscope. These effects were first analysed by THEODOR FÖRSTER [107].

The acceptor may also be stimulated by luminescence from luciferase (**bioluminescence resonance energy transfer (BRET)**) instead of a fluorescent dye. The main advantage of BRET over FRET is that no excitation light shows up in the measured signal, thus selectivity is higher.

If the rate of energy transfer is k_T and the transfer efficiency E_T, then we get

$$E_T = \frac{k_T}{k_T + k_f^D + k_{ic}^D + k_{isc}^D} \tag{6.5}$$

$$\frac{\Phi_{D/A}}{\Phi_D} = 1 - E_T = \frac{\dfrac{k_f^D}{k_f^D + k_{is}^D + k_{isc}^D + k_T}}{\dfrac{k_f^D}{k_f^D + k_{is}^D + k_{isc}^D}} \tag{6.6}$$

$$\tau_{A/D} = \frac{1}{k_f^D} k_f^D + k_{is}^D + k_{isc}^D + k_T \tag{6.7}$$

$$\tau_D = \frac{1}{k_f^D} k_f^D + k_{is}^D + k_{isc}^D \tag{6.8}$$

$$1 - E_T = \frac{\tau_{A/D}}{\tau_D} \tag{6.9}$$

The most interesting application of FRET is the determination of the distance between two fluorophores d according to the FÖRSTER-equation:

$$k_T = \frac{1}{\tau_D}\left(\frac{r}{d_0}\right)^{-6} \tag{6.10}$$

$$E_T = \frac{k_T}{k_T + 1/\tau_D} = \frac{d_0^6}{d^6 + d_0^6} \tag{6.11}$$

d_0 depends on the degree of overlap between the excitation spectrum of the acceptor and the emission spectrum of the donor, it is the critical distance where the probability of energy transfer and intramolecular deactivation of the donor become equal. This is a constant for each donor/acceptor pair, usually between 1–5 nm. Because the transfer efficiency depends on the sixth power of r, determination of d is very precise if d is near d_0. Since d_0 is close to the diameter of proteins, FRET is a very important method, for example, to measure the distance between the binding sites of an enzyme for its substrates. The longest d_0 are observed in lanthanoid complexes ($\approx$10 nm).

6.4.1 Handling Channel Spillover

These measurements are complicated by the fact that we may know neither the distance of the fluorophores, their relative concentration (*e.g.*, inside a cell in fluorescence microscopy), nor their stoichiometry. In addition, we may not have a true base line, since some of the donor fluorescence may “spill” into the acceptor region, similarly the acceptor may be excited by λ_{ex} of the donor. The latter two problems increase with decreasing STOKES-shift. To solve this problem, MULLER & DAVIS [287] measured the signal in the FRET-channel when only either the donor or the acceptor where present, and divided that by the donor or acceptor fluorescence signal, respectively, to obtain donor and acceptor spillover factors, which are concentration independent:

$$\mathrm{DSF} = \mathrm{FRET}_D / F_D, \tag{6.12}$$

$$\mathrm{ASF} = \mathrm{FRET}_A / F_A. \tag{6.13}$$

Thus it is possible to correct for the unknown donor and acceptor concentrations in the FRET-experiment by calculating the total spillover, *i.e.*, the FRET-signal one would obtain at the given (but unknown) concentrations of donor and acceptor if no FRET were occurring:

$$\mathrm{SF} = (F_D \times \mathrm{DSF}) + (F_A \times \mathrm{ASF}). \tag{6.14}$$

The relative FRET-signal is then simply the measured signal in the FRET-channel divided by the spillover, *i.e.*, how many times stronger is the signal than would be expected in the absence of FRET. In the absence of FRET this is 1.0, even very strong FRET results in a relative signal of ≤ 3.

6.4.2 Homogeneous FRET Assays

FRET has become a favourite detection method for clinical assays (*e.g.*, antigen/antibody binding, no separation of bound from unbound ligand required) and for high throughput screening. An example would be the assay for proteolytic activity, where the two fluorophores are covalently bound to either side of the cleavage site of a protein. If the protein is digested, separation of the fluorophores leads to a decrease in FRET. If several 10 000 compounds a day need to be checked for inhibitory activity on a protease (say, HIV-protease) such **homogeneous assays** save a lot of time and material.

One way to evaluate the results of those studies is to calculate the ratio of acceptor to donor fluorescence, and then the relative energy transfer rate ΔF from the ratios of sample and negative control:

$$R = \frac{F_\mathrm{a}}{F_\mathrm{d}} \qquad \Delta F = \frac{R_\mathrm{sample} - R_\mathrm{blank}}{R_\mathrm{blank}}. \tag{6.15}$$

This procedure eliminates some problems with absorbtion and quenching of fluorescence.

6.4.3 Problems to Be Aware Of

FRET is better at determining relative than absolute distances. Several factors which are difficult to control affect FRET, most importantly the relative steric orientation of the transition dipole moments of the fluorophores. The **orientation factor** is between 0 and $4^{1/6} = 1.26$. Most fluorophores are attached to the protein via a (flexible) **spacer arm**, whose length affects the results. Not all proteins become labelled during a labelling reaction. FRET works best if the fluorophores have both large STOKES-shifts and high quantum yields, these properties are unfortunately mutually exclusive. Because of the sixth-power dependency of FRET on the distance between fluorophores, the distance range is limited, donor/acceptor pairs need to be carefully selected for the measurement problem. Concentration of the fluorophores should be low (absorbance less than 0.02) to avoid **self-absorbtion**.

6.4.4 Fluorescence Complementation

The main disadvantage of FRET and BRET assays is that the difference between fluorescence intensity or lifetime in the presence and absence of interaction is measured. This inevitably reduces the signal-to-noise ratio. In fluorescence complementation the fluorescing species is formed only when the probes interact, without interaction there is no fluorescence. Measurements can thus be made with higher sensitivity.

For fluorescence complementation assays the two proteins to be investigated are produced as fusion proteins with parts of GFP. When the proteins interact, the fragments of GFP come into contact and autocatalytically form over several hours the fluorophore. The location of the complex can be visualised in living cells [194]. This method is, however, not suitable to measure the kinetic parameters of the interaction between the proteins.

6.4.5 Pulsed Excitation with Multiple Wavelengths

Emission spectra of most fluorophores are broad, and in multi-colour fluorescence studies there is always cross-talk from the short- to the long- wavelength fluorophores. This is a particular problem in FRET-studies, but also in multi-colour fluorescence microscopy.

In **pulsed interleaved excitation** (**PIE**) [280] and **alternating laser excitation** (**ALEX**) [188] the sample is illuminated alternately by several laser sources of different wavelength. PIE uses pulsed lasers with pulse-widths of about 1 ns and repetition frequencies of several 10^7 Hz, while ALEX switches between continuous lasers at a frequency of several kHz. In both cases the fluorescence caused by excitation with one wavelength is detected before the next pulse (with a different wavelength) is fired at the sample. Since the various fluorophores are excited and measured at different times, cross-talk is avoided and true FRET measurements become possible. On the other hand, excitation frequencies – especially in PIE – are so high that the changes in the sample during a cycle are minimal. The measurements taken in the various channels can be considered quasi simultaneous.

6.5 Photoinduced Electron Transfer

When a fluorescent dye is activated by light, one of its electrons is pushed to a more energy-rich orbital. So far we have looked at processes where that electron simply falls back to the ground state (fluorescence and phosphorescence). However, some activated fluorescent dyes can donate the electron to an acceptor, the dye becomes oxidised and the acceptor reduced ($F^* + A \rightarrow F^+ + A^-$). Alternatively, the empty space in the ground state of the dye can be filled with an electron from

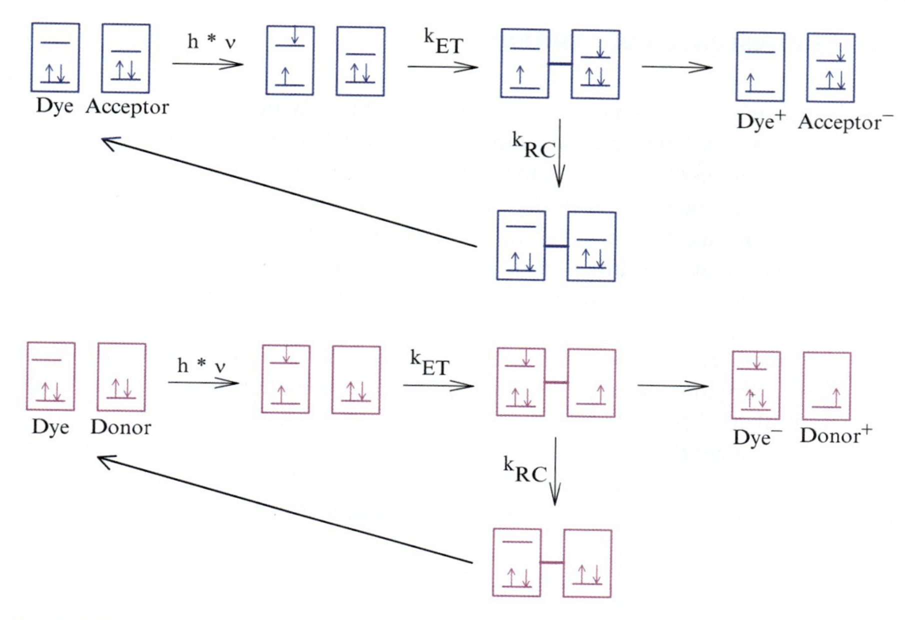

Fig. 6.5 Photoinduced electron transfer (PET). The fluorescent dye can either donate an electron to an acceptor or accept it from a donor. Total charge separation by formation of ion radicals would destroy the dye and result in photo-bleaching. Recombination, however, results only in fluorescence quenching

a suitable donor ($F^* + D \rightarrow F^- + D^+$). In that case, the dye is reduced and the donor oxidised. In either case the excited electron can no longer fall back to the ground state, hence fluorescence is quenched. The formation of such ion radicals would destroy the dye (which is one mechanism of photo-bleaching). Alternatively, charge separation in the complex may be only partial and followed by recombination. In this case, the excited electron returns to the ground state by a different route from that taken in fluorescence, which also results in fluorescence quenching but not in the destruction of the dye (see Fig. 6.5). Efficiency of PET is determined by $\eta = k_{ET}/k_{RC}$.

The most important example for PET is photosynthesis, where the chlorophyll P_{680} is activated and donates its electrons to an acceptor chain which rapidly increases the distance between P_{680} and its electrons. These electrons get replenished by the splitting of water ($H_2O \rightarrow 1/2 O_2 + 2H^+ + 2e^-$). P_{700} is activated by light and transfers its electron to another acceptor chain which eventually reduces FAD ($FAD + 2H^+ + 2e^- \rightarrow FADH_2$). The electrons of P_{700} are replenished by the acceptor chain of P_{680}.

Measuring fluorescence quenching by PET complements FRET since it acts over much shorter distances (0.1–1 nm compared to 2–10 nm). PET is therefore used to measure intramolecular processes like protein folding, rather than the association between proteins, often by fluorescence auto-correlation (*vide infra*).

6.6 Fluorescence Polarisation

If fluorescence is excited with linear polarised light, only those molecules will get excited whose **transition dipole moment for absorbtion** is oriented "nearly" in parallel with the electric field vector of the light beam. This excited state lasts in the order of nanoseconds before light is emitted as fluorescence. If the molecule cannot move during this period (*e.g.*, in the frozen state), the emitted light would also be polarised in the direction of the transition **dipole moment for emission** (in most molecules the transition dipole moments for absorption and emission are within ± 40° of each other). In solution, however, the molecule will undergo random (BROWNIAN) movement during the lifetime of the excited state. Among these movements is rotation around the three axes of the molecule which will change the orientation of its dipole. Thus the emitted light will be less polarised, this effect will be the stronger the longer the lifetime of the excited state is, and the faster the molecule rotates. If the fluorescent lifetime is known, the rates of rotation of the molecule can be determined. Fluorescence polarisation is expressed as **fluorescence anisotropy** A:

$$A = \frac{I_{\parallel} - I_{\perp}}{I_{\parallel} + 2I_{\perp}} \tag{6.16}$$

The factor 2 in the denominator results from the fact that light polarised perpendicular to, say, the x-direction can still have two possible directions (y and z). Both have identical intensities if the rotation of the molecule is random. Turning the polariser by 90° will measure only one of these components. Thus the total fluorescence is given by $I_{\parallel} + 2\ I_{\perp}$. Sometimes you will find polarisation described by the **degree of polarisation** P:

$$P = \frac{I_{\parallel} - I_{\perp}}{I_{\parallel} + I_{\perp}} \tag{6.17}$$

but that is not quite correct from the physical point of view.

It can be shown that in the frozen state, where the molecule can not rotate during the fluorescence lifetime, and if the transition dipole moments for excitation and emission are parallel, $I_{\parallel} = 3/5$ and $I_{\perp} = 1/5$, hence $A_0 = 0.4$. If there is an angle γ between the transition dipole moments, A_0 becomes

$$A_0 = \frac{3cos^2\gamma - 1}{5} \tag{6.18}$$

If the molecule can rotate during the lifetime of the excited state, A becomes

$$A = A_0\left(1 + 3\frac{\tau_F}{\tau_c}\right) \tag{6.19}$$

which is known as PERRIN-equation [323]. τ_F is the lifetime of the excited state and τ_c the rotation correlation time:

$$\tau_c = \frac{1}{2D_{rot}} \tag{6.20}$$

$$D_{\text{rot}} = \frac{k * T}{V_h \eta} = \frac{\frac{A}{A_0} - 1}{6\tau_F} \tag{6.21}$$

V_h is the hydrated molecular volume, k the Boltzmann-constant , T the absolute temperature and η the viscosity of the medium. Thus A will be somewhere between 0 ($\tau_c \ll \tau_F$, molecules have time to assume random orientation during fluorescence lifetime) and A_0 ($\tau_c \gg \tau_F$, molecules do not rotate noticeably during fluorescence lifetime). Thus the fluorophore needs to be selected so that $\tau_c \approx \tau_F$ for maximum sensitivity.

The PERRIN-equation is valid only for molecules that are "nearly" spherical and can rotate freely in all directions. The viscosity in the microenvironment of the molecule must be identical to the viscosity of the bulk solvent and measurements must not be disturbed by processes like energy transfer or reabsorption, which might affect polarisation.

6.6.1 Static Fluorescence Polarisation

If the sample is exited with light of constant polarisation and intensity, the average intensities

$$\bar{I}_\parallel = 1/\tau_F \int_0^\infty I_\parallel(t)\mathrm{d}t, \tag{6.22}$$

$$\bar{I}_\perp = 1/\tau_F \int_0^\infty I_\perp(t)\mathrm{d}t \tag{6.23}$$

are obtained. From these an average $\bar{A}$ can be calculated.

6.6.2 Application

The main application of fluorescence polarisation is the measurement of association between molecules. Such association increases the size of the fluorescing molecule and hence the polarisation of the emitted light. If one measures the concentration dependency of the change, one can determine the number of binding sites and the dissociation constant. Fluorescence may be emitted by Trp-residues of the protein itself, or by attached fluorescent probes. In the latter case one should make sure that labelling does not change the binding properties of the protein: unlabelled protein should compete with the labelled one for the ligand with equal parameters.

6.7 Time-Resolved Fluorescence

To determine the lifetime τ of the exited state of the fluorophore several techniques are available:

Pulse decay Fluorescence is excited with a flash, which is short compared to the lifetime of the fluorophore. The decay of the fluorescence signal is measured directly. In modern instruments the light source is usually a laser, light flashes in the ps and fs range are available.

Single photon timing Excitation is with a high-repetition rate (MHz) laser, which is so weak that for each laser flash at most one fluorescence photon (actually about 1 photon per hundred flashes) is detected by the photomultiplier (operated in photon counting mode). The laser flash starts a timer, which is stopped by the signal from the detector. Thus the time between excitation and emission is measured many times, and the distribution of that time is equal to the decay curve of fluorescence. Advantage of this scheme is that it works entirely digital.

Harmonic response If fluorescence is excited by a light source whose output is amplitude-modulated with a frequency ν of several MHz, the resulting fluorescence will also be amplitude-modulated. However, there will be a phase delay ϕ between exciting and emitted light, caused by the fluorescence lifetime of the fluorophore. Also, the depth of modulation m will be reduced. For the simplest case of single exponential decay, these will be:

$$\tan(\phi) = \nu\tau \tag{6.24}$$

$$m = 1/\sqrt{1 + \nu^2\tau^2} \tag{6.25}$$

If both the light source and the photomultiplier high voltage are amplitude modulated, but with slightly different frequencies, beating of these two frequencies will transform the signal to a lower, more easily measured frequency, without loss of information.

The fluorescence lifetime is sensitive to environmental effects, for example exposure of the fluorophore to quenching agents. Like FRET, time resolved fluorescence has become a favourite method for the development of diagnostic and high-throughput assays.

6.7.1 *Fluorescence Autocorrelation*

If the fluorescence signal of a single molecule is followed over time in a very small volume (say, 1 fl) in a confocal laser microscope, the fluctuation of the fluorescence signal contains information about the diffusion coefficient of that molecule. This becomes accessible from the autocorrelation function (ACF) of that signal (for a more detailed explanation of autocorrelation see p. 264). Since binding of a fluorophore to a large protein molecule reduces the diffusion coefficient, binding kinetics can be followed without the need for separation of bound and unbound ligands. Also, sample size required is extremely low.

6.7.2 Dynamic Fluorescence Polarisation

If $I_{\parallel}$ and $I_{\perp}$ are measured as a function of time, both will decay exponentially, but the decay of $I_{\parallel}$ will be faster. Thus plotting of $\ln(A)$ as a function of t will result in a straight line (if the molecule is spherical), the time required for A to be reduced to 1/e ($\approx 37\,\%$) is the rotation correlation time τ_R. The same techniques as for lifetime measurements may be used.

6.8 Photo-bleaching

Photo-bleaching is based on the fact that very high light intensities can irreversibly destroy fluorophores. Such high intensities can be achieved with a laser beam. If, for example, a membrane is labelled with fluorescent derivatives of lipid, a brief, high intensity laser pulse will bleach all fluorophores in a spot. Over time the bleached molecules will diffuse out of the spot, and non-bleached molecules into it. Thus fluorescence intensity in the spot will gradually approach that of its neighbourhood, the time required is a function of the diffusion coefficient of lipid molecules in a lipid layer. There is a good introduction to **fluorescence recovery after photo-bleaching (FRAP)** on http://asb.aecom.yu.edu/faculty/snapperik/pdf/currentprotocolsFRAP.pdf.

Chapter 7
Chemiluminescence

One of the most sensitive detection methods available is chemiluminescence. Chemiluminescent substrates undergo reactions, where the product is in an excited state, usually the first triplet (T_1) state (see Fig. 7.1). By internal systems crossing (spin-reversal, slow!) it is converted to the first singlet (S_1) state, and reaches the ground state (S_0) by light emission. Even though the quantum efficiency of these reactions rarely exceed 1 %, detection in the amol range is possible with photomultipliers or CCD-detectors.

7.1 Chemiluminescent Compounds

Luminol is a typical example for this type of reaction. Oxidation of luminol by peroxides is catalysed by Fe^{2+}-containing proteins, in particular peroxidases. Horseradish peroxidase is a relatively stable protein that can be easily cross-linked to other proteins like antibodies in the lab. The light-producing reaction requires catalytic co-factors ("enhancers") like *p*-iodophenol or umbelliferol to increase light output by a factor of 100 or so. The reasons, like the reaction mechanism of HRP, are not entirely clear.

Another application of luminol is the detection of body fluids in forensic investigations. These often contain iron, for example in the form of haemoglobin. Spraying a crime scene with this reagent will therefore cause a blue glow of any item covered with even traces of biological fluids. Because of the high sensitivity of this method, detection is possible even after items have been thoroughly cleaned. Luminol-derivatives can be used to label biomolecules for later detection in the pmol–fmol range.

Diphenyl oxalate (Cyalume®) is not by itself chemiluminescent, but after oxidation forms an unstable peroxide which transfers energy to a fluorescent dye, which then radiates light. This is the basis for chemical light sticks whose colour depends on the fluorophore used.

Dioxethane by itself is highly unstable (four-membered ring!), but can be stabilised by appropriate substituents. These also carry a reactive group which allows a protein to be modified with the compound. The labelled protein is then

E. Buxbaum, *Biophysical Chemistry of Proteins: An Introduction to Laboratory Methods*, DOI 10.1007/978-1-4419-7251-4_7,

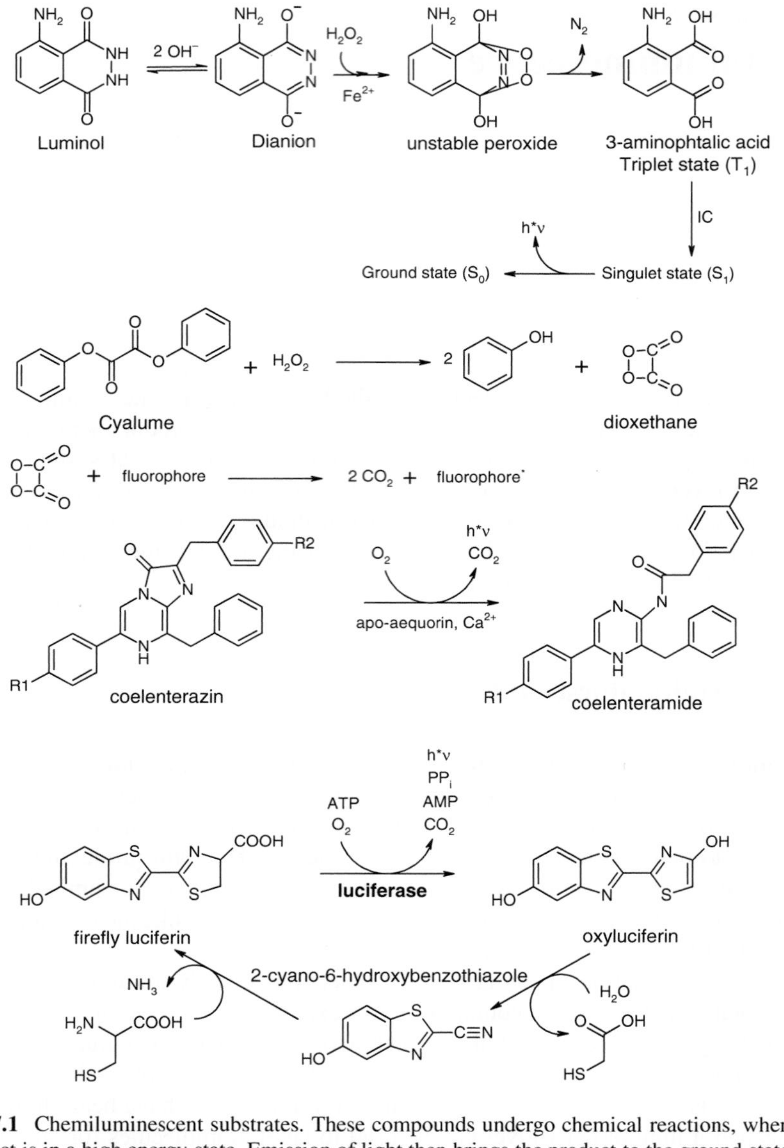

Fig. 7.1 Chemiluminescent substrates. These compounds undergo chemical reactions, where the product is in a high energy state. Emission of light then brings the product to the ground state

used in a bioassay. The amount of protein bound in the assay is detected by a rapid acid/base shift. The free dioxethane rapidly decomposes under light production. Usually only about one dioxethane molecule is bound per protein, quite similarly to labelling with ^{125}I. However, iodine decays with a half life period of 59.4 d, whilst the chemiluminescence from dioxethanes is released within a

fraction of a second. Thus much higher detection sensitivities are possible by chemiluminescence than by radiolabelling.
Other dioxethane derivatives can be cleaved by enzymes like phosphatases. This allows the sensitive detection of alkaline phosphatase-labelled antibodies in a similar way as peroxidase-labelled ones are detected with luminol. These reagents are often patented and hence relatively expensive.

Luciferin is the chemical that glows in the firefly *Photinus pyralis*. To oxidatively decarboxylate it, luciferase initially adds ATP, forming adenyl-luciferin and PP_i. The former then reacts with molecular oxygen to form oxyluciferin under production of blue light. This reaction forms the basis for an extremely sensitive laboratory assay for ATP down to fM concentrations [289].

Aequorin Many marine organisms also produce light, commonly by oxidising **coelenterazine** to coelenteramide. Coelenterazine together with apo-aequorin form the light-emitting protein, **aequorin**. The activity of aequorin is regulated by binding three Ca^{2+} ions. In the lab this is used to assay for intracellular Ca [32]: Apo-aequorin is expressed in the cells, coelenterazine is added to the medium (it is freely membrane permeable).
In the jelly fish *Aequorea victoria* [415] the blue light (469 nm) produced excites fluorescence in green fluorescent protein (GFP) (see Sect. 6.1 on p. 40), so that the animal produces green light. Neither the function of bioluminescence in this organism, nor the reason for wavelength conversion, are known.

7.2 Assay Conditions

For luminol, formation of the luminescent intermediate is a slow process, so light production rate increases for several minutes after mixing peroxidase and substrate, until a plateau is reached. Depletion of substrate and/or poisoning of the enzyme will later lead to a decay of light production rate. This is called **glow kinetics**. Detection can be by photographic film, CCD-camera, or in a misappropriated β-scintillation counter. Chemiluminescence counter for 96-well plates are also available.

Reagents like modified dioxethanes, on the other hand, produce **flash kinetics**, once the reaction has been started, it is over within a second or so. This requires specialised detectors that can inject the necessary reagents and at the same time measure the light produced.

96-well plates used in chemiluminescence assays should be white to ensure that as many of the photons produced as possible reach the detector.

7.3 Electrochemiluminescence

Luminol can oxidise electrically, and then react with hydrogen peroxide, producing light. This can be used as a highly sensitive system to detect hydrogen peroxide produced by oxidases, used in biosensors that detect glucose, cholesterol and other

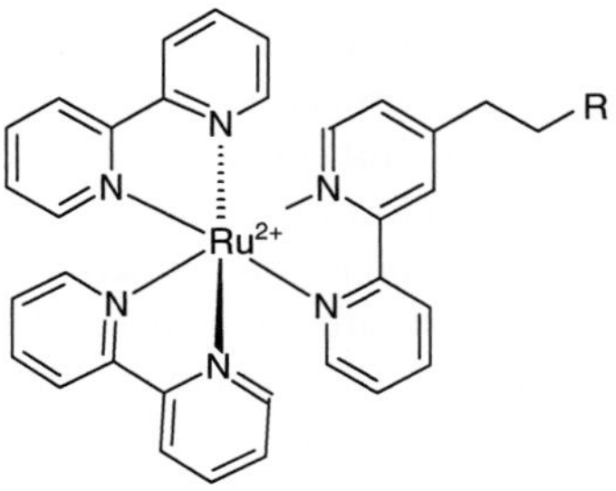

Fig. 7.2 Ru^{2+}-complex $Ru(Bpy)_3^{2+}$ used to label proteins for electrochemiluminescence detection

clinically important metabolites in complex biological fluids like serum or urine. These usually consist of a flow cell, a membrane with the enzyme (about 1 mm^2), and a fibre optics coupled detector.

Alternatively, complexes of ruthenium(II) (see Fig. 7.2) are used as label [29], which are oxidised on a gold anode to Ru(III), at the same time tripropylamine (TPA) is oxidised to $TPA^{+\bullet}$ which looses a proton, creating $TPA\bullet$, a powerful reducing agent which reduces the Ru(III) to Ru^*(II). The latter then decays to the ground state by producing light of 620 nm. Since the ruthenium is not used up in this cycle as long as free TPA is present, the reaction is extremely sensitive. The system is also fast and simple, so it has been developed for detection of biological weapons. Transition metals other than Ru have also been used.

Chapter 8
Electrophoresis

For a more detailed, quantitative treatment of electrophoresis see [60, 428]. Simulations for electrophoretic experiments are available on the web (http://www.rit.edu/~pac8612/electro/E_Sim.html or http://www.jvirgel.de/).

Warning: Electrophoretic experiments require high voltage of direct current which can be lethal (especially with buffers of high conductivity around). Do not interfere with the safety features of the apparatus and remember the old electrician's rule: One hand always stays in the pocket of your trouser!

Since proteins are charged molecules (if $p\text{H} \neq p\text{I}$), they move in an electrical field. Direction and magnitude of the electric force acting on the protein molecules depends on their amino acid composition and the pH of the buffer, the friction experienced depends on the pore size of the medium and the size and shape of the protein:

$$\vec{\mathfrak{F}}_{\mathrm{e}} = Q\vec{\mathfrak{E}} = ze\vec{\mathfrak{E}} \tag{8.1}$$

$$\vec{\mathfrak{F}}_{\mathrm{f}} = f\vec{\mathfrak{v}}_{\infty} \tag{8.2}$$

$$\vec{\mathfrak{v}} = \frac{QE}{f} \tag{8.3}$$

with $\vec{\mathfrak{F}}_{\mathrm{e}}$ and $\vec{\mathfrak{F}}_{\mathrm{f}}$ the electrical and frictional force, Q the charge of the protein, E the potential difference, v_{∞} the velocity at infinite dilution and f the friction coefficient (for a sphere $f = 6\pi\eta r$). Once $\vec{\mathfrak{F}}_{\mathrm{e}} = -\vec{\mathfrak{F}}_{\mathrm{f}}$ the speed becomes constant and

$$ze = 6\pi\eta r\frac{v_{\infty}}{E} \tag{8.4}$$

$$\mu_{\infty} = \frac{v_{\infty}}{E} = \frac{ze}{6\pi\eta r} \tag{8.5}$$

with e the elementary charge, η the viscosity of the medium and μ_{∞} the absolute mobility of the ion (in $\mathrm{m^2\,s^{-1}\,V^{-1}}$). The actual mobility is measured by

$$\mu_j = \frac{Ad_j\gamma}{It} \tag{8.6}$$

E. Buxbaum, *Biophysical Chemistry of Proteins: An Introduction to Laboratory Methods*, DOI 10.1007/978-1-4419-7251-4_8,

with A = cross-section, γ = conductivity, I = current and d_j = distance covered by the ion j in the time t.

Thus electrophoretic separations can be designed to separate by **charge**, **isoelectric point**, or by **molecular size**. Like in chromatography the relative mobility of a species (R_f) is expressed as the ratio of distance moved by the substance (l_s) and the distance moved by a front marker (l_f):

$$R_f = \frac{l_s}{l_f}. \tag{8.7}$$

Electrophoretic methods are normally used for analytical separations, however, equipment for small scale preparative work (up to a few milligrams) is also available.

8.1 Movement of Poly-ions in an Electrical Field

Since most proteins are slightly acidic and electrophoretic buffers usually slightly alkaline, the proteins will form poly-anions. These will be surrounded by positively charged counter-ions. Some of these counter-ions will bind directly to the protein, but because of diffusion and the repulsion of neighbouring counter-ions some will also surround the protein as ion atmosphere (see Fig. 8.1). This ion atmosphere does not exist at the *p*I, where the protein has no net charge. During electrophoresis the counter ions move in the opposite direction from the protein and increase the friction (**cataphoresis**), in addition the separation of protein and counter-ions reduces the ion mobility (**relaxation**).

8.1.1 Influence of Running Conditions

8.1.1.1 Ionic Strength

The electrophoretic mobility μ_i of a (poly-)ion i and its velocity $\vec{\mathfrak{v}_i}$ are related by the strength of the electrical field $\vec{\mathfrak{E}}$ (see Fig. 8.1):

$$\vec{\mathfrak{v}_i} = \mu_i \vec{\mathfrak{E}} \tag{8.8}$$

The protein mobility is calculated from the HENRI-equation:

$$\mu_i = \frac{\zeta \epsilon_r}{1.5\eta} \times \frac{1 + \kappa a_i}{1 + \kappa r_i} \tag{8.9}$$

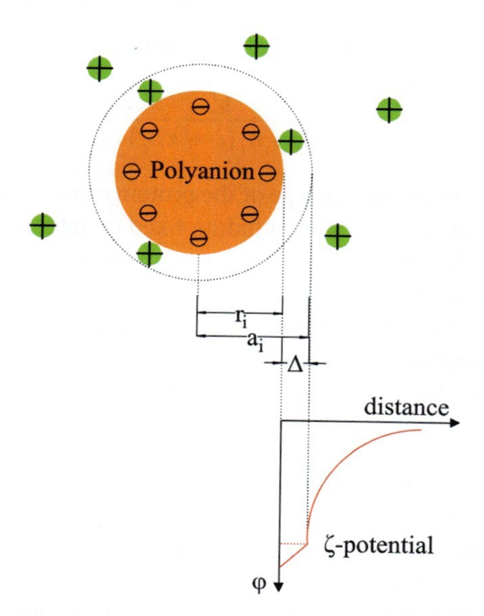

Fig. 8.1 Definition of the ζ-potential: Most proteins are slightly acidic and during electrophoresis in an alkaline buffer they have several negative charges. These are balanced by positive counter-ions. The counter-ions are not only attracted by the negative charges of the protein, but also repulsed by the positive charge of other counter-ions and by diffusion. Hence only some of the counter-ions are bound directly to the protein, the remaining surround the protein as a "ion atmosphere". r_i is the radius of the protein, a_i the radius of the protein plus fixed counter-ions. The thickness of the layer of fixed counter-ions is Δ. The electrical potential of the protein ψ declines linearly with distance in the layer of fixed counter-ions, and exponentially in the ion atmosphere. The potential at a_i is called **electrokinetic (ζ) potential**. This is the potential that interacts with an external electrical field

where κ is the DEBYE–HÜCKEL-parameter (reciprocal of the thickness of the diffuse ion atmosphere):

$$\frac{1}{\kappa} = \frac{1}{F} \times \frac{\epsilon_\mathrm{r} R T^{1/2}}{2 I_\mathrm{c}} \tag{8.10}$$

$$a_i = r_i + (\kappa r_\mathrm{c})^2 \tag{8.11}$$

with F the FARADAY-constant, R the gas constant, T the absolute temperature, r_c the radius of the counter-ion and I_c the ionic strength:

$$I_\mathrm{c} = \frac{1}{2} \sum_{b=1}^{s} z_b^2 c_b \tag{8.12}$$

with s the number of ions formed by a compound i, z the number of proton charges on an ion, and c its concentration.

Then a poly-ion can be viewed as a ball capacitor with

$$\zeta = \frac{Q}{4\pi\epsilon_r a_i (i + \kappa a_i)} \tag{8.13}$$

where Q is its net proton charge (charge of the protein minus fixed counter-ions) and a_i the radius of the protein molecule including fixed counter-ions. The ion mobility μ_i depends on the buffer composition, since as the buffer becomes more concentrated the number of fixed counter-ions will increase, this will decrease Q, ζ, and $1/\kappa$, and increase a_i. All these effects will reduce the mobility. This is described by the **quadratic equation of ion mobility**, which agrees well with experimental findings for $I_c \leq 200\,\mathrm{mM}$:

$$\mu_i = \mu_{i\infty} - \frac{s_i\sqrt{I}}{1 + \kappa r_i + \kappa^2 r_i^2} \tag{8.14}$$

where $\mu_{i\infty}$ is the mobility of the ith ion in an infinitely diluted buffer. A table with important ion properties, including mobilities, is included in the appendix.

High ionic strengths not only reduce protein mobilities, but also increase JOULEAN heating during electrophoresis, which makes runs less reproducible and may lead to additional protein denaturation. On the other hand, a certain ionic strength is required to maintain adequate buffering. From the elementary charge $e = 1.6021 \times 10^{-19}$ A s and the AVOGADRO-constant $N_a = 6.0225 \times 10^{23}\,\mathrm{mol}^{-1}$ it can be calculated that during electrophoresis for 1 h at 1 mA 37.4 μmol of charges, mostly buffer, are transported. Optimal ionic strength for electrophoresis, especially IEF, is in the range of 50–100 mM, the tank volume needs to be large enough to supply the required ions.

8.1.1.2 *p*H

The mobility of a poly-ion will also be determined by the degree of dissociation, which in turn depends on the *p*H:

$$\mu_i' = \mu_i \alpha = \mu_i \frac{K_a}{[H]^+ K_a} \tag{8.15}$$

with K_a the dissociation constant. This is an idealisation since proteins have ionisable groups with different K_a. In practice the mobility of the protein as function of *p*H is a sigmoidal curve, with an inflection point at its *p*I where the mobility is 0.

8.1.1.3 Temperature

Rising temperature strongly reduces the dynamic viscosity of the buffer by about 3 %/K. It also may influence the relative dielectric permeability ϵ_r, but that effect is of far lesser importance.

8.1.1.4 Carrier

In a carrier like polyacrylamide or agar the protein can not go the direct route of length s, but has to take detours around the carrier network of path length s'. Only a fraction f of the gel volume is accessible to the protein. The number of collisions between the gel fibres and a protein molecule is POISSON-distributed, and f is proportional to the probability of a molecule to pass the gel without collision.

The mobility is reduced to

$$\mu_g = \mu \left(\frac{s}{s'}\right)^2 \tag{8.16}$$

The increase in path length is temperature dependent.

In a polyacrylamide gel with constant conditions protein band movement is described by FERGUSON's equation:

$$\log(M) = \log(M_0) - K_R T \tag{8.17}$$

with T the total concentration of the gel (given by the concentration of acrylamide and bisacrylamide). The pore size is proportional to the square root of the total acrylamide concentration.

Y_0 is the **free electrophoretic mobility** at $T = 0$, obtained from extrapolation from a plot of mobility as function of T. These data can be obtained from several gels with different T (but constant C), or from a single gel with transversal acrylamide gradient [294, 348]. The retardation constant K_R depends directly on the molecular radius of the protein $\bar{r}$:

$$K_R = A + B * \bar{r} \tag{8.18}$$

$$\bar{r} \approx \sqrt[3]{\frac{3}{4\pi} \times \frac{M\bar{\nu}}{N_a}} \tag{8.19}$$

with M_r the molecular mass of the protein, $\bar{\nu}$ its partial specific volume (taken as 0.71 ml/g for most proteins, but an elegant way to measure it has been described in [104]) and N_a AVOGADRO's number. However, this calculation neglects the hydration of proteins, for soluble proteins in the order of 0.25–0.35 g water per g of protein.

The determination of the molecular radius from the slopes of FERGUSON-plots [97] is more precise than from a R_f vs. M_r plots and has a basis in first principles. The protein is run in gels of different pore size, and the mobility of sample and reference proteins are plotted as a function of pore size. The slopes of the resulting lines can be plotted against the logarithm of the molecular radius to construct a standard curve. This analysis can also be done using a single gel with a transversal acrylamide gradient [294].

Depending on the nature of the gel, K_R depends on the molecular radius of the protein:

$$\sqrt[2]{K_R} = c(\bar{r} + r) \qquad \text{meshwork of fibres,} \tag{8.20}$$

$$\sqrt[3]{K_R} = c(\bar{r} + r) \qquad \text{meshwork of points,} \tag{8.21}$$

$$K_R = c(\bar{r} + r) \qquad \text{meshwork of planes} \tag{8.22}$$

with r the effective radius of the gel fibres, and c an empirical constant. In other words: Electrophoresis determine molecular radius, *not* molecular mass. Thus changes in the slope of the FERGUSON-plot under different conditions may be used to monitor conformational changes in the protein. Compare the above equation with the SIEGEL-equation for gel filtration $\sqrt[2]{-\log(K_{av})} = c(\bar{r} + r)$.

If the regression line is calculated of $\log(R_f)$ versus gel concentrations T (FERGUSON-plot) the slope gives the retardation coefficient K_R and the extrapolated intercept at $T = 0$ is Y_0.

The radius of an unknown protein can then be calculated from regression lines of $\bar{r}$ vs. $\sqrt[2]{K_R}$.

The free electrophoretic mobility of the protein is calculated from $M_0 = Y_0 \mu_f$, with μ_f the mobility of the front ions which is tabulated for the various gel systems described in www.buffers.nichd.nih.gov. The slope of this regression line is the length of gel fibre per concentration $l' = \frac{dL}{dT} \approx 0.4 \times 10^{12}\ \text{cm/cm}^3$ for polyacrylamide.

Note that in a FERGUSON-plot size isomers give lines which intersect at 0 %T, while charge isomers give parallel lines [149].

If the sample molecules are very long, they may get hooked on the carrier network and stop moving. In this case one can "unhook" them again by changing the direction of the electrical field periodically (**pulsed field electrophoresis**). This method is almost exclusively used with nucleic acids where it allows the separation of entire chromosomes. It will not be discussed any further.

8.2 Electrophoretic Techniques

Electrophoretic techniques can be classified according to several criteria:

- State of protein

  Native: native structure of the protein is maintained.

  Denatured: protein is unfolded with detergents (SDS, CTAB) and reducing agents (DTT, βME). Alkylation of the reduced proteins is sometimes used to prevent re-oxidation.

- Matrix

  Restrictive: separation influenced by size and charge

  Non-restrictive: separation influenced by charge only

  Free flow like non-restrictive, sometimes used for preparative purposes

- Buffer system

  Zone electrophoresis: one buffer, continuous electrical field, continuous *p*H
  Isotachophoresis: Two or more buffers, discontinuous electrical field and *p*H
  IEF, NEPHGE *p*H-Gradient

- Combination methods

  DISC: isotachophoresis/zone electrophoresis
  2-D: IEF/zone electrophoresis

- Gel type

  Round: no longer used since staining and archiving are tedious
  Vertical flat: long separation distances
  Horizontal flat: good cooling

- Gel thickness

  Thick > 1 mm: high sample capacity
  Thin 0.5–1 mm
  Ultrathin < 0.5 mm: low temperature gradient between gel surfaces, low sample volume, fast detection of bands, automatisation possible. Requires use of a stabilising plastic foil or net.

8.2.1 Techniques of Historic Interest

8.2.1.1 Moving-Boundary Electrophoresis

Historically the first electrophoretic method used on proteins was moving-boundary electrophoresis ([412], see Fig. 8.2), invented by ARNE TISELIUS (N.P. 1948). Protein movement was determined only by charge. For a fuller account of this technique see [431].

8.2.1.2 Paper Electrophoresis

Electrophoresis used to be carried out on paper strips soaked in buffer. Because the pores in paper are much larger than even the largest protein molecules, separation is based on charge only. On the other hand, diffusion on this matrix is also significant, thus resolution is lower than in gel-electrophoresis.

Cellulose acetate is often used in clinical laboratories, especially for the analysis of serum proteins or for the diagnosis of haemoglobin variants [204]. This is the cheapest, simplest, and fastest method for electrophoresis. The required apparatus is so simple that it can be home-made even in developing countries (see Fig. 8.3), cooling is not required. Veronal-buffers (=diethylbarbiturate) of *p*H 8–9 is traditionally used, but can be replaced by Tris (280 mM)/taurine (390 mM) buffer *p*H 8.5 where Veronal is difficult to obtain (narcotic!). A voltage of about 250 V is used,

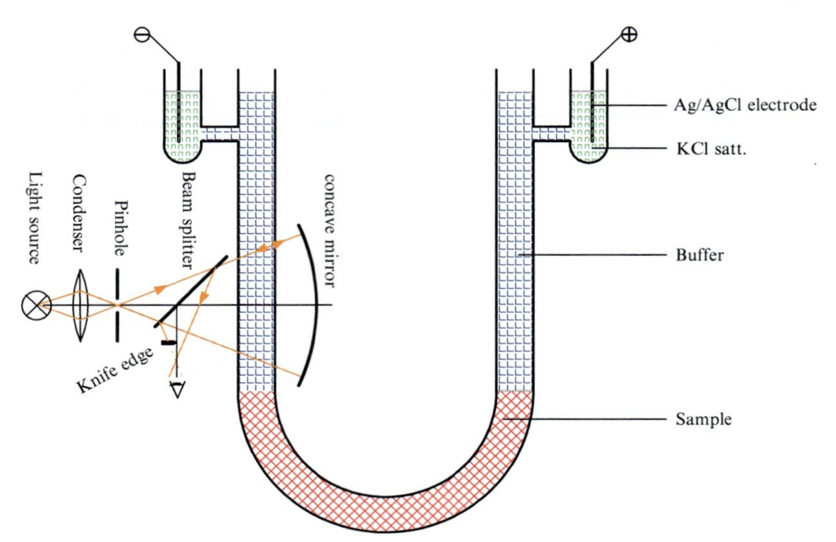

Fig. 8.2 Moving boundary electrophoresis apparatus of TISELIUS: Electrophoresis was carried out in a U-shaped tube of rectangular cross-section. The setup ensured as much as possible that reactive products from electrolytic side-reactions could not come into contact with the protein and that the ionic strength of the buffer could be chosen independently from the electrode solutions. The density difference between buffer and protein solution in buffer ensured a sharp boundary. Movement of that boundary was followed by measuring the refractive index change with a **Schlieren optic**: A knife-edge is introduced between a lens and the image created by that lens, so that light from only half the cone is used for the image. If the medium around the object were homogeneous, this would just result in a darker image. However, if there are changes in refractive index along the optical path, diffraction leads to a Schlieren-pattern (*Schlieren* is the German word for streak), which is the first derivative of the optical density in the direction of the knife edge

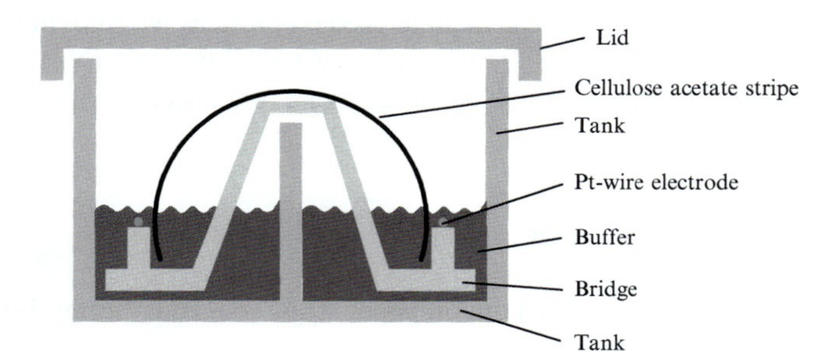

Fig. 8.3 Simple apparatus for cellulose acetate electrophoresis, made from perspex or polycarbonate. The Pt-electrodes end in 4 mm banana plugs mounted at the ends of the bridge with cables from the power supply in banana sockets mounted in the lid. Placing the lid onto the tank closes the electrical circuit. The same assembly may also be used for agarose gels on carrier foils

this results in a current of 3 mA per foil. The foils are stained with Ponceau S and differentiated with 1 % TCA until the background is clear. Then they are placed onto a glass slide and dried for 20 min at 100 °C until they are clear and can be scanned at 530 nm. Immunological detection or zymograms are also possible.

8.2.2 *Gel Electrophoresis*

8.2.2.1 Polyacrylamide

Today gels of cross-linked polyacrylamide [346] are commonly used as a matrix for electrophoresis (see Fig. 8.4). The gel reduces diffusion and prevents convection during electrophoresis. After electrophoresis proteins can be detected by specific staining reactions, this increases sensitivity to about 1 ng per band. Gels may be

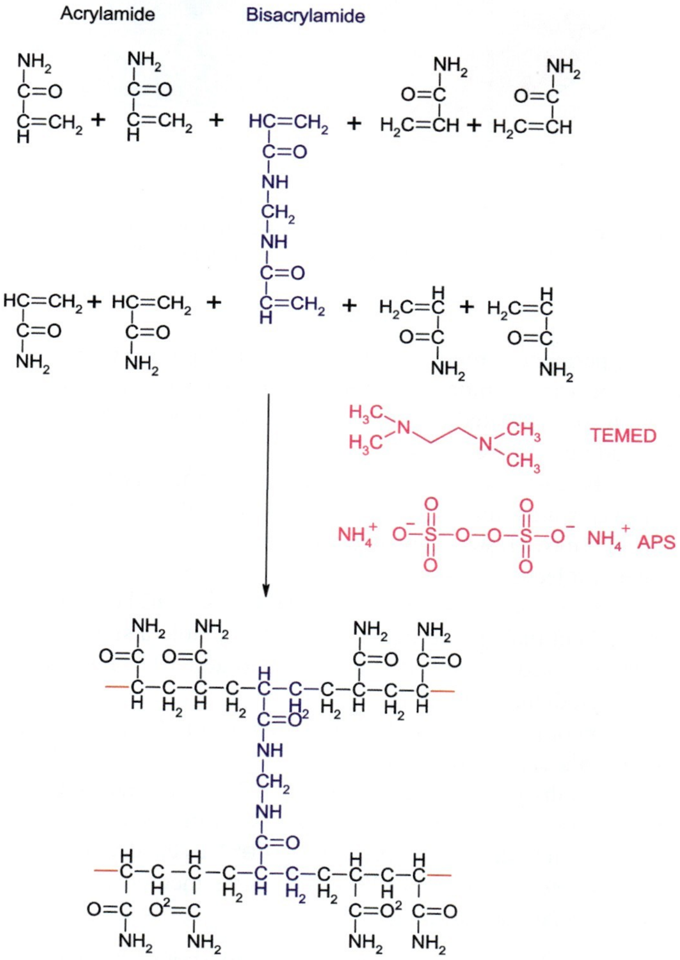

Fig. 8.4 Radicalic polymerisation of acrylamide. Catalysed by *N,N,N',N'*-Tetramethylethylenediamine (TEMED) the initiator ammonium peroxodisulphate (APS) breaks down into two radicals, which start the polymerisation process of acrylamide. *N,N*-methylene bisacrylamide ("Bis") is used to cross-link the growing poly-acrylamide chains into a network

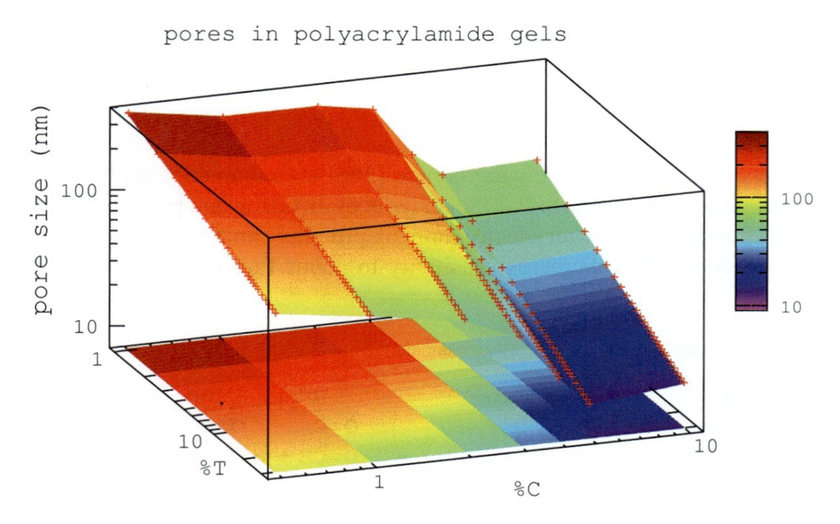

Fig. 8.5 Pore size of polyacrylamide gels as a function of T (in %) (acrylamide + bisacrylamide) and C (bisacrylamide/T), data taken from [390]. The pore sizes were determined experimentally from FERGUSON-plots (R_f vs. T) of DNA molecules of different lengths

dried, giving a permanent record of the results. The pore size of the gels can be adjusted by the polyacrylamide and cross-linker concentration (see Fig. 8.5). If large molecules are to be separated, mechanical strength of polyacrylamide gels can be increased by adding agarose (0.5 %).

The gel may be cast with a concentration gradient of acrylamide [257], this results in narrower bands since leading protein molecules are slowed down while trailing molecules move faster relative to the middle of the band. In addition, a wider range of protein molecular masses can be resolved. SDS–PAGE gels of 5–15 % can resolve molecular masses of about 6–250 kDa, this should be fine for most applications. If a gradient maker (see Fig. 8.6) is not available a step gradient of 10–12 equal steps can be used as well. It is advisable to add 10 % glycerol to the heavy solution to ease gradient formation.

Polymerization of polyacrylamide is achieved by a radicalic mechanism. For alkaline gels radicals are often produced from ammonium peroxodisulphate using *N,N,N',N'*-Tetramethylethylenediamine (TEMED) as a catalyst [187]. An older, riboflavine-based system [75] is tedious and now hardly ever used. Acidic gels are produced either from hydrogen peroxide using Fe^{2+} as catalyst (FENTON-reaction, [96]), or by photo-polymerisation [338]. The latter method gives less brittle and more flexible gels than the former.

Warning: Acrylamide is a potent neurotoxin and should be handled with care. Ready-made stock solutions are commercially available, their use avoids dust formation during weighing. Since polymerisation is always incomplete the gels should be handled with gloves until free acrylamide has been washed out (*e.g.*, during

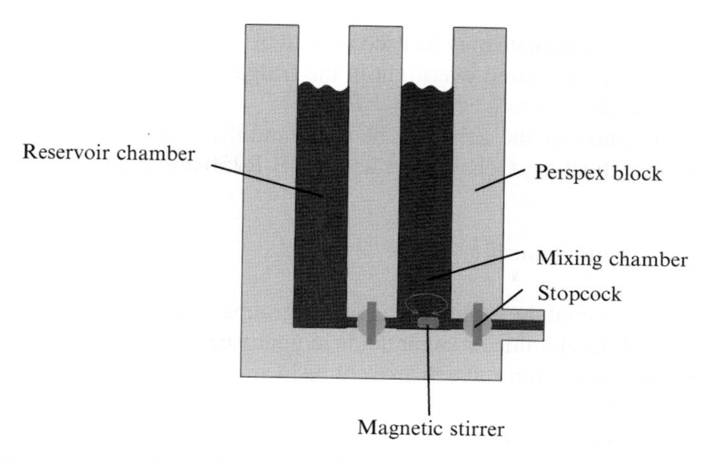

Fig. 8.6 Gradient mixer. The device is made from a perspex block, into which two large holes are drilled which form the mixing and reservoir chamber, respectively. A small bore hole forms channels between the chambers, and between the mixing chamber and the outside. Both channels can be closed by stopcocks. A rapidly spinning magnetic stirrer mixes the content of the mixing chamber with the fluid coming from the reservoir. If operated as depicted, the gradient formed is linear. If the top of the mixing chamber is closed by a plug the gradient becomes exponential. To understand how this works let us assume a step-wise rather than continuous operation. Let the reservoir chamber be filled with 10 ml 5 % and the mixing chamber with 10 ml 10 % solution. Let the 1 ml solution leave the mixing chamber. As a consequence 0.5 ml will pass from the reservoir into the mixing chamber, and the concentration of the solution in the mixing chamber will be $(0.5\ \text{ml} \times 5\,\% + 9\ \text{ml} \times 10\,\%)/9.5\ \text{ml} = 9.74\,\%$. If this procedure is repeated until both chambers are empty, the concentration versus volume plot will be a *straight line*

fixation, staining and differentiation). After washing gels are physiologically inert, but depending on detection method gloves may still be required to prevent fingerprints on the gel.

8.2.2.2 Agarose

If very large pores are required (*e.g.*, DNA molecules up to whole chromosomes or giant proteins like titin [126]), agarose or starch gels are used. Pore size in agarose gels is between 500 nm for 0.16 % and 150 nm for 1 %. Gels with higher concentration are rarely used, as they are no longer transparent. These gels are sometimes also used for enzyme separations in genetics and in clinical laboratories (affinity electrophoresis to separate isoenzymes [9, 284]) since agarose gels are non-toxic, easy to prepare, and run fast. Because of the large pores, agarose gels are also suitable for zymograms and immunoelectrophoresis. 1 % agarose gels can be used for spherical particles up to 50×10^6 Da or 30 nm, sufficient for membrane proteins and even virus. Antibody–antigen complexes can also be separated in agarose, these gels are therefore used for immuno-electrophoresis. As with cellulose acetate electrophoresis tris-taurinate may be used instead of Veronal as buffer. After casting, agarose gels should be stored overnight in the fridge (humid chamber), during this time the gel structure forms.

After electrophoresis the gels may be dried and then the dried gel stained with CBB-R250 or with silver. Gels stained with CBB-R250 are scanned at 605 nm.

8.2.2.3 Electroendosmosis

Most gel matrices contain negatively charged groups (*e.g.*, acrylic acid as a break down product of acrylamide or sugar acids in agarose). Their charge is counterbalanced by small positive ions, which are mobile in an electric field. This movement is accompanied by the movement of water from the positive toward the negative pole (electroendosmosis). This can lower resolution, hence negative charges in the matrix should be kept as low as possible. Acrylamide solutions should be stored over an anion exchanger to remove acrylic acid, and only electrophoresis grade agarose should be used. However, as we will see below, some methods actually use electroendosmosis in their working principle. In those cases, the use of electrophoresis grade agarose must be avoided.

8.2.3 Free-Flow Electrophoresis

Continuous, carrier-free electrophoresis ([143], see Fig. 8.7) is useful for preparation of large particles (virus, cells, organelles) where band spreading by diffusion is limited. The particles are carried by the continuous stream of buffer in y-direction, and separated by an orthogonal electric field in x-direction. Thus the sample particles move at an angle θ to the buffer flow, and

$$\mathrm{tg}(\theta) = \frac{x}{y} = \frac{v_x}{v_y} \tag{8.23}$$

$$v_x = \mu \vec{\mathfrak{E}} = \mu \frac{j}{\gamma} = \mu \frac{I}{q\gamma} \tag{8.24}$$

$$\mathrm{tg}(\theta) = \frac{\mu I}{q \gamma v_y} \tag{8.25}$$

with v_x and v_y the velocities in x- and y-direction, μ the electrophoretic mobility of the particle, I the current and j the current density, q the cross-section of the chamber. γ is the conductivity. Since buffer velocity, conductivity, cross-section and the current are held constant during the run, θ depends only on the mobility of the particles.

The chamber is about 60 by 60 cm^2 and needs to be mounted vertically to avoid sedimentation of the sample. The temperature needs to be kept constant by cooling both plates to avoid convection. The voltage is about 200 V.

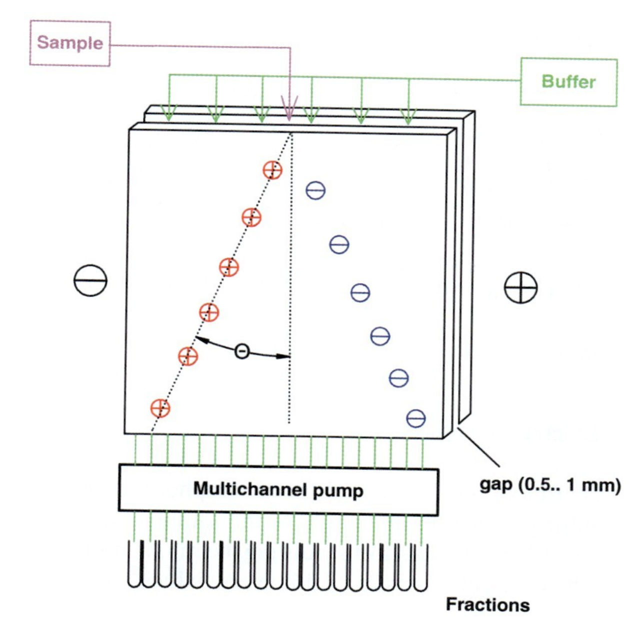

Fig. 8.7 Free-flow electrophoresis. Online detection of separated material is possible with a scanner that moves along the bottom of the chamber

8.2.4 Native Electrophoresis

Native electrophoresis describes conditions, were the secondary, tertiary and quaternary structure of the proteins is maintained. Native electrophoresis separates by both the **charge** (controlled by the *p*H of the buffer), and the **size and shape** of the protein molecules. Because of this, native gels are sometimes difficult to interpret. However, if the gel contains the ligand of a protein, the affinity of the interaction can be determined in a similar way as with gel filtration or ultracentrifugation (affinity electrophoresis, [402]). See Sect. 43.3 on p. 402 for further discussion of these methods.

8.2.4.1 Equilibrium Electrophoresis

This method [222] is used to directly determine the charge of a macromolecule. Electrophoresis is performed in a wide-pored gel (agarose, starch or low percentage acrylamide) under very low electrical field strength (0.2 V/cm) over short distances (a few millimeters). Under these conditions the movement of the protein in the electrical field is balanced by diffusion (in a similar way as in equilibrium centrifugation, see Sect. 25.1.3.2 on p. 241) that is the flux of the macromolecule due to electrophoresis

$$J_e = \frac{c(x)z\vec{\mathfrak{E}}}{f_e} \tag{8.26}$$

is balanced by the diffusion flux

$$J_{\mathrm{d}} = -\frac{kT}{f_{\mathrm{d}}} \frac{\mathrm{d}c(x)}{\mathrm{d}x} \tag{8.27}$$

with f_{e} and f_{d} the translational friction coefficients for electrophoresis and diffusion, with $f_{\mathrm{e}}/f_{\mathrm{d}} \to 1$ for $\vec{\mathfrak{E}} \to 0$. Hence we get

$$c(x) = c_0 \exp\left[\vec{\mathfrak{E}} \frac{z}{kT}(x - x_0) - 2B_2(c(x) - c_0)\right] \tag{8.28}$$

from which z can be determined by curve fitting.

8.2.4.2 Disk-Electrophoresis

If a continuous buffer system were used for electrophoresis, proteins would move with a band width equivalent to the height of the sample loaded onto the gel, and diffusion would broaden the bands further during the run. In normal practice, sample height in the wells of a gel is several millimeters, which would lead to very low resolution.

ORNSTEIN [303] has suggested a solution to this problem: Electrophoresis is started in an open pore stacking gel in the presence of a fully dissociated, fast moving (γ) and a partially dissociated, slow moving ion (α) which form separate zones, both with a common counter-ion (β). The pH (and therefore ionisation of α) is chosen so that the mobility μ_α is less than that of the proteins. Under these conditions, all protein molecules will, given enough time, pass the α ions and then move between the zones of γ and α as sharp bands. Since the amount of protein is not increased, the concentrating effect can only be obtained by reducing the volume, *i.e.*, band height, resulting in a very narrow zone. It is quite instructive to observe this process with pre-stained proteins, available as molecular mass standards from several suppliers.

The classical example used by ORNSTEIN was a system of chloride as leading ion γ, glycine as trailing ion α, and potassium as common counter-ion β:

Ion	Species	Charge z	Mobility μ cm$V^{-1}s^{-1}$	pK$_a$
α	Glycine	−1	−15	9.8
γ	Cl^-	−1	−37	
β	K^+	+1	+37	
π	Albumin (pH 8.3)	−30	−6	

The mobility of an ion (μ) in an electrical field is defined as distance covered (l) per unit time (t) and voltage (U):

$$\mu = \frac{l}{tU}. \tag{8.29}$$

The conductivity (γ) of the solution is given by

$$\gamma = E \times \sum c_i \mu_i z_i \tag{8.30}$$

where c_i is the concentration and z_i the charge of the ith component. Its velocity v_i is then given by

$$v_i = U \mu_i x_i \tag{8.31}$$

where x_i is the fractional ionisation of the ith species, which depends on pH and pK$_a$. Since both solutions are electrically in series, the current (I) flowing through them is identical and we get from OHM's law:

$$U_l = \frac{I}{\gamma_l A} \tag{8.32}$$

and

$$U_u = \frac{I}{\gamma_u A} \tag{8.33}$$

with A the cross-section of the solution normal to the direction of current. In equilibrium $v_\alpha = v_\gamma$ and hence

$$U_u \mu_\alpha x_\alpha = U_l \mu_\gamma x_\gamma \tag{8.34}$$

and

$$\frac{\mu_\alpha x_\alpha}{\sum c_{iu} \mu_{iu} z_{iu}} = \frac{\mu_\gamma x_\gamma}{\sum c_{il} \mu_{il} z_{il}} \tag{8.35}$$

This is a modified form of KOHLRAUSCHS regulating function ([203], experimentally confirmed in [164]):

$$K_E = \sum_{j=1}^{n} \frac{c_j}{\mu_j} = const \tag{8.36}$$

K_E is called the **persistent constant**. Under the conditions stated, the stack of α and γ will move downward with equal velocity, and the border between the ions will be maintained. If an α-ion were to move into the lower phase, it would be subject to a lower voltage and hence move more slowly than the stack until the stack has passed it and it is in the upper phase again. A γ-ion in the upper phase would experience a higher voltage and would move faster than the stack until it has passed the stack and is in the lower phase again.

Taking into account the presence of the β counter-ion we get

$$\frac{\mu_\alpha x_\alpha}{c_\alpha \mu_\alpha z_\alpha + c_\beta u \mu_\beta u z_\beta u} = \frac{\mu_\gamma x_\gamma}{c_\gamma \mu_\gamma z_\gamma + c_\beta l \mu_\beta l z_\beta l}. \tag{8.37}$$

Because of electro-neutrality

$$c_\alpha z_\alpha = c_\beta u z_\beta u, \tag{8.38}$$

$$c_\gamma z_\gamma = c_\beta l z_\beta l, \tag{8.39}$$

hence

$$\frac{\mu_\alpha x_\alpha}{c_\alpha z_\alpha(\mu_\alpha - \mu_\beta)} = \frac{\mu_\gamma x_\gamma}{c_\gamma z_\gamma(\mu_\gamma - \mu_\beta)}. \tag{8.40}$$

The total concentration of a compound is the sum of the ionised and the non-ionised species: $[I] = c_\iota + [H_\iota] = c_\iota / x_\iota$, where x_ι is the pH-dependent fractional dissociation of I. Then

$$\frac{[A]}{[\Gamma]} = \frac{x_\gamma c_\alpha}{x_\alpha c_\gamma} = \frac{\mu_\alpha z_\gamma(\mu_\gamma - \mu_\beta)}{\mu_\gamma z_\alpha(\mu_\alpha - \mu_\beta)}. \tag{8.41}$$

The same equation of course applies to the chloride/protein boundary:

$$\frac{[\Pi]}{[\Gamma]} = \frac{\mu_\pi z_\gamma(\mu_\gamma - \mu_\beta)}{\mu_\gamma z_\pi(\mu_\pi - \mu_\beta)} = \frac{-6.0 \times -1 \times (-37 - 37)}{-37 \times -30 \times (-6 - 37)} = 0.0093. \tag{8.42}$$

If we assume the usual concentration of chloride to be 60 mM, then the albumin concentration once equilibrium has been achieved must be 0.56 mM or 36.3 mg/ml, independent of the protein concentration in the sample or the sample volume. Assuming that we had a sample of 0.1 mg/ml and that the sample in the well was 5 mm high, then we get a concentration factor of 36.3 mg ml^{-1} / 0.1 mg ml^{-1} = 363 and a disk thickness of 5 mm / 363 = 14 μm. It is this concentration effect in the stacking gel that gives DISC-electrophoresis its high resolution. In case of several proteins the same analysis is used repeatedly, the protein with the highest mobility becomes the leading ion for the protein with the second-highest and so on.

Once the stack arrives in the separating gel, the pH increases and with it the dissociation of α, so that $\mu_\alpha x_\alpha$ becomes larger than the mobility of the fastest protein. At the same time, the mobility of the proteins is reduced by the smaller pore-size of the separating gel (higher gel concentration, more cross-linking). α passes the protein stack and the proteins are subjected to zone electrophoresis in a constant electric field.

Separations may be performed in the stacking gel alone, this method is called **isotachophoresis**. The disadvantage of this method is that there can not be baseline-separation of neighbouring bands (this would create ion-free zones between bands which would have infinite resistance). Amino acid or ampholyte spacers, however, may be used to achieve base-line separation of proteins. This method has been used to separate immunoglobulin isoforms [449] or serum proteins [51]. The distribution of the concentration of ions in the zones of a stack is not GAUSSIAN, hence sample concentrations of various ions are calculated from the lengths of their respective zones. To standardise, a second run with a known amount of the ion added to the sample needs to be performed.

The calculation of buffer systems by hand is a lengthy enterprise, the task is greatly simplified by the calculator available on http://www.buffers.nichd.nih.gov/, using equations described in [182–185].

8.2.5 Denaturing Electrophoresis

Most common is the method described by LAEMMLI [218]. The protein is heated in a sample buffer which contains a reducing compound (β-mercaptoethanol or better dithiotreitol (DTT, CLELAND's reagent)) to open *–S–S*-bridges in the protein, destroying its tertiary structure. In some cases it is useful to treat the reduced proteins with iodoacetamide or tris-(2-carboxyethyl)-phosphine (TCEP) to prevent re-oxidation by air. Note that for the characterisation of a new protein it is useful to do the electrophoresis both in the presence and absence of thiol, this simple experiment can reveal disulphide linked quaternary structure.

The buffer also contains the anionic detergent **sodium dodecyl sulphate (SDS)**, which binds to the proteins at a fairly constant ratio of about one molecule of SDS per three amino acids (1.4 g SDS per g of protein) [375]. As the charge of the bound detergent is much larger than the charge of the protein's amino acids, the charge to mass ratio of all proteins is almost constant, and proteins all experience the same acceleration in an electrical field, independent of their composition. At the same time, the high negative charge density makes protein molecules repulse each other, preventing aggregation even at very high concentrations. The pores of the gel slow down big proteins more than small ones, separation is therefore by **size only**. Proteins with 2 % size difference can be separated. If the migration distance of a protein is compared to that of proteins of known size, its molecular radius can be estimated (see Fig. 8.8). For a discussion on the history of the use of SDS in protein chemistry see [319].

Sodium tetradecylsulphate has a higher affinity to proteins and stronger dissociating power than SDS. It has been used as a replacement for the latter to achieve separation of Na/K-ATPase isoforms, that run closely together in SDS–polyacrylamide gel electrophoresis (**SDS–PAGE**) [376]. A more systematic study on detergent-protein interactions in electrophoresis has been performed in [242]. Before highly homogeneous detergents became available, contamination of commercial SDS with analogues of different chain length has led to considerable confusion, as results obtained in some laboratories could not be reproduced in others, that used different brands of SDS [376].

A LAEMMLI-gel consists of two separate zones as discussed with Disk-electrophoresis above. The buffer system is that of ORNSTEIN AND DAVIS (Tris–glycine–chloride). This method is called SDS–PAGE. It is quick, relatively cheap, reproducible and is the most often used type of electrophoresis in a protein laboratory.

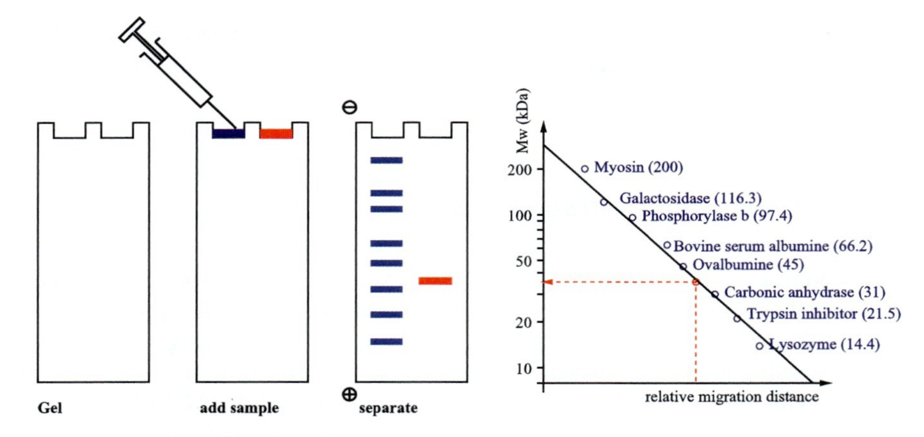

Fig. 8.8 Denaturing gel-electrophoresis. Proteins of known molecular radius are used to establish a standard curve that allows the radius of sample proteins to be determined. The radius is usually taken as an approximation of molecular mass

For low molecular mass proteins (up to 20 kDa) the Tris–acetate–tricinate system with gradient gels of 8–18 % [366, 368] gives higher resolution than the LAEMMLI-system. However, the pK_a-values of the buffers are too far away from the functional pH (7.1), this limits the capacity and resolution of such gels.

Sometimes SDS–PAGE is performed as zone – rather than DISC-electrophoresis with a Tris/acetate [94] or phosphate buffer [424]. Because in such systems no stacking occurs the resolution is limited.

8.2.6 Blue Native PAGE

Sometimes it would be nice to use the high resolution of SDS–PAGE to purify functioning enzymes. However, SDS is a strongly denaturing detergent and enzymatic activity is unlikely to survive the procedure. SDS may be replaced by CBB-G250, which also binds to proteins and gives them a uniform negative charge, but without much denaturation [367]. In addition ϵ-aminocaproic acid is used, which can "salt-in" proteins without protein dissociation. Sulphobetains give a similar effect, but with higher dissociating power.

Membrane protein complexes are solubilised with a mild, neutral detergent of low *cmc* (dodecyl maltoside, Triton-X100, digitonin in order of decreasing dissociating power) and separated by blue-native PAGE. Since the hydrophobicity of some subunits in a complex is averaged out by other more hydrophilic subunits, little loss of the former occurs despite the low solubilizing ability of the method. SDS–PAGE is then used to separate the components of these complexes in a second dimension. Mitochondrial respiratory complexes, solubilised in the same detergent used for the sample, may be used as molecular mass standards (I 1 000kDa, V 700 kDa, III 490 kDa, IV 200 kDa, II 130 kDa).

8.2.7 CTAB-Electrophoresis

SDS–PAGE does not work very well with three classes of proteins:

- Proteins which bind more SDS than the usual 1.4 g per g of protein. This includes very hydrophobic proteins, in particular transmembrane proteins, which can bind 5–7 g/g detergent. Because these protein/SDS-complexes have a higher charge per mass ratio, they run faster and SDS–PAGE gives a molecular size which is too low. Example: the α-subunit of mammalian Na/K-ATPase has a molecular mass of 112 kDa, but runs like a protein of 84 kDa, an error of 25 %.
- Proteins with an unusually large number of charged amino acids also move to positions that correspond to a wrong molecular size. Histone H1 gives a molecular mass of 30 kDa in SDS–PAGE rather than the 21 kDa expected from sequencing.
- Heavily glycosylated proteins contain in their sugar trees negatively charged carbohydrates. Because these sugar trees are highly variable, glycoproteins tend to move not as sharp bands, but as broad smears in SDS–PAGE (but see [331] for the use of borate buffer to reduce the problem).

In the latter two cases the use of the positively charged detergent CTAB (cetyltrimethylammonium bromide) gives better results [43, 45]. CTAB is a milder detergent than SDS, in some cases it is possible to stain enzymes after CTAB–PAGE by their specific reaction (**zymogram**). At the same time CTAB can bring some proteins into solution which are insoluble in SDS.

8.2.8 Practical Hints

For researchers with less than 20/20 vision the refractive index difference between sample wells and their walls is difficult to spot. Loading samples becomes a lot easier if a marker dye is polymerised into the stacking gel. Phenol red is used for SDS-PAGE [381], malachite green for CTAB-PAGE.

The acrylamide solution should be stored refrigerated and over a mixed bed ion exchanger, as over time some acrylic acid may form which can cause electroendosmosis during electrophoresis. Note that presence of acrylic acid is a problem already at very low concentrations, where the reduction of acrylamide concentration does not have a measurable effect.

Dissolved oxygen interferes with the polymerisation of acrylamide since it captures radicals. Therefore the gel mixture must be degassed carefully before polymerisation is started by the addition of APS. During the polymerisation process, contact of the mixture with air must be prevented. Separating gels can be overlayed with water saturated *n*-butanol, for the stacking gel the combs should prevent contact of the mixture with air. However, due to a design error in most commercially available systems this is not the case, resulting in incomplete polymerisation and badly formed wells. For my lab I had better combs made in a craft-shop (see Fig. 8.9), which solved the problem.

Salt concentration should be low, <50 mM. If that is not the case, the sample must be either diluted or the proteins precipitated. Precipitation is most often performed with chloroform/methanol [427] or with TCA/ DOC [23]. However, there may be a problem with re-solubilisation of the precipitated proteins, especially for large, hydrophobic ones.

When gel cassettes are opened, the gel may stick to the surface of the glass plates and tear. This can be prevented by wiping the plates with a tissue soaked in a solution of dichlorodimethylsilane before use (in a fume cupboard!). Once the solvent has evaporated, the plates are rinsed in distilled water and are ready to use. This process creates a monolayer of hydrophobic silicone on the plates, which prevents sticking. The silane solution is available from several manufacturers under trade names (Repelcote, Repel-silane, *etc.*).

Foil-supported gels can be cast without any additives (buffer, detergent, ampholytes, urea), washed twice with distilled water for 10 min each, incubated in 5 % glycerol, and then air-dried over night. The dried gels can be stored for months in the freezer. Before use the gel is thawed at room temperature for 10 min and rehydrated with appropriate buffer/additive solutions for several min to h, depending on thickness. This procedure not only removes unpolymerised acrylamide and catalyst (which can react with proteins, preventing sequencing or mass spectrometric analysis of bands), but it also simplifies gel production in laboratories where different buffer systems are routinely used. The gel may be cut into stripes of required width if a horizontal electrophoresis apparatus is used [201]. The common Tris/glycine/HCl system does not work well with such gels, hence the gel is rehydrated in Tris–acetate. The buffer-stripe at the cathode contains Tris–acetate too, and the buffer-stripe at the anode Tris–tricinate. The tricinate serves as the trailing ion.

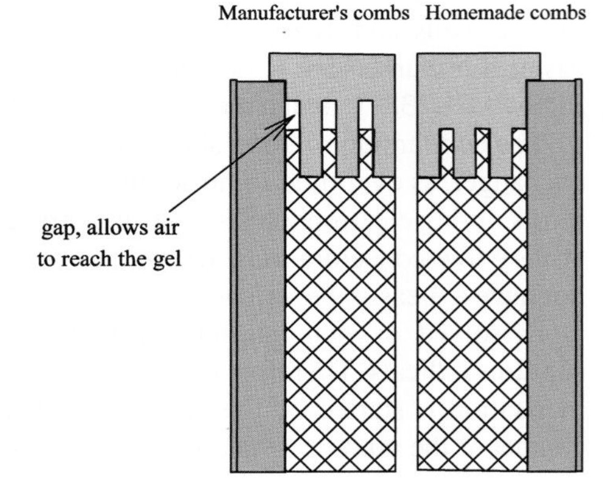

Fig. 8.9 *Left*: Badly designed combs allow the access of air to the polymerising stacking gel mixture. This interferes with polymerisation and results in ill-formed wells. *Right*: Combs as used in my laboratory. There is no gap which would allow air access to the gel. Such combs can be made in any craft-shop from commercially available PTFE sheets

8.2.9 *IEF and 2D-electrophoresis*

8.2.9.1 Ampholytes

Electrolysis of water leads to the establishment of a *p*H-gradient between the electrodes:

$$2H_2O \rightarrow O_2 + 4H^+ + 4e^- \text{anode (positive electrode)} \quad (8.43)$$

$$4H_2O + 4e^- \rightarrow 2H_2 + 4OH^- \text{cathode (negative electrode)} \quad (8.44)$$

Thus the anode becomes acidic, the cathode basic.

Proteins contain carboxylic acid groups, which are uncharged at low, but negatively charged at high *p*H. They also contain amino, guanidino, and imidazole groups that are uncharged at high, but positively charged at low *p*H. Thus proteins will have an excess of positive charges at low *p*H, and an excess of negative charges at high *p*H. There is an intermediate *p*H, at which the number of positive and negative charges on the protein is balanced, that is, the protein has no net charge. This *p*H-value is called ***p*I**, for proteins it ranges from 1.8 (chimp acidic glycoprotein) to 11.7 (human placenta lysozyme). Other compounds with both negatively and positively charged groups (**amphoteric** compounds) behave similarly.

In a *p*H gradient an amphoteric compound will move until it has reached a spot were the *p*H is identical with its *p*I, at this point no electrostatic force will be exerted on it. Should it move from that spot by diffusion, it will attain a charge again and hence move back to its equilibrium position. It will also buffer its surrounding, especially if its pK_a-values are near (within 0.5 *p*H-units) of its *p*I. If the buffer molecules are relatively small, diffusion leads to GAUSSIAN distributions of the concentration of each buffer overlapping with its neighbours, thus the *p*H-gradient is continuous and linear. Such mixtures of buffers are called **ampholytes**. For a quantitative treatment of gradient formation see [163]. To be suitable as carrier ampholyte a molecule must meet the following criteria:

- It must have a high buffering capacity at its *p*I, so that the *p*H-gradient is not flattened by the proteins or air CO_2. This requires that the two pK_a-values surrounding the *p*I are close together (less than 0.5 *p*H-units apart).
- At its *p*I the electrical conductivity must be low, to reduce the current through the gel and hence JOULEAN heating. Even more important, the conductivity over the entire gel must be constant, otherwise "hot spots" would develop where proteins would be denatured.
- The molecular masses of the ampholytes should be between 300 Da (to get stable *p*H-gradients) and 1 000Da (to facilitate their separation from the proteins).
- Optical absorbance in the range from 250–300 nm should be low.
- It must be easy to synthesise a wide variety of molecules with slightly different *p*I, the resulting *p*H-gradient must be smooth.

Amines co-polymerised with epichlorohydrin (1-chloro-2,3-epoxypropane) meet these specifications. This results in straight, branched, and even cyclic molecules of various lengths. In addition epichlorohydrin is stereo-active, so a large number of compounds with slightly different *p*I is generated.

During establishment of the *p*H-gradient, the number of charged species, and hence the conductivity, decreases. Thus the voltage can be increased, specialised power supply units do this automatically in "constant power" mode, so that as much voltage is applied as possible, given the amount of heat that can be conducted away by the cooling system. At the same time they integrate the voltage over time, since the **extend of focusing** (measured in volt hour) is the parameter needed for reproducible experiments.

Salt concentration in the sample should be below 20 mM, if possible even below 5 mM.

Electroendosmosis, combined with some other mechanisms like CO_2 uptake from the atmosphere, ampholyte reduction at the cathode and its oxidation at the anode makes the entire gradient move towards the cathode over time. Because of this **cathodic drift** it is advisable to finish an IEF experiment once all sample molecules have attained their equilibrium position.

8.2.9.2 Immobilised *p*H-Gradients

As an alternative to IEF with carrier ampholytes immobilised *p*H-gradients may be used. These are generated with an immobiline gradient during casting of the acrylamide gel [28]. Immobilines are derivatives of acrylamide with Zwitter-ionic side-chains (see Fig. 8.11). One end of the gradient contains more of the acidic, the other more of the basic immobilines. Due to partial neutralisation of basic and acidic groups a *p*H-gradient is established, which can be calculated using the HENDERSON–HASSELBALCH-equation. Since the immobilines are chemically cross-linked into the gel, the gradient is not subject to cathodic drift by electroendosmosis, in addition the conductivity of the gel and heat generation is lowered and no pre-run is required. The gradient can accept samples with up to 100 mM salt, about five times what is possible with carrier ampholytes. There are no rim effects, so the gels can be cut into narrow strips, each for a single sample. The resolution can be as high as 0.001, 20 times higher than in classical IEF with carrier ampholytes. However, higher voltages and longer separation times are required in immobilised gradients.

8.2.9.3 Isoelectric Focussing

If proteins are added to the gel, they move in the *p*H-gradient until they reach their *p*I, where they are uncharged and do no longer move (see Fig. 8.10). As a result, proteins are sorted by *p*I. Since protein molecules are much larger than the buffer molecules, they show less diffusion and much sharper peaks. Note that the charged groups on the surface of a protein molecule are largely responsible for its *p*I, and that the environment of a group can significantly change its

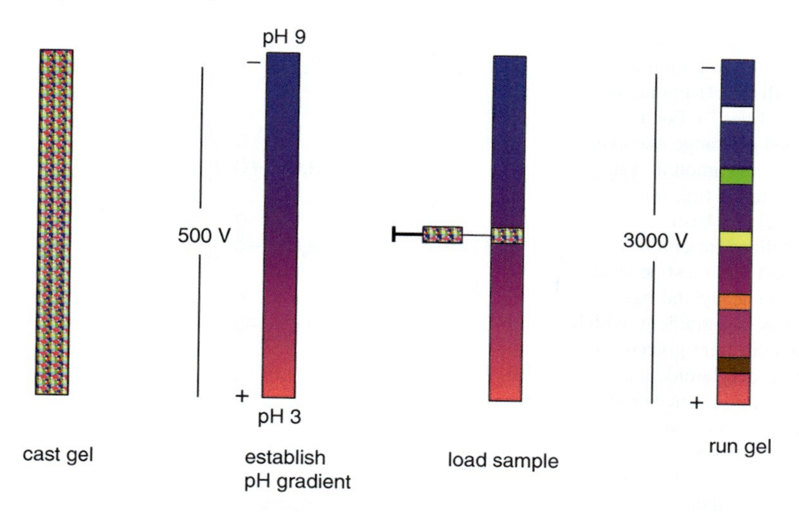

Fig. 8.10 Isoelectric focussing. A pH-gradient is established across the gel by electrophoresis at relatively low voltages ($\approx$ 500 V). The sample is applied to this gradient and run at high voltages until all proteins have achieved their equilibrium position

pK$_a$-value. Thus native and denatured proteins have different pIs. Tertiary and secondary structure has to be destroyed with (nonionic) caotropes like urea[1] or thiourea, and with non-ionic (Triton-X100) or zwitter-ionic detergents (3-[(3-Cholamidopropyl)dimethylammonio]propanesulphonic acid (CHAPS)). A good sample buffer for denaturing IEF is 7 M urea (but see Fig. 8.12), 2 M thiourea, 4 % CHAPS, 65 mM DTT, 1 % ampholyte (of the same pH-range as the gel) and a trace of bromophenol blue. Do not use βME, as it buffers above pH 8!

Isoelectric focusing is an equilibrium-technique, hence it does not matter where in the gradient a sample is added. The sample may even be soaked into the entire gel, increasing the capacity. However, the buffering capacity of the ampholyte at a particular spot has to remain larger than the buffering capacity of the protein after focusing, otherwise a plateau will form in the pH-gradient.

The zone width in IEF depends on electrical field strength $\vec{\mathfrak{E}}$, the slope of the pH-gradient ($\frac{\mathrm{d}pH}{\mathrm{d}x}$), the mobility-dependence of the proteins around the pI ($\frac{\mathrm{d}\mu}{\mathrm{d}pH}$), and the diffusion coefficients D of the proteins. The peak has a GAUSSIAN shape, with a standard deviation of

$$\sigma = \sqrt{\frac{-D}{\frac{\mathrm{d}\mu}{\mathrm{d}pH}\frac{\mathrm{d}pH}{\mathrm{d}x}}}. \tag{8.45}$$

[1] Urea destroys agarose secondary structure and hence cannot be used in agarose gels.

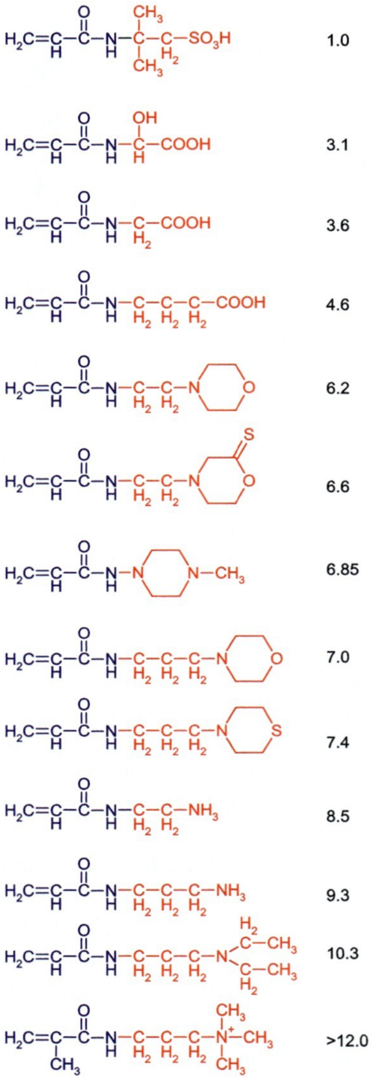

Fig. 8.11 Acrylamido-buffers used to establish immobilised *p*H-gradients (Immobilines™). For a specified *p*H-range one can calculate the amount of the various acidic (mixing chamber) and basic immobilines (reservoir chamber) that must be used [6] in the heavy and light solutions of a gradient, which contain different glycerol, but identical acrylamide and bisacrylamide concentrations. During polymerisation the immobilines get incorporated into the gel, establishing a *p*H-gradient that is not subject to cathodic drift by endosmosis

Proteins are considered resolved if their peak-centres are three standard deviations apart from each other. We get for the resolving power of IEF [419]:

$$\Delta pI = 3 * \sqrt{\frac{D \frac{\mathrm{d}pH}{\mathrm{d}x}}{-\mathfrak{E} \frac{\mathrm{d}\mu}{\mathrm{d}pH}}}. \tag{8.46}$$

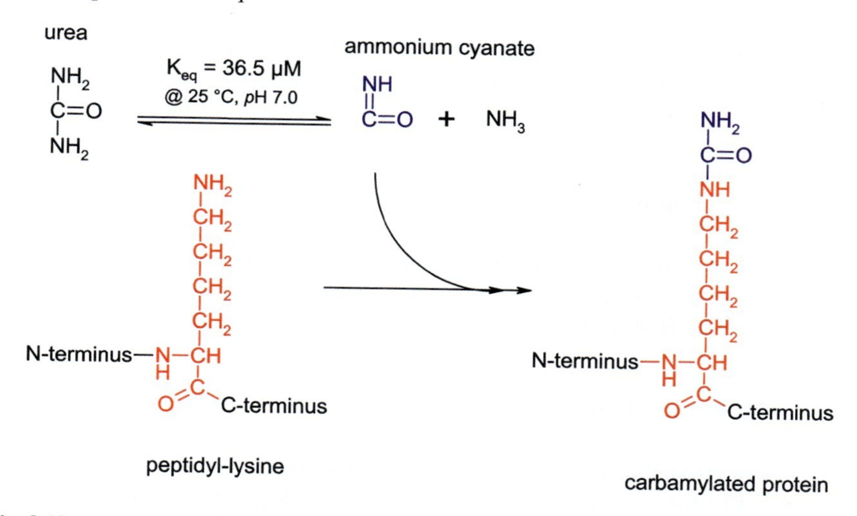

Fig. 8.12 Urea is in equilibrium with ammonium cyanate, which can carbamylate proteins on Lys and Arg residues, and on α-amino groups. This not only changes the *p*I of the protein, but also eliminates proteolytic cleavage sites and changes the molecular mass of peptide fragments. Urea solutions should be prepared fresh. If storage is required, add a little mixed bed ion exchanger to remove the ammonium cyanate formed. Carbamylation of proteins like actin by increased urea concentrations is a proposed mechanism in cataract formation in end-stage renal disease

Under ideal circumstances IEF can separate proteins which differ only by a single charge, resulting in a difference of *p*I-value of 0.001.

8.2.9.4 Preparative IEF

For preparative purposes IEF may be performed in a granulated gel (*e.g.*, Sephadex), the capacity can be as high as 10 mg per ml of gel. Separated proteins are detected by placing a blotting membrane on top of the gel after the run, enough protein molecules will bind to the membrane by diffusion to locate the bands of interest. The bands are transferred to small columns and the proteins eluted with a suitable buffer. Ampholytes may be removed by dialysis or ultrafiltration. For better handling Sephadex Gels may be prepared containing acrylamide. After the run the acrylamide is allowed to polymerise by spraying the gel with catalyst [448].

A different approach has been described in [450]: The sample is enclosed in a cell closed by glass-fibre membranes soaked in immobiline-containing polyacrylamide. The membrane on the cathode side has a *p*H lower, the one on the anode a *p*H higher than the *p*I of the target protein(s). During IEF all proteins with a *p*I outside the *p*H-range bracketed by the membranes will leave the cell. The target proteins stay in solution, which reduces problems with aggregation and precipitation.

Compared to SDS–PAGE IEF works with much lower currents, and hence lower heat production. This certainly simplifies scale-up. However, lack of solubility of proteins at the *p*I is a major hurdle.

8.2.9.5 Non-equilibrium *p*H-Gradient Electrophoresis (NEPHGE)

IEF is not well suited to separate basic proteins, because the range of alkaline ampholytes is limited and cathodic drift further reduces the useful range of IEF. The technique of NEPHGE [297] was introduced to deal with this difficulty.

8.2.9.6 2D-electrophoresis

If after separation the IEF-gel is mounted across an SDS–(or CTAB–)PAGE-gel (see Fig. 8.13), the proteins which have already been separated by *p*I can be separated again by molecular mass. This "two-dimensional" electrophoresis can, under carefully optimised conditions, separate about a 10 000 different proteins from a cell lysate [202]. Conventionally, the gels are presented with low *p*I and low mass on the bottom left, high *p*I and high mass on the top right. Following this convention makes comparing gels from different laboratories much less difficult.

Comparing the protein expression profiles from cells of different developmental stages, or of healthy and diseased cells, can turn up proteins involved in developmental regulation or in disease processes. With luck, such a protein may be a useful drug target. Thus the field of **proteomics** (the proteome of a cell is the collection of all proteins it contains at a given physiological state (defined by time and the

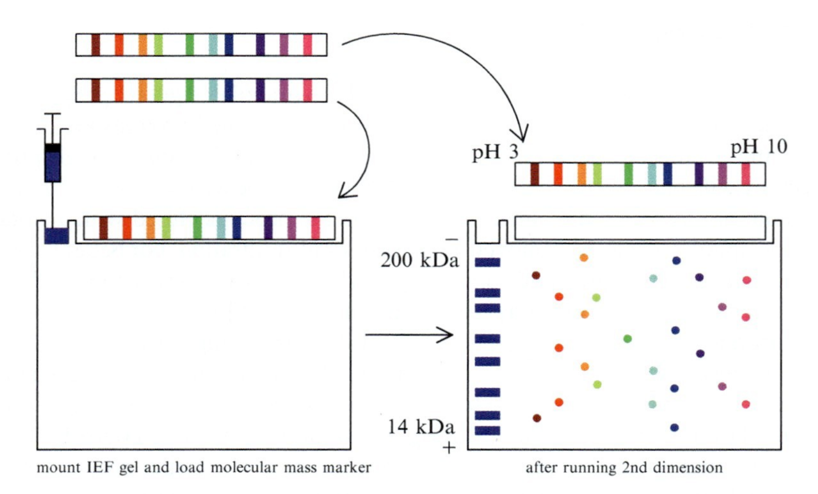

Fig. 8.13 2-D electrophoresis. A protein mixture is first separated on an IEF-gel, which is then mounted across a LAEMMLI-gel. The second electrophoresis separates the bands of proteins with identical *p*I by molecular mass. About 10 000 different proteins can be distinguished on a 2-D gel of mammalian cells

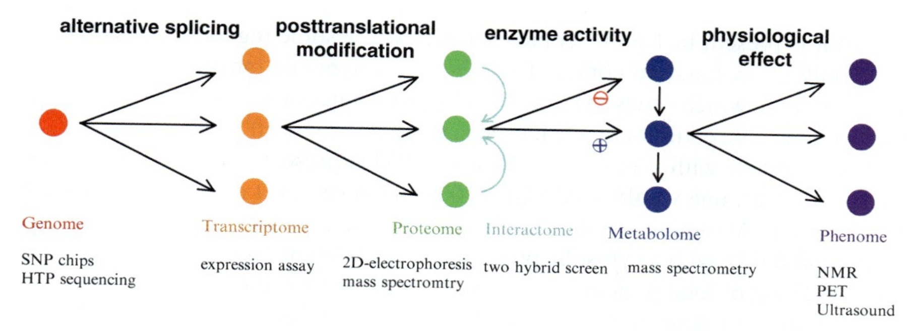

Fig. 8.14 Systems biology. Genome, transcriptome, proteome and metabolome give related, but different information on the state of a cell. Elucidation of the patho-mechanism of complex diseases like metabolic syndrome requires the mining of large data sets that man is no longer able to understand unaided. Ever more complex computer programs handle these data

chemical and physical environment), like the genome is the collection of all genes [423]) has attracted considerable attention in the pharmaceutical industry. It is complemented by investigations into the **metabolome** of cells (see Fig. 8.14) (collection of all metabolites). Identification of protein spots on a 2-Dgel is often achieved by in-gel digestion with proteases followed by MALDI-TOF-MS (see p. 274) analysis of the peptides.

Several programs are available to evaluate 2-Dgels, for example `Msight` (http://www.expasy.ch/msight). The results of these studies should be submitted to international databases (for a list see http://www.expasy.ch/ch2d/2d-index.html) to aid further research.

The methods used for 2D-electrophoresis and spot identification are far from satisfactory, however. For example, in an effort to determine the brain proteome, five laboratories identified 551 differentially expressed proteins, but of these 436 proteins (79 %) were identified only in one lab each [138]. If the detection events are supposed to be Poisson-distributed ($f(k) = \frac{n\,\lambda^k}{k!\,e^\lambda}$), then the system of equations $f(1) = 436$, $f(1) + f(2) + f(3) + f(4) + f(5) = 551$ has the solution $n = 1517$, $\lambda = 0.45$ and $f(0) = 966$. This would mean that only 36 % of all proteins in the proteome were actually detected at all. The problem will particularly affect proteins outside the *p*H-range of IEF (4–11) and very hydrophobic (that is, membrane) proteins. Because of method variations different labs look at different fragments of the proteome, no lab sees all proteins.

It is estimated, that the $\approx 27 \times 10^3$ protein-coding genes in humans give rise to about three times as many mRNA species due to alternative splicing, RNA editing, overlapping units, *etc.*, and that these result in three times as many protein species by post-translational modification, resulting in $\approx 1 \times 10^6$ different proteins in the human proteome. Of those, an average cell may express about 6 000, 80–90 % of which are housekeeping genes expressed in all cells, if perhaps in different amounts (glycolytic enzymes, cell skeleton *etc.*). That would mean that about 1 000 proteins in each cell support its specific function.

Actin is present in 1×10^8 copies per cell, while some regulatory proteins may have only a few hundred copies. Thus there is a dynamic range of 10^6 in protein concentration, which needs to be covered by proteomics methods. In serum or urine, the dynamic concentration range has been estimated to be 10^8–10^{12}.

For a protein with a molecular mass of 50 kDa present in 100 copies per cell (8.3×10^{-18} g), one would need 120×10^6 cells to get the 1 ng of protein required for detection. Assuming a cell diameter of 10 μm that would be 63 μl of cells (plus interstitial fluid); with a cytosolic protein concentration of 400 mg/ml that is equivalent to 25 mg of total protein, way more than can be loaded onto an IEF-gel.

In addition, proteins with a molecular mass outside that covered by conventional SDS–PAGE (10–200 kDa), or a *p*I outside the range covered by IEF (3–10) are likely to be lost. That is also true for very hydrophobic proteins or proteins in aggregates, which may not be solubilised during sample preparation.

Note that the determination of mRNA-concentration, although technically much simpler, can not replace the determination of the concentration of the protein itself because the rates of protein synthesis, modification, and degradation depend on additional factors. In fact, the correlation between the concentrations of mRNA and the protein encoded by it is between 0.4 and 0.7 [128, 136, 245], see also Fig. 9.10 on p. 122.

8.2.9.7 Differential Gel Electrophoresis

The comparison of proteoms by 2D-electrophoresis is time consuming and error prone because no two gels are exactly equal. This results in differences in the spot pattern observed in the two gels, making the alignment of gels difficult. Although consistency is increased by the use of industrially manufactured gels, and there is now software available that can make alignment task easier, the problem still exists.

In differential gel electrophoresis (DIGE) the samples are labelled prior to electrophoretic separation, using different fluorescent [295] or stable isotope labels [135]. The samples are then combined and subjected to electrophoresis on the same gel, thus there is no gel-to-gel variability in separation. After the run, the intensity of the labels are compared in each spot. For this to work the labelling has to meet the following criteria:

- Reaction chemistry and molecular mass of the labels need to be identical, the molecular mass should be low so that labelled and unlabelled protein run identical during SDS–PAGE.
- The label should not change the *p*I of the protein, that is, for each charge destroyed in the labelling reaction (*e.g.*, ϵ-amino group of lysine) the label should contain a charged group of similar pK_a. Thus labelled and unlabelled protein move identical during IEF
- Only about 1–3 % of the protein molecules should be labelled, so that the labelled products are below the detection limit of the mass spectrometer. This in turn requires sensitive detection of the label. Only for very small samples (*e.g.*, from laser capture dissection) one uses saturation labelling

Labelling is performed at *p*H 8.5 for half an hour on ice, then the reaction is stopped by addition of lysine. Note that ampholytes interfere with labelling.

The molecular mass added to the protein by the label is a few hundred Da, SDS–PAGE is able to detect that for proteins smaller than $\approx$ 30 kDa. In such cases, the ratio of fluorescence is determined first by fluorescence scanning, then the gel is post-stained for spot-picking.

Kits working along these lines are commercially available, using different fluorescent dyes (Cy2, -3 and -5 label Lys, CyDye3 and -5 Cys) or activated biotin molecules with either a ^{12}C or a ^{13}C backbone. Alternatively, cells can be grown in the presence of ^{2}H (Leu, Lys), ^{13}C (Lys, Arg) and/or ^{15}N (Lys, Arg) labelled amino acids. In the latter cases mass spectrometry is used to quantitate the labels in each spot (http://msquant.sourceforge.net).

Some cases have been reported in the literature where protein spots behave "funny" in DIGE experiments, therefore reverse labelling controls should be used, *i.e.*, if in the first experiment the standard was labelled with stain A and the experimental with stain B, this should be reversed when the experiment is repeated. The **internally normalised ratio** [8] then becomes

$$R = \sqrt{\frac{\text{Standard}_1}{\text{Experimental}_1} \Big/ \frac{\text{Experimental}_2}{\text{Standard}_2}}. \tag{8.47}$$

This is calculated for each spot, R is the ratio of the analyte in experimentals over standard. Duplicates that vary by more then a preset value (say, 30 %), are discarded before analysis. Samples are considered identical to the reference if $0.7 < R < 1.3$, otherwise a significant difference is assumed.

A mixture of all samples is labelled with a third dye and used as internal standard. Thus comparison of many samples on many gels is possible, signal intensities can be normalised by the intensity of the standard.

8.2.9.8 Other Methods of 2D-electrophoresis

When people speak of 2D-electrophoresis, they usually mean the combination of IEF and SDS–PAGE. However, in some cases it is useful to separate functional protein complexes by electrophoresis in non-denaturing detergents, followed by their analysis by SDS–PAGE. For example, chloroplast photosystem I has been investigated after electrophoresis with oligooxyethylene alkylether sulphate and dimethyl dodecylamineoxide (the latter to stabilise the complex) [162], mitochondrial complexes of endoxidation by "native blue" electrophoresis [367]. With the same method it can be demonstrated that Hsc70 bound to clathrin coated vesicles retains the terminal phosphate from [γ^{33}P]-ATP, while the soluble complex contains only ADP (unpublished observation).

8.2.10 *Elution of Proteins from Electrophoretic Gels*

For small scale preparative work proteins may be eluted from electrophoretic gels, such proteins can be used, for example, to develop antibodies in rabbits or chicken.

The easiest way to elute the protein is to mash up the gel in a suitable buffer as finely as possible. This may be achieved by passing the gel several times through a 27G hypodermic needle, or with a DOUNCE-homogeniser. The gel fragments are removed by centrifugation, the supernatant is collected.

More complete recovery of proteins from the gel can be achieved by electroelution (see Fig. 8.15). The gel fragments are placed on a frit in a small glass tube filled with buffer (see Fig. 8.15). The bottom of the tube is closed by a dialysis membrane. A current is applied over the tube so that the proteins move downward toward the membrane where they may be collected. Equipment for this procedure is available commercially.

8.2.11 *Gel Staining Procedures*

In order to detect the proteins, gels need to be stained. There is a huge selection of literature with staining methods available, but I can cover only the most important methods here.

8.2.11.1 Fixing the Separated Proteins in the Gel

Most often proteins are fixed in a mixture containing 30 % methanol (or *iso*-propanol) and 10 % acetic acid. Both acid and alcohol precipitate the protein, in

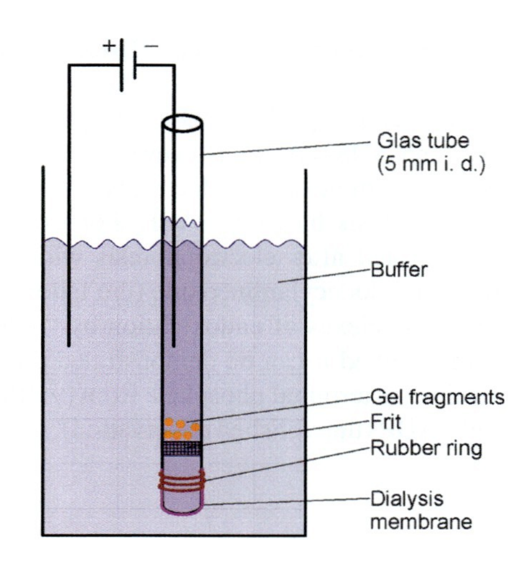

Fig. 8.15 Electroelution of proteins from gel fragments after electrophoresis. The finely minced gel pieces are placed onto the frit, the electric current drives the proteins toward the dialysis membrane at the bottom end of the tube

addition the alcohol helps removing SDS from the gel which would otherwise result in high background staining.

Small proteins and peptides (M_r < 5 kDa) may be washed away in this fixative, in such cases 5 % glutardialdehyde in 30 % methanol may be used instead. Since that solution is not acidic, removal of SDS is faster and more complete and may be further accelerated by heating to 60 °C.

8.2.11.2 CBB-R250

Coomassie Brilliant blue R250 (CBB-R250) is probably the most frequently used stain for proteins on electrophoresis gels. The gel is stained with 0.1 % CBB-R250 in fixative. This dye solution can be used several times, but must be discarded if a precipitate forms. The gel will be stained completely blue, it is differentiated with several changes of fixative. The procedure may be accelerated in a microwave[2] oven and/or by including little bits of sponge, which bind the excess dye. Once most of the background staining has been removed, one can use a weaker destaining solution (5 % acetic acid, 20 % methanol), gels can be left in this solution over night for complete removal of the background.

This procedure is simple, cheap, and works with all types of gel. Other dyes like **eosine** or **calcone carboxylic acid** may also be used in a similar way with somewhat increased sensitivity. Disadvantages of this procedure are that it has only moderate sensitivity (a few μg per band), it is fairly time consuming and proteins are denatured. Thus CBB-R250 stained gels give lower yields in blotting or electroelution. However, the staining is very permanent, gels can be dried and included in lab books for documentation.

With CBB-G250 it is possible to stain only the proteins in the gel, without background. Thus there is no need for destaining [290] and the procedure can be performed over night without human supervision.

8.2.11.3 Silver Staining

There are many different recipes for silver staining, most often used are the procedures of Merril et al. [272] and of Heukeshoven & Dernick [160]. In my experience the latter method is more sensitive and slightly less laborious to perform. Silver staining procedures can be highly capricious and require very pure reagents and scrupulously clean lab ware. However, they are among the most sensitive staining procedures known (about 10 times as sensitive as CBB-R250), and hence have been widely used in proteomics. The silver ions are complexed by Glu, Arg and

---

[2] Microwave radiation can accelerate many lab procedures, and they do so independent from their heating effect. This requires low wattage ovens (< 100 W, "cold microwave") with accurate temperature control. Incubation times may be reduced by 90 % and more.

Cys-residues of the protein, unbound Ag^+ is washed away and the bound Ag^+ reduced with formaldehyde.

If the gel has become too dark it can be differentiated by an ammoni-alkaline solution of 0.185 % NaCl, 0.185 % $CuSO_4 \times 5\,H_2O$ and 3.42 % $Na_2S_2O_3 \times 5\,H_2O$. The washed gel is placed into this solution and transferred into 5 % acetic acid when the desired tone has been achieved.

8.2.11.4 Staining of Glycoproteins

If the sugar residues of glycoproteins are oxidised with periodic acid (H_5IO_6), the C–C-bond between neighbouring ("vincinal") alcohol-groups is opened and a dialdehyde produced. This reacts with SCHIFF'S reagent forming a bright red colour

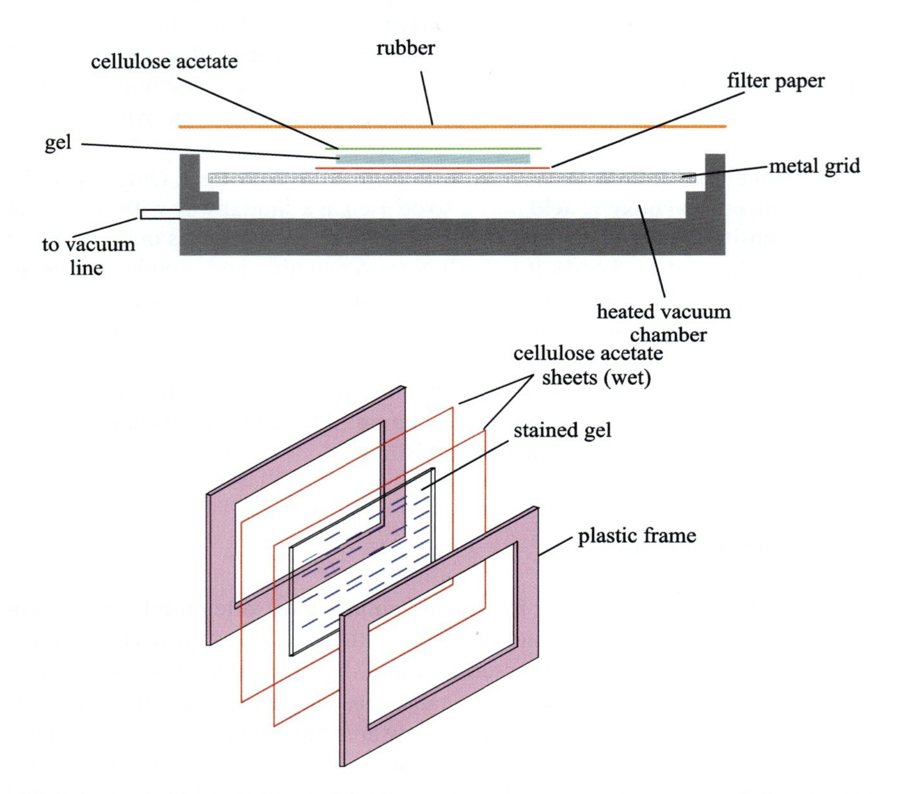

Fig. 8.16 *Top*: The gel can be placed on a sheet of filter paper and exposed to heat while compressed in a special vacuum-appliance. The water is collected in a cold trap in the vacuum line. *Bottom*: Alternatively, the gel is placed between two sheets of cellophane which are spanned into a plastic frame. The assembly is held together with clamps and allowed to air dry over night. The latter procedure has the advantage that the gel can still be used for spot-picking and MS-analysis. In either case, soaking the gel in 5 % glycerol before drying helps to prevent cracking

(periodic acid/SCHIFF (PAS)-reaction). In order to reduce the incubation times with the various solutions it is commonly performed on Western-blots [77]. PAS-staining can be followed with CBB-R250 as total protein stain, bands appear red, purple or blue depending on their sugar content. **Stains All** (3,3'-Diethyl-9-methyl-4,5,4',5'-dibenzothiacarbocyanine) gives similar results in a single step.

8.2.11.5 Fluorescent Staining of Protein Bands

Proteins in SDS–PAGE gels are associated with detergent micelles, which provide a very hydrophobic environment. Certain compounds are virtually non-fluorescent in water, but show bright fluorescence when they partition into a hydrophobic environment. If SDS–PAGE gels are incubated with aqueous solutions of such dyes, protein bands can be made visible by the fluorescence. The first such compound described was **Nile red** [73], the company *Molecular Probes* (now part of Invitrogen, www.invitrogen.com) offers various proprietary **Sypro**®-stains for this purpose. Such stains can be almost as sensitive as silver staining, yet the procedure takes only a few minutes to perform. Additionally, some procedures involve no denaturation of proteins, thus staining does not interfere with blotting and elution. On the down side staining is non-permanent, the gels need to be photographed for documentation. Also the proprietary stains are quite expensive. The latter disadvantage can be overcome by using a homemade dye: ruthenium(II) tris(bathophenantroline disulphonate) [220, 339] is as sensitive as Sypro® Ruby, cheap and easy to prepare. In addition it has a known composition, unlike the proprietary Sypro-stains. Non-disclosed proprietary reagents and kits make repetition of experiments impossible when a vendor changes their product or leaves the market. That attacks the very foundation of science.

8.2.11.6 Negative Staining of SDS–PAGE Gels

Various metal ions (Cu^{2+}, Zn^{2+} and K^{+}) give precipitates with SDS, which turn a SDS-gel opaque. Since the SDS in protein bands is bound to the proteins, no precipitation occurs and the bands stay clear. If the gel is held above a dark background and illuminated from the top, bands can be photographed. They may also be read into a computer, using a simple flat bed scanner (with a black cardboard on their reverse side). This allows quantitation of band intensity using programs like `ImageJ` (rsb.info.nih.gov/ij, the `Java`-based, operating system independent successor of the Macintosh-based `NIH-Image`).

The most sensitive of these staining methods appears to be Zn/imidazole staining [144]. Detection limits rival those achieved with silver and fluorescent staining (a few ng per band). With suitable modification this method may also be used for native gels and even for DNA-gels.

Apart from speed and sensitivity, reverse staining methods have the advantage of being reversible by incubation with chelating agents (EDTA, citric acid). They can then be used for blotting or elution without loss of efficiency.

Gels stained in this way may be stored wet at 4 °C for years, as the metal ions prevent diffusion of proteins. Drying is not possible, as gels would become clear. It is, however, possible to tone the gel by oxidation of *o*-tolidine with potassium hexacyanoferrate(III), which is Zn-catalysed. This gives a deep blue background with clear bands, which survive drying [98]. However, from such gels proteins can no longer be eluted with good yield.

8.2.12 Capillary Electrophoresis

Electrophoresis may also be performed in buffer-filled fine capillaries, the narrow bore (<150 μm) prevents convection (hence no gel is required, unless its sieving effect is used) and dissipates heat very well. Filling of the capillaries is most conveniently done by application of vacuum to a microfluidics-system. Voltage is very high (1 kV/cm), but because of the small cross-sections of the capillaries, currents are low (μA). *Warning*: The high voltages used in capillary electrophoresis are potentially lethal, due care is required in such experiments.

The small cross-section also allows for small sample volumes (nl). Capillaries are several dm long, separation times about 10–20 min. Detection is by absorbance (fmol sensitivity) or fluorescence (zmol sensitivity, *i.e.*, a few hundred molecules). The light is transported to and from the capillary by glass fibres, thus no focusing optics is required. The number of theoretical plates (see Fig. 14.6 on p. 155) obtainable in capillary electrophoresis is given by

$$N = \frac{\mu U}{2D} \tag{8.48}$$

with μ the electrophoretic mobility, U the voltage applied and D the diffusion coefficient. N may be in the order of several 10^5. Ornstein already described such a system in his classic paper [303], using interference microscopy for detection. However, widespread use of this technique came after the publication of [181].

Fused silica capillaries have a layer of positive ions bound to their wall, those in turn bind water. During electrophoresis the ions, and hence the bound water, move toward the negative electrode. Since the bound water constitutes a major fraction of the total water present in the thin capillaries, **electroendosmosis** is much more significant in capillary electrophoresis than in other electrophoretic techniques. In fact, many techniques would not even work without electroendosmosis.

In standard mode (**free solution capillary electrophoresis, FSCE**) sample molecules may be both positively and negatively charged, the electrophoretic movement of sample molecules is vectorially added to the solvent stream caused by electroendosmosis. Thus negatively and positively charged molecules and uncharged ones move towards the negative electrode, but with different velocities.

If molecules with positive charge are covalently bound to the capillary wall the direction of endosmotic flow is reversed and all sample molecules move toward the positive pole (**FSCE in coated capillaries**).

IEF may be performed in ampholyte-filled capillaries, the sample is introduced as a small zone in the middle of the capillary. There are also systems available for isotachophoresis and SDS–PAGE.

In **micellar electrokinetic capillary electrophoresis (MECC)** the (electrically neutral) sample molecules partition between the buffer and SDS-micelles , which move towards the positive electrode (mobile phase). The higher the partitioning coefficient into SDS, the faster the molecules move against the electroendosmotic flow.

Protein–ligand interactions may also be studied and the dissociation constant determined if protein and protein/ligand-complex have different electrophoretic mobilities.

If the buffer in capillary electrophoresis contains serum albumin, **chiral** molecules can be separated due to their different affinity for the protein.

Detection is usually online, which makes such assays quick. Only small sample volumes (about 0.1 μl are required. However, the equipment is quite expensive.

Protein binding to the capillary wall introduces streaking, which lowers resolution. This problem has yet to be overcome.

Chapter 9
Immunological Methods

The following methods rely on the specific recognition of ligands by a protein. Although they are described here (as in most other textbooks) for antibody-antigen interaction, you should keep in mind that they can also be used for other specific interactions, like ligand-receptor. [146] should be consulted first for experimental details, despite its considerable age.

9.1 Production of Antibodies

9.1.1 Isolation from Animals

Originally, antibodies were produced by immunising an animal (in the lab most often rabbits, but mice, rats and chickens may also be used. Large animals like donkeys, sheep or horses are used for commercial scale production) with several monthly injections of antigen. This allowed for the development of a highly specific antibody response.

9.1.1.1 Native or Denatured Antigens?

One question to be answered at the beginning of such a study is whether native proteins are to be used as antigens or denatured proteins isolated from a gel after SDS–PAGE. Because denaturing electrophoresis destroys the 3-D structure of a protein, antibodies raised against native proteins may not be able to detect proteins after electrophoresis (on Western blots, see later). On the other hand, antibodies against denatured proteins may not detect native proteins in immuno-histology, may not neutralise the antigen in biochemical assays, or work in immuno-precipitation studies.

Use of gel-purified proteins for immunisation is technically very simple: the gel is fixed, stained (with any of the common methods), and the interesting band cut out. This is then minced and homogenised with a little phosphate buffered saline (PBS)

E. Buxbaum, *Biophysical Chemistry of Proteins: An Introduction to Laboratory Methods*, DOI 10.1007/978-1-4419-7251-4_9,

and FREUND'S adjuvant by passing it through a 27G needle several times. The mixture is injected into the animal. For the first injection, complete FREUND'S adjuvant is used (a suspension of *Mycobacterium ssp.* cell walls in mineral oil) to stimulate the immune system of the animal. For all booster injections, *incomplete* FREUND'S adjuvant (just the mineral oil) *must* be used to prevent an allergic reaction against the bacterial lipopolysaccharides (granuloma formation).

Note that although acrylamide is very toxic, polyacrylamide is not. Thus this procedure is not harmful for the animal, provided that all unbound acrylamide was washed out, which is the case after fixing and staining. The procedure is very effective, as the gel particles release the antigen slowly, ensuring lasting contact of the animal with the antigen.

9.1.1.2 Adjuvants

For vaccination of humans FREUND's adjuvant is too strong. However, other substances like $Al_2(OH)_3$-gel, squalene (intermediate of steroid and cholesterol biosynthesis), and detergents (*e.g.*, saponin) are used to stimulate the immune system. Unfortunately, we have little idea about their mode of action. Another ingredient of some vaccines is 2-(ethylmercuriomercapto)benzoate (Thimerosal®, Thiomersal®, Mersalyl®). It is used as a preservative when several doses of a vaccine are packed into a single bottle. The increasing use of single-dose packages has reduced the need for this rather ill-reputed substance. Note, however, that the mercury dose contained in a vaccine shot (about 5 μg) is much lower than that contained in a meal of tuna (about 500 μg/kg in many samples). In addition, Thimerosal is degraded to ethyl-mercury ($CH_3CH_2Hg^+$). This is excreted relatively quickly, unlike the methyl-mercury found as environmental contamination (CH_3Hg^+, half-life periods 3.7 and 50 d, respectively), which tends to accumulate in our body and is the causative agent of Minamata-disease. It is therefore not surprising that several large-scale clinical studies found no ill-effect by Thimerosal. In particular, a previously suspected link between Thimerosal and autism must now be viewed as red herring (see http://www.who.int/vaccine_safety/topics/thiomersal/en/index.html for a WHO-statement on this issue).

9.1.1.3 Haptens

A hapten is a small molecule which by itself is not immunogenic. If, however, the hapten is conjugated to a foreign protein ("carrier), antibodies produced against this complex may also recognise the isolated hapten. Once an immune reaction against the complex has been established, boosting with the hapten alone is sufficient to stimulate antibody production.

This is the basis for immunological assays for drugs (legal or otherwise). Injecting therapeutic monoclonal antibodies against drugs may also be used in treatment of addiction [53]. The method may also be used to produce antibodies against short peptides, oligosaccharides or other molecules of biological interest.

The most well-known carrier is haemocyanin from the giant keyhole limpet *Megathura crenulata* (SOWERBY 1825), a Pacific gastropod of impressive size. The protein has a molecular mass of about 5 MDa and consists of multiple subunits that dissociate at alkaline *p*H, each with two copper ions that serve as oxygen transporters. Because it is a protein from a marine animal it requires high salt concentrations to stay in solution (0.9 M ≡ 5.3 %). Since KLH is soluble in up to 50 % DMSO, it can be used for the conjugation of even very hydrophobic haptens. The protein has N-linked oligosaccharides in addition to several hundred each Cys, Lys and Tyr.

As the ratio of hapten to carrier increases, the antibody titre is likely to increase, but the affinity of the antibodies will decrease [318].

9.1.1.4 Isolation of Antibodies

If mammals are used, blood samples are taken in a procedure approved by the local **animal use committee** and allowed to clot over night in a cold room. The serum (which should be clear and yellowish) is decanted and heated to 56 °C for 15 min to destroy complement. After cooling on ice the immunoglobulins are isolated by fractionated ammonium sulphate precipitation (25–50 % saturation) followed by hydrophobic interaction chromatography or dialysis and ion exchange chromatography on a DEAE-matrix.

9.1.1.5 IgY from Egg Yolk

Birds, in particular chickens, have the advantage that antibodies (so called IgY, but actually just IgG) can be isolated from egg yolk, without bleeding the animals. Also, because of the greater evolutionary distance it is often possible to raise antibodies against mammalian proteins in chickens, which would not be immunogenic in mammals. To isolate IgY the egg yolks are separated from the egg white and diluted with ten volumes of phosphate buffered saline (PBS). This mixture is then subjected to fractionated precipitation with polyethylene glycol (molecular mass 4 000 or 6 000, 35–120 mg/ml). The crude IgY preparation can then be further purified by ammonium sulphate precipitation (25–50 % saturation, 144–302 mg/ml). The main disadvantage of IgY is that it does not bind to protein A nor to protein G.

9.1.1.6 Antibody Storage

Antibodies may be stored sterile-filtered dissolved in PBS or as ammonium sulphate precipitate at 4 °C, or in 50 % glycerol at −20 °C. Freezing in PBS and storage at −80 °C is sometimes also recommended. Different antibodies react differently to those conditions, but the storage in 50 % glycerol should be safe for most.

9.1.1.7 Isolation of Specific Antibodies

Such crude preparations of antibodies are often sufficient. However, if there are unwanted cross-reactions, or if the antibody is to be used *in vivo*, further purification may be required. This is usually done by affinity chromatography with a column that contains immobilised antigen. Specifically bound proteins can be eluted with mild denaturants, most commonly 0.1 M glycine/HCl *p*H 2.8. The tubes into which the eluate is collected should contain buffer (0.5 M potassium phosphate *p*H 7.4) to neutralise the solution. Desalting of relevant fractions should be done as soon as possible.

The disadvantage of this procedure is that it selects for mediocre antibodies. Weakly binding antibodies are removed by the washing steps (which is good), medium affinity antibodies are collected, and high affinity antibodies either stay on the column or are eluted with so high a denaturant concentration that they are permanently damaged. Some researchers report better results with electroelution of the bound antibodies. Similar observations have been made in phage display [249], where desorption of the ligand with ultrasound is required to get high affinity virions.

9.1.2 Monoclonal Antibodies

Antibodies isolated from serum have the disadvantage that their specificity and affinity changes from animal to animal, and even from batch to batch for the same animal. For medical applications (diagnostic and even more therapeutic) antibodies are required with consistent, well characterised properties. Those can be obtained if plasma cells from the spleen of an inbred mouse, immunised against the antigen in question, are fused with lymphoma cells from the same mouse strain. The resulting fusion product (**hybridoma**) inherits immortality from the cancer cell and antibody production from the plasma cell.

The fusion products are selected and cloned, the clones are tested for antibody production against the antigen in question. For well characterised monoclonals the precise binding site on an antigenic protein are mapped, and the influence of binding to the enzymatic properties of the target is determined. Hybridomas can be grown in chemically defined medium without the addition of the animal serum commonly used in cell culture. Such media are much more expensive than conventional ones, and give lower growth rates. However, therapeutic antibodies for use in humans have to be guaranteed free of extraneous animal proteins and, even more important, free of animal pathogens.

Another way to produce monoclonal antibodies is to inject the hybridoma cells into the body cavity of mice. There they grow to sizable tumors secreting the antibodies into the peritoneal fluid (ascites). This method avoids the complicated and expensive growth of cells in culture, but it is not very nice on the animals. It is difficult now to get ethics board approval for it.

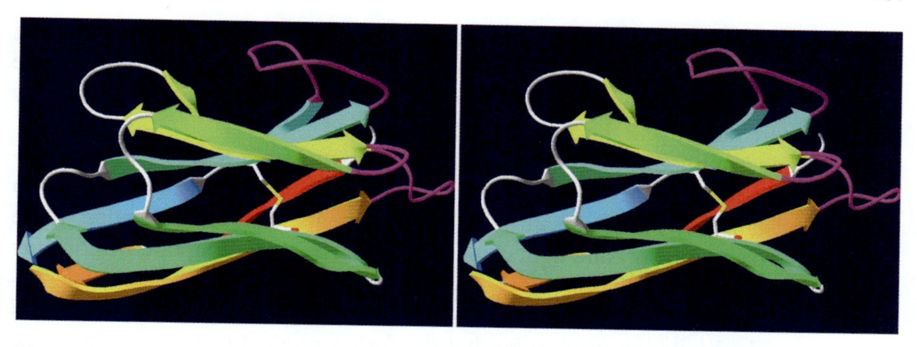

Fig. 9.1 Crystal structure of a nanobody (PDB-code 3eak, N-terminal red to C-terminal blue). Note the sandwich of two anti-parallel β-pleated sheets which are held together by a disulphide bond. This structure is called an immunoglobulin domain. The purple coils to the right are the hypervariable domains that form the antigen binding site

9.1.3 Artificial Antibodies

The most modern approach to antibody generation no longer relies on immunised animals, but on genetic engineering. Since the gene sequence for all antibody domains is known, the process of recombination can be carried out *in vitro*, and the resulting antibodies expressed as part of the coat protein of genetically engineered phages (genotype/phaenotype coupling by **phage display**, [380]). Thus a test tube of these virus contains an entire library of about 10^8 different antibodies.

For such purposes an ideal class of antibodies is found in *Camelidae* (camels, dromedaries and llamas), which consists only of heavy chains, no light chains [288]. The variable segment is called $V_H H$ or **nanobody** (see Fig. 9.1). Sharks also produce single-chain antibodies, **IgNAR** (new antigen receptors), the antigen binding domain is called V_{NAR}. Compared to conventional antibodies, nanobodies and V_{NAR} have considerable advantages. They are small, with a molecular mass of 12–13 kDa. Therefore, they can enter living cells without permeabilisation of the membrane, can cross the blood-brain barrier, and they can sneak into the substrate binding sites of enzymes, thereby inhibiting enzymatic activity. This makes them potential drugs because they can be given orally as they are not destroyed by gastrointestinal enzymes and they are highly soluble in water. Because they are smaller than the kidney threshold for excretion, their serum half life period is short, but can be increased by PEGylation. In addition, nanobodies and V_{NAR} are highly stable molecules, they survive boiling and can work in urea and detergent solutions. Recently, an anti-fungal nanobody was described that can be used in anti-dandruff shampoo [79]. If packed aseptically, they do not require refrigeration for storage and transport. Once developed, they can be produced by genetically altered yeasts with a yield of up to 1 g/l culture medium, and at a cost of only about 1 €/g.

From phage libraries (which are commercially available) it is possible to select those virus which carry antibodies against the antigen in question by repeated "panning". The virus are isolated on immobilised antigens, the bound ones are then multiplied. The process is repeated several times under conditions of increasing stringency. It is even possible to include rounds of mutations in the hyper-variable sections of the gene to increase affinity. Finally, the virus expressing high-affinity antibodies are cloned and the antibodies characterised.

Once good antibodies have been obtained, it is possible to use genetic engineering to create designer molecules for particular applications: antibodies with human constant part for therapeutic use, F_{ab}-fragments, or even molecules where the variable parts of heavy and light chains are linked by a short peptide without any constant part at all.

In a similar way it is possible to create other proteins, *e.g.*, enzymes with novel properties. The limitation used to be that the fusion product of target protein and virus coat protein needed to be transported from the cytosol to the periplasm of the bacterium through the Sec-system in an unfolded state, and then folded in the oxidising environment of the periplasm without the aid of chaperons. In many cases, this did not lead to functional products. A more recent approach is to express the coat and the target protein separately, both labelled with Leu-zipper domains (Jun and Fos, respectively). The coat protein is transported through Sec and folded in the periplasm, however, the target folds in the cytosol and is exported to the periplasm by the **twin arginine translocon (Tat)** [314]. In the periplasm the Leu-zippers interact, the result is stabilised by a disulphide bond. That way the entire chaperon and quality control system of the bacterium is available to ensure efficient folding of the target protein.

9.1.4 Aptamers

In recent years the use of aptamers, single-strand nucleic acid molecules that bind to a target molecule with high affinity (nM) and selectivity, has become more common [40]. Aptamers are selected in a panning procedure similar to phage display, with binding, washing, and amplification steps (**SELEX**: systematic evolution of ligands by exponential enrichment). The selection process can be automated and results in a final product within a few days, much faster than possible with antibodies. Compared to antibodies, aptamers are more stable during storage and less immunogenic. That makes them particularly useful for *in vivo*-applications. The biological half-life period of aptamers is short (elimination by kidney, nucleases), but can be increased by chemical modification if desired. A database of aptamers is available under http://aptamer.icmb.utexas.edu/. In nature, riboswitches are similar to aptamers.

Protein-aptamers have also been described; these are short peptide loops (10–20 amino acids) introduced in a small, stable protein like thioredoxin. They are usually generated in a yeast two hybrid system and can have affinities similar to antibodies and nucleic acid aptamers [169].

Aptamer development can be against unidentified biomarkers on, say, cancer cells. Once an aptamer that binds only to diseased, but not to normal cells, has been isolated, it can be used to enrich the target which may then be identified using proteomics techniques (**AptaBiD**: aptamer-facilitated biomarker discovery, [25]).

9.2 Immunodiffusion

For immunodiffusion experiments agarose gels (1–2 %) are used which have pores large enough to let immunoglobulins and other large molecules pass unhindered. Two little wells are punched into the gel a few millimeters apart, one for the antibody and one for the antigen. The gel is placed into a humid chamber and allowed to stand over night. During this period the molecules will diffuse toward each other, and where they meet a precipitate of antigen-antibody complex will form. After washing away soluble proteins and drying the gel, these precipitates are stained, for example with Coomassie Brilliant Blue or Amidoblack.

Alternatively, one can use a gel that contains the antibody throughout and fill the sample into a single well. The area in which precipitation occurs is then proportional to the concentration of the antigen in the sample.

More informative is the OUCHTERLONY double diffusion technique [307] which works with seven holes, one in the middle and six surrounding (see Fig. 9.2). One can use the middle one for antibody and the surrounding ones for different antigens or *vice versa*. The double diffusion technique results in characteristic precipitation bands, the shape is indicative of the relative affinity of the interaction with the different samples. Thus it is possible to distinguish identical, partly identical, or non-identical antigens. The technique may be used to determine the species from which *e.g.*, a blood spot on a crime scene comes by using antibodies against the serum proteins of different species. This technique is suitable for screening of large sample numbers.

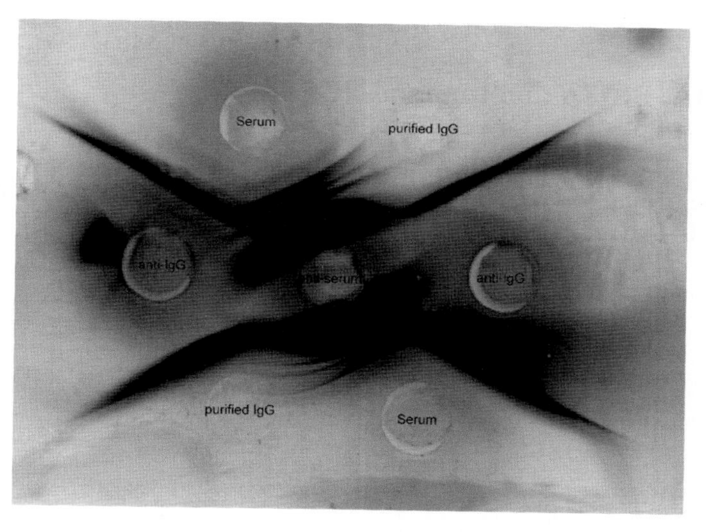

Fig. 9.2 OUCHTERLONY double diffusion test to follow the purification of rabbit immunoglobulins by protein A affinity chromatography. In a rabbit serum sample many proteins react with donkey anti-rabbit serum proteins, but only one of them reacts with donkey serum against rabbit IgG. After purification of the rabbit IgG on a protein A column, only one band is seen after detection with anti-rabbit-serum, this band still reacts with anti-IgG

9.3 Immunoelectrophoretic Methods

Counter-stream immunoelectrophoresis [92] is a native electrophoresis in agar gel carried out at *p*H 8.6 at which antigen moves toward the positive electrode, the antibodies are near their *p*I and transported in the opposite direction by electroendosmosis (beware: do *not* use "electrophoresis grade" agarose, which is selected for a low concentration of the charged groups which cause electroendosmosis!). Precipitation bands are formed at positions that depend on the mobility of both antigen and antibody.

In the **rocket technique** [224], the antibody is present in the entire gel, electrophoresis of the antigen is carried out at the isoelectric point of the antibody, usually around *p*H 8.6. Antigens will form precipitation lines with characteristic shape, which gives this technique its name. The length of these "rockets" is proportional to the antigen concentration.

In **immunodiffusion electrophoresis** [124] complex mixtures of antigens are separated by native electrophoresis in agar gels. In a second step, a slit is cut into the gels parallel to the sample lane, this is filled with antibody solution. Immunodiffusion then leads to the formation of precipitation lanes around the protein spots (see Fig. 9.3).

In all cases non-precipitated proteins are washed out with buffer, the gel is dried and then stained with CBB-R250 or with silver.

The disadvantage of immunoelectrophoresis is the generally high consumption of (expensive) antibodies. Hence they have been largely replaced by the more economic blotting techniques and are of historic interest only.

9.4 RIA, ELISA and Immuno-PCR

Enzyme linked immunosorbent assay [89] and **radioimmunoassay** [439] are techniques that allow the quantitative determination of minute amounts of antigen (10^{-15}–10^{-18} mol). Immuno-PCR is even more sensitive.

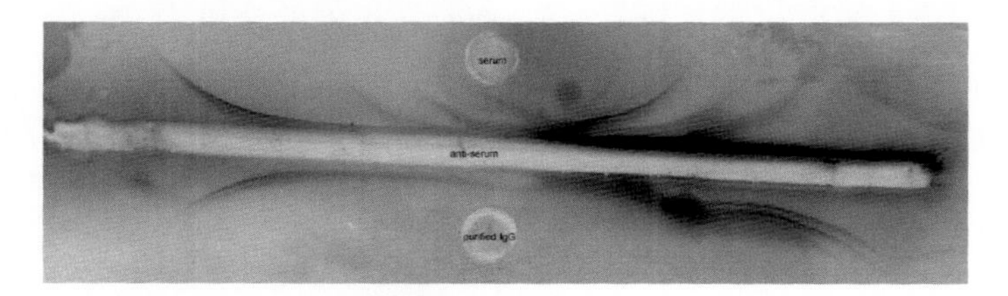

Fig. 9.3 Immune electrophoresis of rabbit serum and purified rabbit IgGs, detected with donkey antibodies against total rabbit serum proteins. Affinity chromatography on a protein A column significantly reduces the number of bands. The samples used are the same as in Fig. 9.2, the higher resolution of immune electrophoresis shows several bands in the purified IgG preparation which appeared homogeneous in OUCHTERLONY double diffusion

9.4.1 RIA

In a radioimmunoassay (RIA) a known amount of radio-labelled antigen (often ^{125}I, a soft γ-emitter) is mixed with a fixed amount of antibody and the sample. The antibody, with any bound antigen, is then separated from the mixture and the bound radioactivity counted. Any unlabelled antigen in the sample would compete with the labelled antigen for the limited number of antibodies, and the measured radioactivity would be lower. By using reference samples with known concentrations of antigen. one can obtain a standard curve.

Several options exist for separation of antibody-bound and unbound radioactive antigen, for example:

- Certain bacteria express proteins that specifically bind the F_c-end of antibodies, for example *Staphylococcus aureus* (**Protein A**) or *Streptococcus ssp.* (**Protein G**). The bacteria use these bound antibodies to mask themselves from the immune system of their host. Preparations of such bacteria (killed in such a way as not to interfere with antibody binding) can absorb the antibody, and can be removed from the rest of the sample by centrifugation. More expensive are the purified proteins bound to agarose gel beads.
- If the antigen has a low molecular mass, separation can be achieved by gel filtration. This is often carried out in **spin columns**, 1 ml syringe bodies filled with about 0.2–0.5 ml of gel. The sample is carefully layered on top and then spun through the gel in a table top centrifuge. The small molecules will enter the gel, while the big antibodies and antibody-antigen complexes can be collected in the flow through.
- Hydrophilic antigens can pass hydrophobic filtration membranes (polyvinylidene fluoride (PVDF) or nitrocellulose), while proteins with their hydrophobic parts get trapped.

9.4.2 ELISA

In ELISA [89] either the antibody or the antigen is bound to a plastic 96-well test plate (see Fig. 9.4). Originally, binding was by hydrophobic interactions with the polystyrene of the plate, but now one can buy plates whose surface has been "activated" with reactive groups which bind proteins covalently.

If the antibody has been bound to the plate, a known amount of labelled antigen is added, quite similar to RIA. However, instead of radioactivity, the label is an enzyme. Usually this is horseradish peroxidase, which is small (40 kDa) and hardy, but alkaline phosphatase, glucose oxidase and β-galactosidase are also used. If the sample contains unlabelled antigen, less of the labelled antigen is bound. After incubation, any unbound antigen is washed away and the remaining antigen is detected by adding the substrate for the marker-enzyme. Detection can occur either photometrically by formation of a coloured product, or by detecting the chemiluminescence emitted during the reaction. This procedure is called a **competitive ELISA**.

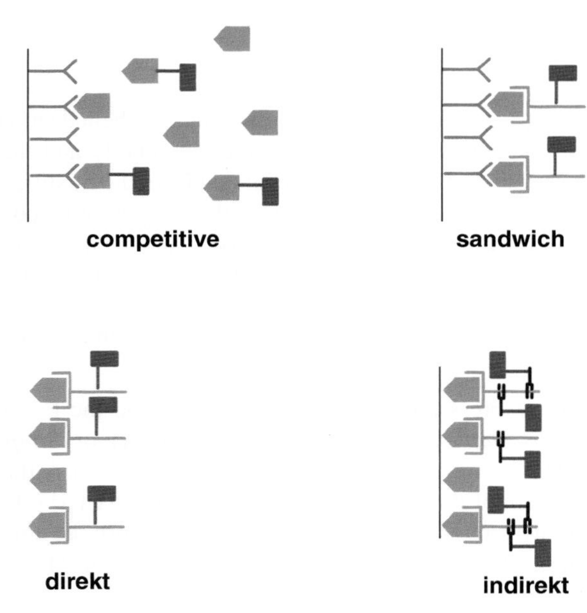

Fig. 9.4 Basic principles of immunological assays. (**a**) In a competitive assay, the antigen to be determined competes with a known amount of labelled antigen for a known amount of antibody. Increasing antigen concentrations lead to a reduction in signal intensity. Competitive ELISA and RIA are examples for this type of assay. (**b**) In a sandwich assay, the antigen is captured by an immobilised antibody and detected by a labelled antibody against a different epitope. (**c**) In a direct assay, the antigen is captured on a solid support and detected by a labelled antibody. ELISA, blots, or immunofluorescence microscopy may be performed like this. This is the standard principle used for FACS. (**d**) More common than the direct is the indirect assay, where the detecting antibody is unlabelled. The label is attached to a second antibody directed against the constant part of the detecting antibody. Note that this assay may also be used to detect the presence of an antibody against the antigen in a serum sample

If the antigen was bound, labelled antibody is added, the remaining steps are the same as described above. The amount of bound antibody, and hence of antigen present, is measured by the strength of the enzymatic reaction. This is called **direct ELISA**.

It is also possible to use an unlabelled antibody bound in the above procedure. After washing away any unbound antibody, a second antibody, labelled and directed against the F_c-part of the primary antibody, is used for detection. For example, if mouse antibodies are used for binding to the antigen, a labelled goat-anti-mouse-IgG antibody can be used for detection. This is called an **indirect ELISA**. There are three major advantages of indirect over direct detection:

- Since several molecules of secondary antibodies can bind to every molecule of the primary antibody, indirect detection is more sensitive than direct. Sensitivity could be increased further by using tertiary antibodies, but as the noise tends to increase too, little is gained by that.

- Indirect detection is more economic, as the complicated labelling procedure needs to be carried out only with the secondary antibody, which is produced in large amounts in big animals (goat, sheep, donkey, horse). A lab may use several different primary antibodies which can be detected by a single secondary one.
- The labelling procedure can, in some cases, lead to a reduction or even loss of antibody activity. One would not like that to happen to a carefully and labouriously isolated primary antibody which is often available only in small amounts. Secondary antibodies, labelled with enzymes or fluorescent dyes, and with exactly specified binding properties, are commercially available, any mishaps during the labelling procedure are the problem of the manufacturer.

Both ELISA and RIA can also be used to detect the presence of antibodies against a certain antigen in serum samples. This is an important method to detect infections with pathogens in clinical diagnostics. If antibodies against the antigen are present, the patient must have been infected with the pathogen (for example HIV).

The biggest disadvantage of ELISA is that during the first binding of antigen or antibody to the plate, only a small percentage of the material available is actually bound. The rest is thrown away, hence the procedure is not as sensitive as it would otherwise be. Additionally, washing of a 96-well plate is a tedious procedure if automatic washing stations are not available (if you are an academic, rather than an industrial, researcher). For these reasons, I have switched from ELISA to dot blots for the detection of protein antigens (see below).

9.4.3 Immuno-PCR

In immuno-PCR [363], either the primary or the secondary antibody is labelled with a specific DNA-molecule (see Fig. 9.5). The assay is performed like an ELISA, but a real-time PCR-step is used to detect the bound DNA. Because of the extreme sensitivity of the PCR-reaction, the detection sensitivity can be increased by a factor of 10^2–10^4 compared to ELISA. In addition, parallel assays to several antigens are possible if the respective primary antibodies are labelled with different DNA sequences. Given the high sensitivity, care must be taken to limit unspecific binding of the labelled antibody, which would decrease the signal/noise-ratio.

9.5 Methods that Do Not Require Separation of Bound and Unbound Antigen

These methods can be performed much faster, and are used in clinical laboratories and for large scale screening for new drugs in industry.

One example is fluorescence polarisation immunoassay: Binding of a small fluorescently labelled antigen to a big antibody molecule increases fluorescence polarisation (see Sect. 6.6 on p. 52 for more detail).

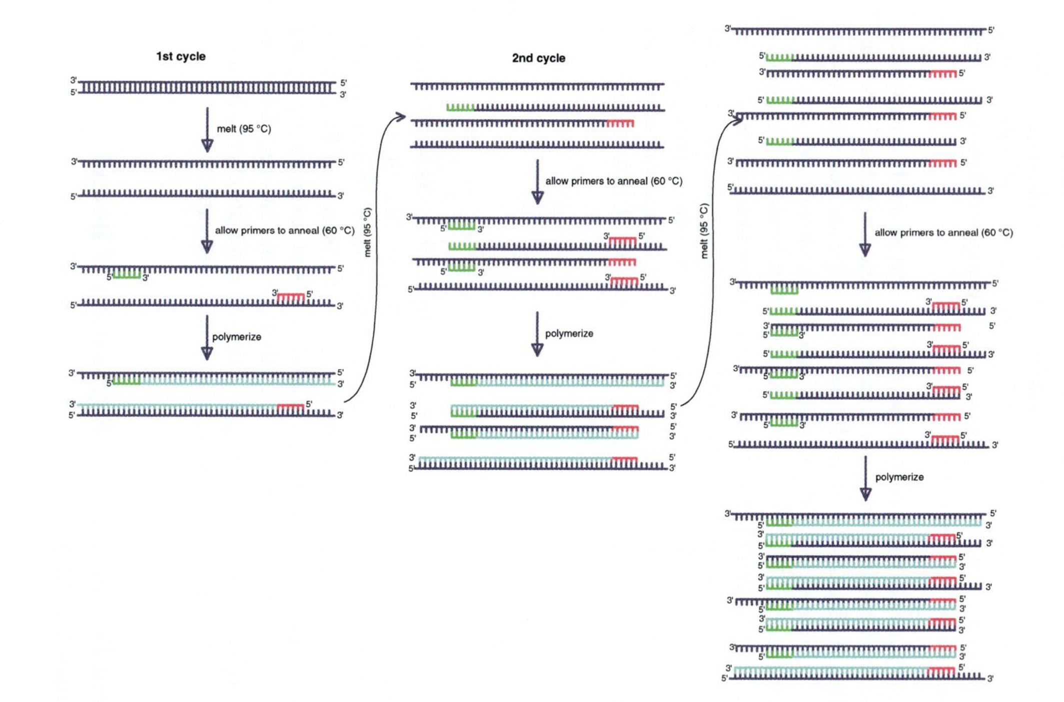

Fig. 9.5 Principle of PCR. The DNA is melted at high temperature, then short DNA fragments ("primers") are allowed to bind at intermediate temperatures. Note that two different primers (purple and green) are required, one each for the coding and non-coding strand. The DNA is then replicated with a temperature-stable DNA-polymerase. Once this has been accomplished, the cycle is repeated several times. Note that from the second cycle on we get DNA-molecules which contain only the stretch between the two primers. Further repetition of the cycle amplifies this product exponentially, so that after, say, 30 cycles a few DNA molecules originally present have been amplified to a detectable quantity. Because the polymerase is heat-resistant (from thermophilic archaeans). the cycles can be repeated fully automatically by changing the temperature without the need to open the reaction tube between cycles. In **immuno-PCR** the original DNA molecule was bound to the detecting antibody. Incorporation of labelled primers into DNA can be followed in real time by hypo- or hyperchromic effects on the label (see p. 37). The number of cycles required to reach a certain signal is a measure of the number of DNA-molecules originally present in the sample

Another example would be radioactive proximity assays, where the antigen is labelled with a weak β-radiator (usually tritium). The antibody is immobilised to plates or beads that contain a scintillator dissolved in the plastic. Only bound antigens are close enough to the plate for the β-rays to be detected by scintillation counting.

In **enzyme multiplied immunoassay technique (EMIT)** the sample is mixed with antibody and enzyme-labelled antigen. Binding of antibody to the labelled antigen reduces enzymatic activity, if the sample contains the antigen it competes with the labelled antigen for the antibody, resulting in higher enzymatic activity. This type of assay is used, for example, in clinical and forensic laboratories to screen for illegal drugs like heroin or cocaine .

The **lateral flow test** or **immuno-chromatographic strip test** requires considerable expertise to manufacture, but is extremely simple and quick to perform. It is therefore used for point-of-care or even home testing (example: home pregnancy tests). The strip looks about like this:

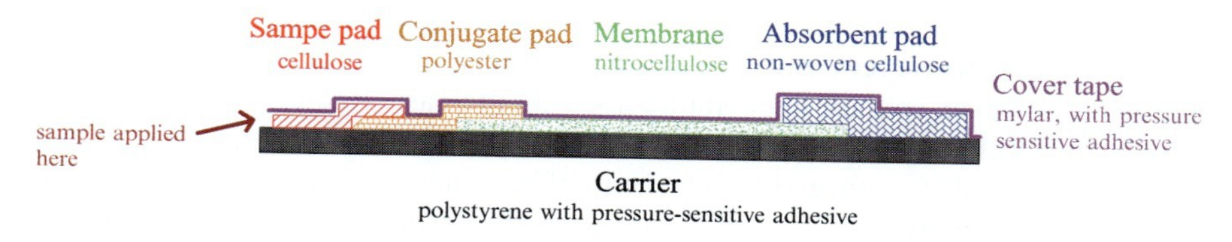

- The sample is applied to the sample pad where it may be modified as needed by dried chemicals (*e.g.*, buffering, removal of blood cells, addition of detergent for better flow). During manufacture of the various pads, the chemicals are added by dipping or by spraying followed by drying. Spraying (similar to ink jet printing) offers better control of the process.
- The preconditioned sample then enters the conjugate pad where it mixes with the detection reagent (*e.g.*, antibody), conjugated on nano-particles of 10–100 nm (dyed latex, colloidal gold). The dried antibodies are protected from damage by polyols (trehalose or similar).
- The mixture of conjugate and sample then moves by capillary action through the membrane, which was designed to carefully control the time that the sample stays in contact with the conjugate. If contact times are too short, sensitivity is lost. If it is too long, background becomes too high and specificity is lost.
- As the mixture moves through the membrane, it encounters the **test line** where an antibody is immobilised, which captures the conjugate/analyte complex. If the analyte is present, some of the coloured conjugate will remain at this line, making it visible to the naked eye. If the test line does not become visible, the test is scored negative (analyte not present).
- Any conjugate without bound analyte (or conjugate/analyte complex in excess of the binding capacity of the test line) gets captured at the **control line**, which simply contains immobilised analyte. This makes the control line visible too, indicating that the test is complete. If the control line does not become visible the test can not be scored (too little sample fluid, not enough time...).

- The absorbent pad binds excess fluid. Its capillary suction moves the reagent/sample mixture through the assembly.
- The cover tape prevents evaporation of liquid, and the carrier holds everything together. The entire assembly may be encased in a plastic housing with windows for sample application and reading.

Development of such tests is a healthy industry.

9.5.1 Microwave and Surface Plasmon Enhanced Techniques

Colloidal gold particles of 20–40 nm in solution change their colour from red to purple when they are brought close together due to interaction of their surface plasmons. This changes the absorbance at 650 nm and can be used to quantify proteins, nucleic acids, and small molecules by bio-recognition. However, the binding kinetics is slow, assays take about 30 min.

If the sample is exposed to low power microwave, the solution temperature increases by several degrees. Because of the skin effect, the radiation can not enter the particles, however, so their temperature does not increase. Rather, the particles as a whole are accelerated. At the same time, the temperature gradient between particle and solution leads to convective flow. Both effects increase the chance for binding of sample molecules to the capturing agents on the surface of the particle, reducing assay time from 30 min to 10 s (*sic*!). Since protein molecules do not absorb much at the usual microwave frequencies of 2.45 GHz, they are not denatured during this process [13].

If silver nano-particles are fixed to a glass surface, they enhance the fluorescence, or chemiluminescence, of molecules bound to them (microwave-accelerated metal-enhanced fluorescence (MA-MEF), and microwave-triggered metal-enhanced chemiluminescence (MT-MEC), respectively). This can be used as fast and sensitive replacement for Northern, Southern and Western-blots, for example for the detection of agents of bio-terrorism where speed is of the essence.

Fluorescence of molecules bound to thin silver or gold layers on a coupling prism is enhanced by surface plasmon resonance, in addition the fluorescent light is sent in a specific direction. For both reasons, measuring fluorescence in an SPR setting (see Sect. 38.8 on p. 374) is more sensitive than in a conventional cuvette. In this case too, binding kinetics can be accelerated with microwaves.

9.6 Blotting

Blotting refers to the detection of biomolecules bound to a membrane:

Nitrocellulose (NC) is the most commonly used material. However, both its chemical and mechanical resistance are limited, as is its protein binding capacity. Proteins bind by hydrophobic interaction. Fibre-reinforced NC-membranes are commercially available, and they have a much lower tendency to fragment during the multiple handling steps of blotting. The membrane is manufactured by dissolving nitrocellulose in organic solvent, mixing the lacquer with water, and evaporating the solvent. Pore size is controlled by the amount of water added and is usually 0.22 or 0.45 μm.

Polyvinylidene fluoride (PVDF, $-CH_2-CF_2-$) has a higher protein binding capacity and strength than NC combined with high mechanical and chemical resistance, making it the ideal material for most applications. Proteins blotted onto PVDF can be used directly even for gas phase sequencing [262]. Note that because of the hydrophobicity of PVDF, it is usually necessary to pre-wet the membrane with methanol, wash with water, and then place it into the desired buffer. There are two exceptions to this rule:

- Total protein on blots may be detected by incubating the air-dried blot with 30 % methanol. This will wet only the protein covered areas and make them translucent. The pattern can then be photographed on a transilluminator [312], no staining is required.
- The dried blot may be used directly for immuno-detection, the reagents come in contact only with the proteins on top of the membrane, but not the membrane itself. Thus, no blocking of unspecific sites is required, and washing steps can be very fast [357].

Nylon (cationised polyhexamethylene adipamine) is positively charged and binds proteins by both hydrophobic and ionic interactions. The ability to undergo ionic interactions unfortunately leads to high background staining in many applications. Nylon is therefore more useful for blotting nucleic acids.

Chemically modified paper was used for gas-phase sequencing of blotted proteins before PVDF became available. Proteins were covalently bound to the paper via azo-groups. This technique is of historic interest only. It is, however, possible to perform blotting onto conventional xerographic paper [442] with some loss of resolution and sensitivity compared to PVDF membranes, but at much lower costs. For student labs this may be an alternative.

Glass fibre filters after chemical activation were sometimes also used for gas-phase sequencing of blotted proteins; they too are obsolete.

There are two ways to get proteins onto the membrane, called Western and dot blots, respectively.

9.6.1 Western Blots

The proteins are first subjected to electrophoresis, most commonly SDS–PAGE. The gel is then placed on top of the blotting membrane and the protein bands are transferred from the gel onto the membrane where they may be detected. Transfer may be by:

Diffusion: Proteins move from the gel to the membrane by random BROWNIAN motion. Only a small fraction of the protein in the gel is transferred, limiting detection sensitivity on the blot. This method is usually used in preparative electrophoresis to locate the band of interest. This may also be done with tissue slices and similar samples, rather than with gels. Apparently, the process can be accelerated by exposure to ultrasound [207].

Capillary force: A pile of dry filter paper is placed on top of the membrane, these suck liquid from the gel through the membrane. This method is more effective with the open pores of agarose gel and hence applied most often in blotting nucleic acids. The biggest advantage is that replica blots can be made easily from the same gel [215, 300].

Vacuum: similar to capillary blotting, but a vacuum pump is used to provide the suction ($\approx$ 200–400 Pa for 30–40 min). This has to be done carefully as not to rupture the gel and membrane.

Semi-dry electrotransfer: The gel and membrane are placed between filter papers soaked in buffer, the pile (see Fig. 9.6) is then placed between flat electrodes made from graphite or carbon glass. A current (0.8–1.0 mA/cm^2) is applied which transfers the proteins from the gel to the membrane. For a 0.5 mm thick gel blotting takes $\approx$ 1 h, for gels of different thickness the time is scaled. The electrodes also serve as heat sinks.

Tank electrotransfer: The sandwich of Fig. 9.6 is placed in a tank filled with transfer buffer with electrodes on both sides of the pile. At least with large proteins transfer tends to be more complete than with semi-dry blotting, but it takes longer and requires larger volumes of buffer.

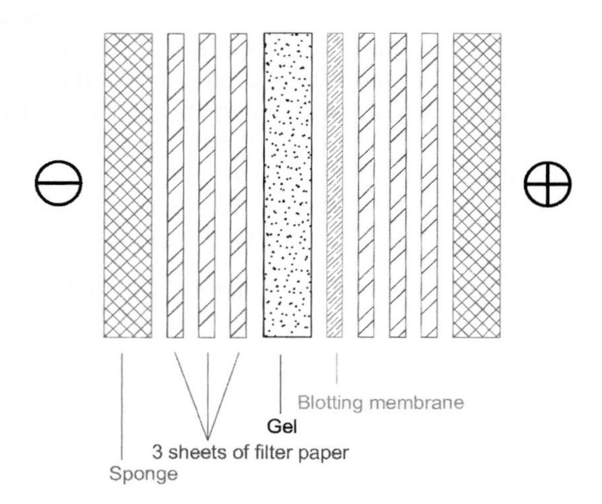

Fig. 9.6 "Sandwich" for Western blotting. The blotting membrane (nitrocellulose or PVDF) and the gel are surrounded by three sheets of filter paper on each side, in tank blotters also by sponges (not required for semi-dry blotters). All of these need to be equilibrated with blotting buffer, and mounted without air bubbles in between. Proteins carry a negative net charge from the bound SDS and move in the electric field from the gel towards the membrane, where they are bound by hydrophobic interactions

Care must be used in electroblotting to orient the sandwich such that the protein moves toward the membrane, not to the other side of the sandwich where it would be lost in the buffer. If SDS–PAGE was used for separation, the proteins have negatively charged detergent bound to them, hence they move toward the positive pole. Paranoid researchers use two membranes for valuable samples, one on each side of the gel, to avoid possible mishaps.

Two buffer systems are most commonly used for transfer, suggested by DUNN [84] and TOWBIN [413], respectively. Of those, the TOWBIN-system (25 mM Tris, 192 mM glycine) is used more often, but that of DUNN (3 mM Na_2CO_3, 10 mM $NaHCO_3$) is cheaper and usually gives better results. For glycoproteins one may use Tris-borate buffer *p*H 9.2, as borate forms negatively charged complexes with the sugar side chains. For sequencing, MATSUDAIRA's buffer (10 mM CHAPS, 10 % methanol, *p*H 11) [262] is used.

It is possible to increase the mobility of large proteins, especially membrane proteins, by adding 50–100 μM SDS to the transfer buffer. On the other hand, for small proteins, interactions with the membrane are increased by reducing the mobility with up to 20 % methanol. These additions have opposite effects, hence there is little point in adding both, as is commonly seen. However, in a semi-dry blotter it is possible to have an SDS-containing buffer on the gel and a methanol-containing buffer on the membrane side [225]. Gels may shrink or swell when transferred into blotting buffer, they should be incubated with the buffer for 10–15 min before blotting to avoid smeared bands.

If a positively charged detergent like CTAB is used for electrophoresis instead of SDS, proteins move toward the negative electrode, hence this procedure is called **Eastern blot** [43, 45].

In both Western and Eastern blots it is possible that detection of an antigen with an antibody fails, because denaturing PAGE and protein binding to the blotting membrane have destroyed the epitope against which the antibody was directed. This risk is obviously greater with monoclonal antibodies, but it can happen occasionally with antisera as well. One way to avoid such problems is to use proteins isolated from a band in an electrophoretic gel for immunisation.

It is possible to quantify protein amount on Western blots by densitometry of the bands. Document scanners like those used in offices are convenient, precise (especially the LED-based ones, as they work with monochromatic light [405]), and cheap but dedicated laboratory instruments offer a wider dynamic range. However, the densitometric signal $\mathcal{E}$ depends on the protein amount m in an exponential manner:

$$\mathcal{E} = a \times m^b \tag{9.1}$$

It is not possible to determine the ratio of amounts in two samples simply from the ratio of the densitometric values, since

$$\mathcal{E}_1/\mathcal{E}_2 = m_1^b/m_2^b \tag{9.2}$$

Hence standard curves are required to accurately measure even relative amounts of protein by Western blot [328]. Such curves can be obtained by using several lanes with different volumes of the reference sample. However, since the exponent *b* may vary from gel to gel, all samples must be run on the same gel. To avoid band warping the sample wells should be filled to the same volume with sample buffer.

Far westerns are made by overlaying a Western blot with a specific ligand that binds to some of the proteins. The bound ligand is then detected. Protein-binding to blotted nucleic acids may be detected in a similar fashion.

9.6.2 Dot Blots

Dot blots are made by spotting samples directly onto the membrane, either with a pipette ("spot blot", use about 2 μl sample, which is best for detecting a protein in the fractions from a chromatographic separation) or with a 96-well vacuum manifold (used if exact quantification is required). For the latter method, good manifolds with gaskets of silicon rubber should be used, those that simply use filter paper are less satisfactory. Suppliers also sell disposable manifolds with membranes sealed to the bottom; this avoids cross-contamination.

A good PVDF membrane will bind more than 95 % of the proteins in the sample (up to about 100–200 $\mu g/cm^2$ (nitrocellulose about 60 % up to 100 $\mu g/cm^2$), hence dot blotting can be 10–100 times more sensitive than ELISA. PVDF-membranes are very hydrophobic, they need to be wetted with methanol, rinsed with distilled water and then equilibrated with buffer. Nitrocellulose membranes can be wetted directly with buffer, they dissolve in methanol.

Native proteins may be used for dot blotting, but sometimes better results may be obtained after denaturation with 15 volumes of 40 % methanol, 3 % trichloroacetic acid (TCA) and 0.2 % SDS [293].

9.6.3 Total Protein Staining of Blots

It is often desirable to use a reversible stain for total proteins on the blot to check that proper transfer has occurred. 0.01 % **Ponceau S** or 0.1 % Fast Green FCF in 1 % acetic acid for 5 min is frequently used, but not very sensitive (5 μg detection limit). Fluorescent dyes like **Sypro Ruby** or **Sypro Rose** interact with the SDS-micelles surrounding the proteins, they achieve detection limits in the order of 1 ng. If PVDF membranes are used, transillumination is a rapid procedure with intermediate sensitivity. The dry membrane is immersed in 20 % methanol which wets the areas covered by protein, but not the rest of the membrane. On a transilluminator the wet bands appear bright on a dark background [347].

If total protein staining is supposed to be permanent, CBB-R250 may be used as with gels. Nice staining is also obtained with 0.1 % India ink in PBS with 0.05% Tween-20 [140]. The soot particles of the ink preferentially interact with the protein bands. This is much cheaper, but about as sensitive as commercial colloidal gold solutions. Note that all permanent protein stains interfere with immune detection.

9.6.4 Immunostaining of Blots

For immune-detection any unspecific protein binding sites must be blocked by incubation with a solution of a non-interfering protein. Low fat milk powder (5 %, spin the solution 20 min at 25 000 g to remove undissolved material) is the cheapest option, and should be used unless there is a reason not to. Specifically, milk powder contains phosphoproteins which react with anti-phosphotyrosine antibodies. Milk powder may also contain biotin and interfere with biotinylated probes. BSA can be used, but is quite expensive. Also, because of its homogeneous nature, it sometimes does not cover all non-specific sites, resulting in high background. Fish skin gelatine [230] looks messy, but is extremely effective as a blocking agent. Polyvinylpyrolidone or polyvinylalcohol have also been suggested. Usually a mild detergent is added (0.05 % Tween 20 or Triton-X100). Due to the hydrophobicity of PVDF these membranes do not need to be blocked, they are dried completely instead. If they are inserted into reagent, only the proteins on the surface interact with the aqueous solution, not the "bare" membrane [357]. In my experience, about 15 min are sufficient for blocking; longer blocking times may lead to a reduction in assay sensitivity by forming a protein layer on top of the antigens. All operations from here on are preformed on a shaker.

After blocking, the blot is incubated with the primary antibody. The antibody is dissolved in blotting buffer, usually PBS (alternatively Tris buffered saline (TBS)) with 0.1 % of the protein used for blocking (to prevent unspecific precipitation of the antibody) and 0.05 % Tween 20 or a similar mild, non-ionic detergent. 0.1 % Thimerosal may be used as bactericidal agent. If phosphatase conjugated antibodies are used, sodium azide works too, but it inactivates peroxidase. This solution is called **blotto**.

Incubation with primary antibody can be for 1 h at 37 °C, or over night in a cold room. Of these possibilities I prefer the latter, because of increased sensitivity and the ability to reuse the (expensive) antibody solution many times.

After incubation with primary antibody, the blot is washed three times 15 min with blotto, and incubated with the secondary, enzyme-conjugated antibody for 1 h at room temperature. The manufacturer will usually recommend a dilution for the antibody, which serves as a starting point for optimising assay conditions. This incubation is followed by a further three washing steps as above. The secondary antibody solution should not be reused.

Detection of enzyme activity is either by forming a coloured precipitate or by detection of the chemiluminescence emitted during the reaction. The following recipes can be used for detection in the various immunological procedures:

- Horseradish peroxidase

  ELISA 100 mM Na-citrate *p*H 5.0, 6 mM H_2O_2, 40 mM *o*-phenylenediamine, incubate 30 min in the dark, then stop with an equal volume of 3 M HCl and read at 492 nm.

  Staining 25 mM Tris-HCl *p*H 7.6, 125 mM NaCl, 2 mM $NiCl_2$, 0.25 mM diaminobenzidine and 0.002 % H_2O_2 give a nice black stain which is easy to photograph and does not fade. Cobalt instead of nickel may also be used [226].

Chemiluminescence Various commercial reagent mixes are available, but homemade reagent works as well [156]: 1 ml luminol (4 mg/ml in DMSO), 1 ml *p*-iodophenol (1 mg/ml in DMSO), 0.6 ml Tris-HCl *p*H 7.5 (1 M), 5 μl 30 % H_2O_2 and 7.4 ml water. Alternatively use 200 μl 250 mM luminol, 89 μl 90 mM *p*-coumaric acid, 2 ml 1 M Tris-HCl *p*H 8.5 and 6.1 μl 30 % hydrogen peroxide for 20 ml reagent.
Commercial reagents are often not pure enough and need to be re-purified for consistent results. The p-iodophenol should be sublimated at 85 °C in vacuo, the crystals should be colourless, not brownish.
Luminol can be recrystallised: dissolve 3 g in a warm mixture of 90 ml *i*-butanol + 40 ml 1 M NaOH, add 30 ml methanol, cool and precipitate with 40 ml HCl. Leave at −20 °C for 30 min, filter through Whatman No. 1 paper and wash with dichloromethane. Dry in vacuo and store protected from light in a desiccator.

- Alkaline phosphatase:

ELISA 1 mg/ml *para*-nitrophenyl phosphate (pNPP) in 1 M diethanolamine* HCl *p*H 9.8, 0.5 mM $MgCl_2$. Incubate 30 min at room temperature, then stop with two volumes of 1 M NaOH and read at 410 nm.
Fluorescence detection Replace the pNPP above with umbelliferyl phosphate.
Staining 100 mM Tris-HCl *p*H 9.5, 100 mM NaCl, 5 mM $MgCl_2$, 0.3 mg/ml nitroblue tetrazolium (NBT) and 0.2 mg/ml 5-bromo-4-chloro-3-indolylphosphate (BCIP, from 50 mg/ml stock in DMF), incubate for 3–4 h in the dark, then wash with 10 mM Tris-HCl *p*H 7.5, 1 mM EDTA. If maximum sensitivity is not required, but cost and time are issues (student practicals), blots may be developed with 100 μM 4-methylumbelliferylphosphate in TBS for 10 min without agitation [426]. Blots can then be photographed on an UV-transilluminator using a fluorescein-filter.
Chemiluminescence is best performed with commercial kits.

Chemiluminescence is more sensitive than coloured precipitates and can be done with X-ray film or with electronic detectors. Dot blots generated with a 96-well vacuum manifold are best counted in a 96-well chemiluminescence detector. This assay is extremely sensitive and has a linear relationship between antigen concentration and signal over 4–6 orders of magnitude (compared to about one order in ELISA). Chemiluminescence can be followed by conventional staining of the blots to obtain a permanent record. Since the substrates used for conventional staining also inactivate the HRP, one can re-probe the blot with new antibody and detect the next antigen first by chemiluminescence and then with a different chromogen. Vector-SG (black), diaminobenzidine (brown), 3-amino-9-ethylcarbazole (brick red) and tetramethylbenzidine (green) have been used to create **rainbow-westerns** [210].

Multi-colour detection of different antigens is also possible with fluorescently labelled antibodies. Modern instrumentation usually scan fluorescence with deep red or near IR diode lasers (680 or 785 nm), where auto-fluorescence of samples is no problem and photo-bleaching is reduced. Either chemiluminescence or IR-fluorescence are detected by CCD-cameras with cooled detectors (down to −40 °C), which allows long signal integration times and improves the signal to noise ratio.

9.7 Immunoprecipitation

To enrich a specific protein, for example in a cell lysate, the protein may be reacted with an antibody. The antibody is then captured with dead Protein A (or G) carrying bacteria as described for RIA. The pellet is resuspended in SDS-sample buffer for analysis. If the cell lysate was metabolically labelled (produced from cells incubated with ^{35}S-methionine/cysteine), bound radioactivity can be counted directly.

In either case, careful washing of the bacterial pellet is required to remove any unbound proteins. Washing solution contains high salt concentration, non-ionic detergents like Triton-X100, and often sodium dodecylsulphate. Increased stringency of washing reduces background, but may also destroy some of the immune complexes formed (see Fig. 9.7).

9.8 Immunomicroscopy

Binding of labelled antibodies can be used to localise proteins inside a cell under the microscope (see Fig. 9.8). Again it is common to use an indirect assay where a secondary antibody carries the label. Extracellular epitopes can be visualised directly, but since antibodies can not penetrate a cell membrane it is necessary to permeabilise the cell membrane by incubation with cold (−20 °C) methanol or methanol/acetone mixtures. ZENKER**'s solution** permeabilises only the plasma, but not the nuclear membrane to immunoglobulin. This can be used to selectively image M-phase cells with antibodies against nuclear antigens [196].

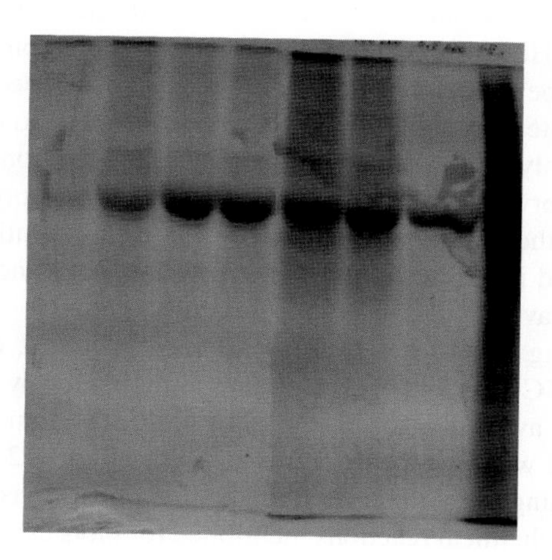

Fig. 9.7 Immunoprecipitation of tubulin from a *Tetrahymena* total cell extract (metabolically labelled with ^{35}S-methionine) with rabbit anti-tubulin antibodies. From the left: 1, 2, 4 and 6 μg of antibody at 0.1 % SDS, 0.05, 0.1 and 0.5 % SDS at 4 μg antibody and total cell extract. Amount of precipitated antigen increases with antibody amount until saturation is achieved. Increasing SDS in the washing buffer increases the specificity of the precipitation, at the expense of yield

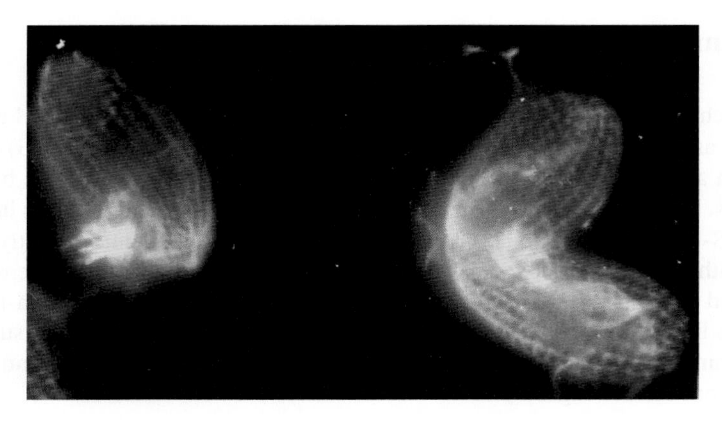

Fig. 9.8 Conjugating cells of the ciliate *Tetrahymena ssp.* by indirect immunofluorescence with rabbit anti-tubulin antibodies and fluorescein-labeled goat anti rabbit-IgG. Cilia, basal bodies, mouth field mebranelles, longitudinal and latitudinal fibrillae, and the mitotic spindle of the micronuclei are visible. Note the haze caused by fluorescent light from outside the focal plane, confocal microscopy would yield much clearer pictures, but had not yet been invented when these pictures were taken

Although immunomicroscopy is sometimes performed on unfixed material, better preservation of structure is achieved by **fixation**, usually with glutardialdehyde and/or methanal (formaldehyde). Both chemicals work by cross-linking cell proteins together into a stable, 3-D network (see Figs. 19.2 on p. 202 and 19.3 on p. 203). Cross-linking of antigens is associated with the risk of blocking the antigenic site. This risk is low to non-existent with antisera, which react with many different epitopes on the antigen. Even if the odd site is blocked, many others are still available. However, with monoclonal antibodies one might be unlucky. Problems seem to be relatively rare however, presumably because not all antigen molecules participate in cross-linking in the same manner, so enough binding sites remain. It is probably worthwhile to try aldehydes as fixation agents first because of their good preservation of structural details. Should problems arise, you can always change to other fixatives, for example ZENKER's solution. An extensive list of fixatives is found in [286]. In any case, material should not stay in fixative for longer than 24 h to avoid excessive cross-linking.

Antibody binding can be detected by fluorescent labels, of which fluorescein isothiocyanate (FITC) is probably the most common. However, a large number of different labels are available, for example for co-localisation studies. Highest resolution is obtained with confocal microscopy (see Fig. 1.12 on p. 16). A second possibility is labelling with enzymes like alkaline phosphatase or horseradish peroxidase. The same chemical reactions mentioned for blots can also be used to detect enzymes on microscopic preparations. Of course much finer details are detectable with fluorescent labelling, but enzyme detection is more sensitive and requires less sophisticated microscopes. It is most commonly used to detect the presence of an antigen in a cell (for example in patho-histology), rather than its subcellular localisation. For electron microscopy, antibodies are labelled with gold particles of defined diameter (few nanometers). It is possible to use different antibodies on the same sample for co-localisation studies if these are labelled with gold particles of different diameter (see Sect. 23.4.1 on p. 226).

Proteins form well-defined networks with other proteins, which depend not only on the cell-type, but also on the status of the cell. These interactions are lost when the cell is homogenised, making immunofluorescence microscopy an important technique to study them. A large number of different antibodies, labelled with different fluorescent probes, is used to stain the cells. Evaluation of the resulting images requires computerised picture analysis tools.

9.9 Fluorescent Cell Sorting

A fluorescence activated cell sorter (FACS) is used to sort living cells by the presence or absence of certain membrane proteins (see Fig. 9.9). The cell mixture is incubated with antibodies which have been labelled with fluorescent dyes. The cells

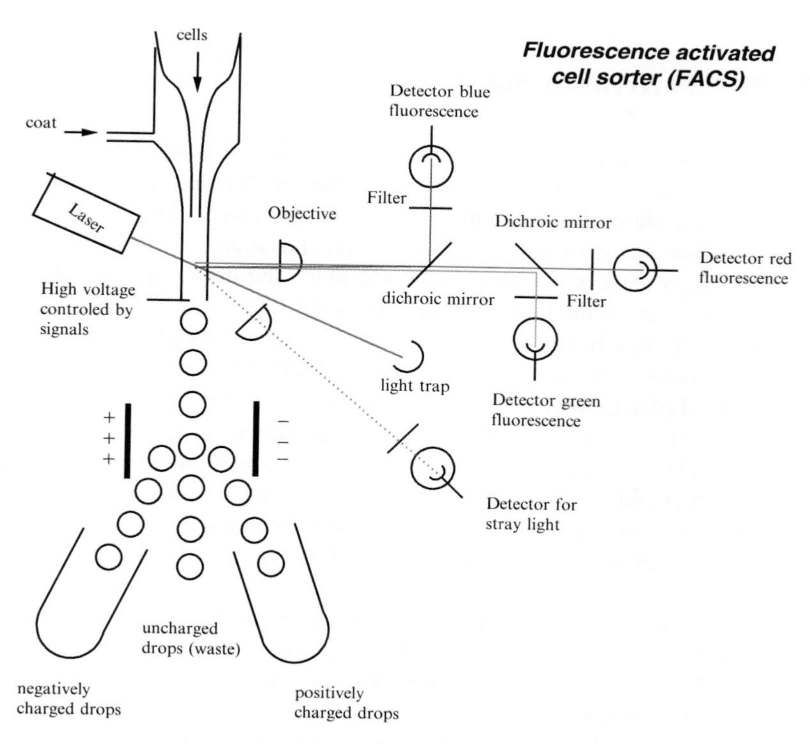

Fig. 9.9 Fluorescence activated cell sorter (FACS). The recent development of relatively cheap diode lasers has made possible the introduction of 10-laser instruments that can measure up to 20 parameters in a single experiment. This has been aided by the development of many new fluorescent dyes with different emission and absorbtion lines

are suspended in a liquid which enters the instrument through a nozzle. A mantle stream of faster flowing buffer ensures that the cells enter the optical plane of the instrument singly and carefully aligned. Fluorescent light is collected by an objective and separated by colour with dichroic mirrors and filters. The intensity of fluorescence associated with each cell is measured separately for each wavelength, stray light (indicating cell size) is also measured. When the cell comes to the outlet of the instrument, a high voltage electrode gives the droplet containing the cell an electric charge, whose polarity and size depend on the presence of several surface markers and cell size. An electromagnetic field then directs the charged droplets into different bins. The entire operation is carried out sterile so that interesting cells can be cultured and further investigated.

One example would be the isolation of pluripotent stem cells from blood samples that may be used one day to create new organs for the patient from whom the blood was collected. Another use for FACS is diagnostics: The concentration of certain cells in blood can be diagnostic for a disease, or can be used to monitor the progress of a disease. The decline of $CD4^{+}T_{h}$-cells in AIDS-patients is an example.

Hint: In discussions with a prospective employer on how your lab is to be equipped, make sure they understand that you are talking about a FACS, not a FAX.

9.10 Protein Array Technology

In the course of the discussion of 2-D electrophoresis, we have already addressed the topic of proteomics, the measurement of the proteins expressed in a certain cell in a specific state of development or disease. 2D-electrophoresis, however, is a tedious and somewhat irreproducible technique, unsuited for routine diagnostics. Array technology usually works as “reversed blot”, the detecting molecules (*e.g.*, antibodies, aptamers or the like) are fixed on a solid support, the labelled analyte is passed over them and bound analyte then detected (**forward phase assay**). However, it is also possible to spot target proteins, for example from lysed cells, onto an array (**reversed phase assay**). These are then detected by labelled antibodies.

Binding of proteins to the surface is most conveniently done by expressing the proteins with a His_6-tag binding to a modified surface containing immobilised Ni^{2+}-ions, or by streptavidin/biotin. Other options include nitrocellulose or acrylamide-coated surfaces. Covalent attachment to surfaces functionalised with aldehyde-, epoxy- or active ester groups is also possible, but may lead to inactivation of the protein.

With currently available technology 10 μl samples are enough to make 200 identical arrays. Each array may contain 500 different samples, (*e.g.*, from different patients), each as quadruplicate. Thus it becomes possible to quantify the immunoreactivity of many samples with a particular antibody in parallel to determine not only the variation of the total concentration of a particular protein, but also certain regulatory modifications. Sensitivity is a few fg. Afterwards the spots may be stained with

Fast Green FCF to measure total protein in each spot. With this method it is possible to detect increased phosphorylation of Akt and Erk in samples from prostate tumors [316]. Label is usually an attached fluorescent dye.

One detection method is **image spectroscopy**, where the fluorescence of bound material is detected by a charge coupled device (CCD)-camera. The CCD-camera is cooled to −20, or even −40 °C to reduce noise. The sample is illuminated with monochromatic light of different wavelengths, fluorescence as a function of x- and y-position is detected by the CCD-chip and converted into a 12-bit (4095 different values) or 16-bit (65,535 different values) digital signal. The main problem is the construction of a light source which can illuminate a relatively large area homogeneously. Laser scanning may also be used.

Radioactivity or surface plasmon resonance and surface enhanced laser desorption ionisation (SELDI)-MS (similar to MALDI-MS, see Sect. 30.2 on p. 274) can also be used. These methods even produce quantitative results.

DNA-arrays [296] have already found a place in diagnostics and research, representing a market of about 7×10^8 US$ in 2001. Such an array contains several thousand gene fragments, it is reacted with fluorescently labelled cDNAs from a tissue, and those genes that are expressed will be highlighted.

By comparison, protein arrays are currently less routinely used, but are expected to catch up in the next few years (see [251] for a recent review). Alternative splicing results on average in three different mRNAs per gene, post-translational protein modification produces about ten different protein species per mRNA. Thus one can estimate that the 27×10^3 human genes correspond to $\approx 1 \times 10^6$ different proteins, all with different physiological roles. Also, because of regulation at the level of translation and protein degradation, and by movement of proteins between cell compartments, the concentration of a mRNA is not a reliable indicator of concentration of its active protein in the cell [128, 135, 245] (see Fig. 9.10). Thus measuring the cDNA-profile of a tissue may not give relevant answers.

A protein chip with antigens from various pathogens could detect binding of antibodies from a patient's serum, and hence contact of the patient with the respective pathogen. A similar chip with human proteins could detect autoimmune-antibodies and help in diagnosing autoimmune-diseases. Detection can be by sandwich assay with labelled detection antibodies or by competition with labelled antigen. Detection times can be as low as 20 min, such chips can therefore be used even in emergency departments. While the detection of serum enzymes (*e.g.*, CK, GOT, AGT) by classical methods requires several successive tests, their detection on a protein chip would be done in parallel, potentially saving time and costs. Protein chips with pollen antigens are already in clinical use to characterise the antibody response of patients with hay-fever. Antibody-based chips may also be used for the detection of small molecules (drugs of abuse in emergency patients, antibiotic residues in foodstuff and the like).

Also in use is the mapping of antibody binding or phosphorylation sites by solid-phase synthesis of peptides on a membrane support. The first spot may contain a peptide corresponding to amino acids 1–10 of a protein, the next spot amino acids 2–11 and so on. Once synthesis is completed, the membrane can be incubated, say,

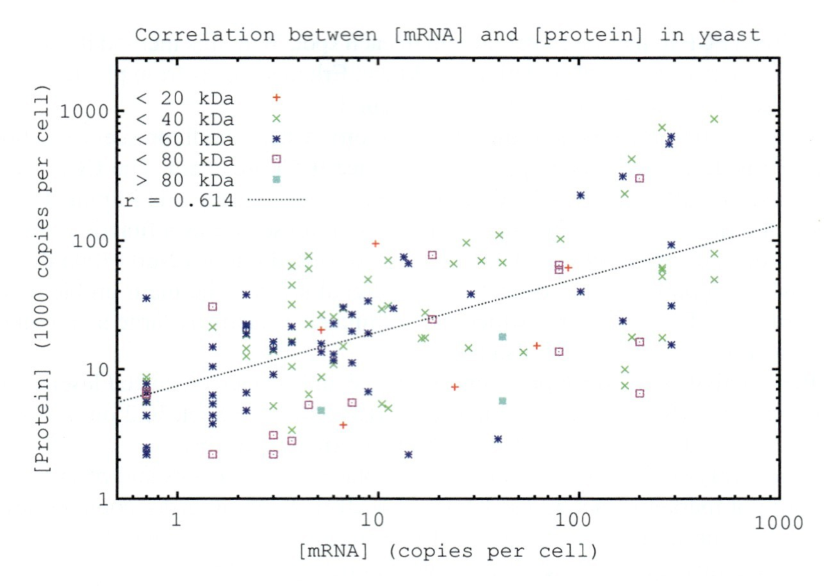

Fig. 9.10 The correlation coefficient between the concentration of proteins and their mRNAs in yeast is 0.614, in other words $r^2 = 37\,\%$ of the variability in the concentration of different proteins is explained by the variation in the concentration of their respective mRNAs. This correlation coefficient is independent of molecular mass as well as *p*I and codon bias (not shown). Data replotted from [136]

with a labelled antibody. Only those spots containing the antibody binding site will be labelled. Because the synthesis of those peptides occurs automated and in parallel, this method is economic.

One problem frequently encountered with protein chips is lack of stability. Proteins denature easily and in particular usually do not survive drying. This seriously limits their application, for example in diagnostics, especially under field conditions. One way around that problem is to spot mRNA encoding for the protein (and a binding tag) and to translate this *in vitro* immediately before use.

Chapter 10
Isotope Techniques

10.1 Radioisotopes

Radioactivity is a phenomenon that can be detected with high sensitivity which can be matched only by chemiluminescence and mass spectrometry. This makes radio-labelling a favourite technique in biochemistry. For this reason the required equipment is almost universally available.

However, there are also some disadvantages to this method which relate to the possible dangers inherent in the use of radioactive material. Although the amounts of radioactivity handled in biochemical experiments is very small (in the order of 10^2–10^3 Bq per sample, 10^7 Bq for a large synthesis), legal complexities are considerable. Since radioactive isotopes can not be destroyed, a considerable part of the costs for each experiment is for disposal. A conscious effort has to be made to keep the radiation dose to the experimenter and to the environment **as low as reasonably achievable** (ALARA-principle), even within the legally permissible values. Each institution has its own safety rules for handling radioisotopes based on local, national, and international regulations. A radiation safety officer will train you on those procedures before you are allowed to work with radioactive materials. The following rules form the basics of safe working practices:

- Lab coats and disposable gloves should be worn whenever handling radioactive materials.
- Place appropriate shielding between yourself and the radioactive material. For low (^{3}H) and medium (^{14}C) energy β-radiation air provides sufficient shielding, for high energy β-radiation (^{32}P) 10 mm Perspex is effective to shield the β-radiation, however, some *Bremsstrahlung* (X-rays) is produced. γ- and X-radiation can be shielded with lead bricks, low energy radiation also with "yellow perspex" which contains chemically bound lead.
- No mouth pipetting, use mechanical pipetting aids.
- No food or drink are allowed in the lab, nor should cosmetics be applied. No smoking either.
- To contain spillages, all work is performed in a tray which should be lined with absorbent tissue.

E. Buxbaum, *Biophysical Chemistry of Proteins: An Introduction to Laboratory Methods*, DOI 10.1007/978-1-4419-7251-4_10,

- Volatile radioactive material (iodine, sulphur) is handled in fume cupboards which are monitored regularly for flow rate.
- Collect all waste that is radioactive. Separate liquid waste, incinerable solid waste, and non-incinerable solid waste. Dispose according to regulations.
- Before and after working with isotopes, check the workplace and yourself for contamination. A GEIGER-counter may be used for γ-radiation and high energy β, but low energy β-radiation can only be detected by swipe tests. Do not forget the soles of your shoes!
- Pregnant females should not work with radioisotopes, nor in labs where such isotopes are handled, to protect the embryo which is particularly sensitive to radiation.
- Any accidents or contamination must be reported immediately. The trouble you get after reporting it is nothing compared to what would happen if you do not!

10.1.1 The Nature of Radioactivity

The nuclei of atoms consist of positively charged protons and neutrons which, as the name would suggest, are electrically neutral. Neutrons are required to prevent the breakdown of nuclei due to the electrostatic repulsion of protons.

Nuclei with the same number of protons (called atomic number Z) are said to belong to the same chemical element. If they have different numbers of neutrons N, they are said to be **isotopes** (from Gr. equal place, namely in the periodic system). The atomic mass is determined by the sum of protons and neutrons because the mass of electrons is so small that it can be ignored. Nuclei with the same mass, but different atomic number, are called **isobars** (from Gr. equal weight). Nuclei with the same number of neutrons are called **isotones** (derived from isotope by replacing the p(roton) with n(eutron)). You can order nuclei in such a way that isotopes are in horizontal rows and isotones in vertical rows, the isobars are on diagonals. This is called a **Karlsruhe Nuclide Chart**[1] [327], also available as a program for Windows www.iaea.or.at/programmes/ripc/physics/faznic/nuchart.zip or directly on the Internet (www-nds.iaea.org/relnsd/vchart/index.html). In the nuclide chart isotopes are given different colours depending on whether they are stable or which mode of breakdown they have (see Fig. 10.1).

Nuclei are stable only at a certain ratio of protons to neutrons. If a nucleus has too many neutrons, the excess neutrons will disintegrate into a proton (increasing atomic number by 1) and an electron, which is sent out at high velocity as **β^--radiation**: $\mathsf{n} \rightarrow \mathsf{p}^+ + \beta^-$. Very few nuclides send out two electrons while converting two neutrons into two protons (double β^--decay). In some neutron rich isotopes the neutron is sent out by **spontaneous emission**, reducing the atomic mass by one, the atomic number stays constant.

[1] Karlsruhe is a city in Germany with a large nuclear research facility.

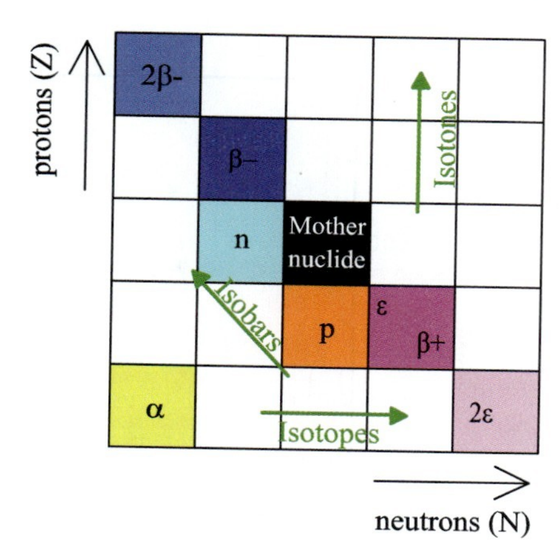

Fig. 10.1 Nuclides produced by various modes of radioactive decay. For details see text. Isotopes are in horizontal, isotones in vertical rows. Isobars are on diagonals

If a nucleus has too many protons, the excess protons can be converted into neutrons, the positive electrical charges are sent out as positrons (β^+**-radiation**): $\mathsf{p}^+ \rightarrow \mathsf{n} + \beta^+$. This process decreases the atomic number, the atomic mass stays constant. Spontaneous emission of protons occurs in a few isotopes under reduction of both atomic number and mass. The other alternative is to send out two neutrons and two protons together, that is a helium nucleus (α**-radiation**). This process decreases atomic number by two and mass by 4. If larger nuclei than He are produced, we call this **cluster decay**.

Some nuclei also convert a proton into a neutron by **electron capture,** ϵ**-decay**, which is the reversal of β-decay: $\mathsf{p}^+ + \beta^- \rightarrow \mathsf{n}$. Double electron capture, where two protons are converted into two neutrons at the same time, has only recently been observed in some long-lived nuclides.

Very heavy nuclei may also break apart by **spontaneous fission**, creating two daughter nuclei of approximately half the atomic mass of the parent nucleus and some excess neutrons. Technically, isotopes undergoing spontaneous fission are used as neutron sources (*e.g.*, ^{252}Cf).

In any of these processes excess energy in the nucleus is emitted in the form of high-energy photons (γ**-radiation, isomeric transition**) . Note that γ- and **X-rays** (also known as RÖNTGEN-rays) are both high energy photons, but have different origin: γ-rays come from processes in the nucleus, X-rays from processes in the electron cloud of an atom. Excess energy may also be used to propel an electron from its orbit (**internal conversion**). In contrast to β^--decay, the resulting electrons are mono-energetic.

Radioactive breakdown of unstable isotopes is a process of first order: the speed of breakdown depends on the amount of isotopes present:

$$v = \frac{\mathrm{d}n}{\mathrm{d}t} = -kn \tag{10.1}$$

$$n_t = n_0 \mathrm{e}^{-kt} \tag{10.2}$$

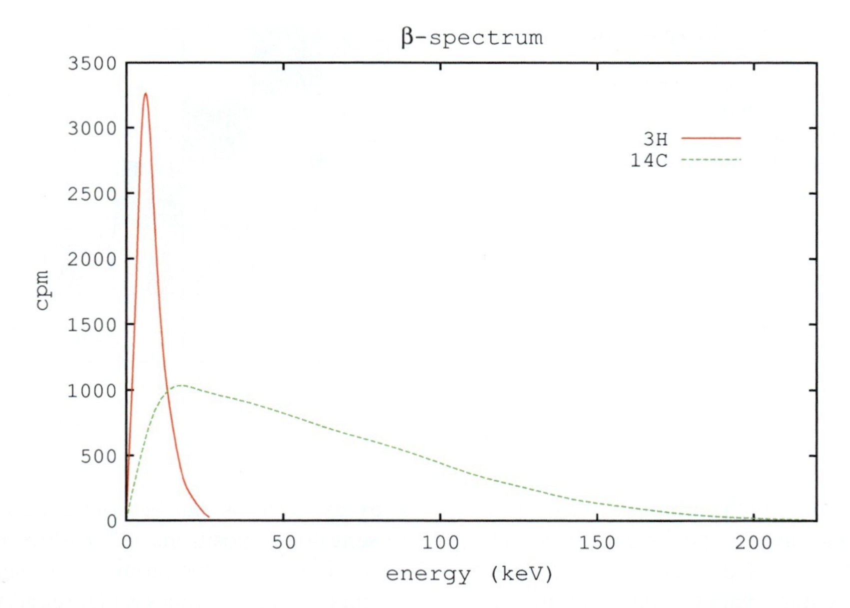

Fig. 10.2 β-spectrum of ^{3}H and ^{14}C. Note the higher energy associated with radiation from ^{14}C and the broad energy distribution of the radiation from both isotopes

k is a constant for a given isotope. More common is to specify the half life period $t_{1/2} = \ln(2)/k$.

In biochemistry you will mostly work with β- and γ-radiators, the instrumentation required to measure α-radiation, positrons and neutrons is usually available only in specialised centres. An example application would be **positron emission tomography (PET)** to measure metabolic activity in various tissues, for example to locate a tumor.

Each nuclear process, just like a chemical reaction, is accompanied by the release of a specific amount of energy. In α- and γ-radiation this energy is entirely in these rays, therefore a spectrum has distinct lines characteristic for each isotope. In β-radiation on the other hand, only part of the energy is released with the electron, the rest as γ-radiation. If the spectrum of β-decay is measured, a broad distribution is found, up to a characteristic maximal value (see Fig. 10.2).

10.1.2 Measuring β-Radiation

10.1.2.1 The β-Counter

The most sensitive way to detect and quantitate β-radiation is to measure the luminescence that is created when β-radiation interacts with certain chemicals

Table 10.1 Properties of some β-emitting isotopes important to biological research. The photon yield is the average number of photons produced by a β-ray as it passes through the scintillator, and in turn is proportional to the signal produced by the photomultiplier. The penetration depth is an approximate value for photographic emulsions

Isotope	^{3}H	^{14}C	^{32}P
Maximum energy (keV)	18.6	156	1 710
Photon yield	30	250	3 300
Penetration depth (μm)	2	100	3 200

(fluors). Toluene, xylene, and other aromatic hydrocarbons give off ultraviolet light when hit by β-rays. This far-UV light is used to generate near-UV fluorescent light in 2,5-Diphenyloxazole (PPO), which in turn induces blue fluorescence in 1,4-Bis-2-(5-phenyl-2-oxazolyl)-benzene (POPOP). This blue light is detected with photomultipliers or solid-state detectors. Since each photomultiplier produces some noise signal, β-counters use 2 photomultipliers both facing the sample vial. The signals from these tubes are sent through a coincidence detector, a count is registered only if both tubes produce a signal at the same time. This lowers noise to about 30–50 events per minute.

The number of photons produced and hence the signal from the photomultiplier is proportional to the energy of the β-ray (see Fig. 10.2 and Table 10.1). The signal is fed into an analyser, and from there into a computer. Isotopes with very different β-spectra, like ^{3}H and ^{14}C, can be measured in parallel in the same sample. This is, however, not quite as easy as with γ-radiation, as β-radiation does not occur in discrete lines. Thus some of the events from the high-energy isotope have the same energy as those coming from the low energy one. The computer programs that come with modern counters can correct this, provided that the ratio of high to low energy activity is within a certain range. This range needs to be determined for each isotope pair and each counter type (as different manufacturers use different programs).

The sample is mixed directly with the scintillator as low-energy β-radiation is screened easily, even by a sheet of paper. If the sample contains water, detergents like Triton-X100 must be added to create a stable emulsion of the sample in the scintillator. Modern scintillator cocktails offered by supply companies are biodegradable, a considerable advantage compared to the messy chemicals used in the home-made cocktails of yesteryear. Solid samples are digested in suitable chemicals, like quaternary ammonium bases, before they are neutralised and mixed with the scintillator. Samples should be of neutral *p*H, as both high and low *p*H can induce chemiluminescence in scintillation cocktails, which is difficult to distinguish from the signal caused by β-radiation. Chemiluminescence is also reduced by low temperatures (about 10 °C). Phosphorescence is another possible problem, samples should be stored under reduced light for about 1 h before counting.

Counting efficiency depends on the energy of the β-radiation, medium and high energy radiation from isotopes like ^{14}C and ^{32}P can be counted with efficiencies of 90 % and better, while for low energy radiation, for example produced by ^{3}H, counting efficiency may be less than 30 %.

Radioactivity is a stochastic phenomenon, the number of events per unit time is given by a POISSON-distribution. This means that for a counting error of less than 2 %, the counter has to detect 10 000 events. Counting time required depends on the activity in the sample, becoming shorter as activity increases. On the other hand, for reasons of cost and possible exposure risks, one would like to keep the activity as low as possible. Thus activities between several hundred and several 10 000 counts per minute (cpm) are a useful compromise (counting time 1–30 min per sample). The counter should be programmed to reject samples with less than 100 cpm, as these cannot be measured in a reasonable time.

The luminescence can be **quenched** by certain chemicals, which may be present in the sample. Quenching leads to both a reduction of counting efficiency and a shift of the detected spectrum to lower energy (see Fig. 10.3). This shift (**sample channel ratio** (**SCR**) can be used to correct for the reduced counting efficiency.

A second method to detect quenching is to use an **external standard**. This is a small amount of americium or a similar γ-radiator, which is placed next to the sample. The γ-radiation produces COMPTON-electrons in the sample, which can be detected by the counter. So in the absence of quenching, the count rate induced

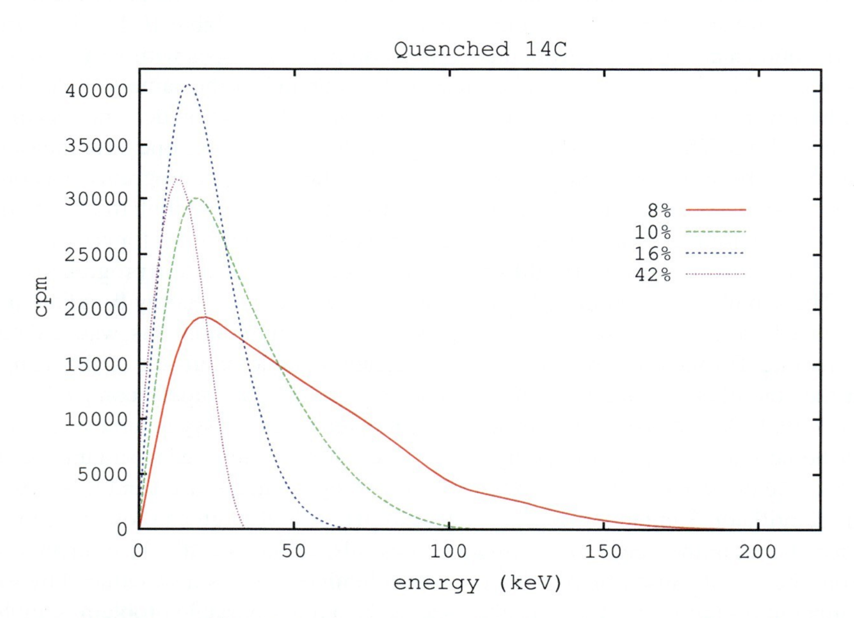

Fig. 10.3 Quenching of ^{14}C radiation by addition of chloroform to the sample. The peak becomes narrower, the maximum is shifted to lower energies, and the integral under the curve (total counts) becomes smaller

by that standard should be identical for all samples, reduction indicates quenching. β-counters will list this reduction as **quench-identification parameter** (**QIP**).

Again, the programs that come with modern instruments can do quench correction for you automatically, but a healthy scepticism towards this form of data manipulation is in order. In particular, at high quenching the multiplication with high correction factors can seriously affect the statistical error of the results. Make sure that your samples quench as little as possible, and that all samples of an experiment quench to about the same degree. In particular, samples should be clear and colourless (yellow is most harmful as it is complementary to the blue light counted). Also, make sure that your standards have a similar composition to your samples. Calibration of the quench correction is possible by making a series of samples with the same amount of radioactivity, but different concentrations of chloroform, which is a very effective quencher.

10.1.2.2 Cherenkov-Counting

High energy β-radiation causes oxygen molecules in air to dissociate, the resulting radicals recombine under emission of blue light (CHERENKOV-radiation). Isotopes like ^{32}P can be counted in a β-counter without the use of scintillation liquid. Counting efficiency is about 25 %, the spectrum is similar to ^{3}H. This option is attractive from both the economical and ecological point of view and also because the sample is not destroyed but may be recovered for other measurements.

10.1.2.3 Autoradiography and Autofluorography

Radioactivity was detected by HENRI BECQUEREL by its ability to blacken a photographic plate in the absence of light. This property can be used to locate radioactivity in a sample. Typical applications include the location of bands in an electrophoresis gel or the location of binding sites on a histological section. For gels and the like X-ray film is most often used. The gel is stained and dried as usual, then placed against the film in a light-tight box and incubated at −20 °C (to reduce background) for several hours or days, depending on activity. Then the film is developed. There are pens with radioactive or phosphorescent ink available to place marks on the gel which afterwards allow alignment of gel and film.

The blackening of photographic film is proportional to the radiation intensity, but only if the intensity is high enough. At low intensities the sensitivity is reduced (SCHWARZSCHILD-effect[2]). This can be counteracted by flashing the film with a small amount of light. Pre-flashed films are commercially available.

With low-energy β-radiation gels can screen much of the radiation from the film. This reduced sensitivity can be overcome by incubating the gel with a fluor like PPO

[2] Named after the photographer KARL SCHWARZSCHILD, 1873–1916.

or sodium salicylate before drying. The film then registers the light coming from the fluor rather than the β-radiation directly (**autofluorography**).

High energy β-radiation may penetrate the film without blackening it. In this case an **intensifying screen** may be placed behind the film where the radiation induces light flashes in heavy-metal based fluors. This light is registered on the film.

For histological sections, the photographic emulsion is poured onto the slide and dried. Thus the silver-grains develop directly on the slide and can be viewed together with the tissue.

10.1.2.4 Phosphor-Imaging

The use of X-ray film for autoradiography involves a tedious procedure, and linearity is limited to about 1 order of magnitude. This makes quantitative assays difficult.

Certain ceramic crystals react to radiation by displacement of their components in the crystal lattice. When the atoms return to their proper position, energy is released in the form of light. However, because the mobility of atoms in a solid is limited, the energy-rich state is meta-stable, activation energy in the form of heat or intense light is required to induce recombination. These physical principles have been used for a long time for the dating of ceramics from archaeological sites by thermoluminescence. In the last couple of years instruments have come on the market where a plate covered with very small ceramic crystals is exposed to radioactivity on blots, gels, histological sections and the like. Afterwards, the plate is scanned by a laser beam to induce recombination. The resulting phosphorescence is measured quantitatively. The procedure is more sensitive than autoradiography on X-ray film, and the response of signal intensity as a function of radiation dose is linear over 4–5 orders of magnitude. The required instruments are very expensive, but because the ceramic plates can be reused, no consumables are needed which makes phosphor-imaging cost-effective if large numbers of autoradiographies are needed.

10.1.2.5 Geiger-Müller Counters

For detection of contamination, the classic GEIGER-MÜLLER-counter is used, but this works only for radiation with high enough energy to penetrate the window of the counting tube. For low-energy β-radiation a methane flow-through counter may be used, this is effectively a GEIGER-MÜLLER-tube without window. However, wipe tests are more common, the surface to be checked is wiped with a filter paper soaked in ethanol (or another solvent suitable for your sample), the paper is then placed into a counting vial with scintillation fluid and counted.

10.1.3 *Measuring γ-Radiation*

10.1.3.1 The Na(Tl)I Scintillation Counter

γ-Radiation interacts less with matter than β, thus it is not necessary to mix the sample with scintillator. Instead, the sample is brought close to a NaI-crystal doped with Tl which emits light-flashes when hit by γ-radiation. Those are detected by a photomultiplier. The resulting signal is proportional to the energy of the γ-quant. Since γ-spectra consist of discrete lines, determination of different isotopes in parallel is easier than with β-radiation.

10.1.3.2 Other γ-Detectors

Ge-based solid state γ-detectors are usually found only in specialised centres, they work only at cryogenic temperatures.

GEIGER-MÜLLER counters work very well with γ-radiation, but exact quantitation is not possible with them.

It is possible to use a **β-scintillation counter** to count γ-radiation. For this purpose the scintillation cocktail is mixed with a solution of 10 % $ZnCl_2$ in water. γ-radiation interacts with the Zn^{2+}-ions, producing COMPTON-electrons, which then cause light flashes in the fluor. For low energy γ-radiation like ^{125}I counting efficiency is about 80 %, for high energy isotopes like ^{22}Na efficiency is lower, about 20 %. The sample need not be mixed with the scintillation fluid but can be placed inside an Eppendorf vial or the like, hence the method is very economical.

10.1.3.3 Photon Activation

High energy X-rays can kick protons or neutrons out of a nucleus, thus creating radioactive elements. The intense X-rays required are produced in a linear accelerator by directing fast electrons onto a water-cooled gold or platinum target. Within a few hours after irradiation the daughter-nuclides of light elements (first two rows of the periodic system) have decayed, then the samples can be placed into a γ-spectrometer for analysis. This allows the sensitive determination of elemental composition even in complex samples, *e.g.*, for environmental monitoring.

10.2 Stable Isotopes

Many biologically relevant elements have not only one, but several stable (non-radioactive) isotopes, for example hydrogen (atomic mass 1 and 2), carbon (12 and 13), nitrogen (14 and 15), and oxygen (16 and 18). Compounds enriched in a rare stable isotope can be used to follow metabolic pathways, the products are detected in a mass spectrometer (see Chap. 39 on p. 383). This has uses in routine clinical diagnosis. For example, if patients are given deuterium oxide (heavy water), it will distribute throughout the body. If then a urine or saliva sample is analysed, the amount of water in the body can be calculated from the volume of deuterium oxide given, the deuterium concentration in that volume and the deuterium concentration in the sample obtained. Since the lean body mass consists of 73.3% water, lean body mass can be determined and the fat content of the body can be calculated as difference between total and lean body mass.

The specificity of enzymes is so high that they can distinguish between different isotopes of the same element. For example, the carbon fixing enzymes in C3- (ribulose bisphosphate carboxylase (Rubisco)) and C4-plants (phosphoenolpyruvate carboxylase) both prefer $^{12}CO_2$ over $^{13}CO_2$, so that the carbon in all living organisms contains less ^{13}C than the carbon dioxide in air. This depletion of heavy carbon is quantified by the $\Delta^{13}C$-value which is several ‰. Interestingly, it is somewhat higher in C3- than in C4-plants. Thus, although sugar produced by sugar beet (a C3-plant) and by sugar cane (a C4-plant) are chemically identical, they can be distinguished in a mass spectrometer (see Sect. 30.2 on p. 274 for details). The use of stable isotopes to determine the structure of proteins in NMR-experiments is described on p. 316, their use for determination of reaction mechanisms on p. 383. In geochemistry and ecology, stable isotopes are even more important than in the biomedical sciences. However, that is beyond the scope of this book.

Part II
Purification of Proteins

Each cell contains several thousand different proteins. If we want to find out their properties, we have to purify them first. How else could we be sure that a particular reaction is really caused by a particular protein?

The basic principles behind protein purification are easily understood, however, it is difficult if not impossible to predict which methods are most suitable for the purification of a given protein. It is rarely possible to purify a protein by a single method; usually the judicious combination of several purification steps is required. Protein purification is much more art than science. Several volumes of *Methods in Enzymology* cover protein purification techniques. For a recent review on recombinant protein expression in *E. coli* see [379]. Protein purification is not only important in the laboratory, but also in industry. The source of your protein should be, whenever possible, an organism in biological safety class I (generally recognised as safe).

Chapter 11
Homogenisation and Fractionisation of Cells and Tissues

If you want to study, say, the enzymes present in the liver, the first thing you have to do is obtain fresh liver tissue. This is done at the abattoir immediately after an animal has been slaughtered. The tissue is transported into the laboratory on ice to minimise proteolytic damage.

Once in the lab, the tissue needs to be disrupted. This is a critical step: Cells should be broken open, but cell organelles should remain intact. Usually the tissue is minced first by hand, then cut into a fine pulp by rotating knifes in a kitchen blender, and finally homogenised by the application of ultrasound or shearing forces. During homogenisation **foaming** should be avoided, as foam inactivates proteins. Best results are often obtained with POTTER–ELVEHJEM-homogenisers (see Fig. 11.1).

Somewhat stronger than the POTTER–ELVEHJEM-homogeniser is the **French pressure cell**: The sample is pressed at a high pressure (15–30 MPa) through the narrow bore of a needle valve. The flow rate in the middle of the hole is higher than that at the rim, so the resulting shearing force will break even cells with a cell wall (bacteria and yeasts).

Freeze–thaw cycles are sometimes used to homogenise small amounts of cultured cells, but may denature proteins. Plant tissue and yeasts may be ground with sand and a little buffer using a mortar and pestle. Animal tissue may be frozen in liquid nitrogen and then ground.

All these steps are performed on ice. Buffer solutions are used to keep the *p*H at the required value; they usually also contain antioxidants (β-mercaptoethanol or DTT) and sucrose or mannitol to keep the osmotic pressure in the solutions at the same level as in the cell ($\approx$ 300–350 mosm). Some very sensitive enzymes also require the addition of their substrate to stabilise them. Buffer composition should approximate the natural surrounding of proteins.

For extracellular proteins of mammals, phosphate buffered saline (DULBECCOS PBS, 1.5 mM KH_2PO_4, 2.7 mM KCl, 137 mM $NaCl$, 8.1 mM Na_2HPO_4) is a good starting point, with the addition of 0.9 mM $CaCl_2$ and/or 0.5 mM $MgCl_2$ as needed. Similar isotonic solutions for species from different taxa are found for example in [286].

For intracellular proteins try 80 mM KCl, 17 mM $NaCl$, 140 mM sucrose, and about 20 mM HEPES or Tris to adjust the *p*H to the desired value. 1 mM EGTA can

E. Buxbaum, *Biophysical Chemistry of Proteins: An Introduction to Laboratory Methods*, DOI 10.1007/978-1-4419-7251-4_11,

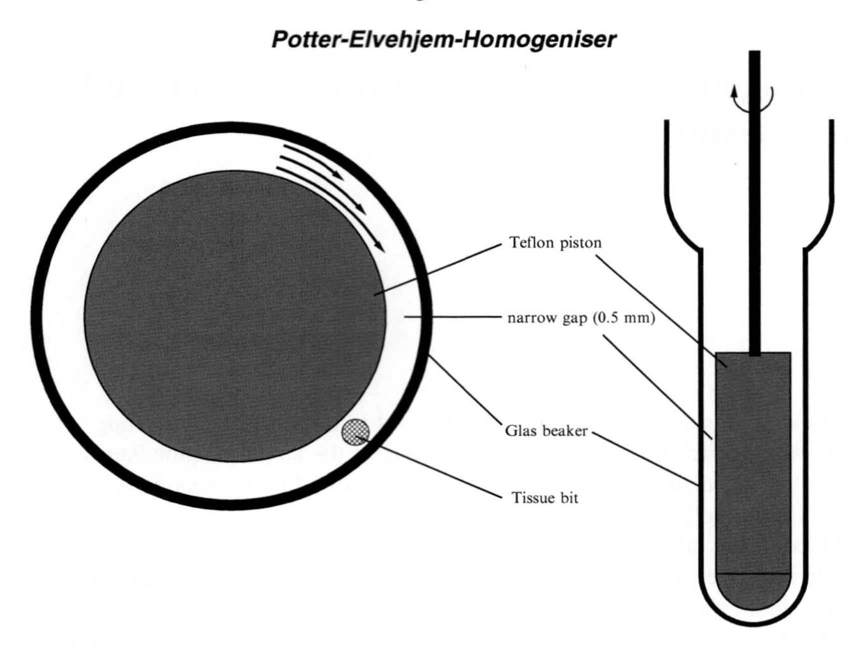

Fig. 11.1 POTTER–ELVEHJEM-homogenisers [336] consist of a stationary glass beaker and a rotating pestle (for safety reasons they are now made from teflon rather than the original glass) with a narrow gap between them in which a gradient in flow velocity develops. Any tissue bit that gets into that gap will be torn apart by that gradient. This method is quite gentle and will vesiculate cellular membrane systems. Do not use the pestle to grind tissue at the bottom of the beaker, rather, move the pestle up and down continuously. A DOUNCE-homogeniser [80] works similarly, but the pestle is a ball rather than a cylinder so it can be hand-operated

buffer free Ca^{2+} to the low values required by some enzymes . Some enzymes need 1 mM $MgCl_2$ for stability.

1 mM NaN_3, or Thimerosal (an organic mercury compound also available under different trade names like Mersalyl), can prevent the growth of bacteria in buffers if long-time incubations, in particular at higher temperatures, are required.

11.1 Protease Inhibitors

Protease inhibitors (see Fig. 11.2 and Table 11.1 for examples) prevent the degradation of proteins; this is essential if reproducible results are to be obtained. Some proteins can change their behaviour considerably if only a small peptide is cleaved off! If you have an antibody against your protein, it is useful to check for proteolytic damage in a Western-blot; you should see only the protein and no degradation products.

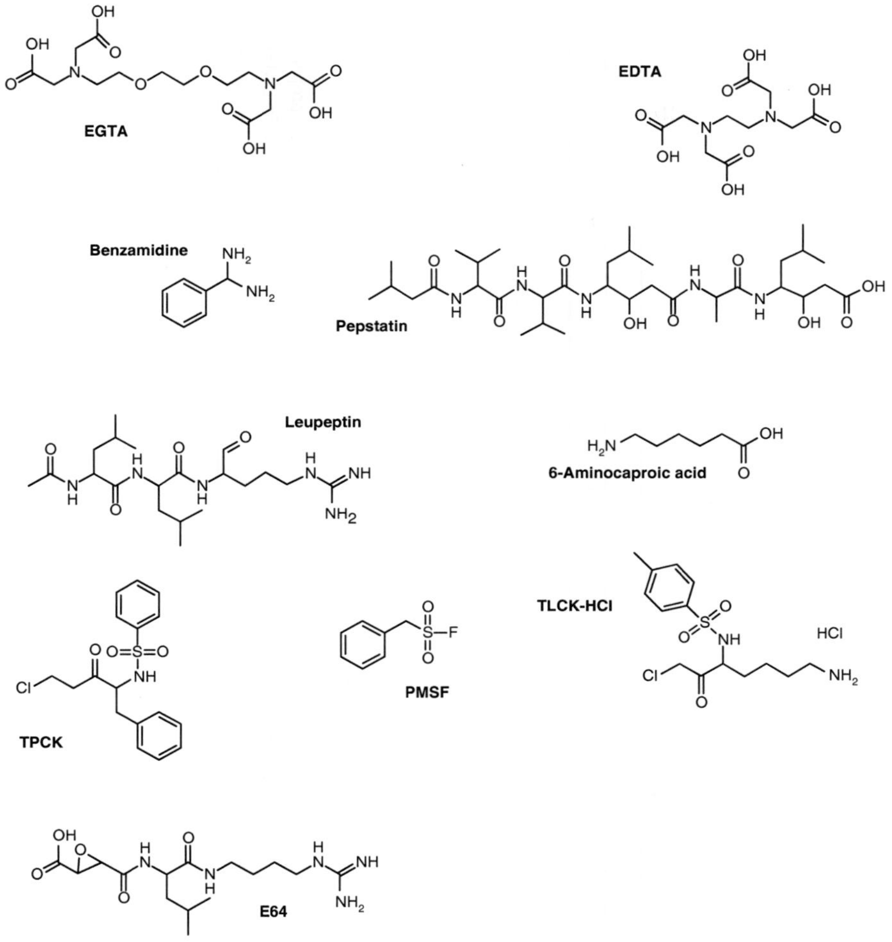

Fig. 11.2 Chemical structure of some important protease inhibitors and inactivators

Metalo-proteases are inhibited by the presence of 0.1–1 mM **EDTA** or **EGTA**, which strip the bivalent metals from their active centre. EGTA has a higher affinity for Ca^{2+} than Mg^{2+}, and can be used when Mg-ions are required for the stability of the enzyme to be isolated. EDTA has higher affinity for Mg^{2+} than for Ca^{2+} and is used in the opposite situation.

In my experience, 5 mM 6-aminocaproic acid, 100 μM EGTA, 100 μM benzamidine, 1 μM leupeptin, 1 μM pepstatin, and 100 μM PMSF are a good starting point. **Safety note**: In many references stock solutions of PMSF in dimethyl sulphoxide (DMSO) are used. This is dangerous as DMSO can make the skin permeable for the highly toxic (acetylcholine esterase inactivator!) PMSF. Use i-propanol instead.

Table 11.1 The most important inhibitors and inactivators of proteases

Substance	Active against	Formula	M_r (Da)	Conc.	Stock
Inhibitors					
6-Aminocaproic acid	Chymotrypsin, fibrinolysin	$C_6H_{13}NO_2$	131.2	5 mM	–
Benzamidine-HCl	Thrombin, trypsin	$C_7H_8N_2$	174.6	100 μM	–
EDTA-Na_2	Metalloproteases	$C_{10}H_{16}N_2O_8Na_2$	372.3	100 μM	1 M in water *p*H 8
EGTA	Metalloproteases	$C_{14}H_{24}N_2O_{10}$	380.4	100 μM	–
Leupeptin	Ser/cys-proteases	$C_{20}H_{39}N_6O_4$	427.6	1 μM	1 mM in methanol
Pepstatin	Asp-proteases	$C_{34}H_{63}N_5O_9$	685.9	1 μM	1 mM in methanol
Inactivators					
E64	Cys-proteases	$C_{15}H_{27}N_5O_5$	357.4	1–10 μM	i-propanol
PMSF	Ser/cys-proteases	$C_7H_7O_2SF$	174.2	0.1–1 mM	1 M in i-propanol
TLCK-HCl	Trypsin-like proteases	$C_{14}H_{21}N_2O_3SCl * HCl$	369.3	0.1–1 mM	50 mM in DMSO
TPCK	Chymotrypsin-like proteases	$C_{17}H_{18}NO_3SCl$	351.9	0.1–1 mM	50 mM in ethanol

Add PMSF-stocks to the buffer only immediately before use, as PMSF degrades in water within 20 min.

Beyond the compounds discussed here, there is a large number of protease inhibitors available which interact with particular proteases. Some of those have pharmaceutical use, for example inhibitors of HIV-protease. Others are used in research to find out which protease is responsible for a particular reaction.

Protease preparations may be freed from contaminating activities by treatment with inactivators, for example, trypsin is treated with TPCK to destroy any chymotrypsin present.

Chapter 12
Isolation of Organelles

After homogenisation the various cell organelles need to be separated from each other. This is done by fractionated centrifugation [103]. First, connective tissue, undamaged cells, and other debris are removed by a brief spin at low speed (10 min at 500 g). In the next step nucleï and plasma membranes are spun down (10 min at 3 000 g). Mitochondria, plastids, and heavy microsomes require about 30 min at 20 000 g to be spun down, 1 h at 100 000 g is required for light microsomes. The remaining supernatant contains the cytosolic proteins.

These crude preparations are then subjected to further purification steps, usually by centrifugation in sucrose gradients [103]. Gradient materials with higher molecular mass like metrizamide or nycodenz (see Fig. 12.1) are expensive, but can be used to make isotonic gradients to preserve the activity of some sensitive membranes.

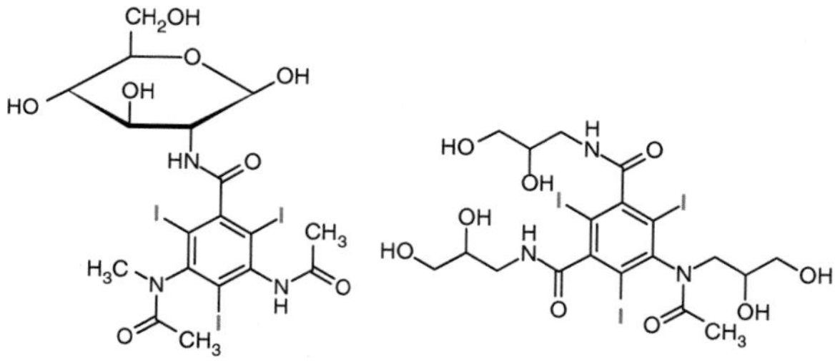

Metrizamide ($M_r = 789$) **Nycodenz** ($M_r = 821$)

Fig. 12.1 Metrizamide and nycodenz, with their high molecular mass and excellent water solubility, can be used to make isotonic gradients for centrifugation of cells and sensitive membranes. In particular, stripping of membrane-associated proteins, which may occur in high sucrose concentrations, is avoided. The main advantage of Nycodenz is that it can be autoclaved without decomposition

E. Buxbaum, *Biophysical Chemistry of Proteins: An Introduction to Laboratory Methods*, DOI 10.1007/978-1-4419-7251-4_12,

Chapter 13
Precipitation Methods

Protein precipitation is a very crude method for purification and it rarely achieves enrichment by more than a factor of 2–3. However, it is quick and cheap. If applied to crude homogenisates, it may remove material that would interfere with later purification steps. It is essential to perform these reactions in the cold because proteins are rapidly denatured by precipitating agents at room temperature. Mixtures of ice and salt are more effective in keeping samples cold than cold rooms, which only chill the scientist.

The precipitating agent is slowly added to the well-stirred protein solution so that the local concentration of precipitant does not exceed the desired one. When the precipitant concentration is so high that the desired protein is just soluble, precipitated material is removed by centrifugation (the pellet may be washed with precipitant solution of the same concentration to recover the desired protein included in the pellet). More precipitant is then added until all desired material has precipitated. It is separated from the solution by a second centrifugation step. The pellet is taken up in buffer and excess precipitant is removed by gel filtration, hydrophobic interaction chromatography, or dialysis.

13.1 Salts

Proteins interact strongly with water, and only the hydrated form is soluble. If the available water concentration is reduced by the addition of salts, proteins precipitate out of solution. Proteins can be separated into two groups depending on their behaviour towards salt, the albumins and globulins (see Fig. 13.1, but compare to Fig. 13.2).

Ammonium sulphate is most often used for precipitation, as it is cheap, non-toxic, highly soluble in water, and strongly ionised. Purified proteins can be crystallised by slow addition of ammonium sulphate; such crystals, suspended in the mother liquor, tend to be very stable if kept refrigerated. Many enzymes are sold in this form by suppliers. Traditionally, ammonium sulphate concentration is given as percent saturation, a saturated ammonium sulphate solution (100 % saturation) is about 4 M at 20 °C (solubility is almost identical to that at 0 °C). The amount of

E. Buxbaum, *Biophysical Chemistry of Proteins: An Introduction to Laboratory Methods*, DOI 10.1007/978-1-4419-7251-4_13,

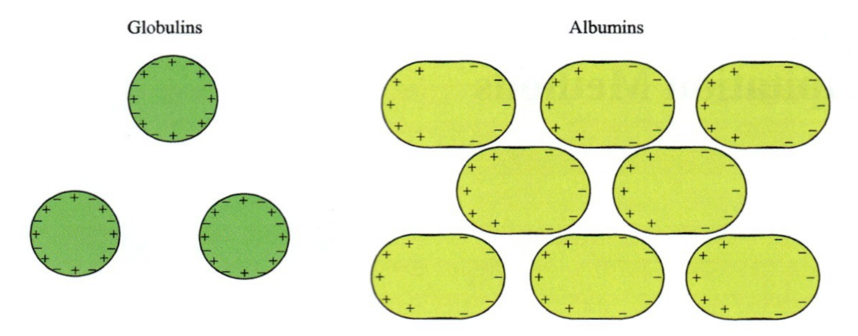

Fig. 13.1 Globulins are, as the name implies, spherical molecules with an even charge distribution. Such proteins are soluble in distilled water and low concentration salt solutions, they are precipitated by high salt concentrations. Albumins, on the other hand, have an asymmetric shape and charge distribution, they are held together by ionic bonds. As a consequence they are not soluble in distilled water and low salt concentrations are required to break these bonds (**salting in**). High concentrations of salt precipitate the protein again

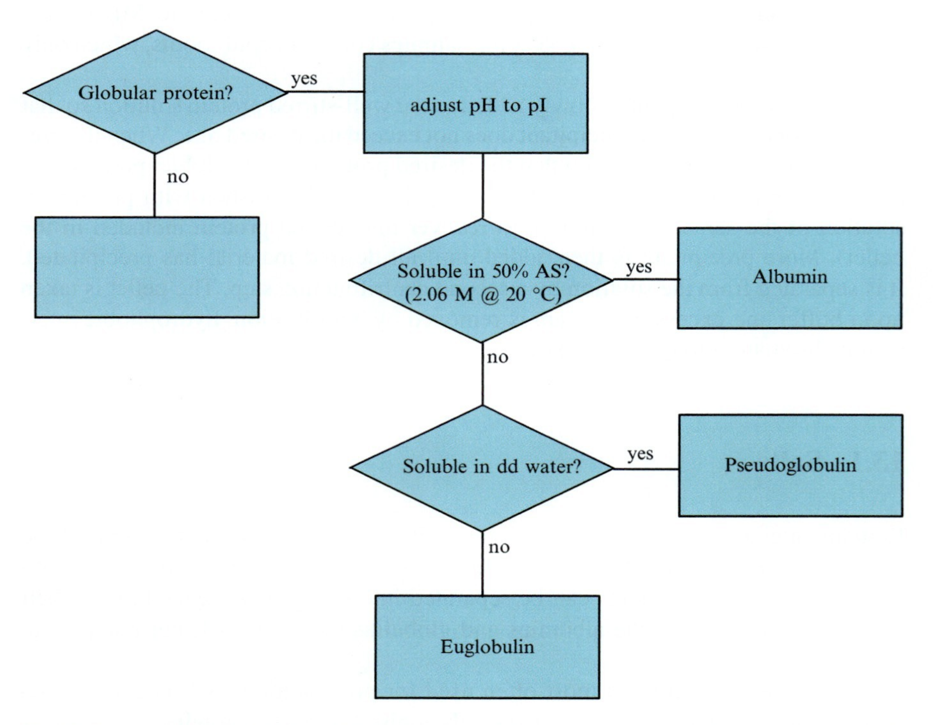

Fig. 13.2 Alternative definition of albumins, euglobulins and pseudoglobulins used especially in the literature on serum protein purification. The distinction is based purely on solubility, without attempt of a molecular explanation. Classification of a given protein may be different from that in Fig. 13.2

solid ammonium sulphate (m in g/l) that needs to be added to a solution of concentration S_1 in order to bring it to the concentration S_2 (both in % saturation) can be calculated by the empirical equation:

$$m = \frac{533(S_2 - S_1)}{100 - 0.3S_2} \tag{13.1}$$

again at 20 °C.

The solubility of proteins S in salt solutions of concentration c is described by the following equation:

$$\log(S) = \beta - \alpha c, \tag{13.2}$$

where α is the salting-out constant, which depends on the nature of both protein and salt, but is largely independent of pH and temperature. β, on the other hand, describes the solubility of a protein as function of pH. If the constants α and/or β of two proteins are sufficiently different, they can be separated by salt precipitation.

Ions can be sorted by their ability to denature (**caotropic effect**) or precipitate proteins (**kosmotropic effect**) into so called HOFMEISTER-series [166]. For anions the caotropic effect decreases in the order trichloroacetate > thiocyanate > iodide > perchlorate > nitrate > bromide > chloride > bicarbonate > tartrate > citrate > acetate > phosphate > sulphate and for cations guanidinium > Ca^{2+} > Mg^{2+} > $N(CH_3)_4^+$ > NH_4^+ > K^+ > Na^+ > Li^+. For protein studies, the anions have stronger effects than the cations.

As HOFMEISTER already realized, one key factor controlling the position of an ion on this scale is its "affinity for water" (modern: hydration energy). Strongly water binding ions or molecules (glycerol, sucrose) will reduce the available water concentration for the protein: this will stabilise protein conformation (higher denaturation temperature, lower enzymatic activity) and reduce solubility, resulting in crystallisation. Weakly water binding ions or molecules (urea) will interact with the protein instead, resulting in denaturation. This can be quantitated by the JONES–DOLE equation [179], which links relative viscosity η/η_0 with the concentration c of a salt:

$$\eta/\eta_0 = 1 + ac^{1/2} + bc + dc^2, \tag{13.3}$$

where a represents the strength of electrostatic effects (relevant only up to 100 mM), and d becomes relevant only for concentrations larger than 1 M. For our purposes, b is most important, as kosmotropes have large hydration shells and hence positive b, whilst caotropes have small hydration shells and slightly negative b. The range for b is about −0.01 to +0.25. There is a linear relationship between the logarithm of the protein denaturation rate constant k_u and the b-value. See p. 257 on methods for viscosity determination.

13.2 Organic Solvents

Water miscible organic solvents like methanol, ethanol, and acetone need to be used even more carefully than salts to prevent irreversible denaturation of proteins. Precipitation is usually done at subzero temperatures. Note that mixing of solvents with water generates heat. After precipitation the solvent is usually removed by **lyophilisation** (freeze–drying). Powders obtained this way are quite stable and can be kept refrigerated for years. This can be used to make an entire investigation from a single batch of protein. **Acetone powders** from many organs of various species can be commercially obtained.

13.3 Heat

Proteins are irreversibly denatured by heat and precipitate out of solution. However, some proteins are more resistant than others, especially in the presence of substrates or other ligands. If crude protein extracts are heated to a carefully chosen temperature for a carefully chosen time, some of the extraneous proteins precipitate while the relevant one stays intact. Precipitated material is removed by centrifugation. It is essential that heating be performed in a reproducible way, especially when scaling up a method. The sample is best held into a boiling water bath under constant swirling until it has achieved the correct temperature, then it is transferred to a second water bath kept at the temperature of the experiment for the desired time. Then it is quickly chilled in an ice/water mixture. This should be done in small aliquots (200 ml) in large ERLENMEYER-flasks (1 l) for effective heat transfer.

Chapter 14
Chromatography

14.1 Chromatographic Methods

In chromatography, analytes are bound to a solid support and then specifically eluted (see Fig. 14.1). Depending on the type of support we distinguish:

Paper chromatography is the oldest method, the sample is spotted onto a paper strip, which is hung with its lower end into a solvent. Capillary action sucks the solvent upward past the sample. Separation usually occurs by partitioning of sample molecules between paper-bound water and the mobile organic phase.

Thin layer chromatography uses a solid support (glass plates, plastic or aluminium sheets) covered with a thin (0.1–2 mm) layer of the separation matrix. This layer is prepared by grinding the matrix material to a fine powder, mixing it with water to a thick slurry, and pouring it over the plates. Because different materials can be used as a matrix, this method is more versatile than paper chromatography.

Column chromatography uses columns filled with the matrix instead of thin sheets. The solvent is pushed through the column by increased pressure to increase flow rate. It is important that the matrix particles are as homogeneous in size as possible to reduce the back pressure of the column. The finer the particles are, the higher their resolving power, but the higher the back pressure of the column. Fine matrix particles ($\leq 5\,\mu$m diameter) are used in **high performance liquid chromatography (HPLC)**, this requires specific instrumentation because of the pressures involved (100 MPa, hence the popular but historically incorrect expansion "high pressure liquid chromatography"). The newest development in this area are bead sizes of less than $2\,\mu$m, this is referred to as **ultrahigh performance liquid chromatography (UPLC)**. For industrial preparation, columns are sometimes replaced by stacks of filter paper, up to about 10 mm thick. This allows very high flow rates, resolution and capacity, with low back pressure. The hydroxy-groups of the cellulose are modified with the separating groups. Even higher flow rates are obtained when the flow is between closely packed filters, rather than through them ("cross-flow").

Gas chromatography Stationary phase is a thin film of liquid on a solid carrier or on the wall of the column (capillary gas chromatography). Mobile phase is

E. Buxbaum, *Biophysical Chemistry of Proteins: An Introduction to Laboratory Methods*, DOI 10.1007/978-1-4419-7251-4_14,

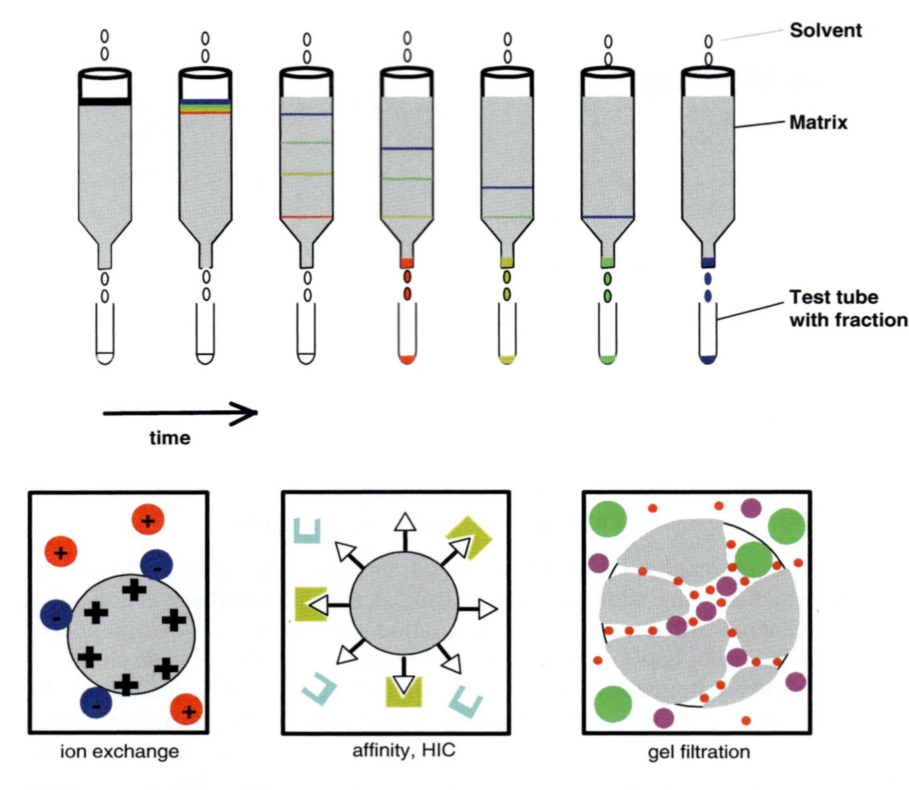

Fig. 14.1 *Top*: The sample is placed onto a separation matrix and eluted with solvent. Different molecules in the sample have different partition ratios between the stationary and the mobile phase. Molecules that bind weakly to the matrix leave the column quickly, those that bind more strongly are delayed. *Bottom*: Separation principles of chromatographic methods

a stream of gas (usually hydrogen). Sample compounds are separated by vapour pressure, slowly heating the column accelerates desorption. At the exit of the column the hydrogen is burned, if sample compounds are present, ions are produced which increase the conductivity of the flame (flame ionisation detector (FID)). WOLLRAB has described cheap gas chromatographs that students can build themselves and use in student labs [433].

Of these methods, only column chromatography is used for isolating proteins, however, the other methods can be useful for separating substrate and product in enzymatic reactions.

Several modes of interaction can be used for chromatography (see Fig. 14.1, bottom):

Ion exchange The electrical charge of proteins depends on their amino acid composition and the *p*H of the medium. If the *p*H is lower than the isoelectric point, the protein will have a positive net charge; if above the *p*I the net charge will be negative. If a protein is passed over a support with charged groups (for example sulphopropyl $-CH_2-CH_2-CH_2-SO_3^-$ or diethylaminoethan $-CH_2-CH_2-NH^+(CH_2-CH_3)_2$) it may or may not bind to it, depending on *p*H. Binding strength depends on the number of charges on the protein, weakly bound protein can be eluted with salt solutions of low concentration, for strongly bound proteins high salt concentrations are required (10 mM to 1 M). This is probably the most often used method for protein purification. It is also possible to use a *p*H gradient for elution (**chromatofoccusing**); this method has higher resolving power, but is much more expensive. Ion exchange is actually a very old technique, first mentioned for water purification in Exodus 15_{23-25}: Bitter tasting (Mg-containing) water is made palatable by bringing it into contact with tree-trunks whose lignin has been digested by certain fungi (white rot). The remaining matrix of cellulose and acid polysaccharides acts as an ion exchanger and binds the Mg-ions.

Gel filtration, also called gel chromatography or size exclusion chromatography, is based on the different size and shape (STOKES -radius) of protein molecules [335]. The matrix contains pores of different sizes. Small proteins can diffuse into all of them and are delayed, while the largest proteins do not fit into any pores and pass the gel without delay. Since no binding of proteins to the gel occurs, this method is very gentle. However, it is applicable only to small volumes of concentrated samples.

Partitioning between hydrophobic and hydrophilic phase is the basis of paper chromatography, the sample partitions between the water bound to the cellulose molecules and the more hydrophobic solvent. It is also possible to have a hydrophobic stationary phase (for example bound octadecyl- (C_{18}) residues) and a hydrophilic solvent, this is called **reversed phase chromatography** (**RPC**). These methods are usually used for small molecules, not so much for proteins.

Hydrophobic interaction Proteins contain a variable amount of hydrophobic amino acids. These are usually shielded by the hydration sphere of the protein. Salts like $(NH_4)_2SO_4$ (0.5 to several M) can strip proteins of their hydration sphere and expose some hydrophobic groups which can bind to hydrophobic groups of the column. Proteins are eluted with a gradient of decreasing salt concentration, restoring the hydration shell and decreasing hydrophobic interaction with the column. Note that proteins can not be eluted from C_{18}-columns used for reversed phase chromatography, at least not in a native conformation. Shorter chains which interact less strongly with proteins are used, for example butyl- or phenyl-groups. Less frequently, hydrophobic displacers like mild detergents, butylamine or butanol are used for elution. Strength of binding may also be controlled by *p*H, which influences the overall charge of the protein.

Ion pairing chromatography Here it is not the sample that interacts with hydrophobic matrix, but its salt with a hydrophobic counter ion. For example, nucleotides can be separated on a C_{18}-column as triethylammonium-salts. The ethyl-groups interact with the C_{18} groups, the retention of the nucleotides depends on the number of TEA-groups bound (ATP > ADP > AMP = P_i). Since TEA-HCO_3 is volatile, chromatography in this buffer is suitable for purification of nucleotide analogues, the buffer is later removed by lyophilisation.

Affinity Many proteins show specific interactions with some other molecules: enzymes with their substrates, receptors with their ligands, antibodies to antigens, or glycoproteins with lectins. If such molecules are chemically bound to a support (see part III on p. 191 for techniques used), those proteins that interact with them will be retained on the column, while all other proteins pass the column unhindered. The bound proteins can then be eluted with a ligand solution or by changing *p*H or ionic strength in such a way that the protein-ligand interaction is weakened. Because of the specificity of ligand-protein interactions, affinity chromatography can sometimes lead to one-step purification protocols. Biological ligands are often expensive and unstable, however, organic chemistry has produced a lot of compounds which, to a protein, look like its ligand, even though the structures may be quite different. Such compounds can be used with advantage for affinity chromatography; the use of Cibacron blue (see Fig. 14.3) for the purification of NAD(P) dependent enzymes is probably the most well known example (see Fig. 14.3).

Immobilised metal affinity chromatography (IMAC) is used mostly to purify proteins genetically engineered to contain a poly-histidine (6–12 residues) tag. The His-residues bind to metal ions bound to the chromatographic medium via complexons (usually nitrilotriacetic acid (NTA) $N(CH_2COOH)_3$, see Fig. 14.2). The protein is eluted either with an imidazole gradient or by stripping both protein and metal from the column with EDTA. The poly-His tag must be exposed on the surface of the protein and in a position where it does not interfere with protein folding. Most frequently, the tag is attached to either the N- or C-terminal end of the protein.

Another example is the capture of phosphoproteins on Fe^{3+}- or Ga^{3+}-columns. This phospho-protein enriched fraction can then be characterised by proteomics.

Hydroxylapatite chromatography (HAC) [72] The theoretical basis is much less clear than for the above methods. Hydroxylapatite (syn. hydroxyapatite) is a crystalline form of calcium phosphate ($Ca_5(PO_4)_3OH$). Depending on chromatographic conditions, the surface of these crystals bears an excess either of

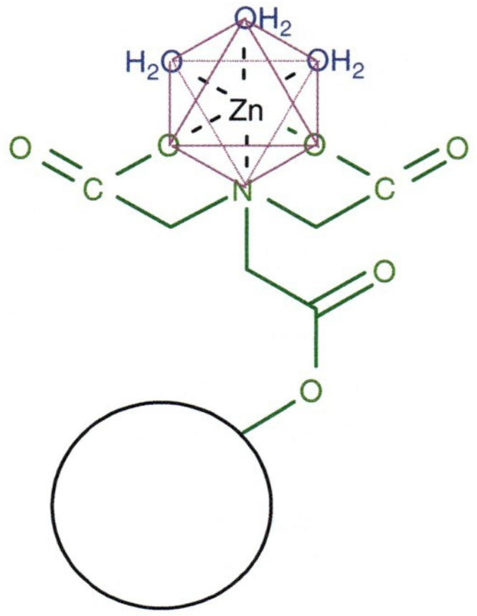

Fig. 14.2 Nitrilotriacetic acid (NTA)-gel with bound Zn^{2+}-ion. Zn^{2+} can accommodate 6 complex ligands in a octahedral configuration. 3 positions are occupied by the gel, the remaining contain water, which can be replaced by protein side chains (especially His). Other ions like Cu^{2+}, Ni^{2+} or Co^{2+} may also be used

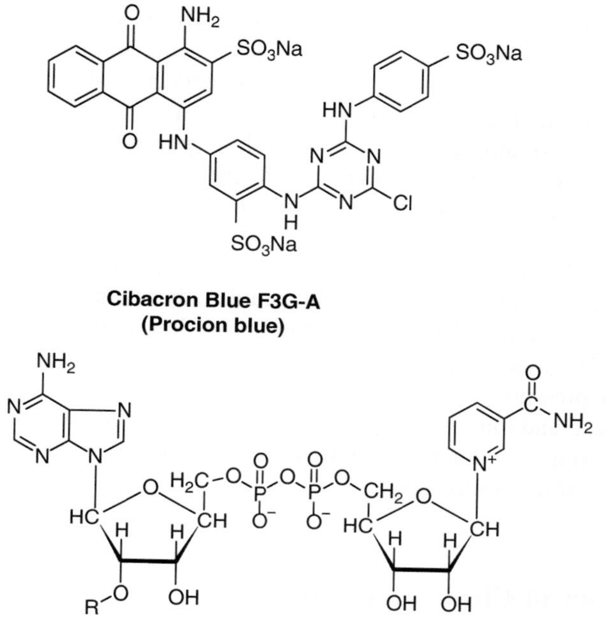

Cibacron Blue F3G-A
(Procion blue)

NAD(P)$^+$

Fig. 14.3 Cibacron blue is used as an affinity matrix for NAD(P)$^+$-dependent enzymes

positively charged calcium or negatively charged phosphate ions. HAC is probably a combination of ion exchange and affinity chromatography: $-NH_3^+$-groups bind to phosphate groups by electrostatic interactions, $-COO^-$-groups to Ca^{2+} by both electrostatic interactions and by chelation. Phosphoproteins show strong chelation to Ca^{2+}, uncharged $-OH$ and $-NH_2$-groups form water-bridged hydrogen bonds. Bound proteins are eluted with a potassium (or sodium-) phosphate solution of increasing concentration (10–500 mM). Note that anhydrous phosphates contain small amounts of pyrophosphate and higher polyphosphates which affect the elution profile of proteins. It is therefore better to prepare phosphate buffers from the hydrates, as otherwise one would have to ensure that the polyphosphate content remains constant from batch to batch. To prevent dissolution of the hydroxylapatite, phosphate concentration in all buffers should be larger than 10 mM. Although results with HAC are even more difficult to predict than with other methods, it is an essential tool for the isolation of some proteins. Hydroxylapatite can be produced in the lab, but industrially manufactured material with ceramic backbone allows much higher flow rates.

Chromatographic methods are the workhorses in protein purification. Columns can be constructed for sample volumes from a few nl in proteomics to several l for industrial scale preparative purification. The equipment for protein chromatography needs to be constructed from biocompatible materials (glass, certain plastics), steel is unsuitable as metal ions can inactivate proteins (steel is commonly used in organic chemistry). Sample vials, columns, and fraction collectors should be kept at 4 °C during chromatographic runs. Fouling of the columns is prevented by filtering all samples and buffers through low-protein-binding (polysulphon) membrane filters of 0.45 or 0.22 μM pore size. Anybody embarking on chromatographic separations is advised to study the Pharmacia handbooks now freely available at http://www1.gelifesciences.com/aptrix/upp01077.nsf/Content/orderonline_handbooks first, the detailed method descriptions are invaluable. If large sample volumes need to be processed in industrial applications, fluid, rather than packed beds, is used with advantage (**expanded bed absorption**). The flow is from the bottom of the column to the top, the matrix particles are kept in suspension by the sample flow. Thus the back pressure of the column is minimised, flow rates can be increased by a factor of about 5 and fouling is prevented. Gel particles need to be heavy to prevent them from leaving the column with the sample flow; a core of dense, inert material like tungsten carbide ensures this.

14.2 Theory of Chromatography

We will now look at factors affecting the performance of chromatographic columns and ways to characterise and compare columns (see Fig. 14.7). Similar considerations would apply to other separation methods where equilibria are established.

In the following we will look at partitioning methods like gel filtration. For chromatographic materials that bind the analytes (ion exchange, affinity), the additional factor to consider is the density of the functional groups on the gel surface. Higher density means higher binding capacity, hence smaller column volumes may be used. On the downside, steric hindrance may occur and binding may be so strong that analytes can not be eluted without denaturation.

14.2.1 The CRAIG*-Distribution*

Before dealing with the continuous process of chromatography, let us look at its discontinuous equivalent, the CRAIG-distribution ([70], see Fig. 14.4): In this type of experiment n separation funnels with a constant volume V of a solvent (for example, water) are placed in a row. The first funnel also contains dissolved sample. Now we add to the first funnel a second solvent, which is immiscible with the first (say, chloroform). We shake the mixture and then allow phase separation. If both phases s (aqueous, stationary) and o (organic, mobile) are in equilibrium, then the activities of a substance X in both phases must also be in equilibrium, and NERNST's distribution law applies:

$$\mu_X^s = \mu_X^o \qquad \Rightarrow \alpha_X^s \propto \alpha_X^o \qquad \Rightarrow \qquad c_X^s \propto c_X^o,$$

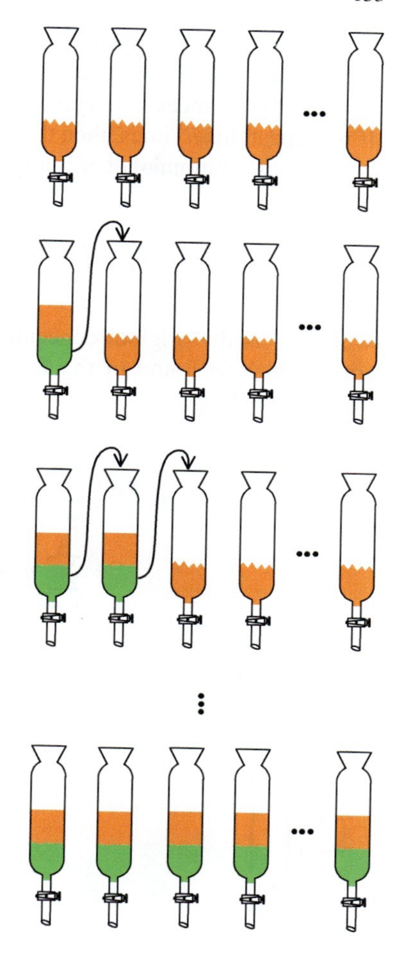

Fig. 14.4 CRAIG-distribution. Analyte distributes between a stationary (*orange*) and a mobile (*green*) phase. During each step, the mobile phase from funnel n is transferred to funnel $n + 1$, and fresh mobile phase is added to funnel 1. The analyte is transported from funnel to funnel with a velocity determined by its partition ratio

where μ, α and c are the chemical potentials, activities, and concentrations, respectively; and the proportionality constant is the ratio of the probability for a molecule to be in either the aqueous or the organic phase:

$$K_{\mathrm{p}} = \frac{P(\mathrm{s})}{P(\mathrm{o})} \approx \frac{\alpha_{\mathrm{X}}^{\mathrm{s}}}{\alpha_{\mathrm{X}}^{\mathrm{o}}} \approx \frac{c_{\mathrm{X}}^{\mathrm{s}}}{c_{\mathrm{X}}^{\mathrm{o}}}. \tag{14.1}$$

This is called the partition ratio. In the following, we will call $P(\mathrm{o})$ simply p. Since $P(s) + P(\mathrm{o}) = 1$, this automatically defines $P(\mathrm{s})$:

$$\begin{aligned} p = P(\mathrm{o}) &= \frac{c_{\mathrm{X}}^{\mathrm{o}}}{c_{\mathrm{X}}^{\mathrm{o}} + c_{\mathrm{X}}^{\mathrm{s}}} = \frac{K_{\mathrm{p}}}{K_{\mathrm{p}} + 1} \\ P(\mathrm{s}) &= \frac{c_{\mathrm{X}}^{\mathrm{s}}}{c_{\mathrm{X}}^{\mathrm{s}} + c_{\mathrm{X}}^{\mathrm{o}}} = \frac{1}{K_{\mathrm{p}} + 1} \end{aligned} \tag{14.2}$$

If we now let the chloroform-phase flow from the first into the second funnel and add the volume V of chloroform to the first, then upon renewed shaking the dissolved substances will again distribute between the aqueous and organic phases. If we repeat this n times, then the concentration of X in the mth funnel ($m \leq n$) will be given by a binomial distribution:

$$P_{\text{m,n}} = \frac{c_{\text{X}}^{\text{o}}}{c_{\text{X}}^{0}} = \binom{n}{m} * p^m * (1-p)^{n-m}, \tag{14.3}$$

where c_{X}^{0} is the original concentration of X (see Fig. 14.5). The highest concentration of X is found in the funnel y, which corresponds to the average of the distribution $P_{\text{m,n}}$:

$$\begin{aligned} y &= \sum_{m=0}^{\text{n}} m P_{m,n} \\ &= \sum_{m=0}^{\text{n}} m * \binom{n}{m} * p^m * (1-p)^{n-m} \\ &= n * p. \end{aligned} \tag{14.4}$$

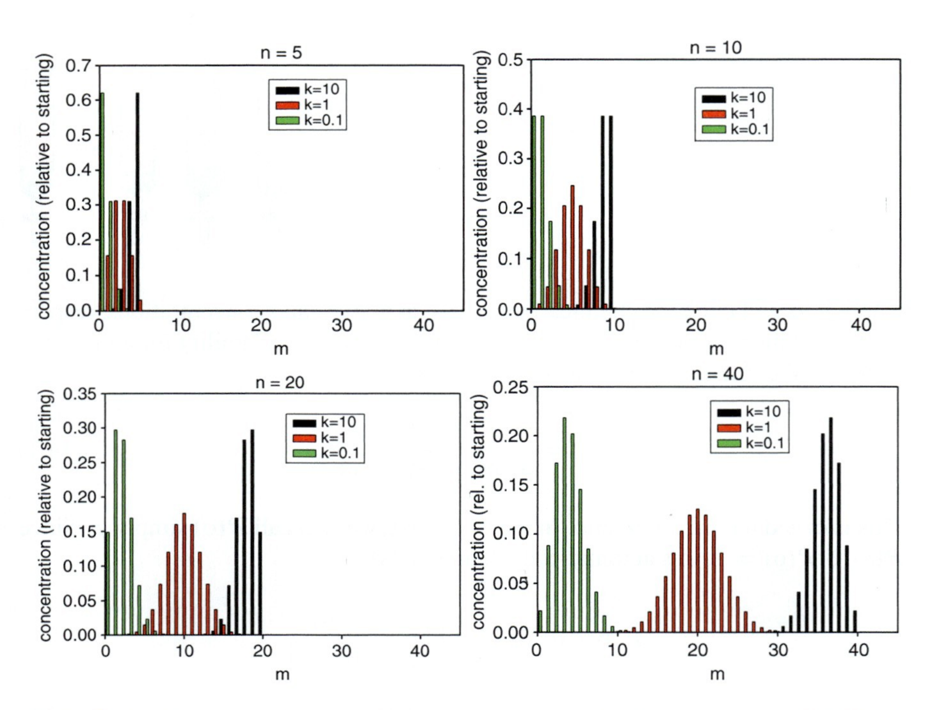

Fig. 14.5 Simulation of a CRAIG-distribution according to (14.3). Three substances with different K were used

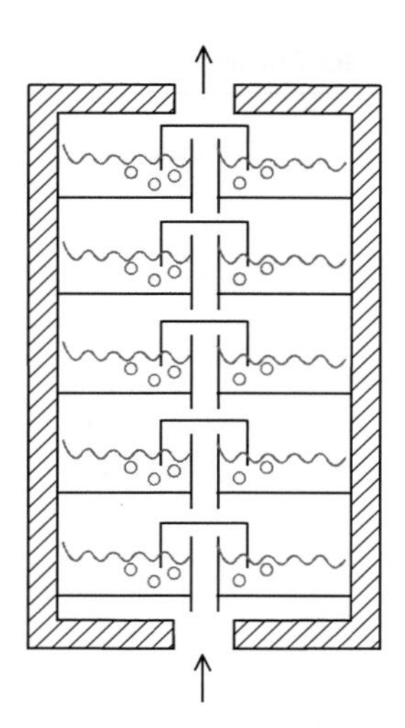

Fig. 14.6 A distillation column with five plates as a model for other equilibrium separation methods. On each plate steam enters from below, and an equilibrium with the liquid phase is established. Substances with high boiling point condense from the steam, in the exchange substances with low boiling point evaporate from the liquid. As the liquid volume on a plate gets larger, some of the liquid drips down onto the next lower plate. The column is well insulated, so that no heat is lost during the process

To change from discontinuous CRAIG-distribution to continuous chromatography, we make the number of funnels n very large (in the order of 10^3–10^4). Each of these conceptual funnels is equivalent to a **theoretical plate** (see Fig. 14.6 for an explanation of that name). At the same time, p is made very small. Then the binomial distribution can be approximated by a POISSON-distribution:

$$\frac{c_X^o}{c_X^0} = \frac{e^{-y} * y^m}{m!} \tag{14.5}$$

for each theoretical plate m.

14.2.2 Characterising Matrix–Solute Interaction

We will do the derivation for the example of gel filtration. Similar considerations also apply for other modes of separation.

The total volume of a chromatographic column V_t is given by its bed height h multiplied with its cross-section, the later depends on the radius r:

$$V_t = hA = h\pi r^2 \tag{14.6}$$

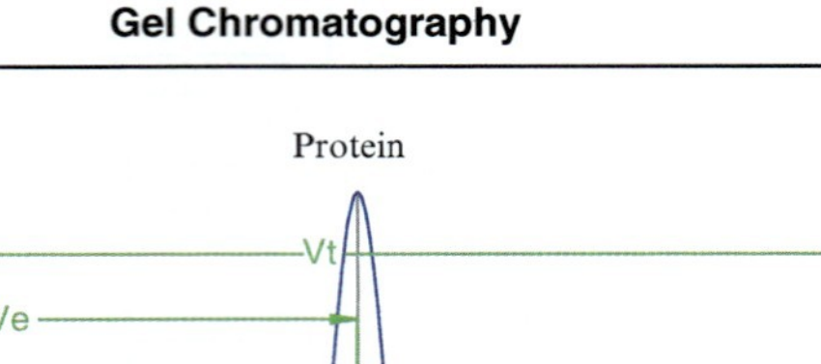
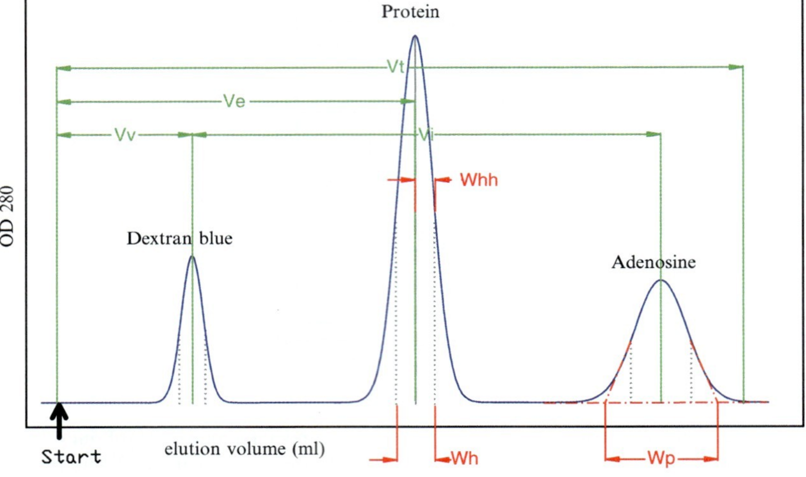

Fig. 14.7 Schematic example of a size exclusion chromatogram. The sample contained Dextran Blue 2000 as very large molecule to mark the void volume V_v and adenosine as a very small molecule to mark the included volume V_i. The protein is assumed to have a size in the separation range of the gel. A substance which binds to the gel (for example by ionic or hydrophobic interactions) would have an elution volume V_e larger than the total column volume V_t. As the elution volume increases, peaks become broader due to diffusion. The peak width W_p is measured by approximating the peak with a triangle reaching to the null-line, alternatively the width at half the peak hight W_h can be used

This total volume consists of three fractions (see Fig. 14.7):

Void volume V_v, which is the space in between the gel particles. This volume is available even to very large molecules, like Dextran blue 2000, a polysaccharide of about 2 MDa molecular mass. V_v is equal to the volume of the mobile phase.

Included volume V_i, which is the volume inside the pores of the gel which are available to very small molecules. This is equal to the volume of the stationary phase. V_i can be approximated as the difference between the elution volumes of a very small molecule like nucleotides and a very large molecule like Dextran blue.

Matrix volume V_m, which is equal to the volume of the molecules forming the gel matrix, this volume is not available to any sample molecules.

Since the method of determining V_i experimentally is somewhat imprecise, $V_t - V_v$ is often used instead, which is equal to $V_i + V_m$.

The partition ratio is then $K_p = (V_e - V_v)/V_i$, which can be approximated by $K_{av} = (V_e - V_v)/(V_t - V_v)$. For a given gel K_p/K_{av} is a constant. One interpretation of the partition ratio is then the part of the included volume of the column that is accessible to the protein. If K_p obtained with a gel is plotted as a function of the logarithm of the STOKES-radius $\bar{r}$ (selectivity curve of a gel), a straight line is formed for a certain size range. Thus the size of proteins can be determined from gel filtration data by comparing their K_p values with that of proteins with known sizes. In many cases, size is taken as approximation for molecular mass M_r.

14.2.3 The Performance of Chromatographic Columns

Ideally, the peaks of the components of a sample should be completely separated after chromatography. Alas, that is often not the case. There is a target conflict: On the one hand, one wants to achieve good separation, on the other hand the costs of the procedure are to be kept as low as possible. Large sample volumes, high flow rates, small columns, and cheap (and hence polydisperse) gels all conspire to reduce the resolution obtained. The narrower the peaks and the larger their distance from each other, the better the resolution of the column.

Peak width W_p is obtained by approximating the peak shape with a triangle (see Fig. 14.7). The width is then measured at the base line. Alternatively, the width at half the peak height W_h is used.

A sample volume V_s is added on the top of a column of cross-section A, its length will be $l_s = V_s/A$. When this sample enters the gel, it will spread to a length of $l'_s = V_s/\alpha$, with α = void volume per unit length. Thus

$$\frac{l'_s}{l_s} = \frac{A}{\alpha} = \frac{V_t}{V_v}, \tag{14.7}$$

which means that the spread can be minimised by minimising the void volume of the column. Further axial spreading then occurs by:

BROWNIAN diffusion of analyte This is a particular concern in non-binding techniques like gel filtration.

Different width of channels between column particles In broader channels flow is faster (eddy diffusion), in addition interaction of solutes with the matrix is reduced as diffusion paths become longer.

Laminar flow in channels The centre of a channel flow is faster than next to the matrix particles where friction slows flow. Again, analyte molecules next to the matrix have a better chance of binding than those in the centre of the channel.

Stagnant mobile phase In some channels flow is reduced as flow direction is not down the column, but sideways.

These effects are minimised by smaller beads, since these have a larger surface area for exchange to occur and the diffusion distances in the void volume become smaller. Note that decreasing the bead size does not change the void volume!

The **resolution** becomes

$$R_s = \frac{\text{distance between peaks}}{\text{average peak width}} \tag{14.8}$$

Baseline separation of peaks is indicated by $R_s > 1.5$. For $R_s = 1.0$, 98 % of solute A will be in the first peak and the remaining 2 % in the second and *vice versa*. The resolution is determined by **selectivity, efficiency** and **retention** of a column:

$$R_s = 1/4 * \frac{\alpha - 1}{\alpha} * \sqrt{N} * \frac{\hat{k'}}{1 + \hat{k'}}. \tag{14.9}$$

The **retention factor** $k' = (V_e - V_v)/(V_v)$ is a measure of how much the column delays the solute molecules. Since in gel filtration $V_e \leq V_t$ the obtainable resolution is lower than in techniques like ion exchange chromatography which do not suffer from this limitation. In above equation the average of the retention factors of both peaks is used.

The **efficiency factor** $N = 5.54 * (V_e/W_h)^2$ measures the zone broadening of a peak on the column. This is also known as the **number of theoretical plates** of the column, an expression developed in analogy with fractionated distillation (see Fig. 14.6 for an explanation). If packing is done properly, the number of theoretical plates achievable per meter of column length depends on the gel used and can be looked up in its spec sheet.

The **selectivity factor** α is determined by the ability of the matrix to separate the two substances, that is by the distance between the two peaks, with $\alpha = k'_2/k'_1 = (V_e(2) - V_v)/(V_e(1) - V_v)$.

14.2.3.1 Peak Symmetry

In Fig. 14.7 the asymmetry factor of the peak can be defined as

$$A_s = \frac{W_h}{W_{hh}} \tag{14.10}$$

This can not only be done at half-height (as shown in the figure), but at any height. A_s should be independent of the height, a plot of A_s against the height should result in a parallel to the x-axis. If this is not the case, the following causes may apply:

- The column is overloaded, sample size must be reduced.
- The peak is not homogeneous, try to separate its components by changing the chromatographic conditions.
- Column is not well packed, repack.
- Unwanted interactions between column and sample (ionic interactions in RPC, binding in SEC). Change the conditions to minimise such interactions.
- Labile complexes formed during chromatography.

14.2.3.2 Is This Peak Caused by a Homogeneous Substance?

One method to use is to systematically change elution conditions and to plot the resulting peakwidths against the retention time. For homogeneous substances the result will be a straight line. If the peak is not homogeneous then it is unlikely that all components react in the same way to the change in chromatographic conditions, and the plot will deviate from linearity, even if the changes do not result in peak separation.

As discussed above, the asymmetry factor should be independent of height. In addition, the ratio of $\mathcal{E}(\lambda_1)/\mathcal{E}(\lambda_2)$ should be constant over the entire peak if the absorbance can be measured at different wavelengths (diode array detector!).

The first and second derivative of the peak react sensitively to inhomogeneities if the signal/noise ratio is good and the data collection rate sufficient. Chromatographic computer software that comes with your instrument can usually calculate the derivatives automatically. General statistics packages like SysStat or Origin also have this function. CHROMuLAN (http://www.chromulan.org/) is an opensource project.

14.2.3.3 How to Increase Resolution

To increase resolution, peak width should be reduced as much as possible. All molecules of a given species should take similar paths through the column. If the surface of the gel is not a horizontal plane, some molecules take a shorter path than others. The chance for such inhomogeneities to occur increases with the diameter of the column, so the column diameter should be adapted to the sample size.

Also, pack the gel as homogeneously as possible, cracks will allow some molecules to move faster, while air bubbles force others to take longer routes around them.

The size distribution of the gel particles should be as narrow as possible to achieve better column homogeneity. When a column is packed, larger particles will settle first, followed by smaller ones. This gradient is unavoidable even with high resolution gels. However, columns should be packed in one go. If more slurry is added repeatedly after settling of the gel, zones of large and small particles will form. Use a large reservoir to contain all the gel used for packing a column.

Shorter diffusion paths and higher resolution can be obtained by smaller gel particles, resolution is inverse proportional to the square of the particle radius. At the same time zone broadening by longitudinal diffusion of sample molecules is reduced, and the surface area of the gel – and hence its capacity – is increased. However, as the back pressure of the column is increased, stronger pumps and more pressure resistant apparatus are required. The gel matrix also needs to be able to withstand the pressure without compression.

The linear flow rate (cm/h) should be low enough to allow partitioning of sample molecules between stationary and mobile phase. However, it should not be reduced beyond that point to reduce the time available for zone broadening by longitudinal diffusion. In addition, denaturation of sample molecules increases with separation time. 5 cm/h are used for analytical gel chromatography, for most other techniques flow can be considerably faster.

Resolution increases with the square root of column length, because longer columns not only have larger peak distances, but also more zone broadening. The back pressure, the run time, longitudinal diffusion, and the costs of column and mobile phase increase with column length.

The viscosity of sample and running buffer should be as low as possible to reduce column back pressure. High sample viscosity relative to the running buffer can lead to viscous fingering, apparent from non-symmetrical peak-shapes. Relative viscosity of sample and buffer (estimated easily from the emptying time of a pipette) should not differ by more than a factor of 2. Differences in density of sample and buffer should also be minimised. If the sample has a higher density than the eluent, sample flow should be from the bottom to the top and *vice versa*.

In non-concentrating techniques the peak width is also determined by the sample zone width, thus sample volume should not be more than a few percent of the column volume. Thus it is often better to use non-concentrating techniques like gel filtration *after* concentrating ones like affinity or ion exchange chromatography.

If in concentrating techniques the columns binding capacity is not used completely, peak width can be reduced and eluate concentration increased by using different flow directions for sample loading and elution.

Often samples are fractioned during the run, and the fractions analysed later. Fractions which are too large decrease resolution, very small fractions increase the work required for analysis and the losses incurred when combining relevant fractions. Something like 20 to 30 fractions is usually adequate. In preparative chromatography it may be possible to collect fractions only near the known elution volume of the desired component.

14.2.3.4 Quantitation

The area under a chromatographic peak is proportional to the amount of the substance causing it. Peak-integrators, which calculate this area, were once common pieces of equipment. Nowadays chromatography stations are controlled by microcomputers running specialised software suites, including this feature. However, if calculating peak areas for partially overlapping peaks is a problem, an online tutorial is available under http://www.vias.org/simulations/simusoft_peakoverlap.html.

In the usual setup the absorbance of the eluate as it leaves the column is measured as a function of time (or volume), either at a fixed wavelength or at multiple (diode array detector). However, a lot of information about the behaviour of the sample on the column is lost. Attempts have been made to scan the length of the column repeatedly during a chromatographic run (reviewed in [2]). This type of experiment resembles analytical ultracentrifugation, where the movement of the protein is also followed during the entire run. This allows to follow binding equilibria to determine dissociation constants. The main obstacle is the high scattering of the gel beads (highest for gels with the smallest exclusion limit), which results in high background absorbance. This method has therefore never really gained popularity.

14.3 Strategic Considerations in Protein Purification

Purification of proteins can be divided into four steps:

Extraction: the source (cells, tissue) is broken up, the component of interest is stabilised by buffer, protease inhibitors, cofactors, chelators and the like. None of these additives must interfere with later purification. The initial extract is clarified, damaging components and particulates are removed, most often by centrifugation. Fractionated precipitation by ammonium sulphate, solvents, or heat may be used at this stage.

Capture: the protein is concentrated and stabilised, *e.g.*, by removal of proteases. Methods are optimised for speed and capacity. This involves the use of coarse chromatographic media ($\approx 100\,\mu m$) with high binding capacities, high flow rates (several 10^2 cm/h), and single step elution profiles (load, wash, elute, recondition). Conditions should be chosen so that as many contaminants as possible are removed with the flow through, this increases column capacity for the desired protein. Narrow cuts are used to favour specificity over yield. Ion exchange chromatography is often used, for recombinant proteins affinity chromatography using a tag on the protein (*e.g.*, His_6).

Purification: most impurities are removed, the sample is further concentrated. Procedures are optimised for resolution and capacity; continuous or multi-step gradients on moderately fine gel matrices (several $10\,\mu m$) and high flow rates (100–150 cm/h) are used. Hydrophobic interaction chromatography is the ideal method after IEC or affinity chromatography capture, as the sample already contains high salt concentrations.

Polishing: trace impurities are removed and the product is brought into a solution compatible with storage and application. Most contaminants now are quite similar to the desired products (splice and glycosylation variants, aggregates) or present in small quantities (endotoxins, pyrogens, virus). At this step, methods are optimised for resolution and recovery since the product is now much more valuable than in previous steps. Gels have very small beads ($\leq 10\,\mu m$) with narrow size distribution and flow rates are low (≤ 100 cm/h). If gradients are used, they are very shallow. SEC is often the method of choice as it allows buffer exchange at the same time.

14.3.1 Example: Purification of Nucleotide-free Hsc70 From Mung Bean Seeds

The 70 kDa heat shock cognate protein is one of the major molecular chaperons in living cells. Because of its tight interaction with nucleotides (the complex survives dialysis against a suspension of activated charcoal for at least 2 weeks) it is difficult to prepare the protein in a nucleotide-free form, to which nucleotide binding could be measured [47].

Preparation of sample material: 1 500 g mung beans were soaked over night in homogenisation buffer, during rehydration and sprouting Hsc70 is expressed in high concentrations.

Capture: Beans were then twice homogenized in a total of 4 l of homogenisation buffer, the cytosol (2 l with 130 g of protein) was prepared by centrifugation at 6 000 g. A 40–80 % ammonium sulphate cut further clarified the solution, reduced the total amount of protein and volume (50 g in 500ml), and removed an apyrase activity that would destroy the affinity column in the next step. Low molecular weight compounds (*e.g.*, nucleotides) were also removed.

Purification: Affinity chromatography on 50 ml ATP-agarose gave 200 mg of a product that was already >95 % pure. Because of the strong interaction between protein and column, the elution volume was unusually large, 130 ml. This product was further purified and concentrated by hydroxylapatite chromatography to remove some actin that had made it to this step, yielding 8 ml with 182 mg. Free ATP was removed by this step.

Polishing: EDTA was than added to the sample to remove Mg^{2+} ions from the nucleotide binding site and hence weaken enzyme–nucleotide interactions. The protein was precipitated with 80 % ammonium sulphate; this results in mild, reversible denaturation and release of most of the bound nucleotides. The sample was centrifuged, the supernatant removed, the pellet washed with ammonium sulphate solution and finally dissolved in 0.5 ml of buffer, containing 120 mg of protein. SEC then separated the dimeric, nucleotide free enzyme from the monomeric, nucleotide-containing one. The final product contained about 70 mg of protein, which was pure by SDS–PAGE and contained less than 0.004 mol/mol of ATP. The enzyme was stored as ammonium sulphate precipitate.

Chapter 15
Membrane Proteins

A nice introduction into membrane biology is [247].

15.1 Structure of Lipid/Water Systems

Interactions between water and lipids are entropy-driven: Water molecules interact strongly with each other by hydrogen bonds (see Fig. 15.2). In ice, each hydrogen atom in a water molecule forms a hydrogen bond with an oxygen atom from another water molecule. The oxygen atom receives two bonds from hydrogen atoms of neighbouring water molecules. This total of four hydrogen bonds per water molecule represents a binding energy of $\approx 21\,\mathrm{kJ/mol}$. In liquid water on average 3.4 hydrogen bonds are preserved, but partner switching occurs on a timescale of 10^{-11} s. If a hydrocarbon molecule enters such a network of water molecules, these interactions reorient since there is little interaction between water and the non-polar hydrocarbon chain. Rather, the water molecules are forced into an ordered, cage-like structure around the hydrocarbon, preserving water–water interaction as much as possible so that there is little change in enthalpy. However, the mobility of both water and hydrocarbon is reduced. This results in an unfavourable loss of entropy. To avoid this, hydrophobic molecules are forced into clusters so that their surface/volume ratio drops. This is the basis for the **hydrophobic effect**. The ΔG of this process can be calculated from the partition ratio K_p of a substance X between water and a hydrophobic solvent (octanol is used to simulate biological membranes):

$$\Delta G = RT\ln(K_p) = RT\ln\left(\frac{[X]_w}{[X]_o}\right). \qquad (15.1)$$

This energy is about $3.3\,\mathrm{kJ/mol}$ for each CH_2-unit in a hydrophobic chain. Composition of membranes is highly variable, about 15–70 % lipids, 0–25 % sterols, and 20–80 % proteins. Prokaryotic membranes do not contain sterols; yeasts and fungi have ergosterol, plants stigmasterol and sitosterol, animals cholesterol (see Fig. 15.1).

E. Buxbaum, *Biophysical Chemistry of Proteins: An Introduction to Laboratory Methods*, DOI 10.1007/978-1-4419-7251-4_15,

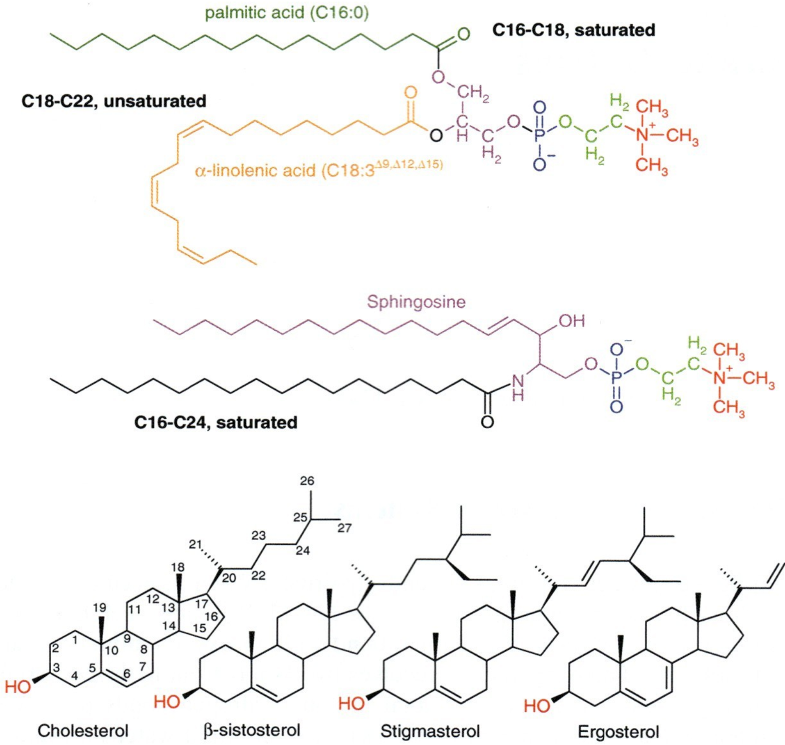

Fig. 15.1 Structure of membrane components. *Top*: Phospholipids, here phosphatidylcholine. Since the fatty acid in position 2 of glycerol is unsaturated, it is kinked and takes up more space than a saturated fatty acid would. This increases the fluidity of the membrane. The fatty acid in position 2 is also often longer than the one in position 1. This allows the phospholipids in the cytoplasmic leaflet to interdigitate with those on the extracellular leaflet. *Middle*: Sphingolipids usually have a saturated fatty acid attached which are straight molecules that can form tight complexes with cholesterol (*Bottom*)

Depending (mainly) on composition and temperature, lipid/water mixtures assume different phases:

Lamellar liquid-crystalline L_α, also called liquid disordered, is the structure we find in a biological double membrane. The lipid molecules are oriented perpendicular to the membrane surface, the hydrophobic tails are flexible, and there is rapid diffusion of lipid molecules (a lipid molecule may reach any point on the surface of an erythrocyte membrane within a few seconds). In experimental systems, the membrane sheets may form stacks, with water layers in between.

Lamellar gel $L_{\beta'}$, also called ordered solid, is the state of biological membranes at low temperature. The tails of the lipid molecules are fully extended, and the molecules closely packed. To achieve this close packing, the molecules are oriented at an oblique angle θ to the membrane. Mobility of lipid molecules is more restricted than in the L_α-phase. For mammalian membranes, the transition

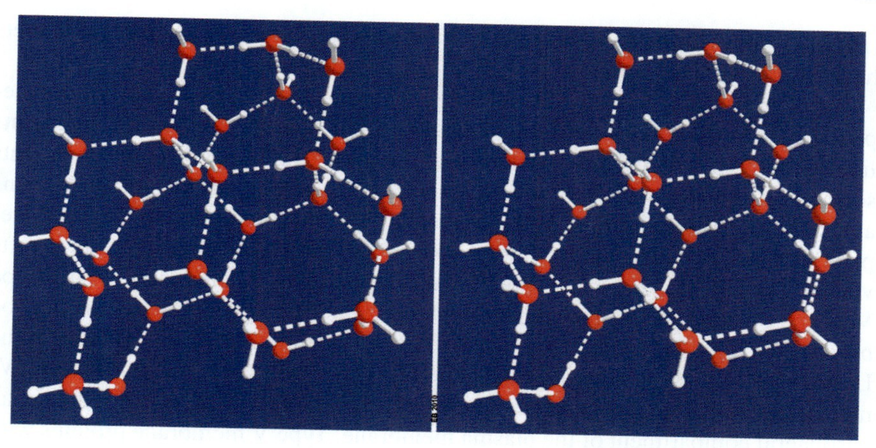

Fig. 15.2 Stereo-diagram of ice 1H (the hexagonal form present under normal atmospheric pressure). Each oxygen is connected to 4 hydrogens, 2 by covalent and 2 by hydrogen bonds. These hydrogens form the vertices of a regular tetrahedron. Note the loose packing of atoms, that leads to the low density (0.91 kg/l) of ice. In liquid water most of the structure is preserved, but hydrogen bond formation is more dynamic, they exchange on a picosecond time scale. At room temperature, each oxygen is surrounded on average by 3.4 hydrogen atoms

temperature between L_α and $L_{\beta'}$ is between 10 and 12 °C. Because of the change in membrane viscosity associated with this transition, membrane enzymes show a kink in the ARRHENIUS-plot in this temperature range.

Ripple phase $P_{\beta'}$ occurs in some pure lipids during the transition between L_α and $L_{\beta'}$. The surface of the membrane is not flat, but rippled. Since biological membranes are mixtures of many different compounds, they do not have a ripple phase.

Liquid ordered L_o forms from mixtures of phospholipids and/or sphingolipids with cholesterol. The acyl-chains of the lipid interacts tightly with the cholesterol, they are straightened out as much as possible. As a result, L_o is ordered like $L_{\beta'}$, but the lipids are more mobile, almost as much as in L_α.

Lamellar crystalline L_c dehydrated, highly ordered lipids on which X-ray crystallography can be used.

Hexagonal the lipids form long, cylindrical tubes, either with the tails sticking in and head-groups out (H_I, high water content), or with head-groups in and tails out (H_{II}, low water content). These tubes are arranged so that each one is surrounded by six others, hence the name. Hexagonal phases are formed by more wedge-shaped lipids, while lamellar phases are formed by cylindrical lipids.

Cubic micellar spherical arrangements (**micelles**) of lipids with the tails inside and the head-groups on the surface where they can interact with the surrounding water.

Cubic bicontinuous network of lipid channels, with water inside and around.

Phase diagrams of phospholipid, sphingolipid, and cholesterol mixtures show non-ideal mixing, with several phases co-existing. This is of high biological significance: cholesterol/sphingolipid rich domains in the outer leaflet of the membrane (**rafts**, L_o) (about 10^6 per cell) float in a phospholipid-rich sea (L_α). Inner leaflets have a different lipid composition and do not form rafts.

Rafts are thicker than the surrounding membrane (48 vs. 39 Å), and can be depicted by atomic force microscopy. They are not solubilised in mild detergents like Triton-X100 (**detergent resistant membranes, DRM**), but different detergents seem to select slightly different membrane populations. Various membrane proteins associate preferentially with either the raft or the phospholipid domains; this depends on the length of their transmembrane helices. Some proteins associate to rafts via acyl- or GPI-linkers, while proteins with farnesyl- or geranylgeranyl-tails (type V, on the cytoplasmic leaflet) prefer the phospholipid-rich membranes. This may be because the highly branched isoprenoids do not fit well into the close packing of the L_0-phase. For reasons not well understood proteins with GPI attachment (type VI membrane proteins) are found in the extracellular leaflet of the apical, rather than basolateral compartment of the plasma membrane. Type V membrane proteins with acyl- or isoprenoid-tails are found on the cytoplasmic leaflet.

Converting an inactive, soluble precursor into an active, membrane-associated protein by attachment of fatty acid, isoprens, or GPI is an important step in cell signalling (Src-like kinases, G_α) and intracellular membrane trafficking (small G-proteins like ARF). One of the reasons is the increase in concentration associated with this process:

$$\frac{\text{Volume of sphere}}{\text{Volume of surface phase}} = \frac{4/3\pi r^3}{4\pi r^2 d} = \frac{r}{3d}, \tag{15.2}$$

where $r \approx 10\,\mu\text{m}$ is the radius of the cell and $d \approx 50$ Å the thickness of the protein layer on the membrane. Thus proteins on the membrane are about 667-times more concentrated than in cytosol. This makes it much easier to find a partner.

A single myristoyl-group (C14:0) on a model peptide leads to membrane binding with an association constant of $10^4\,\text{M}^{-1}$, this is too low to get stable association. Hence acylated proteins usually have 2 such residues, giving an association constant of about $10^7\,\text{M}^{-1}$. In addition, electrostatic interactions between positively charged amino acids (Lys, Arg or, less often, His) and negatively charged lipids stabilises binding.

There are several populations of rafts which differ in size (50–700 nm diameter), lifetime, and composition. **Caveolae** are one subpopulation of rafts, membrane invaginations containing the protein caveolin. Caveolin inserts from the cytoplasmic side into the membrane, oligomerises, and thus forces a concave shape onto the membrane.

15.2 Physicochemistry of Detergents

Before membrane proteins can be separated from each other they need to be taken out of the membrane. This process is called **solubilisation**. Proteins which are only attached to a membrane may be solubilised by elevated *p*H or salt concentration (100 mM Na_2CO_3 *p*H 11.5 [108], 250 mM KI [275], 2 M NaBr [162]), or by caotropic agents like urea. Even the high sucrose concentrations in gradients during preparative ultracentrifugation can strip sensitive proteins of the membrane surface; isotonic metrizamide gradients may be required for preparation in such cases.

Proteins with transmembrane segments can be solubilised only by **detergents**. Detergents form soluble complexes with membrane proteins, any attempt to remove the detergent will usually lead to aggregation and precipitation of the proteins which are not by themselves water soluble. There are several reviews on detergent properties and solubilisation procedures, including [152, 253, 351].

Detergents (also called "surface active agents" or **surfactants**) are molecules with a hydrophobic hydrocarbon, halocarbon or phenyl-chain, and a hydrophilic head-group. The former interacts with lipids and the hydrophobic parts of proteins, the latter with water. Surface activity of a compound can be defined as **surface pressure** $\pi = \gamma_{\mathrm{w}} - \gamma_{\mathrm{soln}}$, the difference between the surface tension of pure water and a solution of the given compound. π reaches a plateau with increasing concentration of surfactant.

There are cationic, anionic, non-ionic and zwitter-ionic head-groups (see Fig. 15.3). Of these the non-ionic detergents tend to be the mildest, which is

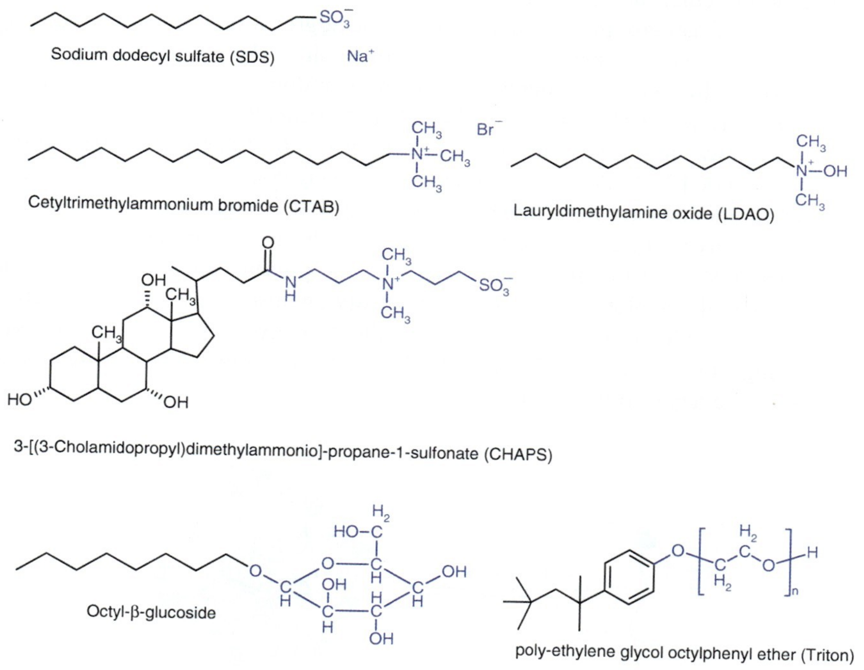

Sodium dodecyl sulfate (SDS)

Cetyltrimethylammonium bromide (CTAB)

Lauryldimethylamine oxide (LDAO)

3-[(3-Cholamidopropyl)dimethylammonio]-propane-1-sulfonate (CHAPS)

Octyl-β-glucoside

poly-ethylene glycol octylphenyl ether (Triton)

Fig. 15.3 All detergents have a lipophilic tail and a hydrophilic head-group. In the case of SDS, the sulphonic acid group bears a negative charge (anionic), while CTAB is positively charged (cationic). LDAO is also positively charged below *p*H 7, but the hydroxy-group looses a proton at higher *p*H, resulting in a zwitter-ion. CHAPS is a zwitter-ionic detergent and at the same time an example for a cholic acid derivative. OG is an example for a nonionic detergent, the hydrophilic head-group is a glucose molecule. Polyethylene glycol based detergents like Triton are probably the ones most often used for protein solubilisation as they are at the same time mild and effective

good for the preservation of protein function, but their ability to dissociate protein complexes, a requirement for the purification of proteins, is limited. Even less dissociating are the fluorocarbons like perfluoro-octanoic acid; they can be used instead of SDS in electrophoresis to study the oligomeric state of proteins [278]. Detergents with strongly polar heads tend to be more dissociating, but also more denaturing.

The hydrophobic chain should have a length of 12–14 carbon atoms, their length then corresponds to the thickness of the hydrophobic part of the biomembrane. Shorter detergents tend to be more inactivating. Detergents with rigid tails are less dissociating than those with flexible tail.

Detergents can be characterised by the following properties:

Critical micellar concentration (cmc) is the concentration where the dissolved monomeric detergent molecules start to aggregate into clusters, the so called **micelles**. It is thus the maximal chemical potential of the detergent that can be achieved (see Figs. 15.4 and 15.7). In the aggregates all hydrophobic tails point to the inside, the hydrophilic heads interact with water. The inside is not water-filled unlike a vesicle. There is a dynamic equilibrium between monomeric detergent molecules and the micelles. Uncharged molecules like sugar, urea, and glycerol increase the cmc; the cmc is also temperature dependent. For ionic detergents, counter-ions decrease the cmc and increase the aggregation number by reducing electrostatic repulsion between detergent molecules. The cmc decreases as the length of the lipophilic tail increases, branching or double bonds increase cmc as they interfere with tight packing of the tails in the micelle.

The easiest way to measure the cmc is to add 5 μM 1,6-diphenyl-1,3,5-hexatriene (DPH) [57] or 10 μM ANS to samples of increasing concentrations of detergent. These compounds are non-fluorescent in water but strongly fluorescent in hydrophobic environments like micelles. Measure base line fluorescence of the dye in water (0 %) and maximal fluorescence in the presence of ≈100 times cmc (100 % signal). A plot of fluorescence intensity against detergent concentration will show a sudden increase at the cmc. Alternatively, surface tension or light scattering can be measured. **Reversed micelles** form in hydrophobic solvents, in that case the head-groups point inward.

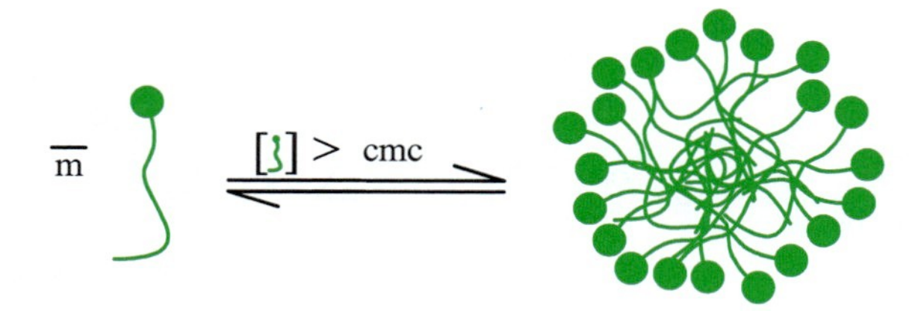

Fig. 15.4 Detergents form aggregates, so called micelles, above a certain concentration. In a micelle the lipophilic tails point to the interior where they are shielded from water, the hydrophilic head-groups are on the surface and interact with water. The concentration at which micelles form (**cmc**) and the size of the micelles $\bar{m}$ depend on both the head-group and the tail length. Note that the interior of the micelle is quite disordered, and the surface rough

Aggregation number $\bar{m}$ is the number of detergent molecules in a micelle . Like the cmc, it depends on temperature and buffer composition, for some detergents also on the detergent concentration. Because of the dynamic nature of the micelle, the aggregation number is best interpreted as size averaged over time.
There is a loose inverse correlation between $\bar{m}$ and cmc (see Fig. 15.5).
Experimentally, the aggregation number may be determined by analytical ultracentrifugation, light scattering, or small angle X-ray scattering. For most detergents it is in the range 50–100. The exception are colic acid derivatives (Fig. 15.6) which have much lower aggregation numbers (two for cholate), the steroid ring systems stick together back-to-back with the hydroxy- and carboxy-groups sticking out at both sides (see Fig. 15.6 for the 3D-structure of cholic acid). These assemblies can form secondary complexes by polar interactions at higher detergent concentrations with $\bar{m}$ between 12 and 100.

Hydrophile–lipophile balance (HLB) expresses the hydrophilicity of a detergent [129, 130]. For oligooxyethylene detergents the HLB can be calculated by HLB = weight % of ethylene oxide/5. The HLB of detergent mixtures can be calcu-

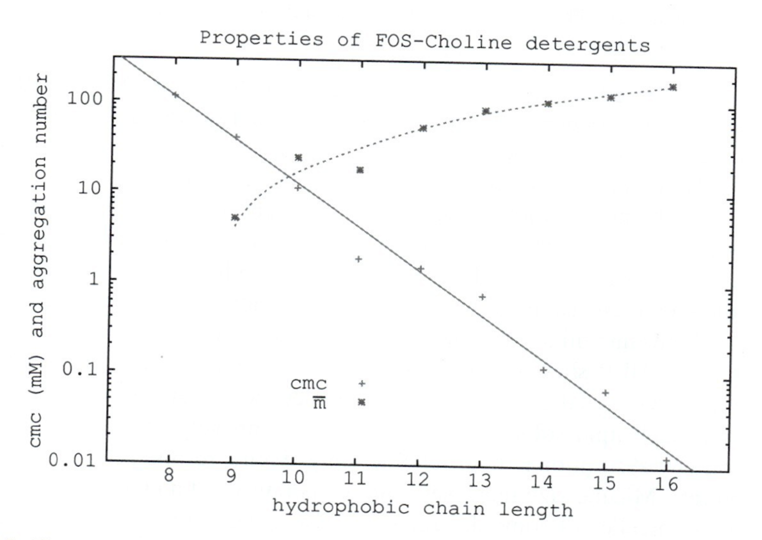

Fig. 15.5 The cmc decreases and the aggregation number ($\bar{m}$) increase with the length of the hydrophobic chain, here for synthetic lysolipids (FosCholine, data from the manufacturer, Anatrace)

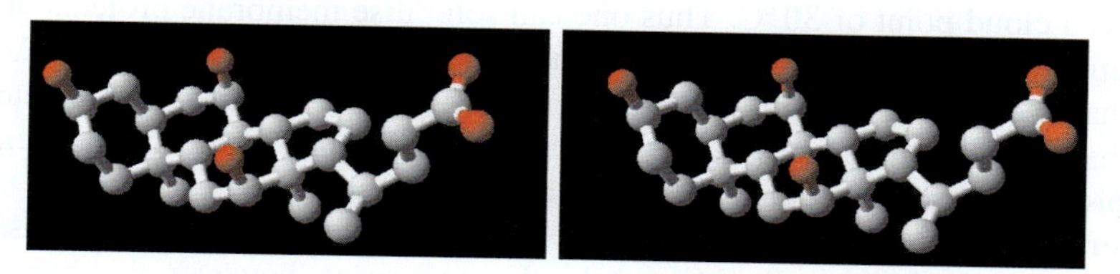

Fig. 15.6 3-D structure of cholic acid (PDB number 2qo4). The molecule has a hydrophilic and a hydrophobic surface, two molecules form a micelle by sticking together with their hydrophobic surfaces

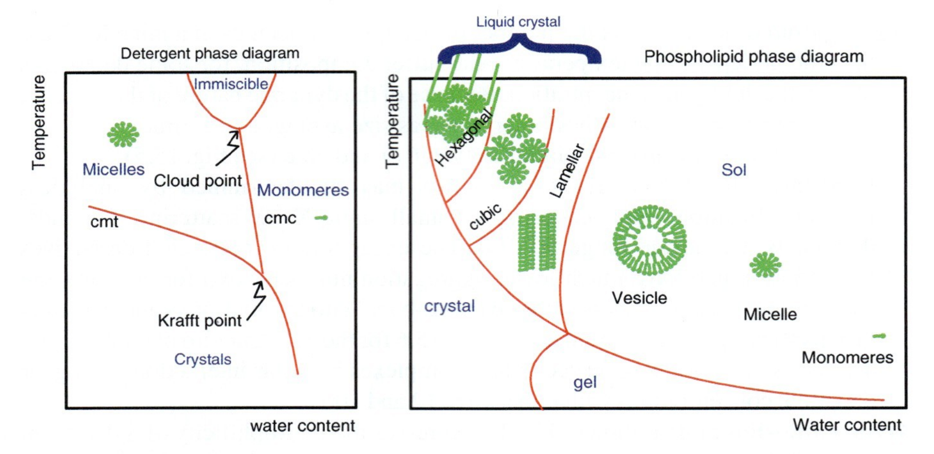

Fig. 15.7 *Left*: Phase diagram of a detergent/buffer mixture. The detergent may exist as either monomer, micelle, or crystal. These phases are separated by the **critical micellar temperature** and **critical micellar concentration**. The lines meet at the triple point, which in this case is called the KRAFFT-point. *Right*: Phase-diagram of lipid-water mixtures. For details see text

lated as weighted average of the component's HLBs. Example: The HLB of a mixture of 25 % Tween 65 (HLB = 10.3) and 75 % Brij 35 (HLB = 16.9) would be $(25\,\% \times 10.3 + 75\,\% \times 16.9)/100\,\% = 15.3$.

The HLB of a detergent influences its rate of flip-flop between the lipid leaflets in the membrane, hydrophilic head-groups prevent it. This slows down the solubilisation of membranes, such detergents will preferentially interact with the protein component as micelles while more lipophilic detergents will partition into the membrane as monomers and, when saturation is reached, preferentially remove lipids into mixed micelles [209]. Thus for solubilising integral membrane proteins the HLB should be between 12–15, for membrane associated proteins detergents with a HLB of 15–20 can abstract the protein from the membrane without much lipid solubilisation. Detergents with very high HLB can not be removed by absorption to hydrophobic (polystyrene) resins.

Cloud point Micelle size increases with temperature. At a characteristic temperature the micelles of some detergents, in particular the oligooxyethylene ones, become so large that phase separation results. For most detergents the cloud point is too high to be useful for protein isolation, the exception is Triton-X114 which has a cloud point of 30 °C. Thus one can solubilise membrane proteins in a solution of Triton-X114 on ice and then gently warm the mixture to 30 °C. After phase separation integral membrane proteins will be enriched in the detergent phase which can be recovered by centrifugation. Despite its theoretical elegance, this method has not found wide application.

Other transitions Many detergent/water mixtures have complicated phase diagrams (including *e.g.*, liquid crystals). In most cases, however, this happens at temperatures too high to be of interest to protein chemists. For a fuller discussion see [223].

Detergents are made in industry for bulk processes and often contain impurities which must be removed if they are to be used for protein solubilisation. In addition, the oligooxyethylene detergents form peroxides when exposed to air, this reaction is catalysed by heavy metals. Purification methods have been described in the literature but are somewhat involved and may use toxic solvents like methylene chloride. Hence it is usually better to purchase biological grade detergents and store them in cold, dark environments protected from air. Purity can be checked by TLC on silica plates with chloroform/methanol/ammonia (63 + 35 + 5 by volume) as mobile phase. Detection is by iodine vapour or by charing with sulphuric acid, in either case only one spot should be visible [90].

The "ideal" detergent has the following properties:

- The cmc is high enough (>1 mM) or the aggregation number low enough (<30) to allow the removal of the detergent by dialysis and ultrafiltration. In addition, a low aggregation number improves the resolution of protein molecular mass determination by gel filtration and analytical ultracentrifugation.
- The partial specific volume $\bar{\nu}$ should be near unity so that the presence of detergents does not influence the results in ultracentrifugation studies. The oligooxyethylene detergents come close to this ideal.
- The average electron density should match that of the buffer used so that small angle X-ray scattering studies can be performed without correction for detergent effects. Such studies are often performed with quaternary ammonium detergents.
- Most important of all, the detergent should efficiently solubilise membrane proteins without inactivation. Such "mild" detergent interestingly do not interact with most soluble proteins (although there are exceptions: serum albumin can bind mild detergents in its hydrophobic pockets). Interaction should be with membrane lipid, not protein.

15.2.1 Detergent Partitioning into Biological Membranes

Lipids and detergents have different molecular shape (see Fig. 15.8); lipids are fairly rod-like while detergents are more conical. A packing factor p can be defined for both lipids and detergents from the molecule volume V_m, the head-group area A_h, and the length of the hydrophobic chain l_c as

$$p = \frac{V_m}{A_h \times l_c}, \tag{15.3}$$

where lipids with their rectangular shape have $p \approx 1$ and in water form double membranes, either as flat sheets or as large vesicles with small curvature. Detergents, on the other hand, have a conical shape, $p \leq 0.5$ and form small micelles with high curvature. It is thus easy to understand that lipid membranes can dissolve only a limited number of detergent molecules before they start to disintegrate into mixed

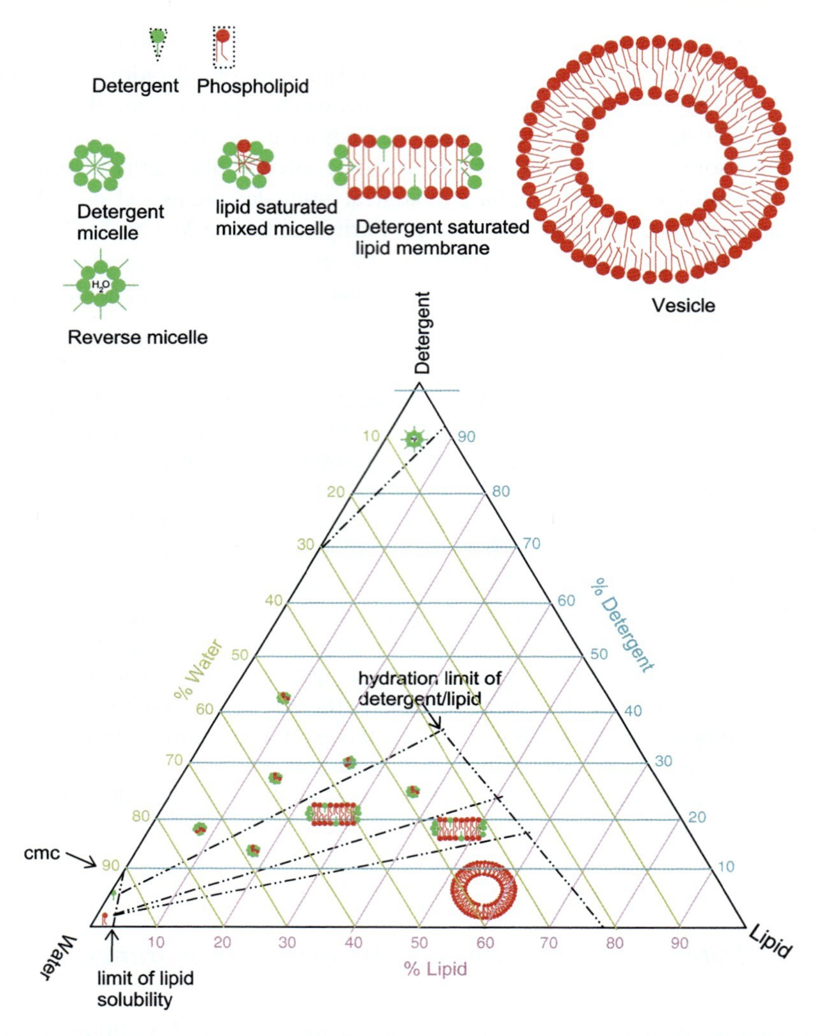

Fig. 15.8 Lipids with their two fatty acid chains have a fairly rod-like shape, in particular when one of the chains is unsaturated and hence kinked. Detergents have only one hydrophobic chain and hence a more conical shape. As a consequence, lipids tend to form flat double membranes or vesicles of large diameter and small curvature. Detergents however, form small, strongly curved micelles.

micelles. On the other hand, mixed micelles can accept only a limited number of lipid molecules. If monomeric detergent is removed from the solution, for example by dialysis, micelles lose detergent molecules, fuse, and eventually form flat double membranes that are protected at the edges by detergent molecules. Once these membranes have reached a critical size they close to form vesicles, initially saturated with detergent. Such vesicles have fluid membranes and are leaky for small molecules.

Further detergent removal from those leads to the formation of tight membranes (liposomes).

The free energy for binding of detergents to lipid membranes and the free energy of micelle formation become equal when $K_p * cmc = 1$ [150]. Large head-groups result in conically shaped detergent molecules; these prefer to form small micelles and the product $K_p * cmc$ is smaller than 1 (about 0.5 to 0.7 for important detergents like SDS, $C_{12}E_8$, Triton or dodecylmaltoside). Such strong detergents will solubilise lipids by their uptake into micelles at a detergent/lipid-ratio of less than 1.

Small head-groups result in rod shaped detergent molecules, these tend to insert into the relatively flat membrane and $K_p * cmc$ will be slightly larger than 1 (2–6 for important detergents like octyl glucoside, decylglucoside or CHAPS). Such weak detergents accumulate in the membrane to a detergent/lipid-ratio larger than 1 before the membrane disintegrates, resulting in solubilisation. In this case we can distinguish high affinity binding of detergent to the membrane at a detergent/lipid-ratio of 1:5 with K_a of $1 \times 10^3 - 1 \times 10^6\,M^{-1}$, and low affinity binding which results in solubilisation.

When detergents are mixed with water at a concentration above the cmc, a phase mixture forms that contains both monomeric (Det_s) and micellar (Det_m) detergent, with

$$[Det]_t = [Det]_s + [Det]_m \tag{15.4}$$

Similarly, the lipid also occurs in monomeric solution and in vesicles. However, since the solubility of lipid monomers in water is very low (200–400 pM for dipalmitoyl phosphatidylcholine depending on temperature), $[Lip]_t \approx [Lip]_m$. Then the molar ratio of detergent to lipid in the membrane is

$$R_e = \frac{[Det]_m}{[Lip]_t} \tag{15.5}$$

and the partition ratio between water and the detergent/lipid mixed micelles becomes

$$K_p = \frac{[Det]_m}{[Det]_s \times ([Det]_m + [Lip]_t)}. \tag{15.6}$$

Since [Water] $\gg$ $[Det]_t$ and [Water] = const,

$$K_p = \frac{R_e}{R_e + 1} \frac{1}{[Det]_s}. \tag{15.7}$$

In a solution of several detergents RAOULTS law applies:

$$X_{i,\mathrm{mon}} = X_{i,\mathrm{mic}} * \mathrm{cmc}_i \tag{15.8}$$

with $X_{i,\mathrm{mon}}$ and $X_{i,\mathrm{mic}}$ mole fraction of ith component in solution and in micelles. cmc_i is the cmc of the pure ith component.

15.3 Detergents in Membrane Protein Isolation

15.3.1 Functional Solubilisation of Proteins

If a detergent is slowly added to a membrane suspension (see Fig. 15.9), detergent molecules dissolve in the membrane plane (detergent solution in lipid, stage I). Even small amounts of detergent have a strong influence on membrane properties like leakiness, fluidity, or the activity of embedded proteins. Detergent binding to membranes is initially hyperbolic, but as saturation is approached it becomes strongly co-operative (stage II, [253]). This concentration dependence is strongly influenced by lipid composition, while the presence of proteins has little influence. During stage II, lipids begin to be removed from the membrane as mixed detergent/lipid micelles. Once the membrane is saturated with detergent, mixed detergent–lipid–protein micelles form (lipid dissolved in detergent, stage III). This last stage of solubilisation requires detergent concentrations higher than the cmc.

Contrary to membrane fragments, the mixed micelles can not be pelleted by centrifugation for 1 h at 100 000 g, which is the operational criterium for solubilisation. Light scattering may also be used to monitor solubilisation. Glycolipids and cholesterol form detergent-insoluble **rafts** [377], some membrane proteins are specifically found in those. These rafts are recovered in the pellet after centrifugation of solubilised membranes.

The structure of the mixed detergent/lipid/protein micelle has been the subject of intensive debate. It now appears that the detergents (and lipids) assemble around the hydrophobic transmembrane section of the protein, forming an **prolate monolayer ring** or **semielliptical torus**. In these, the detergent molecules form a right angle with the protein surface in the middle of the transmembrane segments, and are parallel to the protein surface at the upper and lower ends of the transmembrane segments (see Fig. 15.9). The thickness of the detergent layer parallel to the transmembrane helices in single crystal neutron scattering studies is about 2.5–3.0 nm, which corresponds to the thickness of the hydrophobic part of the double membrane. Perpendicular to the transmembrane helices the thickness is 1.5–2.0 nm, consistent with the size of detergent molecules. This structure satisfies physico-chemical constraints and the calculated binding capacities of proteins for detergent match experimental values. Protein bound detergents behave in this model like detergents at the air-water interface, but quite unlike those in a detergent micelle.

Detergents with very hydrophilic or bulky head-groups can not flip from the outer membrane leaflet to the inner. Thus they solubilise membranes slower, solubilisation occurs by lipid extraction into detergent micelles rather than by partitioning of detergent monomers into the membrane, and requires detergent concentrations higher than the cmc. This may be one reason why structural damage to proteins tends to be higher with detergents like SDS or octyl glucoside than with the oligooxyethylene detergents.

Proteins may stay active in mixed micelles unless the detergent replaces closely bound lipid molecules, which membrane proteins need to maintain their structure (called **annular** or **boundary** lipids as opposed to the **bulk** lipids which may be

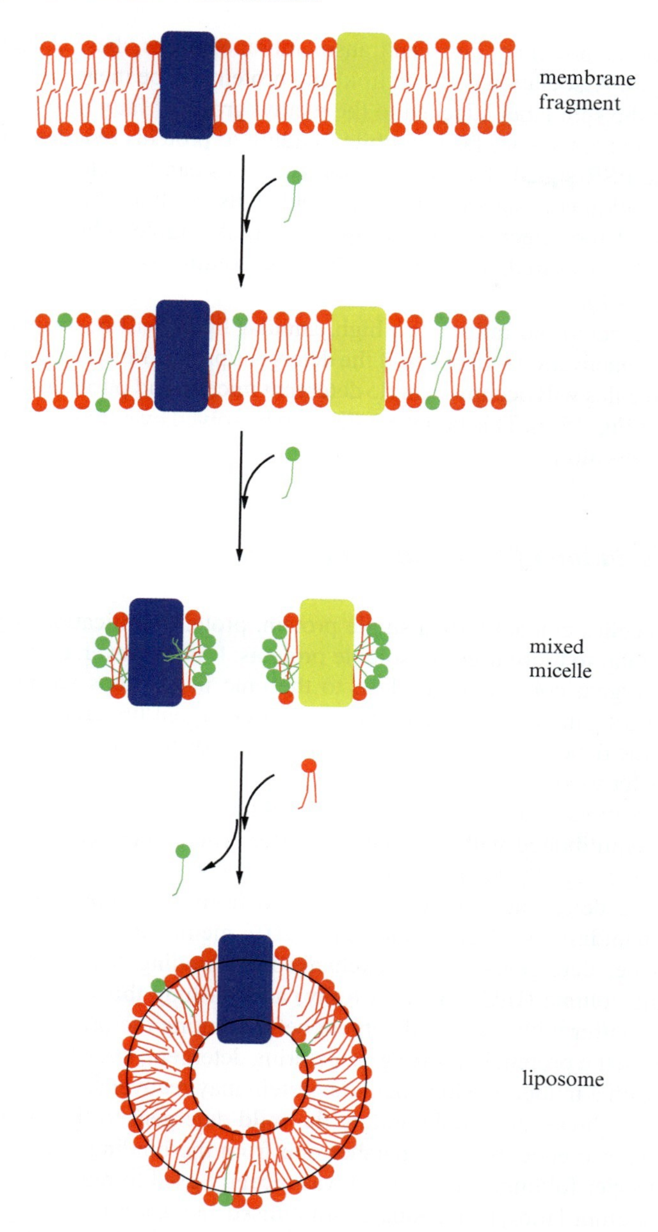

Fig. 15.9 Solubilisation of membrane proteins by detergents (*green*), followed by purification and reconstitution. For details see text

removed with impunity). Annular lipids adapt their conformation to the irregular shape of the protein surface; many detergent molecules can't do this. Note, however, that there is rapid exchange between annular and bulk lipids. **Lipid cofactors** occur in crevices of the protein or between transmembrane helices, these do not exchange with bulk lipid. Association between lipid and protein can be measured by ESR (see p. 303) with the spin label attached to the methyl group of fatty acids. In an undisturbed membrane, they are fairly mobile, binding to proteins reduces mobility and broadens the ESR-signal. Thus association constants can be determined, comparison of association constants for different lipids tells us about the lipid preferences of a protein. If the effect of specific lipids or lipid-soluble substrates on enzyme activity is to be measured, use surface rather than volume concentrations, *i.e.*, mol% ($\frac{[S]}{[S]+[L]} \times 100\%$).

If the detergent/lipid ratio is too high, annular lipids will be stripped from the proteins and inactivation will occur. If the detergent/protein ratio is too small, several protein molecules will occupy a single detergent micelle and purification will not be possible (see Fig. 15.9). This target conflict can be solved by adding extraneous lipid during solubilisation.

15.3.2 Isolation of Solubilised Proteins

If mixed micelles contain only a single protein, protein purification is possible by methods similar to those used for soluble proteins. It is, however, essential to maintain the detergent concentration close to the cmc in all steps so that the mixed micelles do not gain or loose detergent and their detergent/lipid ratio stays constant. Since the cmc depends on buffer composition it needs to be determined separately for each buffer used.

Detergent monomers show unspecific binding to gel filtration columns; these need to be equilibrated with the running buffer (containing detergent at the cmc) for a long time, *e.g.*, by running over night at low flow rates or even with recycling buffer. Bound detergent may be later removed from the column by washing first with 6 M guanidinium chloride followed by 20 % methanol.

If need be, detergents can be exchanged by binding the protein to a chromatographic column (IEC, HIC or affinity) and washing this column with several volumes of buffer with the new detergent before eluting the protein. This way one can solubilise the protein in a strong, denaturing detergent, and after purification replace this with a milder one in which the protein may recover structure and activity. In some cases however, simply adding the mild detergent to the protein preparation may replace enough of the harsher one from the binding sites of the protein to enable proper folding. This method was recently used to renature proteins from sarkosyl (sodium lauroyl sarcosinate) solubilised inclusion bodies by addition of Triton-X100 and CHAPS [407].

Amphipols are hydrophilic macromolecules (polyacrylate) to which hydrophobic side chains are bound. They can not solubilise proteins out of a membrane, but they can keep solubilised proteins in a monomeric state in the absence of detergent. Usually this complex is enzymatically inactive, apparently its structure is too rigid to allow conformational changes during enzymatic turnover. However, the enzyme may become reactivated upon addition of lipid or detergent. Whether or not amphipols can play a role in the purification of membrane proteins remains to be seen. The idea to stabilise the monomeric state and native conformation of a protein during the strains of purification with such a molecule seems interesting enough.

15.3.3 Reconstitution of Proteins into Model Membranes

Once purification has been achieved, proteins are reconstituted into liposomes by the addition of lipid and the removal of detergent. Under suitable conditions, the lipids form closed membrane vesicles (**liposomes**), and the proteins insert into them.

Detergent removal used to be done by dialysis; this was possible only with detergents which have a high enough cmc (>1 mM), so that the concentration of monomeric, membrane permeable detergent allowed efficient dialysis. Many valuable detergents like $C_{12}E_8$ and Triton-X100, however, have a cmc too low for dialysis, and their micelles are too large to pass semipermeable membranes. Gelfiltration is another method, as the mixture of protein, lipids, and detergents passes the gel, small detergent molecules are retained on the gel while the large liposomes leave the column with the void volume. Unfortunately, the resulting liposomes are quite heterogeneous in size.

15.3.3.1 Biobeads SM-2

The introduction of macroporous, hydrophobic polystyrene beads which bind the detergent when added to the reconstitution mixture has given a new momentum to the field. Biobeads SM-2 (BioRad) are most often used for this purpose [353]. Amberlite XAD resins may be an alternative, however, I have no experience with them and will therefore not comment.

Biobeads SM-2 are made from a macroporous styrene/divinylbenzene copolymer with 750 μm beads and pores of 90 Å. The surface area is $300\,m^2/g$. Since the diameter of detergent micelles is 20–40 Å for ionic and 90–100 Å for non-ionic detergents, removal is effective even for low-cmc detergents. Usually less than 20 μM detergent are left after equilibration with a sufficient amount of beads. Lipid/detergent mixed micelles have about twice the diameter of pure detergent micelles, liposomes have a diameter of 800–2 000 Å. Thus lipid losses on Biobeads is small, the binding capacity is about 1 % of that for detergent (80–200 mg/g). Since Protein/lipid complexes are even larger, protein losses on Biobeads are small.

The binding speed of detergents to Biobeads depends on the available surface area, thus one can reduce the rate of detergent removal by adding the beads in several aliquots. This will also further reduce lipid losses, since most of the Biobeads come into contact with lipid only after vesicle formation. Binding velocity doubles with every 15° rise in temperature (equivalent to an activation energy of 29 kJ/mol). Slow removal of detergent results in larger liposomes with a more homogeneous size distribution. 3–5 h seem to be optimal for the entire process which compares favourably with a dialysis time of several weeks.

During detergent removal the mixed detergent–lipid–protein micelles shrink; this increases their curvature and the exposure of hydrophobic domains to the aqueous medium. To avoid the physico-chemical penalties associated with this, micelles fuse into disk-like structures whose edges are covered by detergent. As these sheets grow larger, the amplitude of their random bending motion increases, until closure occurs. These proto-liposomes continue to fuse and to lose detergent. Thus liposome formation is the reversal of membrane solubilisation, a satisfying result. The size of the final liposomes depends on the properties of the detergent which influences the shape and composition of the intermediary stages. Another important factor is the rate of detergent removal which should not be done faster than micelle fusion and post-vesiculation growth can occur.

The most promising way to perform a reconstitution experiment is to add the solubilised protein to pre-formed liposomes to which just enough detergent has been added to get to stage II of the solubilisation process [351]. Under these conditions the liposomes are destabilised enough to allow insertion of proteins but exposure of the protein to large detergent concentration is avoided.

Essential for good results is the preparation of the Biobeads before the experiment [168]: 30 g of beads are stirred with 200 ml methanol, collected on a BÜCHNER-funnel and washed with another 500 ml methanol followed by 1 l of water. Beads are then packed into a column and washed with 2 l of water. The beads are stored under water in the fridge until use.

15.3.3.2 Nano-Disks

Recently introduced as a very exciting tool for membrane protein reconstitution were the so-called nano-disks. These are small monodisperse bilayers (92 Å diameter) whose rim is stabilised by **amphipatic membrane scaffold proteins (MSP)**. These proteins were genetically engineered from lipoproteins by deleting their receptor function, so that only the membrane-organising function remained. Lipid, detergent, MSP, and target protein are mixed with hydrophobic beads, after a few hours the nano-disks can be purified by gel chromatography. That way bacteriorhodopsin could be reconstituted into disks which contained 1 molecule of bacteriorhodopsin, 163 molecules of lipid, and 2 molecules of MSP each [19]. It is obvious that such well-defined structures make the interpretation of physico-chemical experiments much easier.

15.3.3.3 Reversed Micelles

In water, detergents form micelles with the hydrophobic tails pointing inward and the hydrophilic head-groups pointing outward, in contact with the solvent. If, however, detergents are dissolved in hydrophobic solvents like hexane, the structure of the micelles is reversed, with the head-groups pointing inward, solubilizing a small amount of aqueous phase. The tails point out into the solvent.

If non-ionic detergents are used to make reversed micelles, they solubilise little protein from an aqueous sample. However, by addition of a judicious amount of charged detergents one can bring proteins into the micelle by ionic bond formation. Alternatively, one may use a small amount of glucoside-detergent to enrich lectins into such reverse micelles [374]. Although this approach is theoretically attractive, it has seen very little practical application.

15.4 Developing a Solubilisation Protocol

If you need to develop a procedure for membrane protein isolation, you can estimate the amount of detergent required for solubilisation by the following considerations:

There should be at most one protein molecule per detergent micelle. Since the number of protein molecules per micelle will be (in first approximation) POISSON-distributed , you need about 10 detergent micelles per protein molecule. For the average molecular mass of membrane proteins you can assume 100 kDa, thus a 1 mg/ml protein preparation is about 10 μM, and you need a micelle concentration of 100 μM to solubilise them.

One detergent micelle can solubilise about ten lipid molecules. In first approximation, you can assume that a preparation containing 1 mg/ml membrane proteins will be about 1 mg/ml in lipid (actually somewhat more), with an average molecular mass of 750 Da. Thus the 1 mg/ml protein preparation will be about 1.3 mM in lipid and you need a micelle concentration of 130 μM to solubilise them.

To solubilise both lipid and protein from a 1 mg/ml membrane protein preparation, you need a micelle concentration of 230 μM. The next thing you need is the number of detergent molecules per micelle (aggregation number, $\bar{m}$), the molecular mass of the detergent and its critical micellar concentration. These data you can find in the literature of chemical supply companies. For example, CTAB has a molecular mass of 364.5 Da, a cmc of 1 mM and $\bar{m} \approx 170$ detergent molecules per micelle.

To get a micelle concentration of 230 μM, you need $\bar{m} * 230\,\mu\text{M} + \text{cmc}$, that is 40.1 mM in this case, or 14.9 mg/ml.

This calculation uses a lot of handwaving, but gives quite reasonable results (see Fig. 15.10). The next step is to test things in the lab. For this purpose, do small scale solubilisation experiments (about 150–200 μl) with 0.25, 0.50, 1.0, 2.0, and 4.0 times the calculated detergent concentration. Stir the samples on ice for 60 min, take an aliquot for enzyme activity determination and spin the remainder at 100 000 g for 30 min (desktop-ultracentrifuge with rotor for 0.2 ml tubes). Determine the protein concentration and the enzymatic activity in both the supernatant and the pellet. It can be useful to repeat the activity determinations after a weekend to account for the time required later for purification. Repeat this experiment for a number of different detergents.

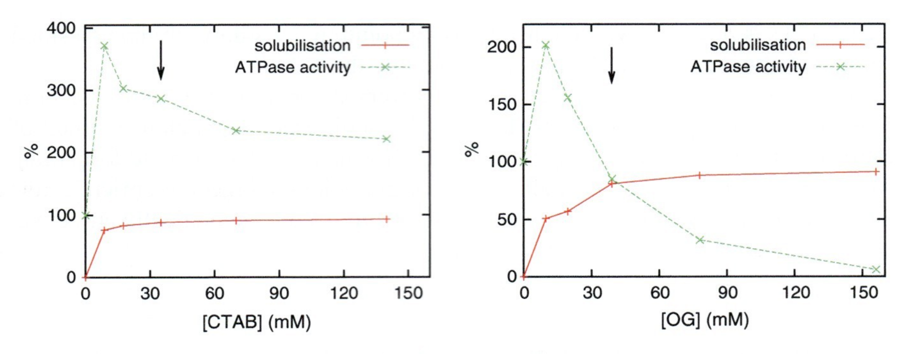

Fig. 15.10 Solubilisation of Mdr1 (P-glycoprotein, an ABC-type transport ATPase) with two different detergents, cetyltrimethylammonium bromide (CTAB, *left*) and octyl glucoside (OG, *right*). In both cases the addition of small concentrations of detergent initially increases enzyme activity (*green dashed lines*). However, increasing OG concentration leads to a rapid inactivation of Mdr1. CTAB leads to a much slower decline in ATPase activity, solubilisation (*red solid lines*) can be achieved at concentrations that do not cause substantial inactivation. The concentration calculated to achieve solubilisation are marked by *arrows*

What you would like to find is a detergent that is able to almost completely solubilise your protein (ratio of protein amount in solution/pellet) without inactivation (enzymatic activity in the aliquot taken before centrifugation). Osmolytes like sucrose or glycerol can protect proteins during solubilisation. Proteins may be more sensitive to proteolytic attack in the solubilised state, appropriate protection is required (*vide supra*). In addition, alkylamines and polyamines can increase the solubilisation yield [441] (about 100 mM), probably by destabilising lipid membranes. They should be removed as soon as possible after solubilisation, however, as they can denature proteins upon prolonged contact.

In practice you will find that small detergent concentrations tend to activate membrane enzymes, this may be caused by increased membrane fluidity. As the detergent concentration increases, inactivation is often seen. You need to find a detergent that can solubilise your protein at a concentration where it is not yet inhibitory.

For example, the following results were obtained with different detergents for the solubilisation of Mdr1 (for other proteins, results may be completely different):

Total inactivation even at low concentration: SDS

Inactivation at solubilising concentration: CHAPS, Zwittergent, cholate, desoxycholate, dodecylmaltoside, octyl glucoside, Hecameg, Lubrol WX, Tween 20, Brij 35

At least 100% activity at solubilising concentration: CTAB, Triton-X100, $C_{12}E_8$, C_8E_5

These experiments have told you which detergents can be used at which concentration to solubilise your protein without (too much) damage. In the next step, try to add lipid (and more detergent) until you can separate your protein from others (*i.e.*: each protein is in a single micelle). Rate zonal centrifugation on glycerol or sucrose gradients is a simple method to use; if you analyse your fractions by SDS–PAGE, your protein should be enriched from proteins of lower and higher molecular mass in some fractions. Alternatively, you can incubate the solubilised protein with glutardialdehyde and look for cross-linking by SDS–PAGE or SEC-HPLC. Cross-linking indicates that there is more than one protein per micelle, and that the detergent/protein-ratio needs to be increased [254]. For 1 mg/ml protein, the addition of 2–3 mg/ml lipid should usually suffice.

Note that in all buffers and gradients you need to include your detergent at the cmc to maintain the protein/detergent/lipid micelles. The cmc, however, depends on the composition of the buffer and should be determined as described on p. 169.

15.5 Membrane Lipids: Preparation, Analysis and Handling

For the solubilisation and reconstitution of membrane proteins, lipids are required. These can be isolated from minced tissues by extraction with chloroform/methanol [31], or alternatively with either petrol ether /i-propanol [340] or with cyclohexane /ethylacetate [18], which are less toxic. The organic phase is extracted with a dilute solution of $MgCl_2$ (0.034 %, FOLCH-partitioning,[105]) and dried with anhydrous sodium sulphate. After evaporation of the solvent the lipid is dissolved in chloroform or cyclohexane and filtered. From this crude lipid preparations the phospholipids are precipitated several times with acetone [11] and extracted with diethyl ether. This preparation can either be used directly as “acetone-precipitated, ether extracted phospholipids”, or individual components purified by column chromatography [189, 356]. Note that all steps need to be done in the absence of oxygen, which would form peroxides with unsaturated fatty acids. These peroxides can inactivate proteins.

Storage should be below −20 °C in solution of no more than 10 % (although chloroform is often used as solvent, cyclohexane is less toxic and easier to remove by lyophilisation). Immediately before use, an aliquot of the solution is lyophilised, the lipids are then taken up in buffer using ultrasound for homogenisation.

15.5.1 Measurements with Lipids and Membranes

15.5.1.1 Analysis of Lipid Composition

Analysis of lipids used to be done by 2-D thin layer chromatography on Kieselgel G plates (activated 1 h at 130 °C) with 1-butanol/acetic acid/water 6 + 2 + 2 for the first dimension and chloroform/methanol/ammonia 50 + 40 + 5 for the second. Spots were made visible successively with iodine vapours (lipophilic substances yellow), spraying with ninhydrin (primary amines red), and ammonium molybdate (phospholipids blue). To determine the nature of the fatty acids present, isolated lipids were hydrolysed and the fatty acids converted to their methyl esters for gas

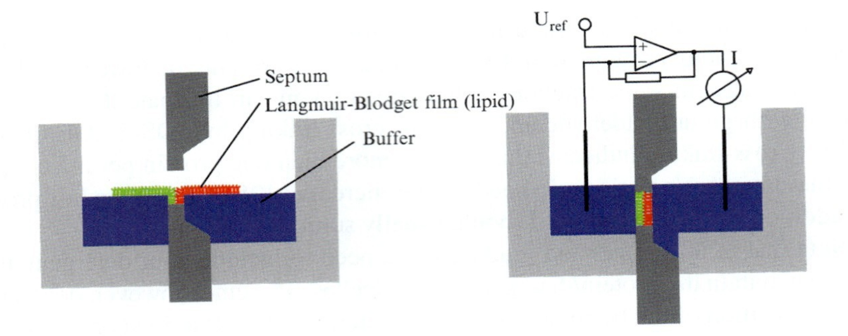

Fig. 15.11 A black lipid membrane is drawn from two LANGMUIR–BLODGETT-films in the orifice of a septum. The procedure allows the membrane to have different composition in both leaflets

chromatography. Today, a lot of the work is done by liquid chromatography coupled mass spectrometry (LC-MS). Blotting of lipids onto PVDF membranes for immunostaining or MS has been described [403].

15.5.1.2 Black Lipid Membranes (BLM)

Black lipid membranes are double membranes in a small (1 mm diameter) orifice of a teflon septum (see Fig. 15.11). Because of destructive interference between the light reflected from the outer and inner leaflet of the membrane, BLMs appear black. The septum separates 2 buffer tanks, the surfaces of the buffers are covered by a single layer of lipid molecules (LANGMUIR–BLODGETT-film) [282]. In such a film the lipid molecules are oriented with their hydrophobic tail into the air and their head-group in the water. As the orifice is moved through that film, a bilayered lipid membrane forms. The experimenter is free to select the lipid composition of both films, and hence membrane leaflets (unlike membranes produced by the earlier "painting" technique [171, 285]). Once the film has formed, its conductivity can be studied by sending a defined current through it and measuring the voltage required (current clamp), or by establishing a defined voltage across it and measuring the current (voltage clamp). This method is used to study the effects of antibiotics and small peptides on the membrane. To increase the signal/noise ratio, patch-clamping can be used (see p. 404).

15.5.1.3 Thermodynamics of Lipids and Lipid–Protein Interactions

Phase changes in lipids can be followed by differential scanning calorimetry (DSC) (see p. 350), interactions between proteins and membranes by DSC, isothermal titration calorimetry (ITC) (see p. 412), or surface plasmon resonance (SPR) (see p. 374).

Chapter 16
Determination of Protein Concentration

Given the importance of proteins, it is perhaps surprising that there is no universally applicable, *absolute* method to quantify protein concentration. The methods in common use are *relative* methods, their result is influenced by the amino acid composition of the protein in question. Quantitation is usually obtained by comparison of the signal with that of a reference protein. The reference protein should be cheap and easy to obtain in consistent quality, and its signal/concentration ratio should be similar to that of the protein to be analysed. Most commonly used are bovine serum albumin (BSA) and immunoglobulin. Simply by choice of assay method and standard protein, one can change the resulting protein concentration by a factor of 20! This needs to be kept in mind when interpreting results. The following methods are in common use:

UV-absorbtion Aromatic amino acids absorb UV-light between 260–290 nm, 280 nm is most commonly used. The method is simple and does not destroy the sample. Results depend on the relative amount of aromatic amino acids in the protein. If the amino acid composition of the protein in question is known, its molar extinction coefficient can be calculated by two methods:

PERKINS The absorbance is calculated from the number of Trp, Tyr and Cys-disulphide residues in the protein molecule: ϵ_{280} = 5550*Trp + 1340*Tyr + 150*Cystine [l mol^{-1} cm^{-1}]. Environmental effects of protein structure on the absorbtion of these residues lead to a small error, usually below 1 % [322].

GILL AND V. HIPPEL This method tries to take account of the environmental effects by calculating the extinction coefficient for denatured proteins (ϵ_{280} = Trp*5690 + Tyr*1280 + Cystine*120) and then comparing the absorbtion of the protein in 6 M Guanidinium*HCl with that of the native protein [115].

In my experience, results of these two methods agree to within 5 %. Samples must be clear (centrifugation, ultrafiltration) to avoid scattering. Other substances with absorbance at 280 nm interfere, in particular nucleotides and nucleic acids (maximum at 260 nm, but the peak is broad enough to cause problems at 280 nm). It is possible to determine nucleic acids and protein in parallel by measuring at both 260 and 280 nm and stray light at 350 nm [422]:

E. Buxbaum, *Biophysical Chemistry of Proteins: An Introduction to Laboratory Methods*, DOI 10.1007/978-1-4419-7251-4_16,

$$D = \epsilon^{n}_{280} * \epsilon^{p}_{260} - \epsilon^{n}_{260} * \epsilon^{p}_{280}, \tag{16.1}$$

$$f^{n}_{260} = \epsilon^{p}_{280}/D, \tag{16.2}$$

$$f^{n}_{280} = \epsilon^{p}_{260}/D, \tag{16.3}$$

$$f^{p}_{260} = f^{n}_{260} * \frac{\epsilon^{n}_{280}}{\epsilon^{p}_{280}}, \tag{16.4}$$

$$f^{p}_{280} = \frac{1}{\epsilon^{p}_{280}} - f^{n}_{280} * \frac{\epsilon^{n}_{280}}{\epsilon^{p}_{280}}, \tag{16.5}$$

$$[N] = [f^{n}_{280} * (\mathcal{E}_{280} - \mathcal{E}_{350}) - f^{n}_{260} * (\mathcal{E}_{260} - \mathcal{E}_{350})]/D, \tag{16.6}$$

$$[P] = [f^{p}_{280} * (\mathcal{E}_{280} - \mathcal{E}_{350}) - f^{p}_{260} * (\mathcal{E}_{260} - \mathcal{E}_{350})]/D. \tag{16.7}$$

The peptide bond absorbs around 215 nm, but so many substances interfere that this wavelength is used only for sensitive detection during chromatography. One problem encountered are chemicals extracted both with water and with organic solvents from plastic tubes that lead to considerable absorbance at 220 and 260 nm [234]. The use of reagent blanks can reduce measurement errors resulting from this.

Biuret is the reaction product of peptides with Cu^{2+} in alkaline solution (see Fig. 16.1). The copper ions are reduced to Cu^{+}, which form a blue-purple complex with peptide bonds. The method is not very sensitive and reducing or complex forming agents interfere. However, in the absence of such interference, colour yield is proportional to the concentration of peptide bonds in the protein.

Lowry et al. Biuret formation is followed by incubation with Folin–Ciocalteu's reagent [243]. This contains Mo(VI), which reacts with reducing compounds like Cu^{+} by formation of molybdenum blue, a huge complex containing 179 410 Mo(IV) and Mo(V) ions [238]. Additionally Tyr (and to a lesser extend Trp, Cys, His and Asn) also react with the reagent, hence colour yield depends on amino acid composition of the protein. Phenols, reducing and complex forming compounds interfere with the assay. Unspecific interference (changes in colour yield or absorbance spectrum of the reaction product) is observed by salts, buffers, sucrose, glycerol, urea, some detergents, and several other compounds that occur frequently.

A further amplification of the signal can be obtained by binding of malachite green or Auramin O to the molybdenum blue [364].

Bicinchonic acid (BCA) This assay too is an amplification of the biuret method. The Cu^{+}-ions react with bicinchonic acid forming a bright red complex [382]. This method is very sensitive and can be used in the presence of detergents. Because it effectively detects peptide bonds, its protein-to-protein variability is lower than many other methods. However, Cys, Trp and Tyr are able to reduce copper under the conditions of the assay. Reducing agents and complex formers interfere, strong oxidising reagent like H_2O_2 bring Cu^{+} back to Cu^{2+}. Unspecific interference is less than with Lowry. The speed of the reaction and the

Fig. 16.1 Determination of proteins by formation of biuret. In alkaline solution, Cu^{2+} is reduced to Cu^{+} by the protein, this forms a purple complex with the peptide bonds of the protein

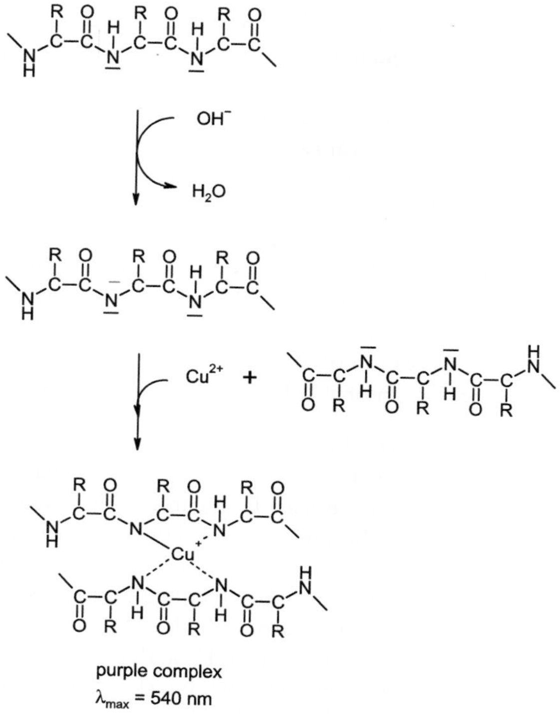

purple complex
λ_{max} = 540 nm

colour yield are sensitive to temperature which must be carefully controlled. Because the colour in the BCA reaction comes from the reaction of the reagent with the Cu^{+} produced by proteins rather than with the proteins themselves, the reaction may also be used to determine proteins bound to surfaces (*e.g.*, microtiter plates or Western blots) or with protein precipitates.

Dye-binding Several hydrophobic dyes bind to hydrophobic patches of the protein, resulting in a change of absorbance or fluorescence spectra. Archetypical for this type of assay is that of M.M. BRADFORD [36], which uses Coomassie Brilliant blue G250 in an acidic solution as reagent. The assay is sensitive and very simple to perform, but interference is caused by any substance that changes protein conformation (urea, detergents). Also, colour yield depends very much on protein composition, apparently Lys and Arg residues form salt bridges with the dye. Protein glycosylation interferes with CBB binding. Because BSA has a lot of hydrophobic patches (for hormone transport), its colour yield is atypically high and it is not suitable as standard protein. Use immunoglobulin instead.
Many other dyes have been suggested for similar photometric assays, but they do not offer a particular advantage. Fluorometric assays use dyes which show no fluorescence in aqueous solution, but strong fluorescence when bound to a hydrophobic patch in a protein, Nile red being one example.

Several proprietary dyes of undisclosed composition have come onto the market for protein quantitation both in solution and in gels. I have a problem with such reagents since the basic requirement for a scientific study is reproducibility. If a reagent of undisclosed composition were withdrawn from the market, reproduction of studies would become impossible.

Amine-reagents react with primary amino-groups yielding fluorescent (*o*-phtaldialdehyde, fluoram) or coloured (ninhydrin) products. Their sensitivity can be markedly increased (and their dependence on amino acid composition lessened) by first hydrolysing the protein into free amino acids. Traditionally, hydrolysis was performed overnight in 6 M HCl at 110 °C, 15 bar. Alkaline hydrolysis, however, does not require such harsh conditions, resulting in better preservation of amino acids [221]. None of these methods has found wide acceptance, probably because they are too labour intensive.

Interference caused by chemicals present in the sample can be avoided by precipitating the protein first. The precipitate is collected by centrifugation, the supernatant with interfering chemicals is discarded. Two methods are commonly used: chloroform/methanol [427] and desoxycholate / trichloroacetic acid [23]. Other reagents (phenol/ether, Coomassie Brilliant blue, calcium phosphate) have not found wide application. Common to all precipitation methods is the problem of re-dissolving the proteins afterwards, especially for high molecular mass and membrane proteins.

One particularly neat method is to dot-blot the protein onto a nitrocellulose membrane. Note that detergents may interfere with protein binding to membranes. The membrane is washed, stained with Coomassie Brilliant blue or Amidoblack and the degree of staining determined either with a scanner, or by extracting the dye from the spot with ethanolic NaOH [157]. Fluorometric determination is also possible after hydrolysis and reaction with o-phtaldialdehyde.

Chapter 17
Cell Culture

Cells from plants and animals – including humans – can be grown in culture if their requirements are met with respect to medium composition and temperature. The following paragraphs will focus on the culture of mammalian cells, even though the culture of insect and plant cells is of considerable economic and scientific interest. More detailed information can be found in [131, 260].

The medium needs to be **isotonic** to the cells and have an ionic composition similar to the interstitial fluid. It also needs to supply nutrients like glucose and amino acids. For **buffering** the medium contains bicarbonate and the incubators have to maintain a partial pressure of CO_2 similar to that in tissue ($\approx 5\,\%$ CO_2). Other media, designed to be used in closed bottles, have a different buffer composition. Phenol red is added to most media so that the *p*H can be checked by eye (red neutral, yellow acidic, purple basic). The addition of 20 mM Hepes can increase the buffering capacity of the medium, but some cell types are inhibited by it.

Some small organic molecules like pyruvate are known to improve cell growth rate. Many hormones, growth factors, and other unknown compounds are required for optimal growth; these are added to the medium in the form of fetal bovine **serum** (FBS, also known as fetal calf serum FCS). Attempts have been made to identify the required compounds and replace the FBS with a chemically defined cocktail. Such serum-free media are available but are more expensive and support less growth than media containing 5–10 % serum.

Apart from such nutrients, **antibiotics** like penicillin, streptomycin, and **antifungals** like amphotericin B are sometimes routinely added to media to prevent yeasts and bacteria from growing. These additions are contentious, however: Sterility should be maintained by aseptic working technique rather than antibiotics.

Incubators maintain a body-like temperature of 37 °C and a high humidity. They also form an enclosed environment where sterility is easier to maintain.

Only few mammalian cell lines can be grown in suspension, most need to attach themselves to a surface. Cell culture dishes are specially treated to make attachment easier, hence they can be used only once. The number of cells grown in culture is limited by the surface area of the dish. Some products are available to increase this surface area, like polysaccharide beads or polyester matrixes. Even using those however the cell density in culture is much lower than that in a body with its efficient supply of nutrients and removal of waste products.

E. Buxbaum, *Biophysical Chemistry of Proteins: An Introduction to Laboratory Methods*, DOI 10.1007/978-1-4419-7251-4_17,

Cells are grown until they cover the available surface area (exchanging the media every 2–3 days). Then they are removed by treatment with trypsin (to digest the extracellular matrix) and EDTA (to remove the bivalent cations necessary for cell attachment). The suspended cells are then seeded into 5–10 new dishes and grown again until confluence. This is called a **passage**. If cells are harvested for studying membrane proteins, the trypsin should be left out (EDTA alone is able to detach cells, it just takes a little bit longer).

Cells which are not needed can be kept frozen in **liquid nitrogen**. For this purpose they are suspended in a special medium with high serum concentration and with 10–20 % glycerol or dimethyl sulphoxide (DMSO) to prevent the formation of ice crystals. Cells need to be frozen slowly (at about 1 °C per min), while thawing should be rapid.

17.1 Cell Types

Cells can be grown directly from tissue samples obtained under sterile conditions from animals or humans. This is called **primary cell culture**. In most cases the cells of interest can be kept alive only for a few days, either because they die or because they are overgrown by fibroblasts.

However, with luck and skill it is sometimes possible to expand an interesting cell type into a stable cell line. These can then be grown for about 20–30 passages until the cells stop multiplying (finite life span of higher organisms, HAYFLICK-limit [148] due to shortening of telomeres during cell division). For that reason it is important to keep frozen stocks at low passage numbers. Stocks of established cell lines can be bought, for example, from the American type culture collection (ATCC), its European counterpart (ECACC), or obtained from colleagues. Scientific ethics requires that researchers make their cell lines available to colleagues, this is best done by submitting them to type culture collections.

Some cancer cells, and established cells treated with certain virus, can be grown indefinitely as they do not stop multiplying at a certain passage number. We speak of **transformed cell lines**. HeLa (from a human cervical cancer) and CHO (from a Chinese hamster ovary carcinoma) are probably the most well known transformed cell lines. Transformation frequently leads to the loss of specialised functions in a cell, which makes them less useful in research. Note, however, that even such immortal cells will change their properties if grown to high passage numbers, results obtained with such cells may differ from those obtained with low passage number cells. In some cases, researchers are interested in immortalising a cell with interesting functions. This can be achieved by fusing the cell with a transformed cell, which needs to be genetically similar (for example from the same mouse strain). One example of this hybridoma technique is the production of monoclonal antibodies from a fusion product of a B-cell with a lymphoma cell (see p. 100).

17.1.1 Contamination of Cell Cultures

Because cell cultures are maintained for a long time with frequent handling, and because in many laboratories more than one cell line is grown, cross-contamination of cell cultures is an all too common problem. In a recent review of 22 studies on 2328 cell lines, 366 (15.6 %) were from a different tissue than stated, 86 (3.7 %) were even from a different species [172], including exotic mix-ups as bird with manta ray. In particular, HeLa cells have a considerable propensity of showing up in places where they do not belong. Therefore, identity of cell lines should be verified (*e.g.*, by short tandem repeat DNA profiling). For research and production of materials regulated by the FDA this is mandatory, some journals now also require it for publication.

In addition, contamination of cell cultures with mycoplasma continues to be a problem; these can be detected easily by PCR. Simpler, but less reliable, is staining of cells with DNA intercalating fluorescent dyes like DAPI, which give infected cells a spotted appearance. Ciprofloxacin or Rifampicin may be used to cure valuable cell lines from mycoplasma infection.

Part III
Protein Modification and Inactivation

A more detailed coverage of protein modification with an extensive bibliography may be found in recent monographs [159, 248]. The Molecular Probes http://probes.invitrogen.com/, Pierce http://www.piercenet.com, and the Thermo Fisher http://www.thermofisher.com catalogues are also useful sources of information. Some of the old tricks may be found in [386], but the experimental conditions employed are often too harsh for enzymological purposes.

Residue-specific reactions are used to introduce labels to proteins and ligands. Examples are radioactive, chemiluminescent, fluorescent or spin labels. They may also be used to bind proteins to a solid matrix, for example, for affinity chromatography (see Chap. 14 on p. 147) surface plasmon resonance measurements (see Sect. 38.8 on p. 374) or to cross-link proteins. Charged, membrane impermeable reagents, and lipophilic, membrane-permeable reagents may be used to study the topology of membrane proteins. On the use of environmentally sensitive probes to follow conformational changes in proteins see the section on fluorescence (p. 39).

If a protein sequence of interest does not contain easily modifiable residues, they can be introduced by site-directed mutagenesis. Note however that even small changes in the size of amino acid residues can potentially have profound structural effects on the protein; the mutants need to be characterised much more thoroughly than is usually done, *i.e.*, by comparing the 3-D- structures of native and mutant proteins.

The thiol-group of Cys, the primary amino-groups of Lys, the protein N-terminus, secondary amines (Trp, His), carboxy-groups (Glu, Asp), guanidino-groups (Arg), phenolic, and alcoholic hydroxy-groups (Tyr, Ser, Thr) are nucleophilic (in decreasing order), and several reagents are available to modify them.

Sugar-residues in glycoproteins may be labelled after conversion of vincinal OH-groups to aldehydes by periodate-oxidation. This reaction is of particular interest because it is less likely to change the properties of the protein than modifications on amino acids, *e.g.*, when coupling antibodies for use as immune reagents. Low concentrations of periodate (<1 mM) and low temperature (on ice) limit the oxidation to sialic acids, at higher concentrations (10 mM) and room temperature other carbohydrates are oxidised also. Sugar oxidases produce aldehydes in specific residues. The aldehydes may then react with primary amine-containing reagents; the resulting

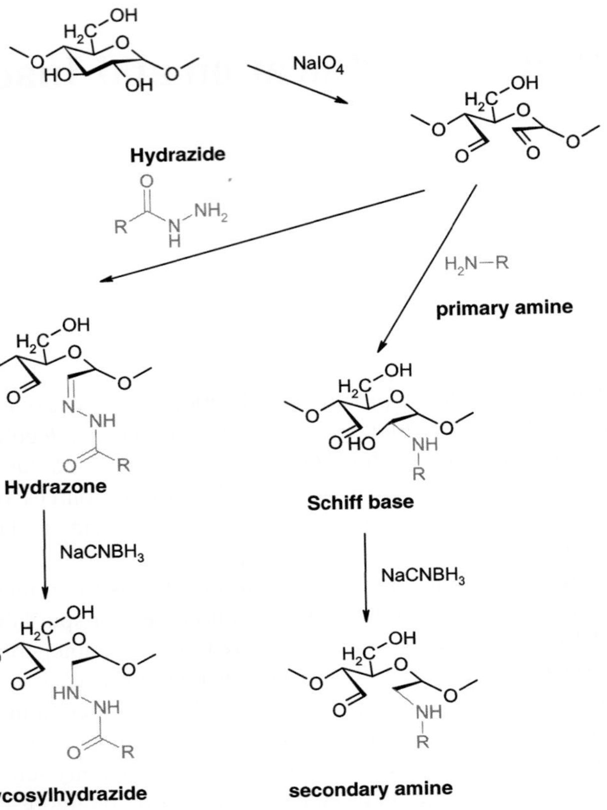

Fig. 1 Labelling of sugar-residues on glycoproteins. The vincinal alcohol-groups are oxidised with sodium periodate to a dialdehyde under ring-opening. The aldehydes can then be labelled using hydrazides or primary amines, followed by reduction of the intermediates with sodium cyanoborohydride

SCHIFF-bases are stabilised by reduction with sodium borohydride (see Fig. 1). Hydrazides may also be used to label aldehydes. Sometimes the reducing ends of oligosaccharides are used for labelling, however, this reaction is very slow due to equilibrium between aldehyde and ring form. Incubation times of a week are not uncommon.

In the following, we will focus mainly on reactions that can be performed under mild conditions which will not lead to gross changes in protein structure.

Chapter 18
General Technical Remarks

An important consideration in protein modification is the length of the spacer-arm. Fluorescence quantum yield of many labels, for example, is reduced if the label is too close to the protein. Steric hindrance may prevent the binding of proteins to affinity matrices if spacer length is insufficient. For special applications labile groups may be introduced into the spacer, for example disulphide-bonds (see p. 219). This allows the label to be removed again during an experiment.

Since some labels are light-sensitive, reactions are performed in the dark, often at 37 °C to accelerate the process. The buffer components need to be chosen so that they are unreactive with the label.

Careful removal of unreacted label from the reaction product is required to prevent it from labelling proteins later. Labelled proteins are usually separated from low molecular mass reagents by size-exclusion chromatography. Often the labelling compound is first reacted with high concentrations (several 100 mM) of a quenching compound like β-mercaptoethanol or ethanolamine.

The modified enzyme may still be able to bind the substrate or even form the product, but affinity for substrate and/or turnover number and/or activation energy may be changed, either positive or negative. Modification of the protein can alter kinetic parameters either

Directly: an amino acid in the binding site is changed, which interferes with (or increases) substrate binding or turnover. This can be by:

- Modification of an amino acid involved in catalysis.
- Steric hindrance of substrate binding or conversion.

Indirectly: binding occurs outside the substrate binding site and changes the conformation of the enzyme and its kinetic parameters.

Addition of substrate to the enzyme/modifier mixture may stop, but not reverse the modification reaction. In some cases the bond to the label can be split by a second chemical reaction, protein activity should be recovered when the label is removed.

Hydrophilic molecules can not pass lipid membranes while hydrophobic molecules can. By the appropriate choice of label it is therefore possible to locate the position of a modified residue in membrane proteins. Also, hydrophobic

E. Buxbaum, *Biophysical Chemistry of Proteins: An Introduction to Laboratory Methods*, DOI 10.1007/978-1-4419-7251-4_18,

molecules tend to bind to the protein core, hydrophilic to the outside. Large and small labels can be used to study accessibility of the labelled site.

18.1 Determining the Specificity of Labelling

Proteins often have more than one binding site for a label. The number of label molecules bound to each target molecule is of critical importance, especially when specific labelling of a particular residue is desired. In the catalytic centre of proteins amino acids like Ser or His may be activated enough to react with reagents that are normally specific for, say, Cys. It is therefore necessary to check specificity of labelling by digestion of the protein and mapping the label to the residue(s).

In order to check whether or not the effort is worthwhile, a TSOU-plot [416] (activity as a function of the number of modified residues) should first be constructed. Let us assume that the enzyme is present initially at concentration $[E]_0$ and that at time t an activity of $[E]$ remains. Then we can define the fractional activity $a = [E]/[E]_0$. Let us further assume that each enzyme molecule contains n groups which can be modified by the reagent used, and that of those s are fast reacting, but non-essential for activity, p react with medium velocity, with i of those essential for enzymatic activity. Then there are $n - s - p$ slow reacting, non-essential groups. Modification of one of the essential groups shall be sufficient for inactivation. The number of groups modified at time t be m, some of those will be essential and some will be not. m can be determined easily if the modifying reagent contains a radioactive marker or a chromophore. If during an inactivation reaction we measure a and m, we get

$$a^{1/i} = \frac{p + s - m}{p}. \tag{18.1}$$

Thus if we plot a, $\sqrt[2]{a}$, $\sqrt[3]{a}\ldots$ against m, in one of the plots data points will be on a straight line, giving us i (see Fig. 18.1). The intersection of that straight line with $a^{1/i} = 1$ gives us s, the slope gives us $1/p$, and the intersection with $a^{1/i} = 0$ gives us $s + p$. If the protein is incubated with high concentrations of the modifier for a long time, a plot of m versus t will approach n as limit for $t = \infty$. For discussion of other inhibition patterns (*e.g.*, with residual activity in the modified enzymes or when modification of more than one essential residue is required for destruction of activity) see the original publication.

18.2 Kinetics of Enzyme Modification

The first to do a thorough, quantitative study on the kinetics of enzyme modification were LEVY, LEBER & RYAN in 1963 [233]. I will use their analysis here, but with a more modern nomenclature.

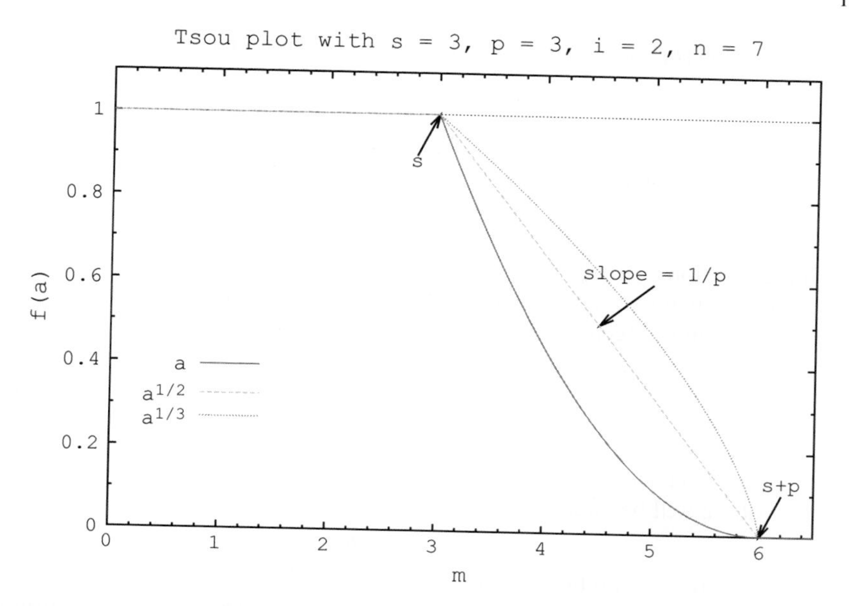

Fig. 18.1 TSOU plot for a hypothetical modification experiment. Since the number of essential residues i was set to 2, plotting $a^{1/2}$ gives a straight line, from the slope and intersection with $a^{1/2} = 1$ and $a^{1/2} = 0$ the number of fast (non-essential) and medium residues, s and p, respectively can be obtained. Note that the number of slow, non-essential residues can not be determined from a TSOU-plot

Let us assume a reaction between an enzyme E and an modifier I:
The modifier first binds reversibly to the enzyme and then a second step irreversibly modifies the protein. The bond formed in the second step is often, but not always, a covalent one. At any given time, there is a certain concentration of unmodified enzyme, $[E]$, which interacts with the modifier, present at concentration $[I]$ according to the law of mass action:

$$[E \cdot I] = K_i * [I]^n * [E], \tag{18.2}$$

where n is the order of the reaction with respect to I. This is the minimal number of modifier molecules bound per molecule of enzyme, the actual number may be higher if some binding sites are saturated throughout the concentration range of I studied.

If we assume the concentration of $[I] \gg [E]$, so that $[I]$ is not changed noticeably during the reaction the rate of modification of the enzyme is

$$\frac{\mathrm{d}[E-I]}{\mathrm{d}t} = -\frac{\mathrm{d}[E]}{\mathrm{d}t} = k_i[E \cdot I] = k_i K_i\, [I]^n\, [E]. \tag{18.3}$$

Since k_i, K_i and $[I]^n$ are constant we can combine their product into a new constant, k' and get

$$\frac{\mathrm{d}[E-I]}{\mathrm{d}t} = k'[E], \tag{18.4}$$

$$[E] = [E]_0 \mathrm{e}^{-k't}, \tag{18.5}$$

which is a reaction of first order.

Before the enzyme was subjected to modification ($t = 0$), the turnover for substrate at saturating concentration was

$$V_0 = -\frac{\mathrm{d}[S]}{\mathrm{d}t} = k_{cat}[E]_0. \tag{18.6}$$

When the reaction between modifier and enzyme is complete ($t = \infty$), the turnover of the substrate will become

$$V_\infty = k^*_{cat}[E]_0 \tag{18.7}$$

with k^*_{cat} the turnover number of modified enzyme molecules. At intermediate times, where only a part of the enzyme is modified ($[E-I] = [E]_0 - [E]$), the turnover of the substrate is

$$\begin{aligned} V_t &= k_{cat}[E] + k^*_{cat}[E-I] \\ &= k_{cat}[E] + k^*_{cat}([E]_0 - [E]), \end{aligned} \tag{18.8}$$

$$V_t - V_\infty = (k_{cat} - k^*_{cat})[E] = (k_{cat} - k^*_{cat})E_0 \mathrm{e}^{-k't} = (V_0 - V_\infty)\mathrm{e}^{-k't}. \tag{18.9}$$

After taking logarithms this becomes

$$\ln(V_t - V_\infty) = \ln(V_0 - V_\infty) - k't. \tag{18.10}$$

Thus plotting $\ln(V_t - V_\infty)$ as a function of t results in a straight line with slope k'.

k' depends on the concentration of the modifier. At a given $[E]_0$ we get

$$k' = k_i K_i [I]^n, \tag{18.11}$$

$$\ln(k') = \ln(k_i K_i) + n \ln([I]). \tag{18.12}$$

Thus plotting $\ln(k')$ (or the half life period of the enzyme $t_{1/2} = \ln(2)/k'$) as function of $\ln([I])$ gives a straight line with slope n.

A substrate may prevent binding of the modifier to the enzyme, and therefore competitively protect against modification. The concentration of free enzyme (without substrate bound) $[E]_f$ can be calculated from the concentration of non-modified enzyme $[E]$ as

$$[E]_f = \frac{[E]}{1 + \frac{[S]}{K_s}}. \tag{18.13}$$

The rate of reaction with the modifier is then determined by $[E]_f$, not $[E]$, thus:

$$-\frac{\mathrm{d}[E]}{\mathrm{d}t} = k'[E]_f = \frac{k'[E]}{1 + \frac{[S]}{K_s}} \tag{18.14}$$

and

$$\ln(V_t - V_\infty) = \ln(V_0 - V_\infty) - \frac{k't}{1 + \frac{[S]}{K_s}}, \tag{18.15}$$

$$t = \ln\left(\frac{V_0 - V_\infty}{V_t - V_\infty}\right)\left(\frac{1}{k'} + \frac{[S]}{K_s k'}\right), \tag{18.16}$$

$$t_{1/2} = \frac{\ln(2)}{k'} + \frac{\ln(2)[S]}{k' K_s}. \tag{18.17}$$

If the half-life periods $t_{1/2}$ are plotted against $[S]$ at constant $[I]$, a straight line is obtained, the x-intercept is $-K_s$.

Alternatively, it is possible to plot $t_{1/2}$ as function of $1/[I]$ at constant $[S]$, this is similar to a LINEWEAVER–BURK-plot. If the experiment is repeated for several $[S]$ the lines all intersect at a common point on the y-axis, indicating competitive inhibition of the inactivation process. It is also possible to have uncompetitive, non-competitive or partial competitive inhibition of the inactivation process, depending on whether the substrate and inactivator can bind to the enzyme at the same time and whether any SE · I-complex can proceed to SE – I.

Chapter 19
Amine-Reactive Reagents

Aliphatic amines (ϵ-amino group of Lys) are moderately nucleophilic when deprotonated (beyond pH 9 for the free amino acid, but often much lower in proteins); on the other hand, the inactivation of the labelling reagent by water increases with pH. The α-amino group has a pK_a of 7.6–8.5, thus the N-terminus of a protein can be labelled specifically around pH 7. Aromatic amines require very reactive probes (thiocyanate, sulphonyl chloride), but can be labelled at any $p\text{H} > 4$. The nitrogen of peptide bonds and Gln, Asn, Arg and His side-chains, and of guanosine and adenosine are almost unreactive. Amino-sugars in glycoproteins may also react.

The buffer used in labelling must not contain free amines (*e.g.*, Tris), ethanolamine may be used to quench unreacted reagent after labelling.

Isocyanates are too unstable to be prepared commercially, however, these compounds can be prepared form the corresponding azides by heating to 80 °C (see Fig. 19.1).

Isothiocyanates are more stable and more commonly used for amine-labelling. However, the thiourea bond is not particularly stable, other chemistries may be preferable, for example in capillary electrophoresis of amines. Isothiocyanates are also used in EDMAN-sequencing of proteins (see p. 271).

Succimidyl esters are quite specific for aliphatic amines in aqueous solutions, hydrolysis by water is low if $p\text{H} < 9$. The resulting amide bond is almost as stable as the peptide bond, they may be used in cases where thioureas are not stable enough. BOLTON–HUNTER-reagent (3-(4-hydroxyphenyl)propionic acid succimidyl ester) is used to introduce phenol-groups in proteins before radioiodination (see Fig. 23.1 on p. 224).

Sulphosuccimidyl esters tend to be more water soluble than succimidyl esters, their higher polarity makes them less likely to react with groups buried in the protein core. Their stability is often too low for commercial preparation, but they can be prepared in the lab by carbodiimide coupling between the corresponding carboxylic acid and N-hydroxysulphosuccinimide (NHSS).

Sulphonyl chlorides are highly reactive and completely hydrolysed by water within a few minutes. Therefore, they are often used at low temperatures. However, once formed the resulting sulphonamide bond is very stable and survives protein hydrolysis (end group analysis with dansyl chloride, WEBER's reagent). In aqueous buffers, reactions of sulphonyl chloride with other reactive groups in

E. Buxbaum, *Biophysical Chemistry of Proteins: An Introduction to Laboratory Methods*, DOI 10.1007/978-1-4419-7251-4_19,

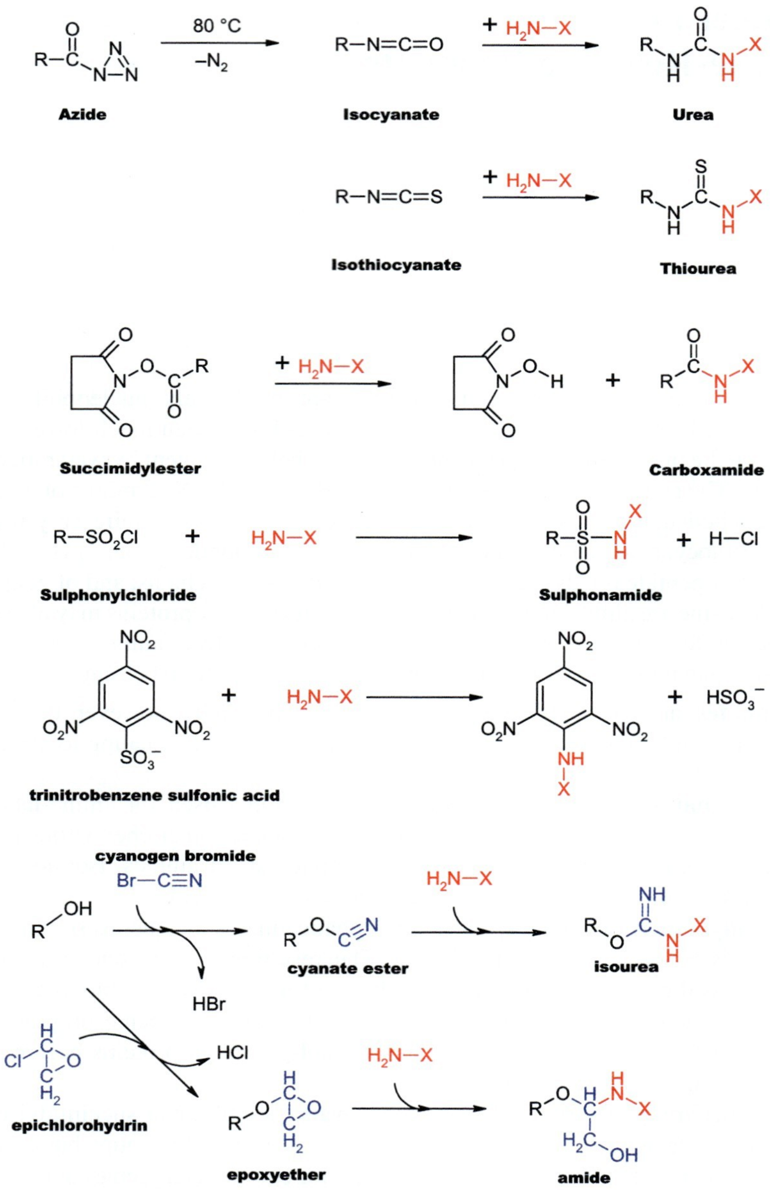

Fig. 19.1 Amine-specific reagents label the ϵ-amino group of lysine and the N-terminus

proteins (Cys, Tyr, His, sugars) is insignificant. Sulphonyl chlorides are unstable in DMSO, hence they can not be stored in that solvent. Sulphonyl fluorides are less reactive than the corresponding chlorides and used for the specific labelling of Ser and Thr groups in the active centre of hydrolases.

NBD-Cl and the more reactive NBD-F are usually used for labelling SH-groups, but also label primary and secondary amines. The fluorescent quantum yield and absorbtion and emission spectra of the resulting compounds depend on the hydrophobicity of the environment.

Aldehydes form SCHIFF bases with primary amines, these can be reduced to stable alkylamine with sodium borohydride or – more specific as it does not reduce disulphide bonds or aldehydes – with sodium cyanoborohydride[1] or dimethylamine borane. The most commonly used aldehyde is **pyridoxal phosphate** , the resulting SCHIFF-base has an absorbtion maximum of 430 nm, after reduction this changes to 325 nm with $\epsilon = 4800\,\mathrm{M^{-1}cm^{-1}}$ for simple model compounds. *ortho*-phtaldialdehyde (OPA), or fluorescamine, form highly fluorescent compounds with amines, which may be used for their quantification and small molecule labelling before chromatography. Acetone can also be used to modify primary amines.

Aldehydes are also used for protein cross-linking. Even methanal (formaldehyde) is actually a bifunctional reagent (see Fig. 19.2). Because of this cross-linking activity it is often used in microscopy as fixative for tissues. Glutardialdehyde (see Fig. 14.3), with two aldehyde groups, is more effective, but also more likely to block antigenic epitopes. Formaldehyde coupling may be used to make antibody-conjugates with peroxidase or phosphatase. ^{14}C- or ^{3}H-Methanal followed by borohydride reduction is used to radio-label proteins (reductive methylation). Heating formalin fixed samples in Tris-buffer with detergent (*e.g.*, SDS) and caotropes (*e.g.*, urea) can partially reverse the cross-linking. After enzymatic digestion the resulting peptides can be identified and quantified by LC/MS/MS [292], allowing proteomics studies on pathology tissue archives.

2,4,6-trinitrobenzene sulphonic acid (TNBS) gives with amino-groups of $pK_a >$ 8.7 an intensively coloured product which can be photometrically determined at 420 nm ($\epsilon = 2.20 \times 10^4\,\mathrm{M^{-1}cm^{-1}}$ for α- and $\epsilon = 1.92 \times 10^4\,\mathrm{M^{-1}cm^{-1}}$ for ε-amino groups in the presence of excess sulphite). There is an isosbestic point at 367 nm with $\epsilon = 1.05 \times 10^4\,\mathrm{M^{-1}cm^{-1}}$. The derivative of α-amino groups is unstable against hydrolysis with hydrochloric acid (8 h at 110 °C) and can thus be differentiated from the ε-derivative. TNBS is not membrane-permeable unlike fluoro-dinitrobenzene.

Ethoxyformic acid (Eth−O−CO−O−CO−O−Eth) has been used to radio-label amino-groups with ^{14}C (100 μM at *p*H 4), binding is reversible by incubation with 100 mM hydroxylamine at *p*H 7 [270].

Cyanogen bromide CNBr is often used to couple amines to hydroxy-groups, for example of agarose. The activated gel is commercially available in lyophilised form, thus no handling of the (very poisonous!) cyanogen bromide is required.

[1] Commercial preparations of sodium cyanoborohydride are often contaminated with interfering material. It should be re-crystallised before use (11 g dissolved in 25 ml acetonitrile, centrifuged and precipitated with 150 ml methylene chloride over night at 4 °C). Store in vacuo and make fresh solutions every day. Nickel or cobalt ions may be used to trap cyanide during reduction with cyanoborohydride.

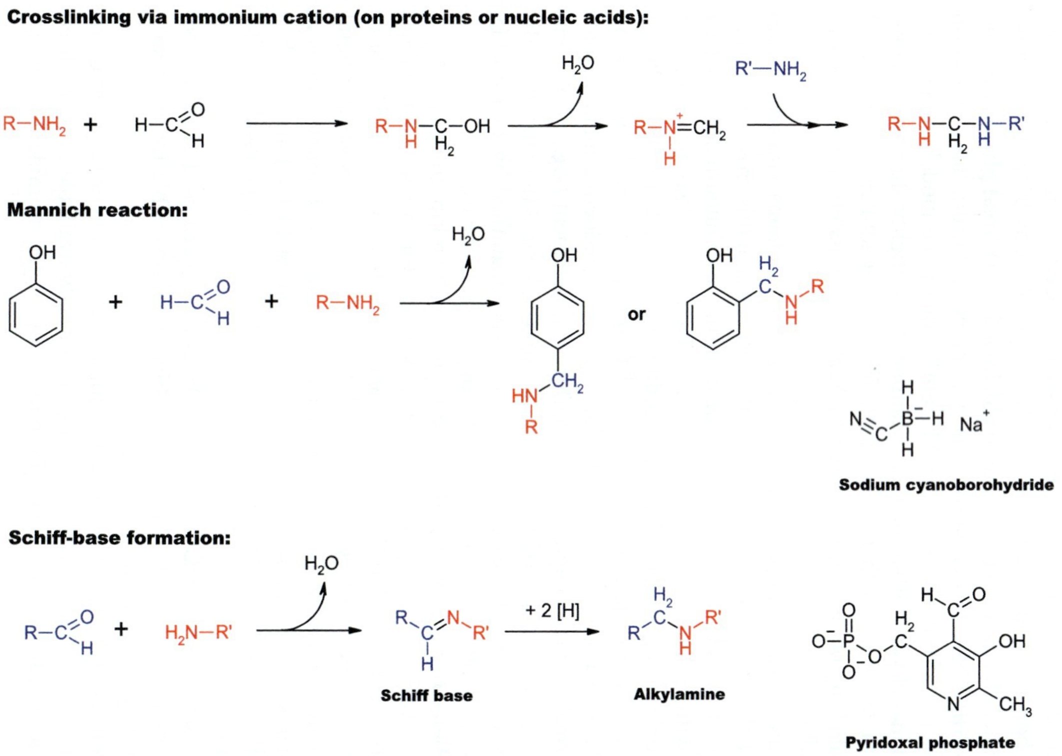

Crosslinking via immonium cation (on proteins or nucleic acids):

Mannich reaction:

or

Sodium cyanoborohydride

Schiff-base formation:

+ 2 [H]

Schiff base

Alkylamine

Pyridoxal phosphate

Fig. 19.2 Methanal (formaldehyde) can cross-link two amino-groups (both on proteins and nucleic acids) via an immonium cation (*top*) or an amino-group with a carbon bearing an active hydrogen (here the *o*- and *p*-hydrogen of phenol) via the MANNICH-reaction (*middle*). This reaction may be used to conjugate compounds that have only an active hydrogen, but no easily modifiable functional groups, to proteins. The reactions can be reversed by heating in Tris-HCl [302]. SCHIFF-base formation (*bottom*) can be used to label a protein with an aldehyde (often pyridoxal phosphate), the bond is stabilised by reduction with sodium cyanoborohydride

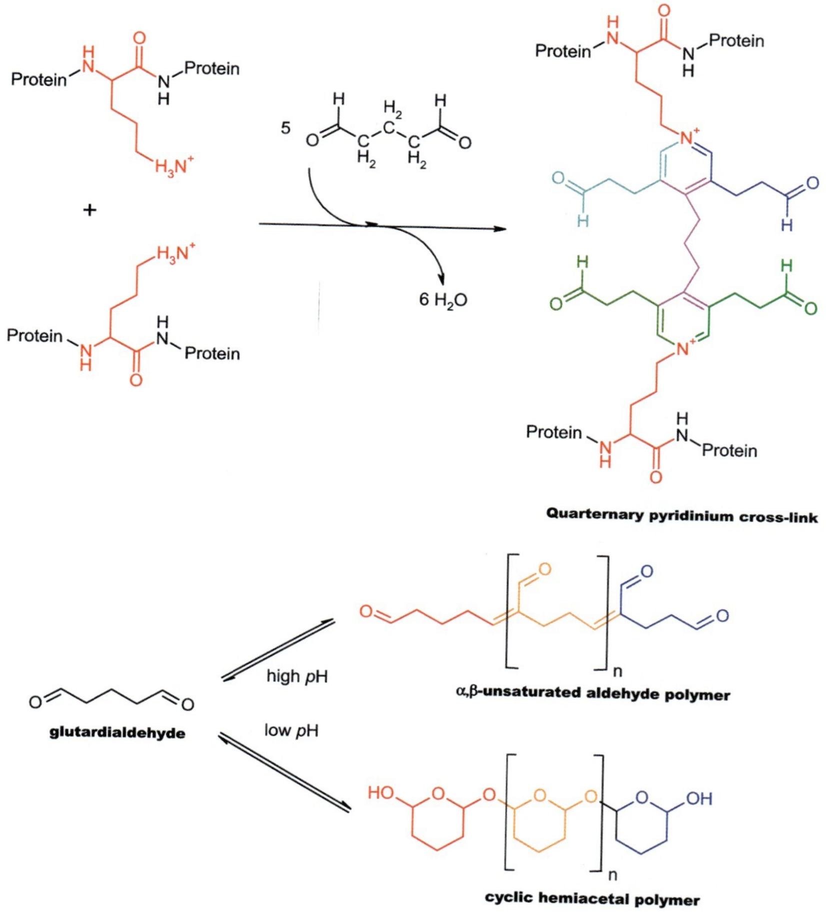

Fig. 19.3 Example for cross-linking of Lys-residues by glutardialdehyde. As a bifunctional cross-linker, glutardialdehyde can form quite complex products in addition to those formed by simple aldehydes

Such pre-activated gels are often used to prepare affinity columns. Although the stability of the isourea bond is not as good as with other chemistries (*e.g.*, epoxy) it is sufficient for most purposes.

Epoxides like epichlorohydrin can also be used to pre-activate agarose gels. They react with any nucleophile ($-NH_2$, $-SH$, $-OH$) to form quite stable bonds.

Dicarboxylic anhydrides react according to $\text{Protein}-NH_3^+ + \text{anhydrid} \rightarrow 2H^+ \text{Protein}-NH-CO-R-COO^-$. The high density of negative charges dissociates proteins, and Lys-labelling protects proteins against trypsin digestion [200].

Chapter 20
Thiol and Disulphide Reactive Reagents

The thiol-group of cysteine is the strongest nucleophile in most proteins, in the active centre of an enzyme the group may be even more reactive. It can be labelled with high efficiency and selectivity under mild conditions (pH 6.5–8.0); the resulting thioethers or thioesters are stable under biologically relevant conditions (see Fig. 20.1). This makes thiol-reactive probes a favourite tool for protein modification. Modified bases like thiouridine also allow their use on nucleic acids. Excess reagent can be quenched with cysteine, DTT, or β-mercaptoethanol before it is separated from the labelled protein by dialysis or gel filtration.

Haloalkanes like iodoacetamides react with thiols, they may also react with protonated amines and hence should be used at $p\text{H} \approx 8$, where aliphatic amines are not protonated. Reactions with His and Met occur only if free thiols are absent. Haloalkanes are light sensitive so the labelling reaction needs to be performed in the dark. However, the labelled protein is stable even when illuminated. If the protein is hydrolysed, the thioether is converted to S-carboxymethyl cysteine under loss of label. Chloroacetamides are less reactive than iodoacetamides and may be used to selectively label particularly reactive SH-groups.

Maleimides, unlike iodoacetamides, cannot react with His, Met and Tyr even under unfavourable conditions. In other respects the two reagents are similar. At $p\text{H} > 8$ maleimides can not only react with amines, but they can also be inactivated to maleamic acid. At $p\text{H} > 9$ ring-opening by nucleophilic reaction with neighbouring amines can cross-link amine- and thiol-groups. Coumarin maleimides are non-fluorescent until they have reacted with thiols and can be used to quantitate free thiols in proteins and small molecules.

Disulphides react with thiols by disulphide interchange ($\text{R–S–S–R} + \text{HS–Protein} \rightleftharpoons \text{R–SH} + \text{Protein–S–S–R}$), this reaction is thiol-specific. Yield is limited since the reaction is fully reversible. The most well known compound of this type is ELLMAN's reagent (5,5-dithiobis-(2-nitrobenzoic) acid, which can be used for the quantitation of free SH-groups in a protein because of the intensive colour of the reaction product ($\epsilon_{412} = 13\,800\,\text{M}^{-1}\text{cm}^{-1}$). The analogous 2,2'-dithio-bis-(5-nitropyridine) is used in the same manner, but has a higher extinction coefficient ($\epsilon_{412} = 19\,800\,\text{M}^{-1}\text{cm}^{-1}$).

E. Buxbaum, *Biophysical Chemistry of Proteins: An Introduction to Laboratory Methods*, DOI 10.1007/978-1-4419-7251-4_20,

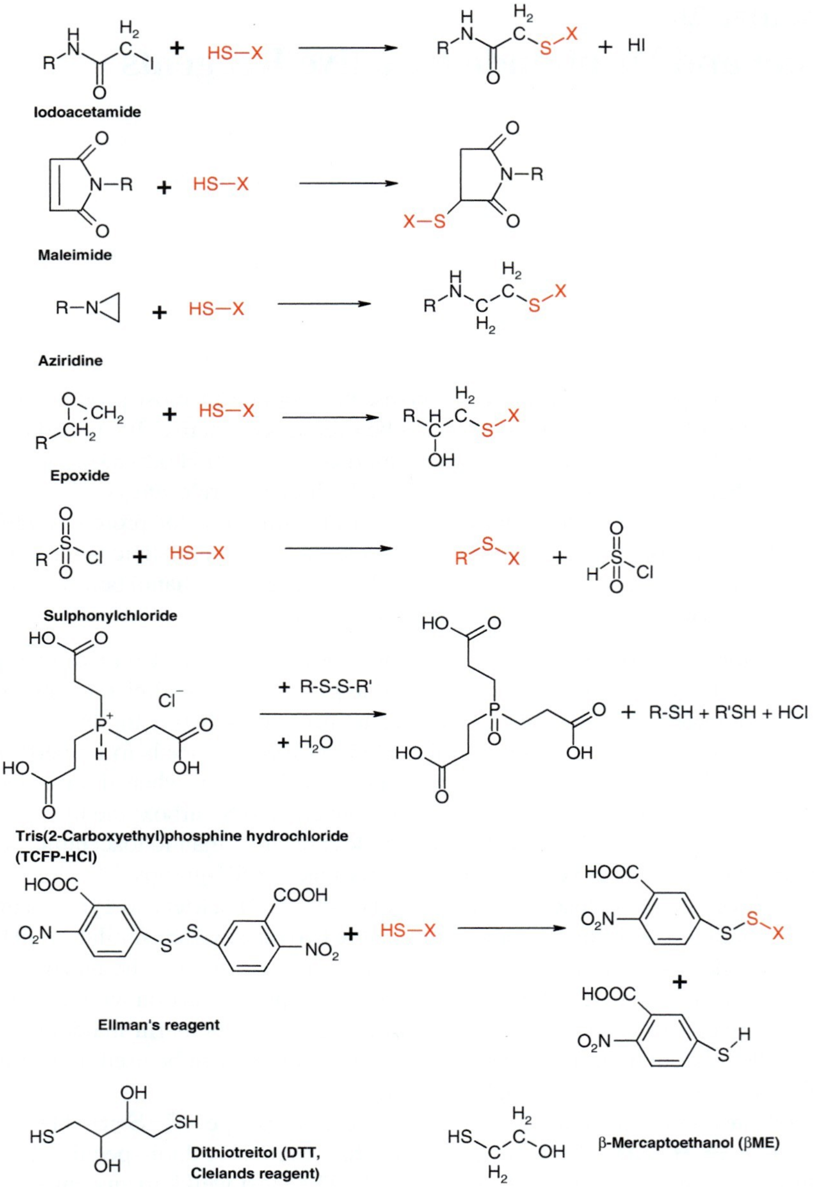
Iodoacetamide
HS–X
+ HI
Maleimide
Aziridine
Epoxide
Sulphonylchloride
Tris(2-Carboxyethyl)phosphine hydrochloride (TCFP-HCl)
+ R-S-S-R'
+ H2O
+ R-SH + R'SH + HCl
Ellman's reagent
Dithiotreitol (DTT, Clelands reagent)
β-Mercaptoethanol (βME)

Fig. 20.1 Thiol-specific reagents used to label cysteine-residues

Because of its reduction potential CLELAND's reagent (dithiotreitol (DTT)) is often used to split disulphide bonds; the initially formed mixed disulphide reacts further so that two Cys are formed. This reaction can be quantified, oxidised DTT has a ϵ_{283} of $273\,M^{-1}cm^{-1}$.

Aziridines Ethyleneimine is used to form *S*-2-aminoethylcysteine in 1 M Tris-HCl *p*H 8.6 by nucleophilic ring opening. This amino acid analogue looks sufficiently like Lys to introduce a trypsin cleavage site into proteins [236].

NBD-Cl and fluoride react with both thiols and amines, the thiol adduct is excited at shorter wavelength. They will react preferentially with thiols if the reaction is carried out at $p\mathrm{H} < 7$, but transfer of probe from the thioether to an amine may occur later. The fluorescence of the label is sensitive to protein conformation and can be used to monitor ligand binding or protein folding.

Epoxides react similar to iodoacetamide. The first step in labelling with iodoacetamide is actually the elimination of HI and formation of the epoxide, which then reacts with the protein.

Methane thiosulphonate reagents change an −SH group into −S−S−R, this can be reversed by DTT.

***p*-Hydroxymercuribenzoate** reacts with disulphides, the resulting compound is readily destroyed with KCN, which forms a strong complex with mercury. 2-Chloromercuri-4-nitrophenol has similar properties, but the modification can be followed easily from absorbance at 405 nm

Phenylarsine oxide (Phe−As=O) can reversibly label two closely spaced SH-groups (6.3–7.3 Å apart). Lonely SH-groups can then be labelled in any way desired. After removal of the protection group with DTT, the closely spaced SH-groups may be labelled with a different reagent [214].

Metals Certain metals like Au, Ag or semiconductors can form dative bonds with the unshared electron pair of sulphur in both thiols and disulphides, leading to the adsorption of sulphur-compounds to metal surfaces.

Several SH-reagents are available in deuterated form for use in differential gel electrophoresis (DIGE) studies (see p. 88 for details on this technique).

20.1 Cystine Reduction

Cystine may be reduced to cysteine by reagents like DTT or β-mercaptoethanol. Of these, DTT has a lower redox potential (−330 mV), less odour, and a higher price. Because of the two −SH-groups, DTT can reduce a cystine in a bimolecular reaction rather than via a mixed disulphide, the reaction is faster and has a better equilibrium position.

TCEP (tris-(2-carboxyethyl)-phosphine) does not contain thiols and does not need to be removed prior to modification of the protein with SH-specific probes. Thus the risk of re-oxidation by air–oxygen is reduced. In addition, it is odourless, stable in aqueous solution and highly soluble (1.08 M at 20 °C), stable to air, and specific for SH-groups. It is, however, unstable in phosphate buffer and since it is charged it can not be used prior to IEF.

Reduction may be performed with reducing agents (homocysteine, TCEP or dihydrolipoamide) immobilised onto agarose beads. Their high local concentration creates a strongly reducing environment, in addition removal of excess reagent is simple.

Chapter 21
Reagents for Other Groups

21.1 The Alcoholic OH-Group

The reactivity of alcohols in aqueous solutions is low, most reagents prefer to react with more nucleophilic groups like amine or thiol.

Tosyl- or diphenylphosphoryl-modified Ser can undergo β-elimination under formation of dehydroalanine, this can be hydrolysed at the $C^{\alpha}-N$ bond under alkaline conditions (0.1 M NaOH) [386]. A similar reaction occurs with *O*-glycosidic bonds on Thr and Ser when treated with alkali, this may be used to identify the sites of O-glycosylation. Phospho-Ser or -Thr also show this reaction (see Fig. 21.2). In general, however, the reaction conditions are probably to rough to be useful.

After tosylation the alcohol may be converted into a thiol by incubation with thioacetate at *p*H 5.5, followed by hydrolysis.

Sulphonyl fluorides and organic phosphoesters can label Ser and Thr OH-groups in the active centre of enzymes, where they are more reactive. Protease inactivation by PMSF is an example (see Fig. 21.1).

Alternatively, periodate oxidation of alcohols leads to aldehydes which may be modified with hydrazines or aromatic amines. This reaction is of particular significance for the modification of sugar residues in glycoproteins, it will also work with Ser but not Thr. Because the sugar residues of glycoproteins are usually not involved in protein secondary and tertiary structure, their modification is unlikely to alter binding and enzymatic properties of proteins; one of the application is the cross-linking of antibodies with horseradish peroxidase via their sugar chains for ELISA [432].

Galactose oxidase can introduce aldehyde groups into Gal-residues in glycoproteins, these in turn may be introduced using Gal-transferase.

N-methylisatoic acid anhydride is used to label molecules like sugars or nucleotides, the small size of this probe and its environmentally sensitive fluorescence allow ligand binding to proteins to be monitored. The Mant-group is probably the smallest fluorescent label available (see Fig. 6.2 on p. 41).

Phenylboronic acid derivatives form complexes with vincinal alcohols, like sugars. They also react with Ser in proteins if there is an adjacent His (catalytic triad!).

E. Buxbaum, *Biophysical Chemistry of Proteins: An Introduction to Laboratory Methods*, DOI 10.1007/978-1-4419-7251-4_21,

Fig. 21.1 PMSF is used to modify Ser

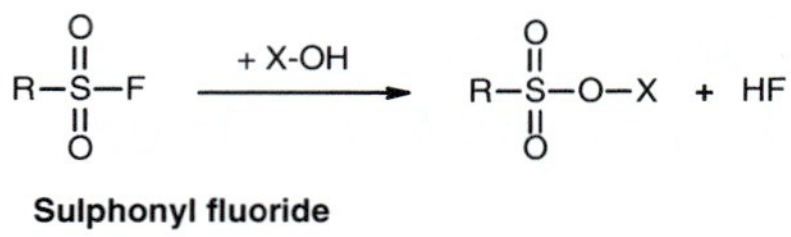

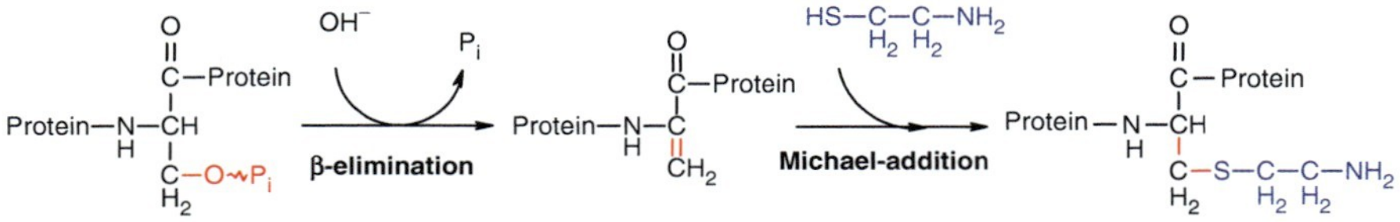

Fig. 21.2 Labelling of phosphoserine residues in proteins. The resulting thioethylamine may be used to identify phosphoserine in mass spectroscopy, or may be used to introduce amine-reactive reagents. Before labelling, phosphoproteins or peptides may be enriched by IMAC on a Fe^{3+} or Ga^{3+} column, which have high affinity for phosphoproteins

21.2 The Phenolic OH-Group

The reactivity of Tyr depends on the ionisation state of the OH-group and on the micro-environment inside the protein. The simplest modification is the introduction of iodine (in particular radio-iodine ^{125}I and ^{131}I). Iodination is often achieved from KI in combination with an oxidant like Chloramine T or H_2O_2. The iodide is oxidised to iodine ($2I^- + 2H^+ + H_2O_2 \rightarrow I_2 + 2H_2O$). The iodine reacts spontaneously with Tyr, forming iodotyrosine and HI, the reaction may involve I^+ as electrophilic intermediate. Pierce offers Iodo-Gen, a hydrophobic oxidant that is used to coat the test tube in which the iodination is performed (see Fig. 21.3). Since the oxidant is not in solution, damage to proteins is limited. Free radio-iodine and protein are separated by gel chromatography on disposable desalting columns. It is convenient to leave the unbound radio-iodine on the column and dispose of it as solid waste. Note that radio-iodine is volatile and must be handled only in well-functioning fume cupboards. As student I once experienced an instructor who routinely did iodinations for immunoassays on the open bench, his attitudes changed when students, just for fun, held a GEIGER-counter to his throat and picked up a significant signal.

Sulphonyl chlorides, iodoacetamides and NBD-Cl sometimes react with Tyr.

Tyr can be nitrated with tetranitromethane (or peroxynitrite) in the *ortho*-position. The o-nitrotyrosine has an absorbance of $4\,100\,M^{-1}cm^{-1}$ at 428 nm in alkaline solution, the pK_a of the OH-group is shifted from 7 to 10 by the strong inductive effect of the nitro-group. This can inactivate enzymes which use Tyr in acid-base catalysis. Absorbance is environmentally sensitive and decreases in a more hydrophobic surrounding. This can be used to measure protein conformational changes after ligand binding. Nitrotyrosine is an effective quencher of fluorescence and has been used as acceptor in FRET-studies.

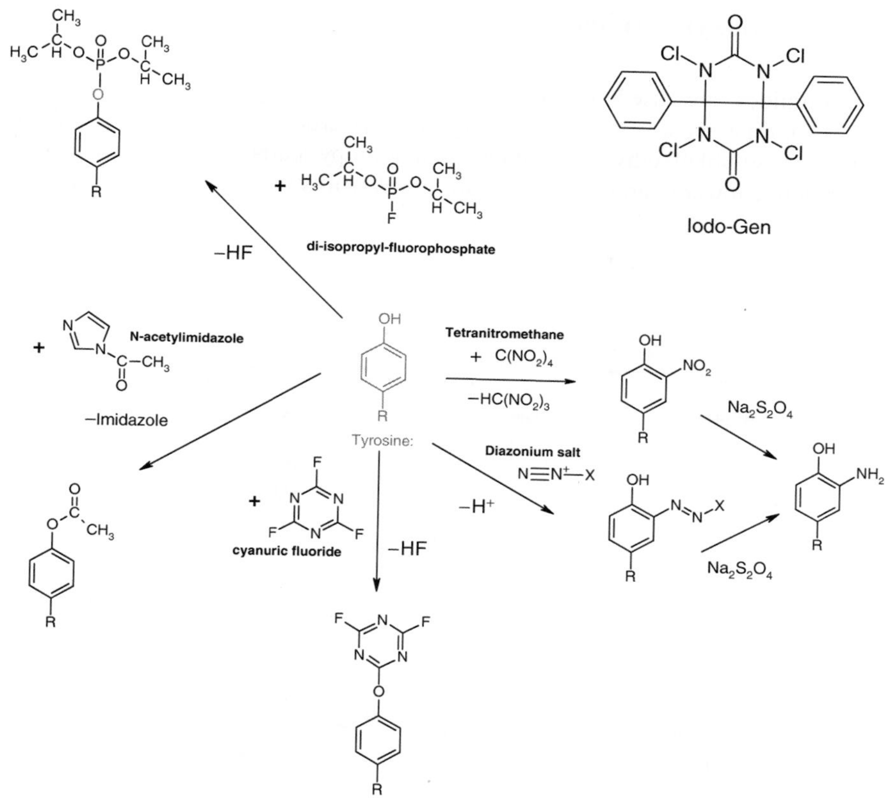

Fig. 21.3 Reagents to modify tyrosine. For details see text

The nitro-group can be reduced to an amino-group with sodium dithionite and then labelled with amine-reactive probes. Tetranitromethane reacts also with SH-groups, His, Met, and Trp. In addition, at acidic *p*H, TNM can cross-link the tyrosine residues of a protein (zero-length cross-linking).

Aromatic diazonium salts react with Tyr to yield azo-dyes. If required, these can be reduced with dithionite to *o*-aminotyrosine and then amine modified. Lys and His may also be labelled.

Acetylimidazole labels at neutral *p*H with a change in absorbance of $\Delta\epsilon_{278} = 1.210\,M^{-1}cm^{-1}$, the reaction is reversible under alkaline conditions in the presence of a nucleophile like Tris or hydroxylamine.

Cyanuric fluoride can label Tyr groups under alkaline conditions, when the OH-group is ionised. Diisopropyl fluorophosphate has also been used, but is highly toxic.

21.3 Carboxylic Acids

Carbodiimides are used for labelling or cross-linking carboxy-groups (see Fig. 21.4). The most common are dicyclohexylcarbodiimide (DCCD) and the more water soluble 1-ethyl-3-(3-dimethylaminopropyl)carbodiimide (EDAC). Also very reactive toward carboxyl-groups are the WOODWARD-reagents (isoxazolium

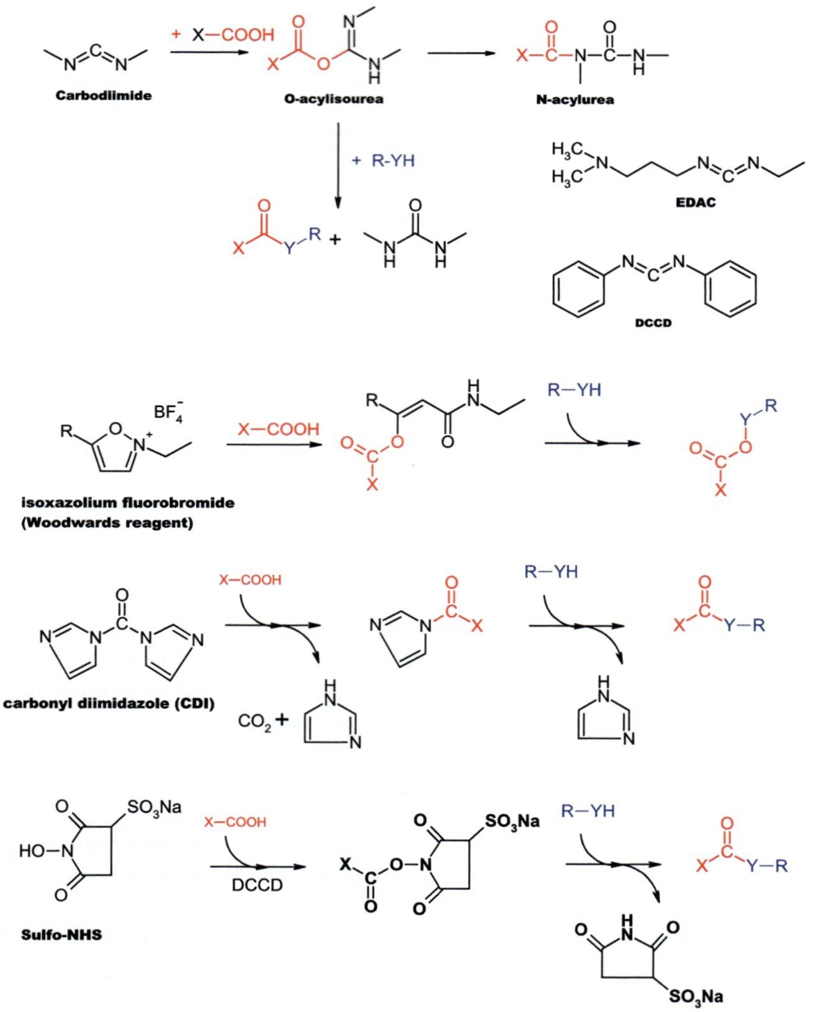

Fig. 21.4 Reagents to modify carboxylic acids. The acid is first activated, then coupled to an nucleophilic reagent (-YH: alcohol, thiol, amine *etc.*)

fluorobromides). Sulpho-NHS esters are sometimes stable enough to be shipped as intermediates. Like amines, carboxylic acid groups are stronger nucleophiles when deprotonated, that is, above their pK_a.

21.4 Histidine

Diethylpyrocarbonate (ethoxyformic anhydride) labels His specifically at *p*H 5.5–7.5 by transfer of a carboethoxy-group to N^3 (see Fig. 21.5), resulting in a change of absorbance at 240 nm, $\Delta\epsilon \approx 3\,200\,M^{-1}cm^{-1}$. The reaction is reversible by incubation with hydroxylamine or other nucleophiles (Tris!) at neutral or alkaline *p*H. Higher DEPC/His ratios result in the modification of both N^3 and N^1, this is not reversible. Instead, incubation with hydroxylamine leads to the opening of the His-ring. In side-reactions DEPC may also react with Cys, Tyr and primary amino groups, Lys ϵ-amino groups may form isopeptides with neighbouring carboxy-groups.

Before diethylpyrocarbonate became available His was modified by photo-oxidation in the presence of methylene blue, rose bengal, or eosine. Electrons in these dyes are excited by light to higher orbitals, from where they may enter the triplet state. The energy is then transferred to nearby oxygen molecules (for which the triplet state is the ground state) which become activated and can oxidise other molecules. Apart from His, photo-oxidation may also affect Trp, Met, and to a lesser extend Tyr, Ser and Thr, especially at higher *p*H.

In microenvironments where the nucleophilicity of His is increased labelling with iodoacetate, iodoacetamide, or *p*-bromo-phenacyl bromide ($\Delta\epsilon_{271} = 17\,000\,M^{-1}cm^{-1}$), is possible. This can be used to label reactive His in the catalytic centre of enzymes.

21.4.1 Tryptophan

Trp can be oxidised on the indole ring with hydrogen peroxide in 1 M Na_2CO_3 *p*H 8.4 (see Fig. 21.6). This results in a reduction of absorbance, $\Delta\epsilon_{280} = 3\,490\,M^{-1}cm^{-1}$.

N-bromosuccimide and 2-hydroxy-5-nitrobenzyl bromide require harsh reaction conditions (10 % acetic or formic acid, 8–10 M urea) and are therefore of limited use. The nitro-group in the latter reagent gives the product strong absorbance in alkaline solution ($\epsilon_{410} = 18\,000\,M^{-1}cm^{-1}$). To avoid reaction with Cys, the protein should be reduced and carboxymethylated.

Formic acid saturated with HCl can formylate the ring nitrogen of Trp, result-ingresulting in an increase in absorbance at 298 nm. The reaction is reversible in a *p*H-stat at *p*H 9.

KOSHLAND's reagent (2-hydroxy-5-nitrobenzylbromide) can be used to quantitate Trp-groups since its reaction product has a strong absorbance

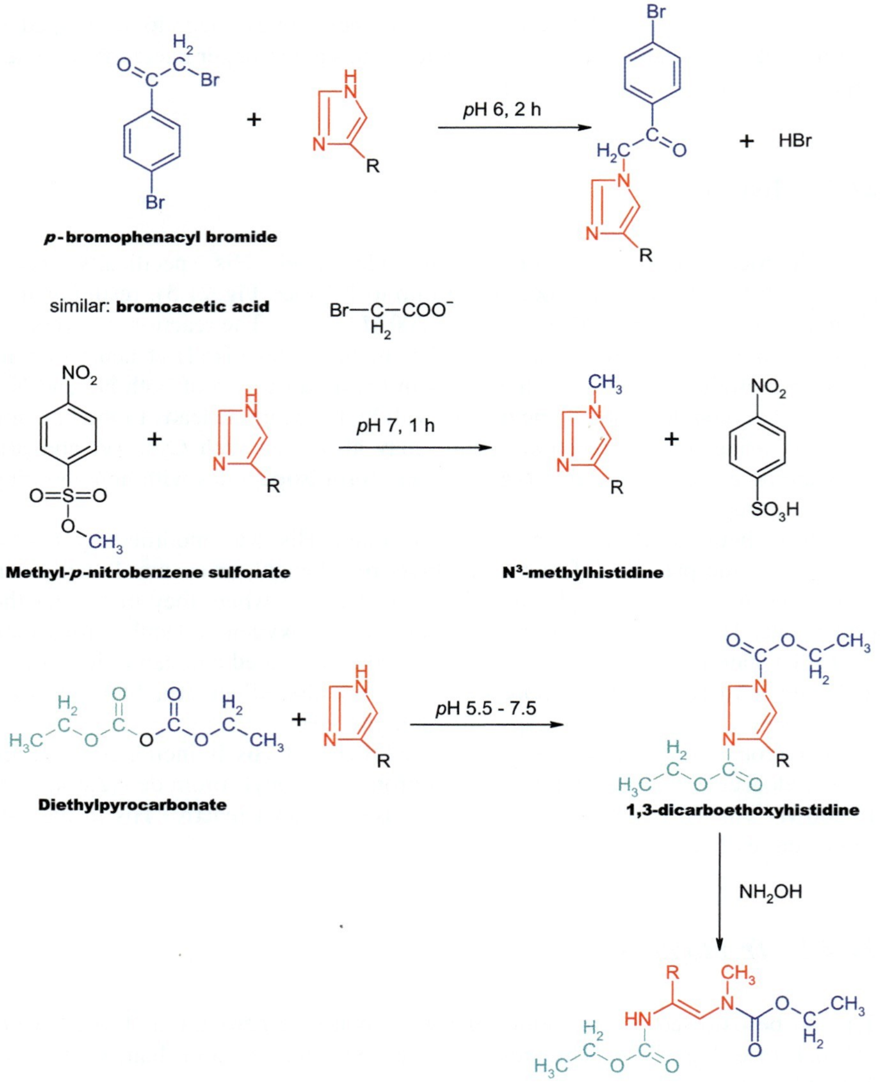

p-bromophenacyl bromide

pH 6, 2 h

+ HBr

similar: **bromoacetic acid** $Br{-}CH_2{-}COO^-$

Methyl-*p*-nitrobenzene sulfonate

pH 7, 1 h

N^3-methylhistidine

Diethylpyrocarbonate

pH 5.5 - 7.5

1,3-dicarboethoxyhistidine

NH_2OH

Fig. 21.5 Histidine modifying reagents

($\Delta\epsilon_{410} = 18\,000\,\mathrm{M}^{-1}\mathrm{cm}^{-1}$ at *p*H 12), however, the reaction conditions are harsh (180 mM acetic acid, 10 M urea, 37 °C). If the protein survives these conditions, the reagent has the advantage that its absorbance spectrum depends on environmental conditions (acidity, hydrophobicity).

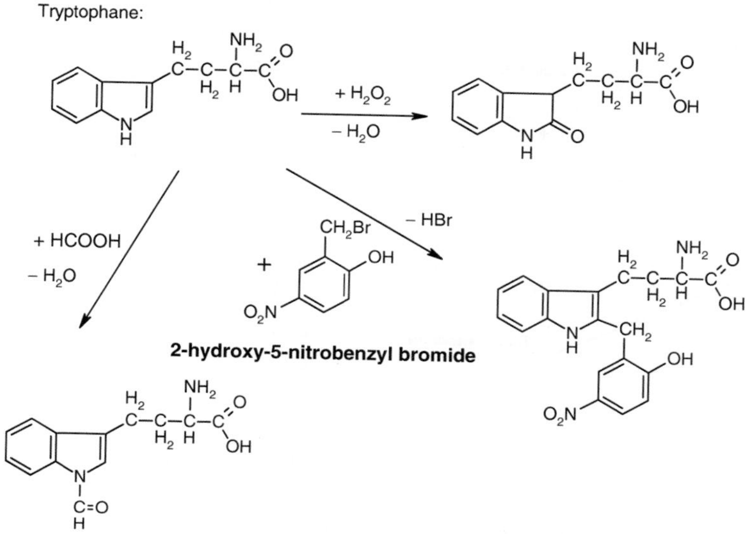

Tryptophane:

$+ H_2O_2$ / $- H_2O$

$+ HCOOH$ / $- H_2O$

$- HBr$

2-hydroxy-5-nitrobenzyl bromide

Fig. 21.6 Reagents to modify Trp

21.4.2 Arginine

Phenylglyoxal [400], 4-hydroxy-3-nitro-phenylglyoxal (chromogenic, $\epsilon_{316} = 10\,900\,M^{-1}cm^{-1}$), 2,3-butanedione [440] and 1,2-cyclohexanedione [315] also label Arg (see Fig. 21.7), but also amino- and SH-groups and His, but with lesser efficiency. Borate (50 mM *p*H 8) accelerates the reaction about 20-fold. The reaction products are generally not stable to acid hydrolysis (6 M HCl at 110 °C for 24 h), but can be protected by mercaptoacetic acid.

Ninhydrin [401] also labels Cys and Lys.

21.4.3 Methionine

Methionine is fairly hydrophobic and often buried in the core of the protein, making modification more difficult than it should be from the chemists point of view.

Of greatest importance is the specific splitting of peptides and proteins at Met-residues by cyanogen bromide [133] (see Fig. 21.8). This compound is highly toxic and should be handled accordingly. Since Met is a relatively rare residue in proteins (abundance 1.7 %), the resulting fragments are much larger than those obtained with trypsin or chymotrypsin. A 100-fold excess of the reagent is commonly used under acidic conditions (70 % trifluoroacetic acid) for 4 h in the dark, followed by lyophilisation (with KOH in the trap!).

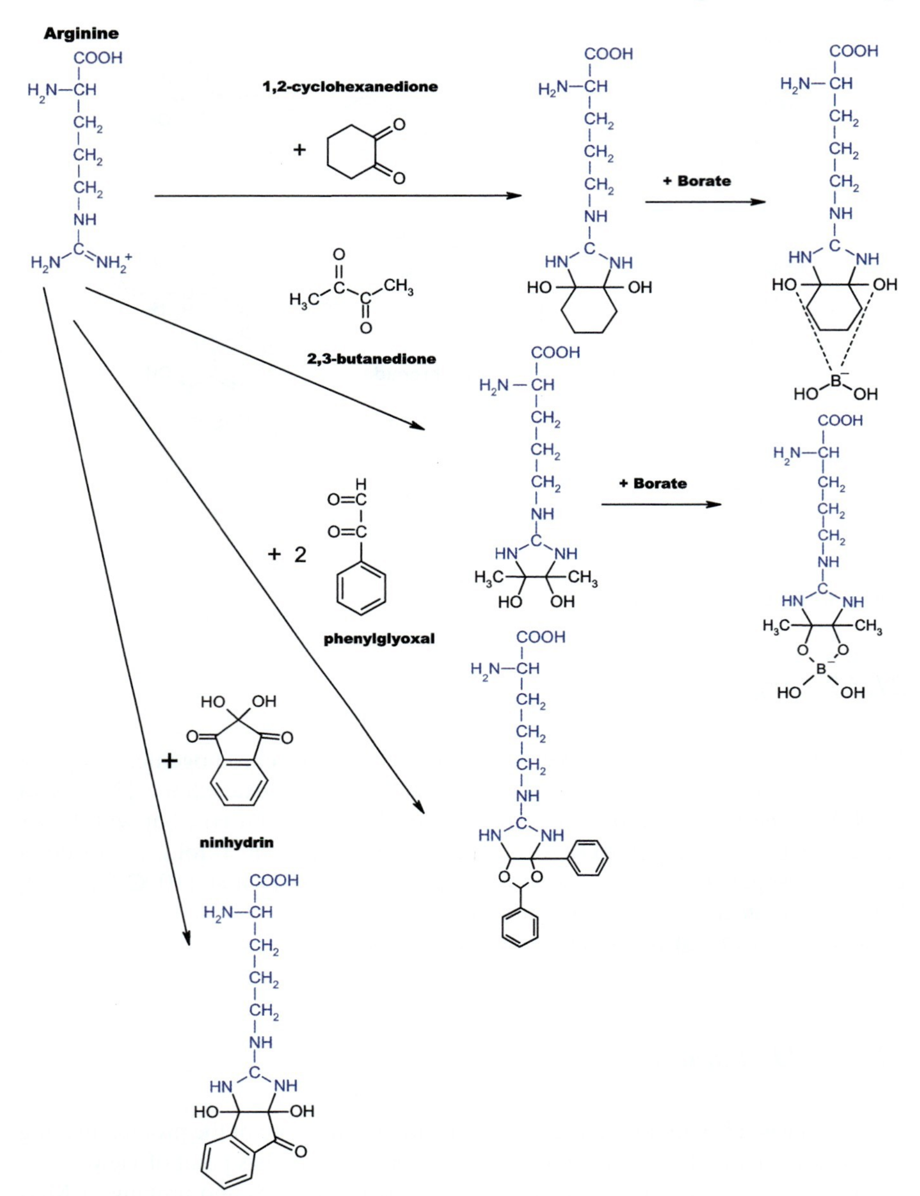

Fig. 21.7 Labelling of Arg with diones and ninhydrin

The sulphur in methionine may be oxidised step-wise with Chloramine T (accompanied by a change in absorbance at 245 nm), sodium periodate, hydrogen peroxide, or the hydrophobic *t*-butylhydroperoxide. Cys interferes with that reaction, it can be reversed with mercaptoethanol, DTT, and especially *N*-methylmercaptoacetamide (1 M, 37 °C, 21 h).

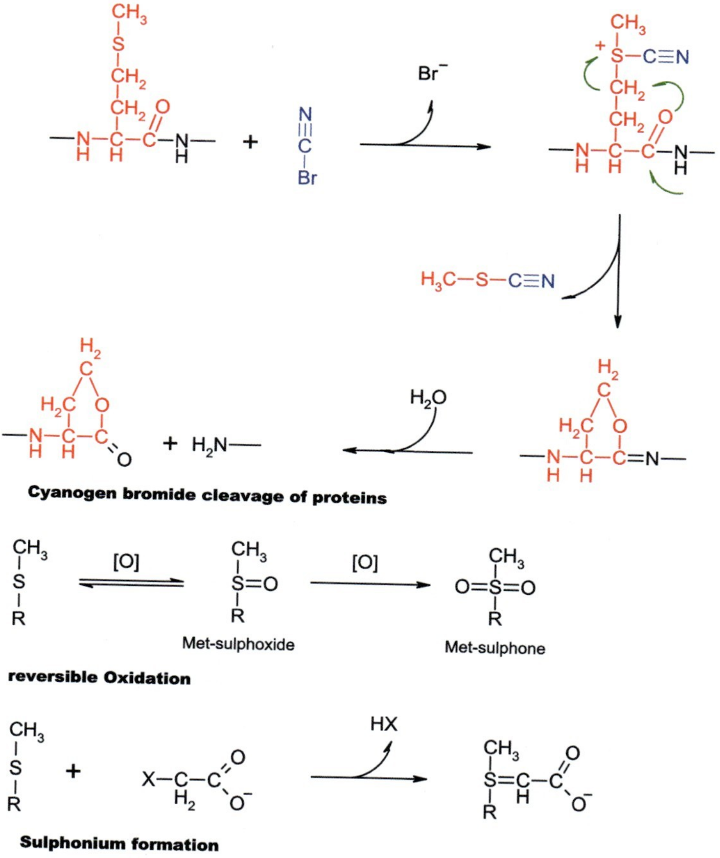

Cyanogen bromide cleavage of proteins

reversible Oxidation

Met-sulphoxide

Met-sulphone

Sulphonium formation

Fig. 21.8 Modification of proteins at Met-residues. Of the reactions shown the cleavage of proteins with cyanogens, bromide is the most important

Halogen-compounds like iodoacetate react much slower with Met than with Cys, and often only at low *p*H after disruption of protein structure with caotropes. The result is a sulphonium-compound.

Chapter 22
Cross-linkers

Bifunctional reagents may be used to cross-link proteins (see Fig. 22.1). They may be homo-bifunctional (two identical reactive groups) or hetero-bifunctional, for example with a thiol- (maleimides) and an amino-specific end (NHS-ester). Intra-molecular cross-links are favoured over inter-molecular ones by low protein concentration, a high net charge of the protein (electrostatic repulsion) and by a high protein-to-reagent ratio. Since the cross-linking of monomers and oligomers follow a different time course the oligomeric state of proteins can be investigated by cross-linking. However, analytical ultracentrifugation and laser light scattering offer much higher precision.

1,3-dibromoacetone can cross-link Cys or His-groups, note that this compound acts as tear-gas.

Glutardialdehyde is used to cross-link amino-groups, and one might expect this to work via a SCHIFF-base. However, cross-linking is stable without reduction, and even the mono-functional methanal (formaldehyde) can cross-link proteins (see Figs. 19.2 and 19.3 on p. 202f.).

22.1 Reversible Cross-linkers

Compounds with disulphide bonds can be used to make cross-linking reversible and distinguish the effects of cross-linking and amino acid modification (see Fig. 22.2). For this, cross-linking is performed under oxidising conditions. At a convenient time the protein is exposed to reducing conditions and the disulphide bond opens. However, the amino acids involved in the cross-link stay labelled with the fragments of the reagent. Methyl-3-(*p*-azidophenyl) dithiopropioimidate (PAPDIP, the imidoester reacts with amino-group, the azido-group is then used to unspecifically label neighbours [74]), 3,3'-dithiobis(succimidylpropionate) (DTSP, LOMANT's reagent [240], membrane permeable or with a sulfonic acid group Sulfo-DTSP not membrane permeable) and *N*-(4-azidobenzoylglycyl)-*S*-(2-thiopyridyl)-cysteine (AGTC, labels Cys reversible via the thiopyridyl-Cys and any neighbours unspecifically via the azido-group [59]) are the most well-known reagents of this type.

E. Buxbaum, *Biophysical Chemistry of Proteins: An Introduction to Laboratory Methods*, DOI 10.1007/978-1-4419-7251-4_22,

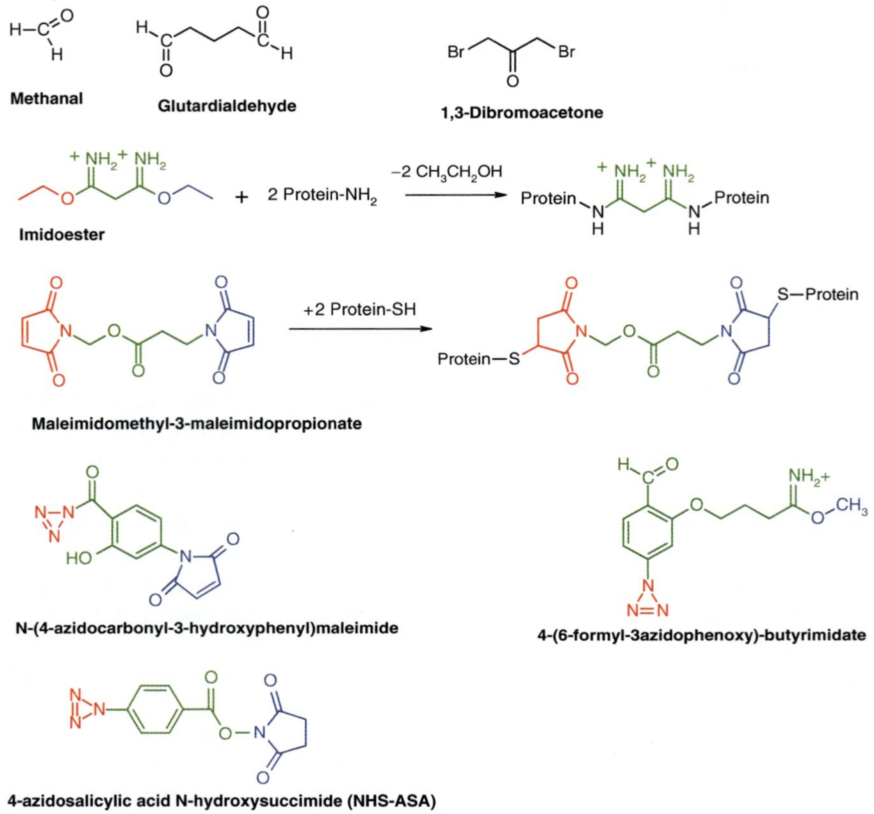
Methanal
Glutardialdehyde
1,3-Dibromoacetone
Imidoester
2 Protein-NH2
−2 CH3CH2OH
Protein
Maleimidomethyl-3-maleimidopropionate
+2 Protein-SH
S—Protein
Protein—S
N-(4-azidocarbonyl-3-hydroxyphenyl)maleimide
4-(6-formyl-3azidophenoxy)-butyrimidate
4-azidosalicylic acid N-hydroxysuccimide (NHS-ASA)

Fig. 22.1 Reagents used to cross-link proteins

Glycols can be cleaved with periodate (in the presence of Tris to block the resulting aldehydes). These compounds are used when reduction of disulphide bonds in the protein must be avoided.

22.2 Trifunctional Reagents

Trifunctional reagents (see Fig. 22.3) work in principle like the bifunctional ones described above, but have three business ends. They are often derived from biocytin (biotin on the ϵ-amino group of Lys) by derivatisation of the C′-carboxy- and α-amino group. One of these ends may be used to label a hormone, the other may have a photoreactive group that cross-links to the receptor after binding. The biotin can then be used to isolate the complex or to find it, *e.g.*, on Western-blots by means of its interaction with avidin or streptavidin. Trifunctional phosphine derivatives have also been described.

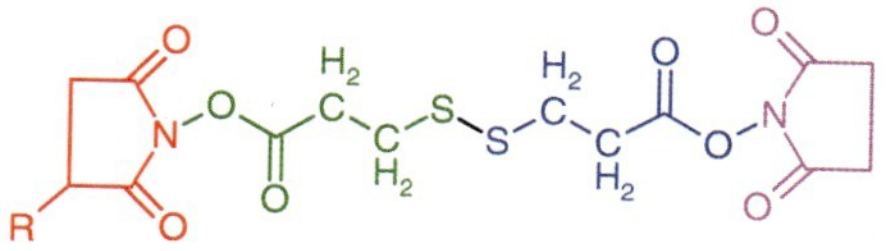

R = H: DTSP, Lomant's reagent
R = SO_3Na: Sulfo-DTSP (water soluble, not membrane-permeable)

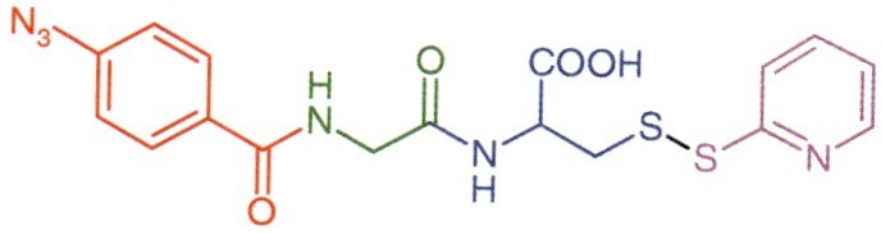

N-(4-azidobenzoylglycyl)-S-(2-thiopyridyl)-cysteine (AGTC)

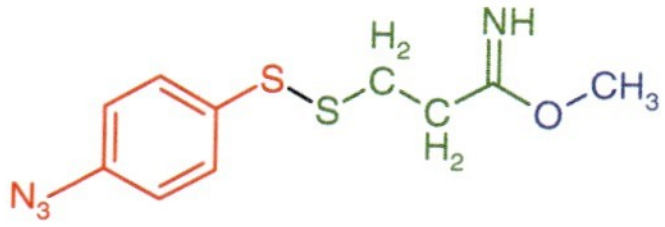

Methyl 3[(p-Azidophenyl)dithio]propionimidate (PAPDIP).

Fig. 22.2 Reversible cross-linkers

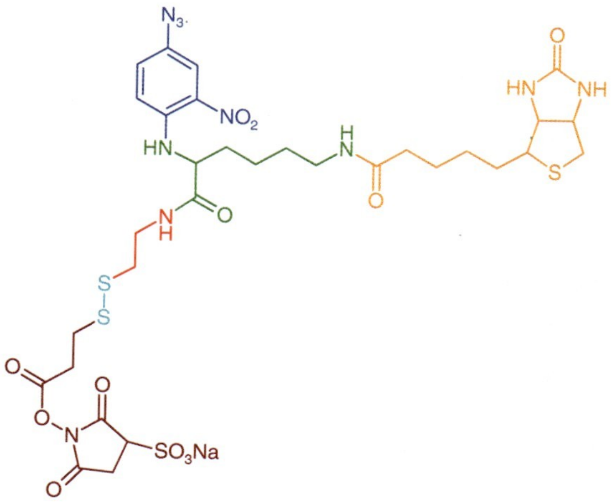

Fig. 22.3 Trifunctional reagents like Sulfo-SBED can be used to label a bait-protein (say, a hormone) via the amine-reactive sulfo-NHS-ester (*brown*). The bait is then allowed to bind to the target (say, receptor), and photo-crosslinked via the azido-group (*blue*, on photoreactive groups see p. 224). The complex is purified by binding the biotin (*orange*) to avidin, and the bait removed by cleaving the S – S-bond (*cyan*). The entire molecule is built on a Lys-backbone (*green*)

Chapter 23
Detection Methods

23.1 Radio-labelling of Proteins

There are two principal routes to radio-labelled proteins: metabolic labelling *in vivo* with labelled amino acids (in particular ^{35}S labelled Cys and/or Met) and chemical introduction *in vitro*. Metabolic labelling has the advantage that it will not interfere with the proteins' function, however, only a small part of the radioactivity used ends up in the protein you are interested in. Some stays in small molecules, some is bound to other proteins. Chemical labelling is performed on an already purified protein, so only unreacted label must be removed by gel filtration.

The most well-known way to radio-label a protein is to introduce ^{125}I to Tyr-residues. For this, $^{125}I^-$ is oxidised to $^{125}I^+$, which readily reacts with Tyr. Oxidation may be performed with Chloramine T or a hydrophobic derivative of this compound which is used as a layer on the reaction tube (Iodogen, Pierce, see Fig. 21.3 on p. 211). Thus the contact of the oxidising reagent with protein is minimised. H_2O_2 produced by glucose oxidase is an even gentler oxidant. *p*H should be below 8.5 to prevent labelling of His. The big advantages of ^{125}I are its comparatively low price and the fact that this γ-radiator can be counted without a scintillator. The half-life period of 59.4 d allows convenient use and disposal by decay.

Proteins without Tyr can be labelled with ^{125}I using the BOLTON-HUNTER-reagent, which reacts with primary amino-groups (see Fig. 23.1). Alternatively, ^3H-succimidylpropionate or ^3H-methanal may be used to introduce tritium. Incubation of a protein over night with ^{35}S Met leads to the incorporation of radioactivity too, apparently radiolysis leads to the formation of compounds which react with the protein (thioaldehyde $-CH{=}S$?) [186].

For applications in whole organisms, *e.g.*, for cancer diagnostics and therapy, proteins are labelled with a large number of isotopes. Some metals may be bound to SH-groups, or to chelators linked to the protein.

E. Buxbaum, *Biophysical Chemistry of Proteins: An Introduction to Laboratory Methods*, DOI 10.1007/978-1-4419-7251-4_23,

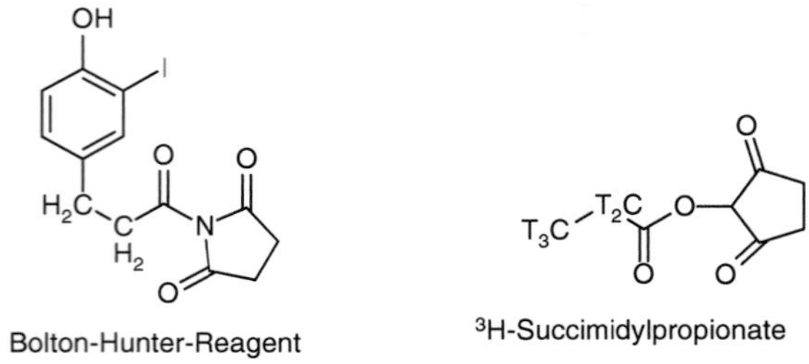

Fig. 23.1 Radio-labelling of proteins

Bolton-Hunter-Reagent

^{3}H-Succimidylpropionate

23.2 Photo-reactive Probes

Photo-reactive probes linked to substrates or other ligands are first bound to a protein in the dark and then photolysed, usually with UV-light. The reactive breakdown-products bind to any nearby protein in a shotgun-like manner. Thus the vicinity of the ligand binding site gets labelled with the (radioactive) breakdown-products and can be mapped onto the primary structure of the protein by proteolytic digestion followed by purification and sequencing of the resulting peptides.

The archetypical photoaffinity label is the arylazide [371]. Photolysis at >300 nm produces a nitrene which can react directly with double bonds, primary amines or active hydrogen bonds. If that has not happened after about 10^{-4} s, ring expansion leads to a dehydroazidopine which reacts with nucleophiles like amino- or sulphydryl-groups (see Fig. 23.2). Per-fluorinating the ring reduces ring-expansion, favouring the other reactions.

Diazirines can be even more reactive, especially with fluorine in the neighbourhood. They are small (about the size of a water molecule) and used when arylazide analogues can not be used due to steric hindrance. Diazirine-derivatives of Leu, Ile and Met have even been used to metabolically label proteins [393]. T_2-diazirine is used to label surface-exposed residues, also in enclosed cavities. Optimum wavelength is 360 nm.

p-Nitrophenyl-3-diazopyruvate [121] is another photoreactive cross-linker of proteins which is now used as surgical glue to close wounds [116].

A little known fact is that nucleotides are photo-reactive, no chemical modification is required. Photolysis occurs when ATP is irradiated with light of 254 nm [443]. The same is true for GTP, CTP, UTP, dTTP [91], and probably other nucleotides.

23.3 Biotin

Biotin (Vit. H) is a small molecule that binds with extremely high affinity ($K_d \approx$ 1.3 fM) to **avidin** (from bird eggs) and **streptavidin** (from *Streptomyces avidinii*). In addition, (strept)avidin is extremly stable against denaturation, to dissociate its complex with biotin requires 8 M guanidinium HCl at *p*H 1.5. Biotin can be linked to macromolecules (*e.g.*, antibodies) often without changing their properties.

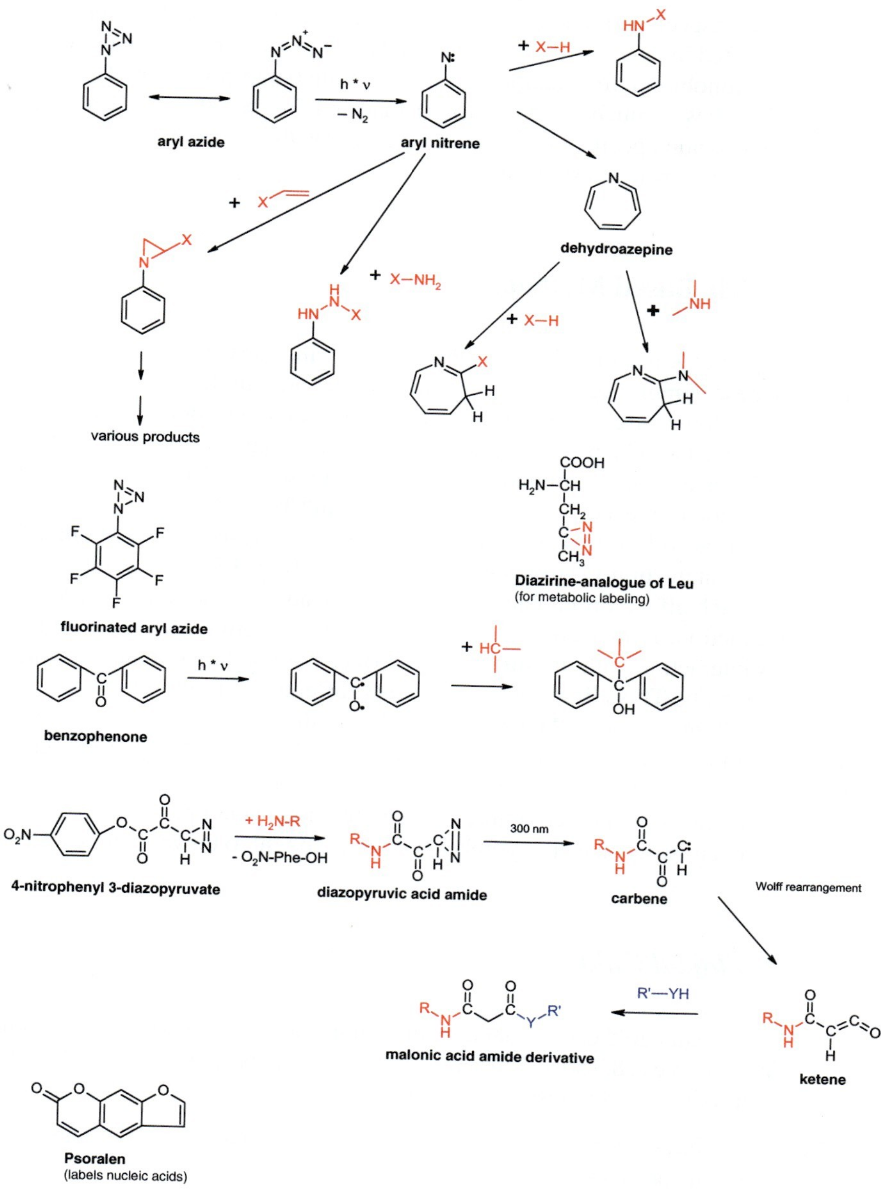

Fig. 23.2 Photo-reactive reagents used

(Strept)avidin labelled with enzymes, fluorescent markers, *etc.*, is then used to detect these macromolecules. Because the biotin binding sites in (strept)avidin are located in 9 Å deep pockets, spacers are required between biotin and macromolecules. Modern biotinylation reagents have hydrophilic (*e.g.*, PEG-based) spacers, which are less likely to cause unspecific interactions. The affinity between biotin and (strept)avidin may be modulated by changes on the biotin molecule to make binding more easily reversible. Imminobiotin, for example, is released from (strept)avidin at $p\text{H} \leq 4$.

Streptavidin has a much lower *p*I (5–6) than avidin ($\approx$ 10) and is not glycolysated. Hence non-specific binding to streptavidin is lower than to avidin. However, its solubility in water is slightly lower.

23.4 Particle Based Methods

The use of particles in bioassays probably started with latex beads as carriers in RIA. From these humble beginnings there have been enormous developments. Some of these technologies are briefly discussed here, but quantum dots are discussed with fluorescence (see Chap. 6.2 on p. 44), radioactive proximity assays in Sect. 9.5 on p. 107. Chromatographic supports are discussed in Sect. 14.1 on p. 147. An expanded discussion of the subject may also be found in Chap. 14 of [159].

Proteins can adsorb to hydrophobic surfaces (latex, polystyrene) at about 2.5–3 mg/m^2, this interaction is strongest near the *p*I. If the *p*H is changed later, the protein may leach off. Hydrophobic adsorption may lead to denaturation of proteins and loss of function. To make more economic use of scarce proteins, and to stabilise the proteins interaction with the surface, the surface may be functionalised to bind proteins covalently. This allows binding at a *p*H different from the *p*I, where proteins are usually more stable. Any of the chemistries discussed above may be used. Cyanogen bromide (BrCN) activated sepharose is stable when lyophilised and is commercially available. It binds proteins via amino-groups. This is the traditional route for home-made affinity columns, although the bond is not particularly stable. Hence other chemistries like epoxy-gels have been brought to the market as well.

23.4.1 Colloidal Gold

Particles of colloidal gold can be manufactured in the laboratory with defined diameter (see for example [78, 159]). These may be used to tag proteins, *e.g.*, antibodies. Binding of proteins to gold presumably occurs by:

- Electrostatic interactions between negatively charged colloid particles and positive sites on the proteins
- Hydrophobic and VAN DER WAALS interactions between proteins and metal surface
- Dative bonds between gold and sulphydryl-groups of the protein

Labelled antibodies have been used to detect their antigens in immuno-electron microscopy. It is possible to use several antibodies labelled with gold-particles of different sizes in dual- or triple-labelling studies.

Gold-labelled antibodies may also be used in Western blotting, the sensitivity can be increased by silver-amplification. For the use of gold and silver nanoparticles to enhance fluorescence and chemiluminescence signals see Sect. 9.5.1 on p. 110.

23.4.2 *Magnetic Separation*

Small magnetic particles with polymer coat functionalised, for example, with (strept)avidin, Protein A, glutathione or titanium dioxide (for phospho-peptides), can selectively bind biomolecules of interest. They are then separated from unbound material by means of a strong magnet. This is a very quick method for immunoprecipitation and similar assays. Re-suspension of the beads is eased when the particles are superparamagnetic. Such particles are drawn into a magnetic field, but – unlike ferromagnetic materials – they do not become magnetised themselves and there is no attraction between them once the external field is removed. *In vivo*, functionalised magnetic beads can be used as contrast agents for NMR-tomography [191]. Since in an electromagnetic field the particles cause heating, they may also be used for cancer treatment.

Beware of the strong magnets, which can stop watches and cause injury from flying steel objects.

Chapter 24
Spontaneous Reactions in Proteins

One would like to think of proteins as relatively stable entities, at least in the absence of modifying enzymes. Alas, this view is mistaken. As time passes, proteins can undergo spontaneous modifications that change their structure and function.

At particular risk for modification are residues that are surface-exposed and flexible. Surface exposure can be estimated from a hydropathy plot [217], as sequences with negative hydropathy (high hydrophilicity) are more likely to interact with the aqueous environment than with the hydrophobic protein core. Flexibility can be estimated from the side chain volume [341], as side chains with large volumes sterically hinder reactions. Thus, protein regions where both the hydropathy and flexibility plot are negative are more likely to be subject to modification reactions.

24.1 Reactions

Deamidation

The quantitatively most important reaction is the **deamidation** of Gln and Asn (see Fig. 24.1), which changes the primary structure from an uncharged to a negatively charged amino acid (Glu or Asp, respectively). It may even introduce non-proteinogenic amino acids like iso-Asp. Polar residues preceding the Asn or Gln, and Gly following it, accelerate deamination; while bulky, hydrophobic residues at either position slow it down. In some cases, this requires removal of sensitive Asn-residues from pharmaceutical proteins or adenovirus-preparations by genetic engineering.

Enzymatic deamidation of Gln in heterotrimeric G-proteins by **cytotoxic necrotising factors** of certain bacteria has recently been demonstrated and leads to activation [305].

24.1.1 Racemisation

Racemisation of amino acids (from L- to D-form, see Fig. 24.2) and Pro **cis/trans-isomerisation** may occur.

E. Buxbaum, *Biophysical Chemistry of Proteins: An Introduction to Laboratory Methods*, DOI 10.1007/978-1-4419-7251-4_24,

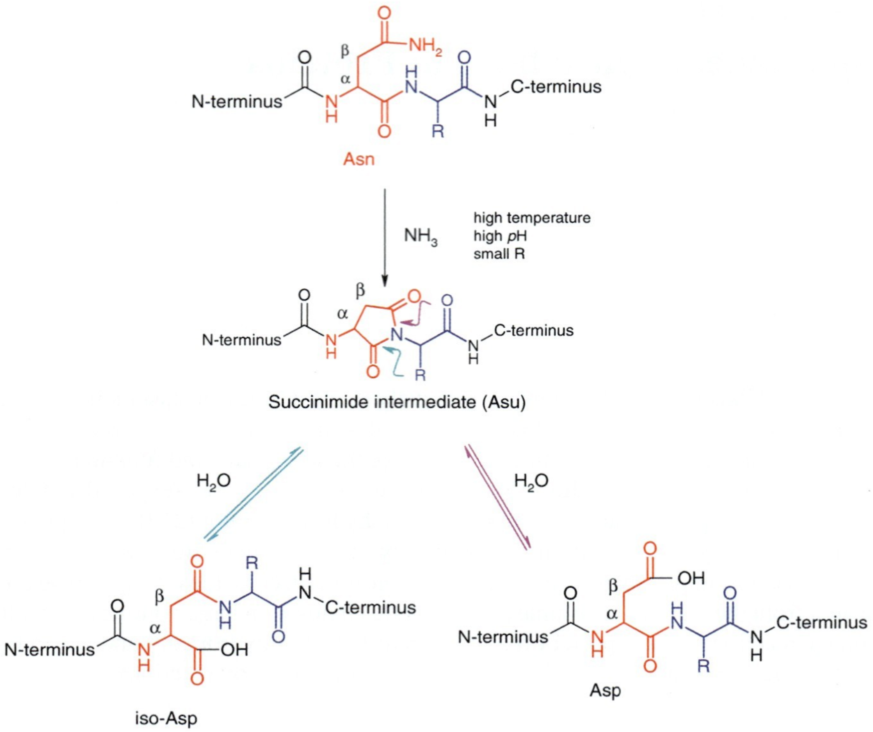

Fig. 24.1 Asn can spontaneously deamidate, producing ammonia and a symmetric succinimide intermediate (Asu). The latter can be hydrolysed at either of the keto-C − N bonds, resulting in Asp or in iso-Asp. Because the hydrolysis of Asu is reversible, Asp racemises too, but slower than Asn. Because during sample hydrolysis Asn is converted to Asp, the two amino acids are usually measured together as Asx. Racemisation of Gln and Glu is analogous, but slower

24.1.2 Oxidation

Cys-thiols may be **oxidised** to disulphide bonds. Met, and to a lesser extend His, Trp and Tyr, are also subject to oxidation (see Fig. 24.4). Oxidation is caused by reactive oxygen species, especially in the presence of metal ions that can catalyse a FENTON-reaction. Peptide bond cleavage may occur. This is a reaction catalysed by both acids and bases, so storage conditions should be within $2 < p\text{H} < 11$. −Asx−Xxx− is particularly sensitive to hydrolysis, especially Asn−Pro− (see Fig. 24.3).

24.1.3 Amyloid-Formation

Proteins may also change their secondary structure, for example from α-helix to β-strands, by amyloid formation. This process is accelerated by multivalent metal

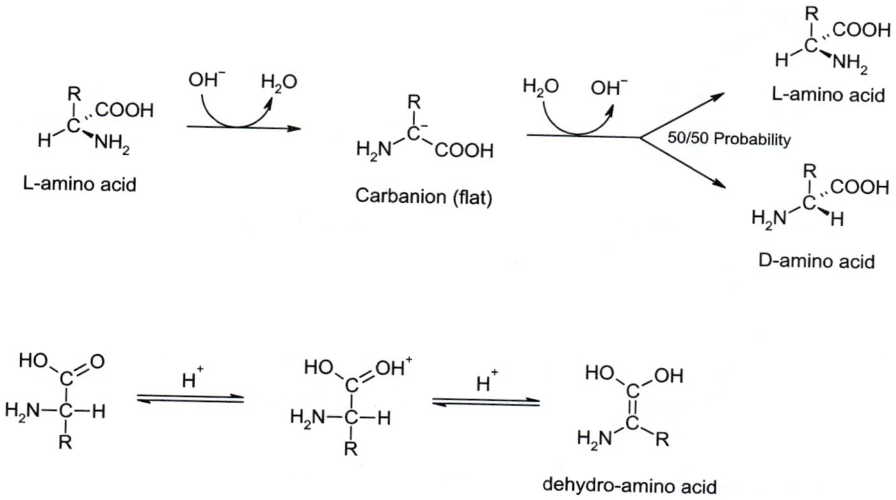

R COOH H C NH$_2$ L-amino acid
OH$^-$ H$_2$O
R C$^-$ H$_2$N COOH Carbanion (flat)
H$_2$O OH$^-$
50/50 Probability
L-amino acid
D-amino acid

H$^+$ H$^+$
dehydro-amino acid

Fig. 24.2 *Top*: Base-catalysed racemisation of amino acids. A hydroxy-ion abstracts the hydrogen from the α-carbon, forming a carbanion, which is flat and not stereoactive. Re-addition of the hydrogen can then occur with equal probability to either side of the carbanion, due to mesomery the free electron pair of the carbon is found both above and below the plane of the molecule. Thus either L- or D-amino acids are formed with equal probability. *Bottom*: Acid-catalysed amino acid racemisation. Protonation of the carboxy-group is followed by elimination of the α-proton. Reversal of this reaction results in 50 % each D- and L-amino acids. Acid-catalysed racemisation is slower than base-catalysed

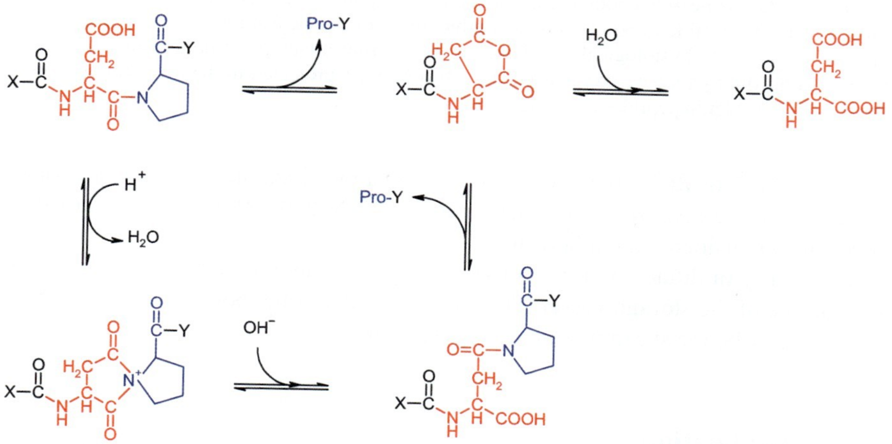

Pro-Y H$_2$O H$^+$ H$_2$O Pro-Y OH$^-$

Fig. 24.3 Hydrolysis of Asp–Pro. Pro is a particularly good leaving group

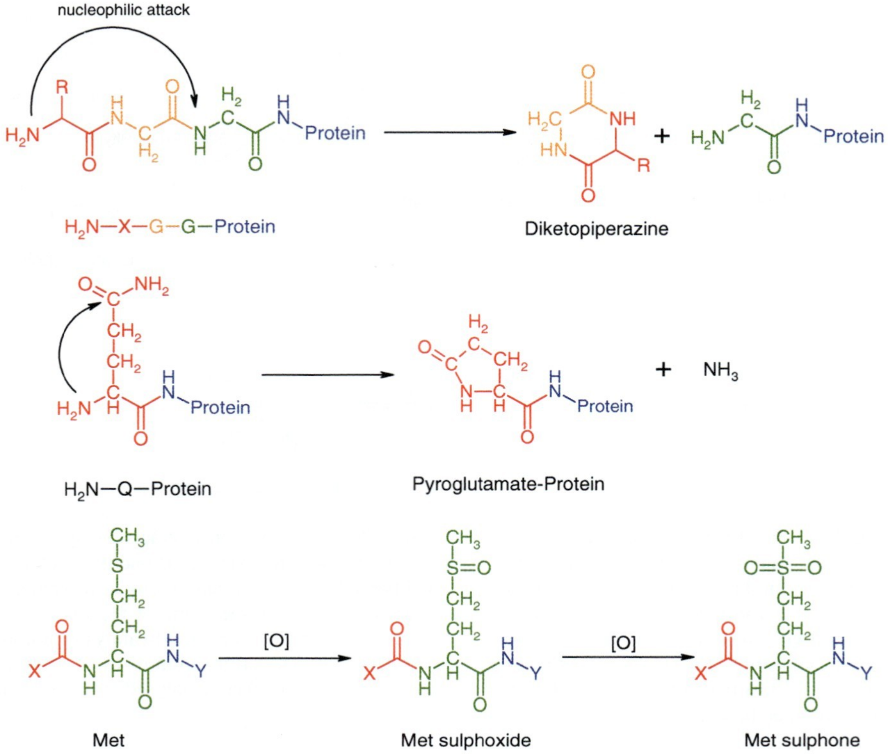

Fig. 24.4 *Top*: Diketopiperazine formation from proteins with $NH_2-X-G-G-$ in the N-terminus. Gly is flexible enough to allow for a nucleophilic attack of the terminal amino group onto the $G-G$ peptide bond. *Middle*: Pyroglutamate formation from an N-terminal Gln. Such proteins also occur physiologically and are protected from attack by aminopeptidases. *Bottom*: Oxidation of Met by reactive oxygen species. Most Met are protected in the hydrophobic core of the protein, but Met in peptides are more at risk

ions like Al^{3+} or Zn^{2+} [61]. This has led to suggestions that Al^{3+} from cooking utensils may be involved in the pathogenesis of ALZHEIMER's disease, however, this is now considered a red herring.

Especially in dilute solutions, proteins tend to adhere to surfaces, for example, the surface of the storage vessel. This not only reduces the potency of the solution, but also may be the starting point of aggregation.

24.2 Applications

Some proteins in our body are very long-lived, they may be produced during embryogenesis and then stay in place our entire lives. This is true for lens crystallins and the collagen in bones and cartilage. Such long-lived proteins are subject to amino acid racemisation, in particular the racemisation of Asn, which is faster than other amino acids [153]. Disease processes that lead to a more rapid turnover of such proteins can be monitored by measuring their D-Asp content [267]. In addition, Asn racemisation in long-lived proteins can be used to determine the biological age of the individual from which they were taken. This is useful in wildlife biology and forensic science.

In archaeology amino acid racemisation can be used to date fossils (*e.g.*, hair, bones, teeth, egg shells or corals), provided that proteins (*e.g.*, collagen) can still be extracted (see [145] for a review). Amino acids are obtained after acid hydrolysis of proteins (6 M HCl at 110 °C over night), they are derivatised and measured by chiral gas- or liquid chromatography. The main problem has been to establish the racemisation rate constants, which depend on secondary structure of the protein (the collagen triple helix, for example, does not allow enough steric freedom to form Asu) and the environment (*p*H, temperature, ions, reactive oxygen species), see for example [246]. Leaching of amino acids and peptides into the sample increases the problem. Therefore this method is most useful to determine relative ages of different finds from the same site, which it can do relatively cheaply. A review on the controversy around amino acid racemisation and other dating techniques on dating Amerindian finds was recently provided in [332].

In the pharmaceutical industry, and to some extend even in the food industry, modification of proteins during storage limit the shelf life of their products. Aim is to find storage conditions that allow a shelf life of at least 2 a. This is often accomplished by addition of excipients, that is, compounds which ease manufacturing, increase stability or improve delivery of the active ingredients. Further specialised literature on this subject can be found for example in [109, 192, 317].

Part IV
Protein Size and Shape

Chapter 25
Centrifugation

Centrifugation is used for preparative and analytical separation of molecules and organelles (see [103]); centrifugation is one of the techniques used to determine the molecular mass from first principle, rather than by comparison with standard proteins. The other two are osmometry (see Chap. 26 on p. 251) and laser light scattering (see Sect. 29.1 on p. 261). For a recent review on modern centrifugation techniques see [229]. Centrifuges can be categorised by their running speed: **clinical** centrifuges allow velocities of a few thousand rounds per minute, **high speed** centrifuges up to 25 000 rpm. Centrifuges which allow even higher speeds are called **ultracentrifuges**. An **analytical** ultracentrifuge has the facility to observe the sample during the run. In a **preparative** ultracentrifuge, analysis of the sample is possible only outside the centrifuge after the run is completed, but rotor capacities can be much larger. Observation of the sample during the run in an analytical ultracentrifuge can be done with a photometer (detection limit several μM), fluorimeter (detection limit for fluorescein 5 nM) or by interferometry with a *Schlieren*-optics (see Fig. 8.2 on p. 68). Interferometry has the advantage that it can be used even in cases where the sample has no suitable absorption lines in the UVIS-range (*e.g.*, polysaccharides). On the other hand it is possible to distinguish different sample species by their absorption spectra.

Several manufacturers (Beckman-Coulter, Thermo) have introduced analytical ultracentrifuges capable of digital data acquisition during the last two decades, resulting in a revival of centrifugation as a tool to study the size and shape of proteins and their association into complexes **in solution** without the need for association to surfaces or matrices. Software for the evaluation of analytical runs is available either from instrument manufacturers or from the Internet, *e.g.*, `Sedfit` for Windows (www.analyticalultracentrifugation.com) or `UltraScan` for Linux (www.ultrascan.uthscsa.edu/).

Digital data acquisition results in large data sets (1×10^5 data points with a signal/noise ratio of 100–1000) allowing global curve fitting with narrow error margins for parameters (0.1 % under ideal conditions). Thus analytical ultracentrifugation has become an important tool to study the **interactome** of cells, or the way in which proteins form complexes to achieve physiological functions. Originally, analytical ultracentrifugation (both sedimentation velocity and sedimentation equilibrium techniques) were developed to study pure, (monodisperse) samples. Modern

E. Buxbaum, *Biophysical Chemistry of Proteins: An Introduction to Laboratory Methods*, DOI 10.1007/978-1-4419-7251-4_25,

approaches, however, allow the de-convolution of the data to obtain parameters of several species. These go beyond the scope of this book, an introductory discussion is available on the Internet (www.kolloidanalytic.de/uz/theoindex.htm).

Warning: Centrifuge, and especially ultracentrifuge, rotors can cause considerable damage to life and property in the event of a failure. Centrifuges have armoured shielding that is supposed to contain the fragments in such cases. Make sure you are properly trained in the operation of such equipment, and never run a rotor in a centrifuge not approved for that rotor type, as the kinetic energy of the rotor may exceed what the chamber can contain in an accident. As rotors and centrifuges get older, they are de-rated to account for possible metal fatigue. This requires complete logging of all runs. Do not exceed the rating of your equipment.

25.1 Theory of Centrifugation

For the purposes of molecular hydrodynamics, proteins are considered rigid, impenetrable, and incompressible bodies. The friction they experience comes only from the exposed surface.

25.1.1 Spherical Particles

The force $\vec{\mathfrak{F}}$ required to keep a particle at a given speed $\vec{\mathfrak{v}}$ depends on the speed and the friction that the particle experiences in its environment (expressed as friction coefficient f):

$$\vec{\mathfrak{F}} = \vec{\mathfrak{v}} * f \tag{25.1}$$

Since force is the product of mass (m) and acceleration ($\vec{\mathfrak{a}}$), replacement yields

$$m * \vec{\mathfrak{a}} = \vec{\mathfrak{v}} * f \tag{25.2}$$

For the ideal case of a sphere (with radius d in a medium of viscosity η), the friction coefficient can be calculated from STOKES' formula:

$$f_s = 6\pi\eta d \tag{25.3}$$

Note that if a protein were a perfect sphere, its radius would be

$$d = \sqrt[3]{\frac{3M_r\bar{\mathfrak{v}}}{4\pi N_a}} \tag{25.4}$$

with the **partial specific volume** $\bar{\nu}$ of a solute defined as the change of volume of a solution that is caused by the addition of a given amount of solute:

$$\bar{\nu} = \frac{\partial V}{\partial m} \approx 1/\rho_{\mathrm{p}}. \qquad (25.5)$$

The surface area of a soluble protein can be estimated by

$$A = \frac{11}{12} M_{\mathrm{r}}^{2/3} \qquad (25.6)$$

which is about twice that of a smooth sphere. Therefore the surface is quite bumpy, with plenty of opportunity for interaction with solutes. The effective mass of a sphere is the product of its volume ($V = 4/3 \times \pi \times d^3$) and its effective density ($\rho_{\mathrm{eff}} = \rho_{\mathrm{s}} - \rho_m$):

$$\vec{v} = \frac{4d^2 \vec{a}\, \rho_{\mathrm{s}} - \rho_{\mathrm{m}}}{18\eta}. \qquad (25.7)$$

Note that the effective mass will be negative if the density of the particle is less than that of the medium; in this case the particle will float rather than sediment. This can be useful to separate particles from contaminants of higher density. The density of large proteins is assumed constant $\rho_\infty = 1.4106\,\mathrm{mg/ml}$, below 30 kDa it can be calculated as a function of molecular mass according to:

$$\rho(M) = \rho_\infty + \Delta\rho_0 \mathrm{e}^{-(M_{\mathrm{r}}/k)} \qquad (25.8)$$

with $\Delta\rho_0 = 0.145\,28\,\mathrm{mg/ml}$ and $k = 13\,400\,\mathrm{Da}$ [101]. Small proteins fold more tightly than the larger ones, this can be confirmed from X-ray crystallographic structures.

According to (25.7) the sedimentation velocity $\vec{v}$ is proportional to the centrifugal acceleration $\vec{a}$, the square of the radius r from the centrifugal axis and the difference between the densities of the particle and the medium. This equation is valid only for spheres, but can be used as an approximation for "similar" shapes.

The centrifugal acceleration is the product of the rotor radius r and the square of the angular velocity $\vec{\omega}$. It is common to give the centrifugal acceleration relative to the gravitational acceleration on earth ($g = 9.81\,\mathrm{m/s^2}$, note that $g = |\vec{g}|$ is the absolute value of $\vec{g}$):

$$\vec{a} = \frac{r * \vec{\omega}}{g} = 11.18 * r * \left(\frac{\mathrm{rpm}}{1\,000}\right)^2. \qquad (25.9)$$

Since $\vec{v} \propto \vec{a} \propto r * \vec{\omega}^2$ SVEDBERG and PEDERSEN [395] defined the sedimentation constant S of a particle as sedimentation velocity relative to centrifugal acceleration:

$$S = \frac{\vec{v}}{\vec{\omega}^2 * r}, \qquad (25.10)$$

S has the unit of 1 s, but since many biological objects have sedimentation constants in the order of 10^{-13} s, this value is defined as 1 S (Svedberg). Salts have sedimentation coefficients of about 0.1 S, virus of more than 1 000 S. However, S depends on the density and viscosity of the medium, thus you will find values for water of 20 °C in the literature. Measured values can be corrected as follows:

$$S_{20,\mathrm{W}} = S_{\mathrm{T,m}} * \frac{\eta_{\mathrm{T,m}} * (\rho_{\mathrm{s}} * \rho_{20,\mathrm{W}})}{\eta_{20,\mathrm{W}} * (\rho_{\mathrm{s}} - \rho_{\mathrm{T,m}})} \tag{25.11}$$

assuming that the density of the particle does not change noticeably between 20 °C and the actual temperature. This correction is easily accomplished by the program SEDNTERP (www.rasmb.bbri.org/), taking into account the properties of many commonly used buffers.

The sedimentation constant of a particle is determined from the integrated form of (25.10):

$$S = \frac{\ln(r_t) - \ln(r_0)}{\vec{\omega}^2 * (t_t - t_0)}. \tag{25.12}$$

25.1.2 Non-spherical Particles

For non-spherical bodies the friction coefficient f can not be calculated easily. However, f can be obtained from the diffusion coefficient D:

$$D = R * T/f \tag{25.13}$$

with R the universal gas constant (8.315 J mol^{-1} K^{-1}) and T the absolute temperature. An elegant method to determine D is laser light scattering (see p. 261).

Since $S_{\mathrm{sphere}}/S_{\mathrm{obs}} = f_{\mathrm{obs}}/f_{\mathrm{sphere}}$, these ratios give the maximum shape asymmetry. Very large sedimentation constants compared to theoretical values for a given protein sequence point to oligomerisation, small values point to an extended shape of the molecule.

Historically, the sedimentation coefficient was determined from the movement of the midpoint of the concentration curve over time, but global fitting to the LAMM-equation, which describes the movement of a molecular species χ as a function of time and radial position under the influence of sedimentation and diffusion, is more precise:

$$\frac{\partial c}{\partial t} = D\left[\left(\frac{\partial^2 c}{\partial r^2}\right) + \frac{1}{r}\left(\frac{\partial c}{\partial r}\right)\right] - s\omega^2\left[r\left(\frac{\partial c}{\partial r}\right) + 2c\right]. \tag{25.14}$$

There is no explicit analytical solution to this equation, both SEDFIT [370] and UltraScan use finite element methods to solve it. These programs can also handle fitting in cases where the sample contains several components of different size. This may or may not be obvious by inspection of the experimental data [229].

25.1.3 Determination of Molecular Mass

25.1.3.1 Sedimentation Velocity

The determination of f and S of proteins is important because they are related to molecular mass M_r. For proteins larger than 30 kDa, $\bar{\nu}$ can be taken to be 0.71 ml/g, for smaller proteins the correction of (25.8) can be applied [101]. It is also possible to calculate $\bar{\nu}$ from the amino acid composition of the protein [444]. Then

$$M_r = \frac{S * f}{1 - \bar{\nu}\rho}. \tag{25.15}$$

For glycoproteins the overall partial specific volume is calculated from the weight fractions of protein and carbohydrate (w_p and w_c) and their respective partial specific volumes, $\bar{\nu}_p$ and $\bar{\nu}_c$. The latter depends on carbohydrate composition, usually determined by mass spectrometry.

25.1.3.2 Sedimentation Equilibrium

As molecules are sedimented by the centrifugal field their concentration at the bottom of the cell increases. This in turn increases their diffusion against the centrifugal field which is concentration dependent according to FICK's second law. If the experiment is performed at low centrifugal speed diffusion, sedimentation will be in equilibrium after some time and the concentration at the top and the bottom of the sample cell will no longer change. The BOLTZMANN-equation holds:

$$\frac{c_1}{c_2} = \mathrm{e}^{-\frac{E_1 - E_2}{k*T}} \tag{25.16}$$

with c_1 and c_2 the concentrations of solute with distances r_1 and r_2 from the rotor axis, E_1 and E_2 the corresponding potential energies and k = BOLTZMANN-constant (1.381×10^{-23} J/K).

$E_1 - E_2 = \Delta E$ is the work that has to be expended to move a molecule against the centrifugal force from r_2 to r_1:

$$\begin{aligned} \Delta E &= \int_{r_2}^{r_1} m(1 - \bar{\nu}\rho) \\ &= 1/2m(1 - \bar{\nu}\rho)\rightarrow^2(r_2^2 - r_1^2) \end{aligned} \tag{25.17}$$

If (25.16) and (25.17) are combined, replacing k by R to get M_r in Dalton rather than m in kilograms we get:

$$M_r = \frac{2\,R\,T}{(1 - \bar{\nu}\rho)\,\vec{\omega}^2} \ln\left(\frac{c_r}{c_a}\right) \frac{1}{r^2 - a^2} \tag{25.18}$$

with $a =$ distance of meniscus to rotor axis and r the distance to a second point in the vial. c_a and c_r are the corresponding protein concentrations. Plotting $\ln(c_r)$ against r^2 yields a straight line, the slope is M_r. If the experiment is performed in both water and $^2H_2{}^{18}O$, both M_r and $\bar{\nu}$ can be determined.

Note that these equations describe the ideal case of non-interacting particles. Experimentally, this requires moderately high salt concentrations ($\approx$100 mM) to suppress the effect of protein molecule charge which otherwise would result in an underestimation of the molecular mass. Protein concentration should be below 1 mg/ml.

If there are several sedimenting species present, the absorbtion $A(r)$ becomes:

$$A(r) = \sum_{i=1}^{n} c_{i,0}\epsilon_i d \times \exp\left[\frac{M_i(1-\bar{\nu}_i\rho)\omega^2}{2RT}(r^2 - r_0^2)\right] + \delta, \tag{25.19}$$

where the summation goes over all n molecular species, $c_{i,0}$ is the concentration of the ith species at the reference position r_0 and ϵ_i its molar absorbtion coefficient. d is the optical pathlength and δ a baseline offset that compensates for small differences in reference and measurement cells. The term $M_i(1-\bar{\nu}_i\rho)$ is the molecular mass of the protein reduced by the mass of the displaced buffer (buoyant molecular mass). For a single sedimenting species (25.18) and (25.19) are identical, except that parameters can be determined more precisely by curve fitting to (25.19) than by straight line fitting to the linearised form. Several programs for handling sedimentation equilibrium data are available, including WinNonlin (www.bbri.org/RASMB/rasmb.html), SEDPHAT (www.analyticalultracentrifugation.com) and UltraScan (www.ultrascan.uthscsa.edu/).

25.1.3.3 Membrane Proteins

Since only solubilised membrane proteins can be investigated by analytical centrifugation, the results of above experiments would be the buoyant molecular mass of the protein/detergent mixed micelles. This is the sum of the protein and detergent components, but if the density of the medium is identical to that of the detergent, the detergent becomes gravitationally transparent and the molecular mass of the protein can be determined [250, 266, 349]. Density can be matched with $D_2{}^{18}O$, sucrose, glycerol or nycodenz.

25.1.3.4 Using a Preparative Ultracentrifuge

Investigations described above are usually performed in analytical ultracentrifuges that allow observation of the sample during the run. For several decades, such instruments were difficult to get, now they are offered by several manufacturers. However, they are expensive and require trained operators. They can be used only for analytical purposes and require highly purified samples.

For these reasons the use of preparative ultracentrifuges, an equipment that is more widely available, for analytical purposes was explored originally by MARTIN AND AMES [258] for **sedimentation velocity** experiments. The samples are spun not in a homogeneous medium, but in a density gradient. The behaviour of particles in a gradient is described by the following equation:

$$\int_{t=0}^{t_a} \vec{\omega}^2 * S_{20,w} dt = \int_{r_0}^{r_c} \frac{\rho_p - \rho_{20,w}}{\eta_{20,w}} * \frac{\eta_{T,M}}{\rho_p - \rho_{T,M}} * \frac{1}{r} dr \qquad (25.20)$$

This equation does not have a closed solution, but can be solved numerically on a computer.

A gradient designed in such a way that the higher centrifugal acceleration a particle experiences as it moves further outward is exactly compensated by the higher density and viscosity of the medium is called **isokinetic**. Under these conditions the distance covered by a particle is a linear function of time, and the ratio of these distances for two molecules is equal to the ratio of their molecular masses. A linear 3–10 % sucrose or 10–30 % glycerol gradient are isokinetic in the Beckman SW60 rotor. For other rotors the gradient needs to be redesigned, but it has become common practice to just use the same gradient. Under these conditions

$$\frac{\Delta r(1)}{\Delta r(2)} \approx \frac{S_{20,w}(1)}{S_{20,w}(2)} \approx \frac{M_r(1)}{M_r(2)} \qquad (25.21)$$

This equation is valid only if both protein molecules have the same shape and partial specific volume. For soluble proteins it is assumed that they have, more or less, a spherical shape and a partial specific volume of about 0.71 ml/g (lower for glyco-, nucleo- and metalloproteins, higher for lipoproteins). You can expect an error of about 10 % from violations of these conditions.

POLLET et al. [333] have described the use of desktop ultracentrifuges for **sedimentation equilibrium** measurements. Centrifugation is carried out with small (100 μl) samples, the resulting gradient is stabilised by inclusion of 0.5 % bovine serum albumin. After the run, samples are fractioned with a capillary pipette controlled by a micro-manipulator.

Since the concentration of the analysed protein can be determined by enzymatic activity or its radioactivity, it can be present in much lower concentrations than those required for *Schlieren*-patterns. This allows the investigation of self-association – in the above paper of insulin – and of ligand binding. These methods do not require the use of sector-shaped cells and even work with fixed-angle rotors.

25.1.3.5 Self-forming Gradients

If a solution of a solute (*e.g.*, CsCl, metrizamide, nycodenz or percoll) is centrifuged, the solute will distribute according to (25.18). Thus a gradient will form under the centrifugal force from a homogeneous solution. Proteins, organelles or nucleic acids will move in this gradient until they have obtained a position where their own density is the same as that of the surrounding solution. This is important for measuring the density of nucleic acids. The classical experiment of MESELSON & STAHL [273], which demonstrated semi-conservative replication, used this method to achieve baseline-separation of ^{14}N- and ^{15}N-DNA. The concentration of the sample molecules as function of radius $c(r)$ follows a GAUSSIAN curve with

$$c(r) = c(r_0) \exp\left(\frac{-(r - r_0)^2}{2\sigma^2}\right), \tag{25.22}$$

$$\sigma^2 = \left(\frac{RT}{\omega^2 r_0 M_\mathrm{r} \bar{\nu} (\mathrm{d}\rho/\mathrm{d}r)}\right), \tag{25.23}$$

where $\mathrm{d}\rho/\mathrm{d}r$ is the slope of the density gradient.

Warning: Self-forming CsCl gradients have to be used with care. If they are spun too hard, the solubility of CsCl is exceeded at the bottom of the gradient, leading to crystal formation. These crystals can puncture the sample tubes under centrifugal force, leading to rotor imbalance and possibly rotor failure.

25.1.4 Pelleting Efficiency of a Rotor

There are many reasons why an experiment that worked in one rotor is to be repeated in another one. Larger sample sizes may be required, or the original rotor may no longer be available. Under these conditions it is useful to calculate how long the centrifugation will take in the new rotor. This is achieved by comparing the pelleting efficiency k of the two rotors, which is the time required to sediment a particle of given sedimentation constant:

$$\begin{aligned} k &= \frac{t}{S} \\ &= \frac{\ln(r_{\mathrm{max}}/r_{\mathrm{min}})}{\vec{\omega}^2} * \frac{10^{13}}{3\,600}, \\ &= \frac{\ln(r_{\mathrm{max}}/r_{\mathrm{min}}) * 253303}{(\mathrm{rpm}/1\,000)^2}. \end{aligned} \tag{25.24}$$

If the k-factors of both rotors are known, a procedure can be transferred between them using the simple relationship

$$\frac{k_1}{t_1} = \frac{k_2}{t_2}. \tag{25.25}$$

If a rotor is not to be run with maximum speed, a corrected k-factor (k') can be calculated as follows:

$$k' = k * \left(\frac{\text{rpm}_{\text{max}}}{\text{rpm}}\right)^2. \tag{25.26}$$

This equation is useful if a rotor can no longer run at its maximum speed due to old age. Such "derated" rotors can still be used, if centrifugation times are increased accordingly.

25.2 Centrifugation Techniques

Fractionated If the sedimentation constants of the particles in a sample are very different (by a factor of 3 or more), they can be separated by spinning the sample at a speed sufficient to pellet one component, but not the other component. This is a crude method usually used only for pre-purification (see Fig. 25.1). The

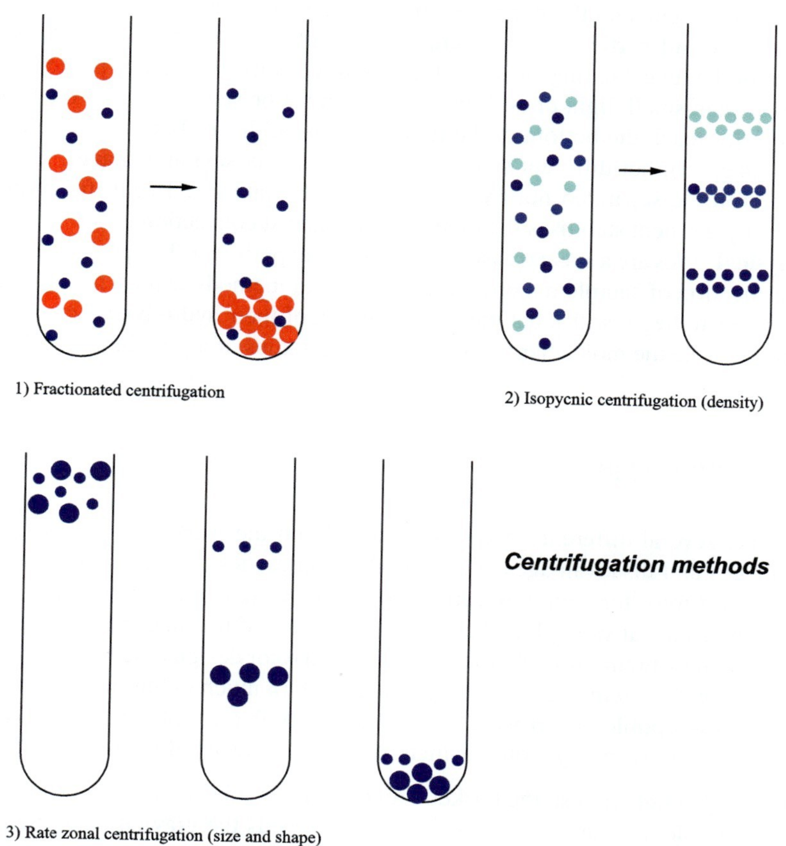

Fig. 25.1 Centrifugation methods. For details see text

pelleted material is always contaminated with the lighter particles, and needs to be resuspended and centrifuged again ("washing").

Isopycnic centrifugation is used to separate particles of different density. The centrifugation tubes are filled with a density gradient that encompasses the whole range of densities found in the sample particles. Centrifugation is continued until all particles of the sample have achieved their equilibrium position, that is until they are in a position were the density of the medium is the same as the density of the particle. Further centrifugation has no effect. Cell organelles tend to have quite different densities and can be separated by isopycnic centrifugation. Note that for cells, vesicles, and other closed membranes the equilibrium density depends on the osmotic pressure of the gradient medium, hence different equilibrium densities are observed with different gradient solutes. Isotonic gradients from metrizamide and nycodenz can be made by mixing isotonic solutions of these materials (290 mOsm) with an isotonic buffer to measure the "true" density of membrane systems.

Rate zonal centrifugation is used to separate particles by the sedimentation constant which is determined by their size and shape. The density gradient in the tube is designed so the density of the medium at the bottom is lower than the density of all particles in the sample. Thus all particles move toward the bottom of the tube, but they do so with different velocity. Big, heavy particles move faster than small, light ones. Centrifugation must be stopped before the heaviest particles reach the bottom of the tube. Proteins all have about the same density (≈ 1.4 g/ml), but different sizes (*i.e.*, molecular masses) and shapes. They can therefore be separated not by isopycnic, but by rate zonal centrifugation. The density gradient stabilises the tube contend against convection .

Analytical runs are a special case of a rate zonal experiment. The purpose is not the purification of sample material (most analytical methods require highly purified samples to begin with), but the determination of their hydrodynamic properties, for example the molecular mass and STOKES-radius of a protein molecule.

25.3 Rotor-Types

There are several different rotor types available on the market, each with its own advantages and disadvantages. Knowing their properties should enable you to select the best rotor for your application. Old rotors were made of aluminum, but the limited mechanical strength and chemical resistance of this material have led to its replacement by titanium steel. Some very modern rotor designs use composite materials (carbon fiber reinforced plastic), which result in much lighter rotors, which are even less susceptible to corrosion. Because of the high price of composite materials they are used only in large rotors which would be too heavy if made from titanium.

Swinging bucket In rest, the bucket with the sample tube hangs downward. As the rotor accelerates and the centrifugal force exceeds the gravitational, the bucket swings into a horizontal position. Then the direction of the centrifugal force and the axis of the centrifugation tube are identical. Particles therefore move through the whole length of the tube, which can take considerable time, but leads to good resolution.

Fixed angle In a fixed angle rotor the tube axis is not identical with the direction of the centrifugal force. Particles are first pressed against the wall of the tube, then slide along the wall to the bottom. This results in short centrifugation times. The rotors are also more robust and easier to set up than swinging bucket ones and they can be run at higher speeds. However, fixed angle rotors are unsuitable for rate zonal experiments.

Vertical tube A vertical tube rotor is the extreme case of a fixed angle rotor, with the tube at 90° to the direction of the centrifugal force. However, from their mode of operation they are more similar to swinging bucket rotors. When the rotor is at rest, the gradient in the tubes is oriented from top to bottom. As the rotor accelerates and the centrifugal force exceeds the gravitational, the gradient re-orientates itself from the inner to the outer wall of the tube. In this gradient the particles move. When the rotor decelerates, the gradient again re-orients itself to the top-bottom direction. One might expect these re-orientations to result in disturbed gradients, but the opposite is true. Because the distance between the inner and outer wall is small, centrifugation times are short; because the gradient after the run has the whole length of the tube, resolution is good. Since there are no wall effects, VT-rotors can be used for rate zonal experiments. Handling and stability of a VT-rotor is similar to a fixed angle one.

Near vertical tube While a VT-rotor combines the advantages of swinging bucket and fixed angle rotor, there is one problem left. If the sample contains precipitating material, this will be distributed along the outer wall of the tube, and upon redistribution of the gradient as the rotor decelerates contaminate all fractions of the gradient. For such cases the NVT-rotor was designed. The tubes stand a few degrees off the vertical, so precipitating material should experience a wall effect and slide to the bottom of the tube to form a pellet. On the other hand the angle of the tube is small enough that non-precipitating particles do not experience noticeable wall effects.

Continuous flow and zonal rotors are designed for large sample volumes. They are loaded and unloaded while the rotor is spinning. Their use requires special training and a modified centrifuge. However, if those are available, these rotors can make available experimental material in larger quantities, and often with higher purity, than other rotor types.

25.4 Types of Centrifuges

There are some centrifuge types which you are unlikely to encounter in a lab like milk centrifuges. Centrifuges driven by a hand-operated crankshaft, once common in chemistry labs, have been largely replaced by electrical instruments (which is a shame, they were fun!). The following types are common:

Clinical centrifuges operate at relatively small speeds (several hundred to several thousand rpm) on small volumes (up to 50 ml per sample). Special rotors are available for 96-well plates and other sample containers. They come with or without refrigeration; if you need to keep your samples cool you need a refrigerated centrifuge! Placing a non-refrigerated centrifuge into a cold-room does not work well: The air resistance of the rotor results in time-dependent heating.

High-speed centrifuges operate with large rotors (up to several l of sample) at moderate speeds. Up to about 20 000 rpm cooling is efficient enough to remove friction heat at ambient pressure, higher speeds require a vacuum. These instruments are workhorses for protein or organelle isolation.

Preparative ultracentrifuges come either as large floor models or small, desktop ones. They are always refrigerated and operate at speeds up to 150 000 rpm corresponding to about 800 000 g. The **airfuge** was a small desktop centrifuge driven by a stream of nitrogen acting on a turbine, rather than an electric motor. Because of the high speeds involved all ultracentrifuges must be operated with the greatest care!

Analytical ultracentrifuges are dedicated instruments for the measurement of hydrodynamic properties. Because of the long path-lengths required, instruments like the "Model E" used to be quite large. In modern instruments the optical path is folded by mirrors and hence fits into the same case as a preparative ultracentrifuge.

25.5 Determination of the Partial Specific Volume

The partial specific volume $\bar{\nu}$ [211] is defined by:

$$\bar{\nu} = \frac{\partial V}{\partial m} \tag{25.27}$$

$$= \frac{1}{\rho_\mathrm{b}}\left(1 - \frac{\rho_\mathrm{solution} - \rho_\mathrm{b}}{c}\right) \tag{25.28}$$

with ρ_b and ρ_s the density of buffer and protein solution respectively and c the protein concentration. If the allowable error for $\bar{\nu}$ is $\epsilon = \Delta\bar{\nu}/\bar{\nu}$, then

$$\Delta(\rho_\mathrm{solution} - \rho_\mathrm{b}) = -\rho_\mathrm{b} c \bar{\nu} \epsilon \tag{25.29}$$

and

$$\frac{\Delta c}{c} = \frac{\bar{\nu}\rho_\mathrm{b}}{1 - \bar{\nu}\rho_\mathrm{b}}\epsilon \tag{25.30}$$

With $\bar{\nu} = 0.75$ ml/g, $c = 5$ mg/ml, and $\rho_\mathrm{solvent} = 1.0$ mg/ml for $\epsilon = 1\,\%$ $\Delta(\rho_\mathrm{solution} - \rho_\mathrm{solvent})$ must be $\pm 4\,\mu$g/ml, and the individual densities must bedetermined to

$\pm 2\,\mu\text{g/ml}$. This requires thermostating the solutions to $\pm 10^{-2}\,^{\circ}\text{C}$ so that results are not unduely influenced by thermal expansion. Fortunately, only relative values for the densities are required.

Density measurements to that precision are possible by determining the natural frequency of a U-shaped glass tube filled with either solvent or solution ($f \approx 500\,\text{Hz}$). The position of the vibration nodes are kept constant by an abrupt change in tube radius. The effective mass of the vibrator is $m = m_0 + \rho V$, with m_0 the mass of the empty tube. For an undamped oscillation

$$2\pi f = \sqrt{c/m} = \sqrt{\frac{c}{m_0 + \rho V}} \tag{25.31}$$

Since m_0 and V are constant and the concentration is fixed, the density is

$$\rho = A((1/f)^2 - B) \tag{25.32}$$

and a density difference

$$\rho_1 - \rho_2 = A[(1/f_1)^2 - (1/f_2)^2] \tag{25.33}$$

The instrument constants A and B are determined by filling the tube with air and water as standards, whose values are fixed by international conventions.

Chapter 26
Osmotic Pressure

Assume two compartments separated by a semipermeable membrane (see Fig. 26.1). One compartment contains pure solvent, which can freely cross the membrane, for example, water. The other compartment contains solvent plus a solute, which is too large to cross the membrane, say, a protein. If both compartments are open to the atmosphere – that is, exposed to the same pressure – then water will pass from the first compartment into the second in an attempt to equalise the chemical potentials on both sides. However, no matter how much water passes from compartment 1 to compartment 2, the protein concentration in the second compartment (and hence its chemical potential) will always be larger than that in the first. To establish an equilibrium, we have to apply an outside force that counteracts the force drawing water into compartment 2. This can be achieved by putting pressure onto compartment 2, which would press water from there into compartment 1. At a certain pressure, the pressure-driven flow of water from 2→1 will exactly balance the concentration driven flow from 1→2. We call this the osmotic pressure π of solution 2. This pressure is related to the solute concentration by

$$\Pi = RT\left(\frac{1}{M_r}c + B_2c^2 + \dots\right), \tag{26.1}$$

where c is the concentration (in g/ml), B_2 the second virial coefficient[1] (higher coefficients are not usually required), R is the universal gas constant, T the absolute temperature, and M_r the molecular mass. Thus, in principle, measuring the osmotic pressure of the solution of a macro-molecule with known mass concentration is a simple way to determine the molecular mass M_r.

[1] The virial equation ($V_m * P = n * R * T + n * B_2 * P + n * B_3 * P^2 + \dots$, P = pressure, V_m = molecular volume) is used to describe the behaviour of non-ideal gases. The ideal gas law is $P * V = n * R * T = n * B_1$. $B_2, B_3, \dots$ = second, third, ... virial coefficients are used to describe the interaction strength of molecules, they are negative for attractive and positive for repulsive forces. In most cases inclusion of the second virial coefficient is sufficient to account for observed behaviour under biologically relevant conditions. A common way to determine the virial coefficients is to measure the osmotic pressure (Π) of a solution as function of its concentration: $\frac{\Pi}{\varrho k_B T} = 1 + B_2\varrho + B_3\varrho^2 + \dots$ with $\varrho = \frac{cN_a}{M_r}$ the density of particles in solution. Other methods to determine virial coefficients include static light scattering and small angle neutron or X-ray scattering.

E. Buxbaum, *Biophysical Chemistry of Proteins: An Introduction to Laboratory Methods*, DOI 10.1007/978-1-4419-7251-4_26,

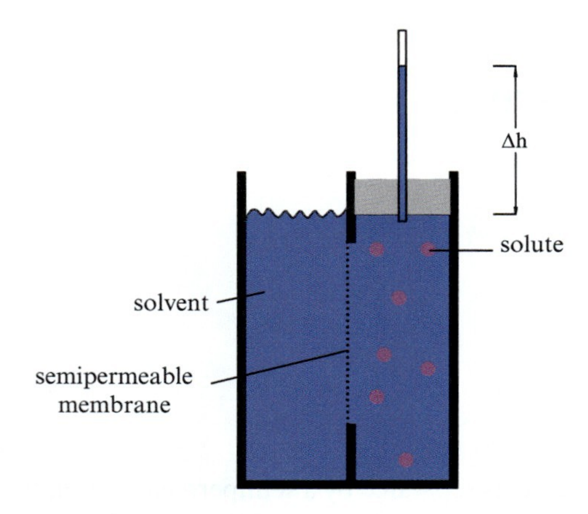

Fig. 26.1 Principle of the PFEFFER-cell. Two solutions with different concentrations of solute are separated by a semi-permeable membrane, through which solvent, but not solute, molecules can pass. The device is named after WILHELM PFEFFER (1845–1920), who built this device based on earlier work by MORITZ TRAUBE (1826–1894). Solvent rises into the pipe until the hydrostatic pressure $p = \Delta h \times \rho$ driving fluid into the compartment with pure solvent exactly opposes the osmotic pressure π, drawing fluid out of that compartment

Above equation may be rewritten as

$$\frac{\Pi}{c} = \frac{RT}{M_r} + B_2RTc + \ldots \tag{26.2}$$

If one plots $\frac{\Pi}{c}$ as function of c, the molecular mass can be determined from the y-intercept and the virial coefficient from the slope of the resulting line. For a mixture of macro-molecules, the number-averaged molecular weight ($\sum c_i / \sum (c_i M_i)$) will be obtained.

Although ENGELHART had already shown the macromolecular nature of haemoglobin in 1825 from the iron/protein ratio [88], most scientists refused to believe that molecules could be that big. The determination of the molecular mass of haemoglobin to 67 kDa by ADAIR [3,4] a hundred years later by osmotic pressure changed this.

The cytosol of a cell has an osmotic pressure of 13 bar!

26.1 Dialysis of Charged Species: The DONNAN-Potential

Above derivation assumes that all species involved in the equilibrium are non-charged. If the macro-molecule were charged, it would have counter-ions. In this situation the counter-ions, but not the macro-molecules, can pass the membrane. The result is an electrical potential difference ψ across the membrane, which can seriously affect measurements of osmotic pressure. The electrochemical potential of an ion of charge z is

$$\mu_i' = \mu_i^0 + RTln(c_i) + zF\psi \tag{26.3}$$

with F the FARADAY-constant of 96 485 J/(mol V). This leads to

$$\Delta\psi = \frac{RT}{zF} \ln \frac{c_1}{c_2} \tag{26.4}$$

The simplest way to prevent such potentials from affecting the results of measurements is to include a sufficient concentration of salts which swamp any effects caused by the macro-molecule. 100 mM NaCl is frequently used.

Chapter 27
Diffusion

Molecules move by random BROWNIAN motion so that over time all concentration differences are evened out and the entropy of the system is maximised. FICK**'s first law of diffusion** (for simplicity here for the 1-D case) states that the net flux J is proportional to the concentration difference and the pathlength x:

$$J = -D\left(\frac{\partial c}{\partial l}\right) \quad (27.1)$$

where D is the diffusion coefficient of the substance in question. Because of mass conservation, a molecule entering a given volume element must either leave again, or increase the concentration in that element, this is expressed in the **continuity equation**:

$$\left(\frac{\partial c}{\partial t}\right) = -\left(\frac{\partial J}{\partial l}\right) \quad (27.2)$$

FICK**'s second law** states:

$$\left(\frac{\partial c}{\partial t}\right) = D\left(\frac{\partial^2 c}{\partial l^2}\right) \quad (27.3)$$

For multi-dimensional diffusion the equations are similar, if more complex. The average displacement $< l >$ of molecules, which spread from an infinitesimal thin starting zone, is time-dependent:

$$< l^2 >= 2Dt \quad (27.4)$$

which is the equation for a random-walk from the origin. The width of the starting zone increases with the square-root of time.

According to EINSTEIN, the diffusion coefficient D is related to the **friction coefficient** f by:

$$D = k \times \frac{T}{f} = \frac{R}{N_a} \times \frac{T}{f} \quad (27.5)$$

STOKES has derived f for a sphere of radius r in a medium of viscosity η:

$$f_0 = 6\pi\eta r \quad (27.6)$$

E. Buxbaum, *Biophysical Chemistry of Proteins: An Introduction to Laboratory Methods*, DOI 10.1007/978-1-4419-7251-4_27,

and the diffusion coefficient of the sphere becomes

$$D_0 = \frac{RT}{6N_a \pi \eta r} \tag{27.7}$$

For other shapes of importance to protein science the friction coefficients have also been derived:

Shape	f/f_0	r_e
Prolate ellipsoid	$\frac{P^{-1/3}(P^2-1)^{1/2}}{\ln[P+(P^2-1)^{1/2}]}$	$(\alpha\beta^2)^{1/3}$
Oblate ellipsoid	$\frac{(P^2-1)^{1/2}}{P^{2/3}\tan^{-1}[(P^2-1)^{1/2}]}$	$(\alpha^2\beta)^{1/3}$
Rod	$\frac{(2/3)^{1/3}P^{2/3}}{\ln(2P)-0.30}$	$\left(\frac{3\beta^2\alpha}{2}\right)^{1/3}$

where α is the radius along the long (major) axis, β the radius along the short (minor) axis, $P = \alpha/\beta$ and r_e the equivalent radius of a sphere of equal volume to that of the ellipsoid or rod.

Chapter 28
Viscosity

Interaction between molecules slows down their movement; this intermolecular friction is observed macroscopically as **viscosity**.

To measure the viscosity of a liquid two parallel glass plates can be used. The space between them is filled with the liquid. If one plate is moved relative to the other, fluid next to the moving plate has to have a velocity near that of the plate, while fluid next to the stationary plate has to have a velocity close to zero. The shear force $\vec{\mathfrak{F}}$ required to move one plate against the other is given by:

$$\vec{\mathfrak{F}} = \eta A \frac{\mathrm{d}\vec{\mathfrak{v}}}{\mathrm{d}y} \tag{28.1}$$

with η the viscosity, A the area, y the distance between the plates and $\vec{\mathfrak{v}}$ their relative velocity. $\frac{\mathrm{d}v}{\mathrm{d}y}$ is the shear gradient in the liquid. In NEWTONIAN fluids η is independent of the shear gradient, in non-NEWTONIAN fluids it is not.

If fluid flows through a long, narrow tube, the flow is fastest in the centre of the tube, and near zero next to the walls. The flow rate is given by HAGEN-POISEUILLE's law:

$$\frac{\Delta V}{\Delta t} = \frac{\pi r^4}{8\eta l} \Delta P, \tag{28.2}$$

where $\frac{\Delta V}{\Delta t}$ is the volume flow rate, r and l the radius and length of the tube, and ΔP the pressure difference over the tube. In OSTWALD's capillary viscosimeter (see Fig. 28.1) the fluid is moved by suction up a U-shaped tube. It is then allowed to flow back by gravity force, the time t required for the meniscus to drop between two marks is measured, the volume between the marks is known. Then

$$\eta = \frac{\pi r^4 t}{8 l V} h \mathrm{g} \rho \tag{28.3}$$

with h the average height difference, g the gravitational acceleration and ρ the density. Their product is, of course, the pressure driving the fluid through the capillary. Although in theory the setup can be used to determine absolute viscosities, in

E. Buxbaum, *Biophysical Chemistry of Proteins: An Introduction to Laboratory Methods*, DOI 10.1007/978-1-4419-7251-4_28,

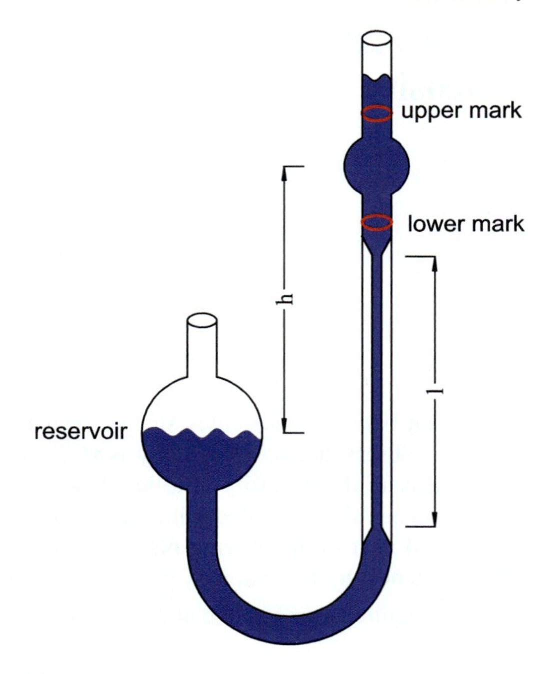

Fig. 28.1 OSTWALD viscosimeter. The time required for the fluid to fall from the upper to the lower mark is measured. The instrument is usually used to measure relative viscosities

practice one usually determines relative viscosities η_r (pure solvent η_0 and solutions of macro-molecules of several concentrations η_x):

$$\eta_r = \frac{\eta_x}{\eta_0} = \frac{t_x \rho_x}{t_0 \rho_0} \tag{28.4}$$

The only data required are the densities of the solutions, which can be determined with high accuracy by the method described on p. 248.

Alternative methods to measure viscosities include the COUETTE and the floating rotor viscosimeter, where the sample is between a static outer and a rotating inner cylinder. The viscosity is calculated from the torque required to rotate the inner cylinder. These work better than the OSTWALD viscosimeter for non-NEWTONIAN fluids.

From the relative viscosity η_r the **specific viscosity** η_s can be calculated:

$$\eta_\mathrm{s} = \frac{\eta_x - \eta_0}{\eta_0} = \eta_\mathrm{r} - 1 \tag{28.5}$$

If the specific viscosity is plotted as a function of sample concentration $[X]$, a straight line is obtained that can be extrapolated to $[X] = 0$ to obtain the **intrinsic viscosity** $[\eta]$ as y-intercept.

The intrinsic viscosity $[\eta]$ of a molecule depends on its size and shape. For a coil protein (in 6 M guanidinium hydrochloride) the empirical relationship between the number of residues n and $[\eta]$ is

$$[\eta] = 0.716n^{0.66}. \tag{28.6}$$

Intrinsic viscosity can be used to monitor any process that changes either size or shape of a protein, *e.g.*, folding and oligomerisation.

Chapter 29
Non-resonant Interactions with Electromagnetic Waves

29.1 Laser Light Scattering

29.1.1 Static Light Scattering

An electromagnetic field of light which is not in resonance with energy levels of the electrons of a sample can induce forced oscillations of these electrons, which then re-emit light of the same wavelength as the incident light, but in a different direction. This is called scattering. The intensity of the scattered light depends on the molecular polarisability α of the molecule

$$\alpha = \frac{n_0 M_r \partial n}{2\pi N_a \partial c} \quad (29.1)$$

where n_0 is the refractive index of the solvent, M_r the molecular mass and N_a AVOGADRO's number. $\partial n/\partial c$ is the refractive index increment of the sample as function of the weight concentration (g/ml), conveniently measured with a differential refractometer that can determine even small differences in refractive index with the required precision (see Fig. 29.1).

Scattering is qualitatively observed by the TYNDALL-effect: If a sample is observed at right angle to an incident light beam, particles present in the sample will scatter light into the eye of the observer. This is a very sensitive procedure, even low concentrations of very fine particles will cause a visible haze. In higher concentrations scattering can also be measured as loss of intensity in the incident light beam (for example measuring the concentration of bacteria in a liquid medium at 600 nm in a photometer).

Light scattering gives information on the size and shape of a particle (*e.g.*, large molecules or liposomes). Today, lasers are almost exclusively used as light sources, their light is linear polarised (laser light scattering (LLS)). As long as the particles are much smaller than the wavelength of the light ($d \leq 0.1 * \lambda$), the intensity of the light scattered is:

E. Buxbaum, *Biophysical Chemistry of Proteins: An Introduction to Laboratory Methods*, DOI 10.1007/978-1-4419-7251-4_29,

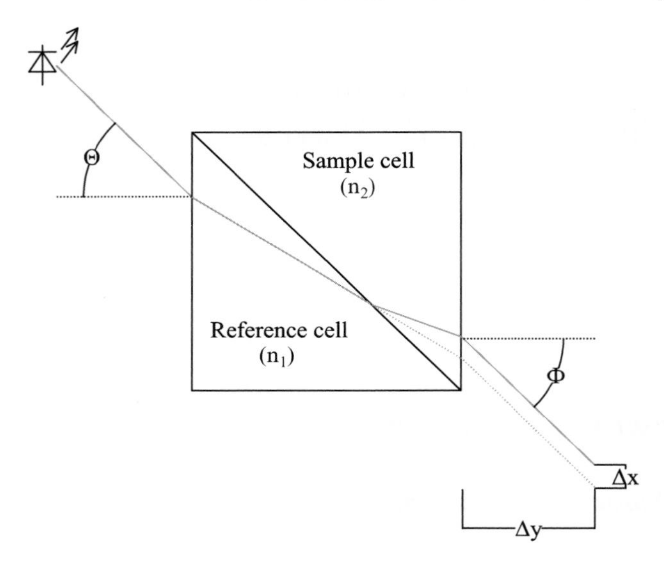

Fig. 29.1 Differential refractometer. The light of a LED is passed through a collimating optics onto the reference and sample cells, and then onto a photodiode array detector. The angle $\Phi = \Delta n / \tan(\Theta)$ is measured as height difference $\Delta x = 2\Delta y \tan(\Theta)\Delta n$, and since Δy and Θ are constant, $\Delta x = c\Delta n$, with c an instrument constant. Sensitive instruments can measure differences in refractive index of 10^{-8}

$$\begin{aligned} R_\theta &= \frac{I_{\mathrm{sc}}}{I_0} \\ &= \frac{16\pi^4\alpha^2}{r^2\lambda^4} \\ &= \frac{4\pi^4 n_0^2}{N_{\mathrm{A}}\lambda^4 r^2} \left(\frac{\partial n}{\partial c}\right) c\, M_{\mathrm{r}} \\ &= K c M_{\mathrm{r}} / r^2 \end{aligned} \tag{29.2}$$

with λ = wavelength of light, r = distance of observer, K = instrument constant, θ angle of scattering, R_θ = RAYLEIGH-ratio.

The refractive index increment is a function of the electron density of the molecule, and very similar for biologically interesting compounds. Hence scattering is a direct function of the particle mass. However, scattered light has a low intensity, high concentrations and high laser intensities are required. As a rule of thumb, for a laser beam of 100 mW and a molecular mass of 100 kDa, particle concentrations of 1 mg/ml are required for reliable results. These high concentrations lead to non-ideal behaviour, thus **virial coefficients** (see Footnote 1 on p. 251) need to be taken into account:

$$\frac{1}{M_{\mathrm{r}}} = \frac{K}{R_\theta} c + 2\, B_2\, c + \ldots \tag{29.3}$$

Scattering Kc/R_θ is measured as function of sample concentrations. If the plot is linear, one can extrapolate to zero concentration, and also determine the virial coefficient as slope of the line. Non-linear plots are found if the sample has a concentration-dependent association-dissociation equilibrium. If the solution contains a mixture of solutes the weight average molecular weight ($\sum c_i M_i / \sum c_i$) is determined if all molecules have similar $\partial n/\partial c$. Dust-particles count as molecules of very high molecular mass: use filtration or centrifugation to ensure that your sample is dust-free!

If the particles are larger than 10 % of λ, destructive interference between light waves scattered on different parts of the particle causes the intensity of scattered light to depend on the scattering angle:

$$\lim_{\theta \to 0} \frac{Kc}{R_\theta} = \left[1 + \frac{q^2 r_G^2}{3}\right] \left[\frac{1}{M_r} + 2\,B_2\,c\right], \tag{29.4}$$

$$q^2 = \left(\frac{2\pi \sin(\theta/2)}{\lambda}\right)^2, \tag{29.5}$$

$$r_G^2 = \frac{\sum_{i=1}^{n} m_i r_i^2}{M_r} \tag{29.6}$$

with r_G = radius of gyration, the mass-weighted average distance of the various centres of mass of parts of the particle from the centre of mass of the entire particle, m_i = mass of the ith element, r_i = distance of the ith element from the common centre of mass.

Thus light scattering can not only determine the molecular mass of a particle, but also its size in terms of r_G. For simple shapes, r_G can be calculated:

Shape	r_G
Sphere of radius r	$r * \sqrt{3/5}$
Stick of length l	$l\ /\ \sqrt{12}$
Prolate ellipsoid of radii α, α, γ	$\alpha\sqrt{2 + \gamma^2/5}$
Coil of n links of length l	$l\sqrt{n}/\sqrt{6}$

29.1.2 Dynamic Light Scattering

BROWNIAN motion of particles leads to a fluctuation of the intensity of scattered light from a small volume element of solution over time. Thus measured scattering intensity has a noise component which depends on the motion of the scattering particles. This noise is caused by interference of light scattered by different particles and constantly changes because of the motion of the particles.

The fluctuation characteristics of the noise signal can be determined by an **autocorrelation function (ACF)** :

$$\mathrm{ACF}(\tau) = \langle I(t) * I(t+\tau)\rangle / \langle \bar{I} \rangle^2, \tag{29.7}$$

where $\langle\rangle$ means averaging over all times t using a fixed time difference τ. $\bar{I}$ is the average intensity. If τ is short, values of intensity can be expected to be close together, *i.e.*, to be highly correlated. As τ becomes larger, the intensities become more likely to have dissimilar values, thus the autocorrelation becomes smaller exponentially:

$$\mathrm{ACF}_{\mathrm{Trans}} = A_0(1 + f * \mathrm{e}^{-(D_{\mathrm{Trans}} q^{-2} t)}) \tag{29.8}$$

$\mathrm{ACF}_{\mathrm{Trans}}$ is the autocorrelation caused by translational movement, which depends on the translational diffusion coefficient D_{trans} and the coherency of the light beam (0..1). Fluctuations may be caused either by simple diffusion, or by chemical reactions. It is thus possible to determine reaction rates from correlation, without perturbation of the equilibrium and without separation of educts and products [86, 252].

Once the size of the molecule becomes comparable to the wavelength, rotational diffusion and internal motion of the molecule also contribute to light scattering.

29.1.3 Quasi-elastic Scattering

So far we have looked at the case of completely elastic light scattering, where the intensity of scattered light depends on the size of the scattering particle. In quasi-elastic scattering DOPPLER-effects lead to a change in the wavelength of the scattered light, depending on the speed of the particle that causes scattering. Since the particles move in random direction, the wavelength of the scattered light can be higher and lower than that of the incident light, the spectral width (width of the peak at half maximal hight $\Delta\tilde{\nu}$) of the scattered light is higher than that of the incident light and is described by a LORENTZ-curve:

$$I(\tilde{\nu}) = \Delta\tilde{\nu} * \frac{1}{(\tilde{\nu} - \tilde{\nu}_0)^2 + (\Delta\tilde{\nu})^2} \tag{29.9}$$

$$\Delta\tilde{\nu} = \frac{D * q^2}{2\pi} \tag{29.10}$$

$$|q| = \frac{4\pi}{\lambda} \sin\left(\frac{\theta}{2}\right) \tag{29.11}$$

$$D = \frac{kT}{6\pi\eta r_s} \tag{29.12}$$

with D = diffusion coefficient and r_s = radius of a sphere with given diffusion coefficient.

29.1.4 Instrumentation

The sample must be completely free of dust and gas bubbles (vacuum filtration), about 1 ml is required. Cuvettes used in light scattering are round to allow the observation of the sample under any angle. Concentration required is higher for lower molecular mass and longer wavelength (which means that the recent appearance on the market of blue-violet laser diodes constitutes a significant progress). To reduce the effect of light refraction, the cuvette is placed in a bath of toluene which has a similar refractive index like glass and a known RAYLEIGH-ratio, thus the instrument constant K can be determined by simply removing the cuvette. The scattered light is detected by a photomultiplier mounted onto a goniometer.

Simple light scattering instruments are used for quality control in polymer chemistry, they measure the size distribution of molecules without much user input. Research instruments on the other hand allow the determination of molecular mass, radius of gyration, and diffusion coefficient. These are usually found in institutes that specialise on molecular hydrodynamics (and which usually also have complementary equipment, like analytical ultracentrifuges). Instruments which measure light scattering online at the exit of a gel filtration column are also available.

29.2 Small Angle X-ray Scattering SAXS

As we have seen, for light scattering experiments the molecule should have a diameter of about $0.1 * \lambda$. For very large molecules, and for particles like liposomes, visible or UV-light can be used. However, biological samples absorb below 300 nm and far-UV light below 200 nm is absorbed by nearly everything. Shorter wavelengths (X-rays below 1 nm) can penetrate samples again, but become short compared to biomolecules. For example, the CuK_α-radiation has a wavelength of 1.54 Å, about one-tenth of the radius of a small protein. Scattering is limited to very small angles (a few degrees). Equation (29.4) then leads to the GUINIER-equation, valid for dilute solutions:

$$\lim_{\theta \to 0} \frac{Kc}{R_\theta} = \exp\left(\frac{16\pi^2 r_G^2}{3\lambda^3}\right) * \sin^2\left(\frac{\theta}{2}\right). \tag{29.13}$$

Scattering data have been used to aid in the reconstruction of the 3-D structure of molecules [396] or those parts of a molecule that are not well resolved in X-ray structures [325].

29.3 Neutron Scattering

Thermic neutrons have an equivalent wave length of 2–4 Å, and can be used for the determination of r_G. However, they are scattered by collision with nucleï rather than by interactions with the electrons of a molecule. The refractive angle of the solvent for neutrons depends on the $^1H/^2H$-ratio, which can be adjusted to mask certain compounds. The weak interaction of neutrons with matter requires high sample concentrations (mg/ml and higher). Thermic neutrons are produced by nuclear fission in research reactors.

29.4 Radiation Inactivation

The probability to hit a target increases with its size. Hitting a molecule with energy rich radiation leads to its fragmentation. Only accelerated electrons or neutrons can be used because they penetrate the sample and the doses received by molecules in the front and in the back of the sample are identical. Available flux densities with electrons are higher, they are produced in a linear accelerator (about 10 MeV).

Samples are prepared under exclusion of oxygen, and snap-frozen in liquid nitrogen. Some cryoprotectant like sucrose or glycerol may be added. Samples are kept deep frozen during the experiment (−60 to −80 °C, this requires cooling with liquid nitrogen since considerable amounts of heat are produced in the beam line). Beam intensity plotted as a function of the distance from the centre of the beam forms a 2-DGAUSS-curve (diameter ≈1 m), so all samples can be irradiated at the same time, the dose of the samples depends on their position and is determined by photo-dosimeters.

Thawing of the sample is done under a protective nitrogen or argon atmosphere, the samples should be analysed immediately after thawing.

Target theory then describes the relationship between surviving protein and radiation dose. The probability $P(n)$ of a volume V_i to receive n hits is given by the POISSON-distribution :

$$P(n) = \frac{e^{V_i} * V_i^n}{n!} \tag{29.14}$$

The protein will survive only if its volume is not hit at all, that is, its survival probability is $P(0)$, so we can write:

$$N/N_0 = e^{-V_i}. \tag{29.15}$$

The molecular mass of a protein is

$$M_r = V_i * \rho * N_A. \tag{29.16}$$

The density ρ of proteins can be assumed to be 1.411 g/ml for large proteins, for proteins <30 kDa the density is higher and can be calculated from (25.8) on p. 239. AVOGADRO's number N_a is 6.022×10^{23} particles/mol.

Thus if the damage of a passing electron were limited strictly onto its path, radiation target analysis would be a method to determine the molecular mass of a protein based on first principles (if the density is determined separately). Unfortunately, a passing electron will produce a spur of long range secondary electrons (**Δ-rays**), which can inactivate proteins that are not directly in the path of the primary electron. Thus the exponent of (29.15) is not directly proportional to the protein volume.

In practice, the logarithm of remaining activity is plotted as a function of radiation dose, the slope of the resulting lines is plotted for a few standard proteins as function of the molecular mass to obtain a standard curve. From this an estimate of the molecular mass of the sample can be obtained.

The advantage of this method is that it can be performed on impure samples.

Because radiation inactivation is not based on first principles; molecular mass determined vary amongst different laboratories, depending on the standard proteins used and the exact procedure followed. Thus radiation inactivation is more suitable for relative than for absolute measurements. For example, P-type ATPases like Na/K-ATPase cycle between the so called E_1 and E_2-conformation. Partial reactions of the enzyme which require only one of the conformations have a radiation target size equivalent to the molecular mass of one catalytic subunit. Those reactions however, which require cycling between both conformations have a target size of about twice that value [46]. This would indicate that Na/K-ATPase is active as diprotomer.

Part V
Protein Structure

Chapter 30
Protein Sequencing

An important step in the characterisation of a new protein is the determination of its primary structure (amino acid sequence). This can be determined either by repeated chemical cleavage of terminal amino acids (either from the N- or C-terminal end, see Fig. 30.1 and Fig. 30.2 respectively), or by MS/MS.

30.1 Edman Degradation

If the protein is hydrolysed, the amino acids can be separated by ion exchange or reversed phase chromatography and quantified. For quantitation the amino acids are converted to coloured (with ninhydrin) or fluorescent (with OPA) derivatives either before or after chromatography (pre- and post-column derivatisation, respectively). This gives the amino acid composition, but not the sequence.

To get the sequence, the amino acids need to be removed one at a time. The classical way to do this is EDMAN-degradation (see Fig. 30.1).

The protein is bound to a matrix and its N-terminal amino acid converted to a phenyl thiourea derivative, which is cleaved off with acid. Because the efficiency of the reaction is only about 80–96 %, the signal diminishes each cycle. Under optimal conditions, up to 50 amino acids can be sequenced, and only about 100 pmol of purified protein is required. To get high sequencing yields from small samples, the use of automated sequencers is required. They also limit the amount of (expensive) staff time required. Two basic principles are now in use:

Gas phase sequencers The sample protein is held on a membrane (glass fibre or PVDF), the coupling base (trimethylamine) and cleavage acid (trifluoroacetic acid) are added to the sample as vapour by an argon stream. No washout of sample can occur. Only the extraction of the reaction products by organic solvents occurs in the liquid phase.

Pulsed liquid phase sequencers The basic principle is the same as in the gas phase sequencer, except that the cleavage acid is added as a small droplet; just enough volume to moisten the sample. During the cleavage step the excess acid evaporates. The time required for the cleavage reaction compared to the gas phase sequencer is reduced, resulting in cycle times of less than 30 min.

E. Buxbaum, *Biophysical Chemistry of Proteins: An Introduction to Laboratory Methods*, DOI 10.1007/978-1-4419-7251-4_30,

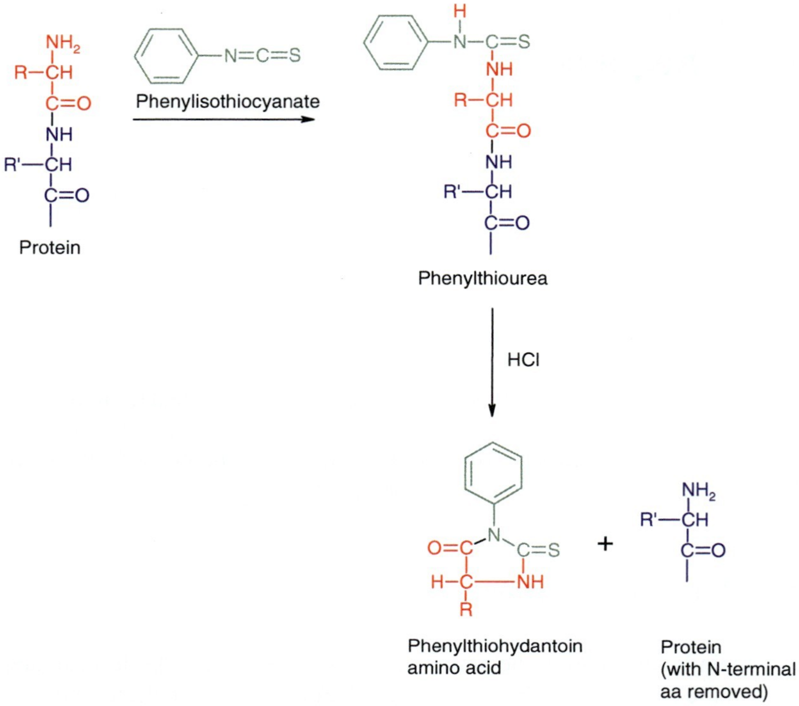

Fig. 30.1 EDMAN-degradation of a protein. In each cycle one amino acid is cleaved of from the N-terminus and identified. The remaining protein can go directly to the next cycle

30.1.1 Problems that May Be Encountered

Great care has to be taken to ensure homogeneity of the sample. If there are two or more different amino acids produced in each cycle, determination of the sequence becomes difficult or – if the amounts of the different proteins are not very different – impossible. Because the initial yield of a protein is unpredictable and can vary widely around an average value of 50 %, EDMAN-sequencing can not be used for quantitation of proteins. About half of all naturally occurring proteins have a blocked N-terminus, for example with N-acetyl or N-formyl groups. Blockage may also occur during protein purification, *e.g.*, by unpolymerised acrylamide during electrophoresis.

Modified amino acids generally cause problems in EDMAN-sequencing because the modified thiohydantoins elute at a different position in the chromatogram (*e.g.*, methyl-His or methyl-Lys) or are too polar to be extracted by ethylacetate (glycosylated Asn, Thr, Ser) or because the modification is unstable and destroyed during the harsh sequencing reactions (phospho-Ser or -Thr). In many cases mass spectrometric approaches can solve such problems, either for analysing the peptide directly

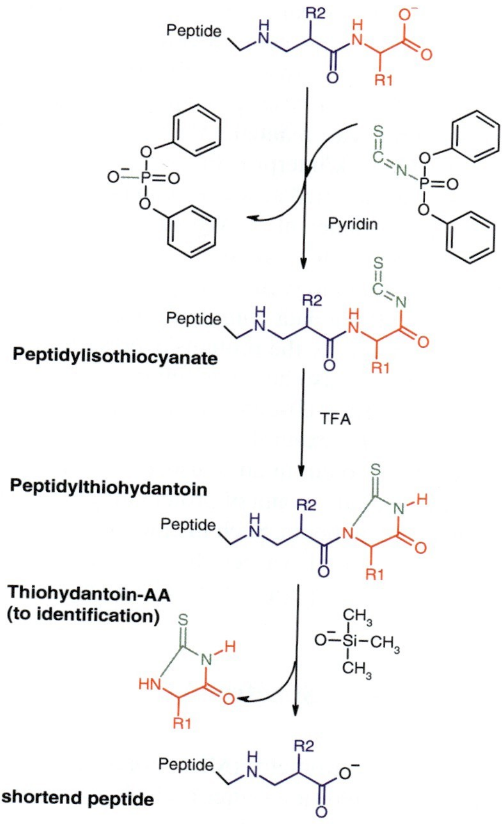

Fig. 30.2 Similar to N-terminal sequencing by the EDMAN-procedure it is also possible to sequence from the C-terminal end [125]. Once the protein has been sequenced at both ends it is possible to find the entire sequence by searching a gene sequence database. Both N- and C-terminal sequencing can be performed in the same instrument

or for analysing unusual PTH-amino acids. More difficult to deal with are modifications that are too unstable to survive even the initial protein purification (*e.g.*, phosphorylated Asn, His, or Cys). It is not surprising that we know very little about the physiological roles of such modifications.

30.1.2 Sequxencing in the Genomic Age

Today protein sequences are usually determined by sequencing their genes, which is quicker and easier. Indeed, great effort has been spent in the last couple of years in sequencing the entire genome of bacteria, yeasts, plants, animals and humans [67, 418]. With these data available it is often sufficient to sequence the first 10 or 20 amino acids of a protein and then look for the gene sequence in a computer data base (NCBI GenBank www.ncbi.nlm.nih.gov/Genbank/index.html, EMBL nucleotide sequence database www.ebi.uk/embl, DDBJ www.ddbj.nig.ac.jp), using the

BLAST (basic local alignment search tool) program. Annotated information containing genetic, protein specific and literature data on a sequence can be obtained from Ensembl www.ensembl.org/index.html. Protein sequences can be found on the Expasy Universal protein knowledgebase (Uniprot) www.expasy.uniprot.org (SwissProt with manual and TrEmbl with automatic annotation) and InterPro www.ebi.ac.uk/interpro. Protein structures are collected on PDB www.rcsb.org, these are classified at SCOP scop.berkeley.edu and CATH cath-www.bio-chem.ucl.ac.uk/latest/index.html. Where structure information is not available the Swiss-Modell Repository swiss-model.expasy.org/repository may help with hypotheses based on homology modelling.

With all the data coming in from gene sequencing attention has now shifted to identify not only the proteins expressed in a cell and their post-translational modifications, but also how they interact with each other (**interactome**). It appears that proteins form well-defined, cell-type specific networks with each other. These can be studied for example by immunofluorescent microscopy of whole cells. Appearance of a protein in an unusual spot can be a valuable diagnostic marker. In such cases the total amount of protein may not change, only the interaction of the protein with its neighbours. Such interactions are lost when the cells are homogenised. This information is also accessible over the internet from MIPS for mammalian proteins mips.gsf.de/proj/ppi and from Cellzome for yeasts http://yeast.cellzome.com.

30.2 Mass Spectrometry

In mass spectrometry (MS), atoms or molecules are ionised and accelerated in an electrical field (ion source). The ions are then separated by m/z ratio (analyser) and detected (detector and recorder). In principle, this can be done with both positive and negatively charged ions, however, in practice positive ions are used almost exclusively. For tryptic peptides the charge +2 is most commonly encountered, the charges reside on the N-terminus and the C-terminal Lys or Arg.

30.2.1 Ionisers

Electron ionisation (EI) The sample is bombarded with electrons from a heated filament. This method is used only for simple applications because of the high thermal stress that the sample is exposed to. It is unsuitable for large molecules. Excess energy absorbed by molecules during ionisation leads to fragmentation, the fragment spectrum can be used for identification.

Fast atom bombardment (FAB) The analyte is mixed with a suitable matrix (like glycerol) and brought onto a cone, where it is hit by a stream of heavy, positively charged ions (Cs^+). If the sample were exposed to the ion beam without the matrix, its surface would rapidly change and the stream of sample ions would stop. This type of ioniser can be coupled to liquid chromatography by connecting

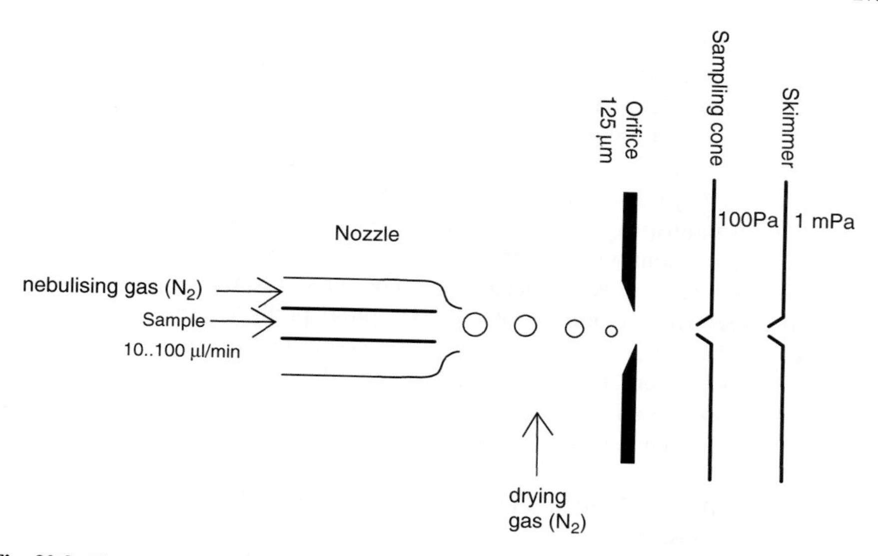

Fig. 30.3 Electrospray ionisation. For an explanation see text, details of the analysers are found in Fig. 30.6

a metal frit to the exit of the column, which acts as target for the ion beam. Excess liquid is constantly drained from that frit by a wick.

Plasma desorption mass spectrometry uses spontaneous fission (see p. 125) in ^{252}Cf to generate two daughter nucleï like Ba and Tc with similar mass and speed, but moving in opposite direction. During decay the nuclei loose their surrounding electrons, *i.e.*, they form a **plasma**. One nucleus is used to desorb an ion from the sample spotted on a aluminised Mylar-target connected to a high voltage source. The other strikes a detector, whose signal starts the timing of a TOF-analyser (*vide infra*). The method is of historic interest only and has been superseded by MALDI.

Electrospray ionisation (ESI) [429] The sample is pressed through a capillary nozzle (10 µm diameter) through a stream of warm nitrogen gas into a vacuum (see Fig. 30.3). The nozzle forms one electrode, the charged droplets move through the vacuum towards a second electrode (with an accelerating voltage of about 9 000 V). As the solvent evaporates, the charge density on the droplets increases until the electrostatic repulsion exceeds the surface tension of the droplet, leading to disintegration (Coulomb-explosion). Repetition of this process leads to smaller and smaller droplets, until each droplet contains only a single sample molecule. Complete evaporation of solvents then results in sample ions. This type of ioniser is particularly gentle, it can be coupled to a chromatography column (LC/MS). Droplet formation at the tip of the capillary may be assisted by a nebuliser gas, ultrasound, or heat. This allows the production of a stable ion current despite varying flow rates and water/solvent ratios, and hence quantitative evaluation of spectrograms. Since ionisation occurs at ambient pressure, maintenance of the ion source is easy and all solvents and volatile buffers can be used. A flow-splitter may be used to connect both the mass spectrometer and other detectors (absorbance, fluorescence, conductivity...) and/or a fraction collector to a column. The total ion current can be plotted over time to get a conventional chromatogram. Very low flow rates (*e.g.*, from capillary electrophoresis) may be increased using a sheath fluid. Flow rates are usually several $\mu l/min$, in **nanospray** instruments down to 1 nl/min. Because the same compound produces ions with different numbers of charges, and hence different m/z, spectrograms are more complicated than those produced with, say, MALDI.

Desorption electrospray ionization (DESI) [404] produces ions by directing a stream of charged droplets from a nebuliser connected to a high voltage source onto a surface. There the droplets kick out ions of the molecules present. The entire process occurs under ambient conditions and can be performed, for example, on the skin or even tongue of a patient or test person, without harming them (see Fig. 30.4).

Matrix assisted laser desorbtion/ionisation (MALDI) [388] The sample is dried with a light-absorbing organic acid which ensures that the protein molecules are protonated and physically separated from each other. If a laser of appropriate wavelength is fired onto this target, a thin layer (a few molecules thick) of the organic acid is evaporated and ionised, taking sample molecules with it. The resulting ions are accelerated in a strong electric field (about 25 kV). Most matrix chemicals absorb UV-light (see Fig. 30.5), but recently the use of IR-absorbing matrices has gained favour since they can be used directly on a blot

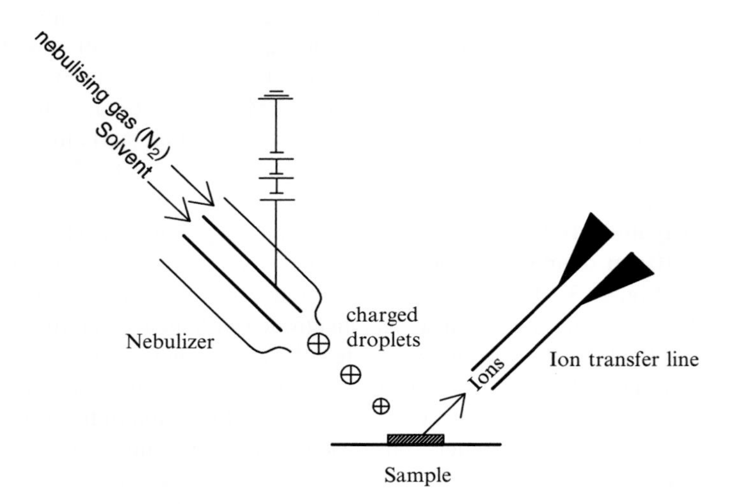

Fig. 30.4 For Desorption electrospray ionization (DESI) a stream of charged droplets is directed against a surface where they kick out ions of the molecules present there. These are collected and directed into a mass spectrometer. Since the process occurs under ambient conditions and is not harmful in any way, it can be used to sample, for example, the molecules on the skin or even tongue of a test person or patient

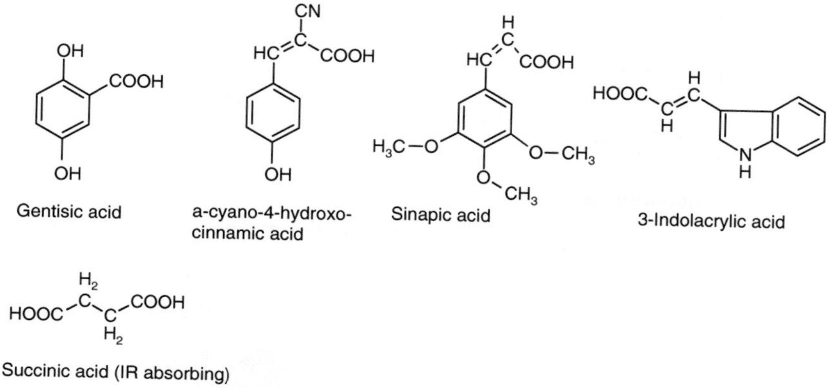

Gentisic acid

a-cyano-4-hydroxo-cinnamic acid

Sinapic acid

3-Indolacrylic acid

Succinic acid (IR absorbing)

Fig. 30.5 Chemicals used as matrix for MALDI

on PVDF. Lasers have a pulse time in the ns-range with an irradiance of several MW/cm^2 for a spot size of about 100 μm. Frequently used are N_2- (337 nm) or Nd-YAG-lasers (1064 nm with frequency tripling to 355 nm). The sample is observed through a video-microscope. A few μl of sample with about 10 ng of protein are required (most of which can be recovered from the spot afterwards), they are spotted onto a stainless steel sample holder and transferred into the vacuum chamber of the spectrometer. Commonly used buffers, detergents or denaturants do not interfere with MALDI in moderate concentrations, depending on the matrix used. MALDI produces mostly single-charged ions, spectrograms are particularly simple to interpret. MALDI is not used if quantitation of ions is required.

Laser induced liquid beam ion desorption (LILBID) The sample is injected into a high vacuum as a jet or stream of droplets against a target cooled by liquid nitrogen. Any excess liquid is trapped on this target as ice and can not pollute the vacuum. On its route, a high-intensity IR-laser beam is fired into this jet, converting some of the sample into a supercritical state which turns into gas. Depending on laser energy, protein complexes may be ionised *in toto* or broken down into subunits. Detection sensitivity is in the amol range [283].

Of these methods, MALDI and ESI are commonly used for the ionisation of proteins.

30.2.2 Analysers (See Fig. 30.6)

Electrostatic/magnetic sector analyser A magnetic or electric field orthogonal to the direction of movement of the ions deflects them from their path. Double focusing by combining both electrostatic and magnetic sectors results in higher

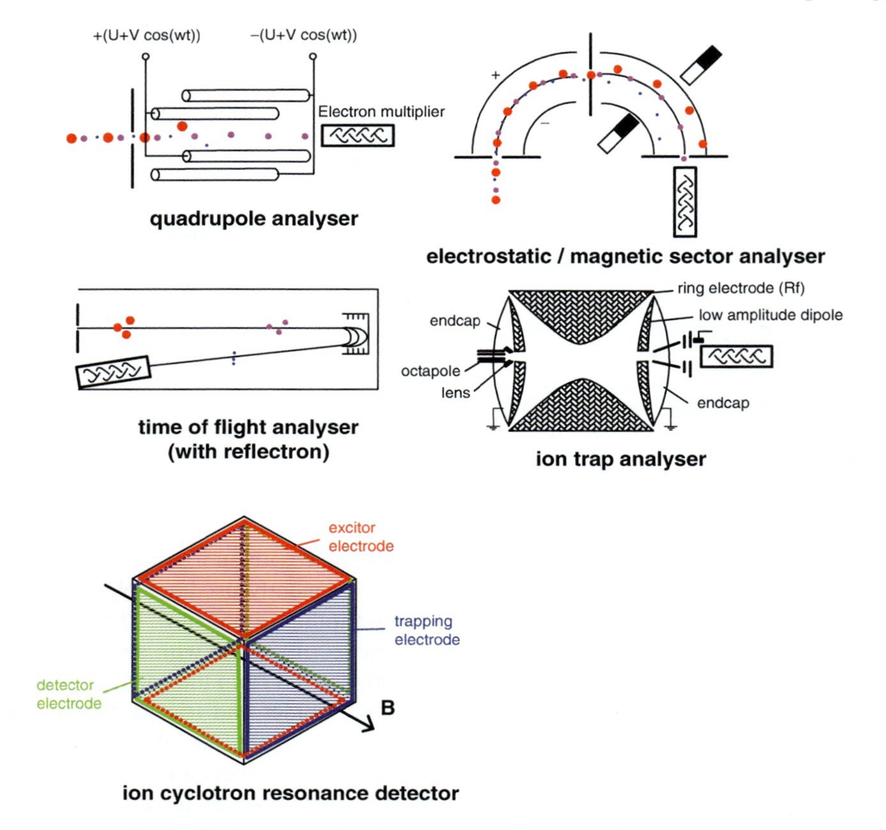

Fig. 30.6 The analysers used in mass spectrometry. For details see text

resolution: The electrostatic sector analyser ensures that all ions entering the magnetic one have the same velocity. Because of their limited molecular mass range ($m/z \leq 1\,000$) these instruments are more useful to chemists and nuclear physicist than to protein scientists.

Quadrupole analyser [177] The four rods of this analyser are electrodes with both DC (U) and radio-frequency ($V * \cos(\omega t)$) voltage between them. Adjacent rods have opposite voltage and a phase difference of 180° for the RF-component. Opposing rods are connected to each other. In this arrangement ions undergo transverse oscillation; only ions of a selected *m*/*z* pass to the detector, ions of different *m*/*z* collide with the rods. By scanning the DC and RF voltages a quadrupole analyser can be adjusted to different *m*/*z* ratios [411]. Instruments often contain three quadrupoles, serving as sample analyser, collision cell (*vide infra*) and fragment analyser, respectively.

Time of flight analyser (TOF) [406] The acceleration of an ion in a constant electrical field depends on its mass/charge ratio, since

$$\vec{\mathfrak{F}} = m * \vec{\mathfrak{a}} = \vec{\mathfrak{E}} * Q \tag{30.1}$$

with $\vec{\mathfrak{F}}$ = force (N), m = mass (kg), $\vec{\mathfrak{a}}$ = acceleration ($\mathrm{m\,s^{-2}}$), $\vec{\mathfrak{E}}$ = field strength (V/m) and Q = charge (C). Hence the time required for an ion to cover a certain distance depends directly on m/z. A TOF-analyser is a long (2 m) tube without magnetic or electric fields, which the ions pass in a few µs. This method requires either a pulsed ion source (like MALDI) or a shutter. About 50 spectra are taken and averaged for noise reduction, which takes only a few minutes.

Resolution of TOF-analyser is limited by the fact that even ions of the same m/z ratio may have different kinetic energy when entering the analyser, and hence leave it at slightly different times. This results in broadening of the peak. In modern instruments this effect is reduced by a **reflectron** which has the same charge as the sample ions and hence redirects them. Ions with higher kinetic energy penetrate further into the reflectron, thus their flight path is slightly longer than that of ions with lower kinetic energy. In addition, the reflectron increases the flight path length without the need for a longer instrument.

To reduce the peak width at the ioniser the application of the electrical field is timed a few ns after the laser pulse. Thus all ions enter the flight tube at the same time (**pulsed extraction**).

A $\frac{m}{\Delta m}$ of 10 000 or so is possible with this type of detector.

Ion trap analyser A 3-D version of the quadrupole analyser, where the ions are contained by an Rf-field in a ring electrode between endcaps at ground potential. After ion collection a smaller Rf-field between the endcap dipoles is used to eject the ions in order of m/z. The ion source must be pulsed so that no ions can enter during the ejection phase. If a constant ion source like ESI is used, gating lenses at the entrance can be used. He gas in the ion trap can be used to fragment an isolated ion species, the fragments are then ejected one by one. At this stage it is possible to select one of the daughter ions for further fragmentation (and so on up to $\mathrm{MS^{10}}$). This type of analyser is very sensitive, but with limited mass range.

Ion cyclotron detector [66] The ions are injected into a cubic analyser and kept on circular tracks by a magnetic field of several Tesla, causing a LORENTZ-force $\vec{\mathfrak{F}}_l = e\vec{\mathfrak{v}}\vec{\mathfrak{B}}$ perpendicular to the direction of their movement and the magnetic field, *i.e.*, towards the centre of the cell. This force is balanced by the centrifugal force $\vec{\mathfrak{F}}_c = m \times \vec{\mathfrak{v}}^2/r$ directed away from the centre of the cell. With $\omega = \vec{\mathfrak{v}}/r = 2\pi f$ the cyclotron frequency becomes $f = \frac{\vec{\mathfrak{B}}e}{2\pi m} = 1.536 \times 10^6 \vec{\mathfrak{B}}z/M$. As a consequence, all ions of the same m/z have the same frequency, ions with different velocities simply have orbitals of different radii. A small potential between trapping electrodes (perpendicular to the magnetic field) that causes the ions to oscillate keeps the ions in the cell for extended periods of time (if necessary, hours). The two pairs of plates that are parallel to the magnetic field serve as excitation and detection electrodes, respectively.

Radio-frequency between the excitor plates elevates the ions to an orbit of higher energy. The movement of the ion packets induces an electrical signal in the detector plates which is amplified in a superheterodyne[1] receiver. Ions of different m/z ratio give signals of different frequencies. The FOURIER-transform of the voltage across the detector plates is the mass spectrum. Since detection does not consume the ions they can be used for additional experiments, *e.g.*, MS/MS.

This type of detector has a very high mass discrimination, $\frac{m}{\Delta m} = 100\,000$ and more, a very high sensitivity (30 zmol $\equiv$ 18 000 molecules [22]), and high mass range (10^3) therefore complete spectra are obtained very fast. This allows for global proteome analysis, where even low abundance species can be detected. It may also be used as collision cell. Since collisions between ions and gas molecules leads to the deceleration of the ions, a very good vacuum (μPa) is required.

The methods for ionisation and analysis can be combined almost at will, so in the literature you find expressions like MALDI-TOF-MS to describe the method used in a particular experiment. The detector is usually an electron multiplier, nowadays linked to a computer via a fast ADC. A setup for the identification of contaminants in biological fluids is depicted in Fig. 30.7.

30.2.3 Determination of Protein Molecular Mass by Mass Spectrometry

A more detailed treatment of this subject can be found in [199].

The *m*/*z*-ratio depends on the molecular mass of the protein, the number of bound protons and the charge added during the ionisation process. Thus a single protein in the sample will give multiple peaks in the mass spectrum, each differing by one charge and the weight of one proton from its neighbours. Molecular mass determination (de-convolution) is done by solving the following system of equations for each pair of peaks:

$$m_1/z_1 = \frac{M_r + n * X}{n}, \tag{30.2}$$

$$m_2/z_2 = \frac{M_r + (n+1) * X}{(n+1)}, \tag{30.3}$$

resulting in:

$$n = \frac{m_2/z_2 - X}{m_2/z_2 - m_1/z_1}, \tag{30.4}$$

$$M_r = n * (m_2/z_2 - X) \tag{30.5}$$

[1] In a superheterodyne receiver ("superhet") the signal of frequency f_s is mixed with a reference signal of frequency f_r. The resulting beat signal of frequency $f_s - f_r$ contains all the information of the original, but at a much lower, and therefore easier to handle, frequency.

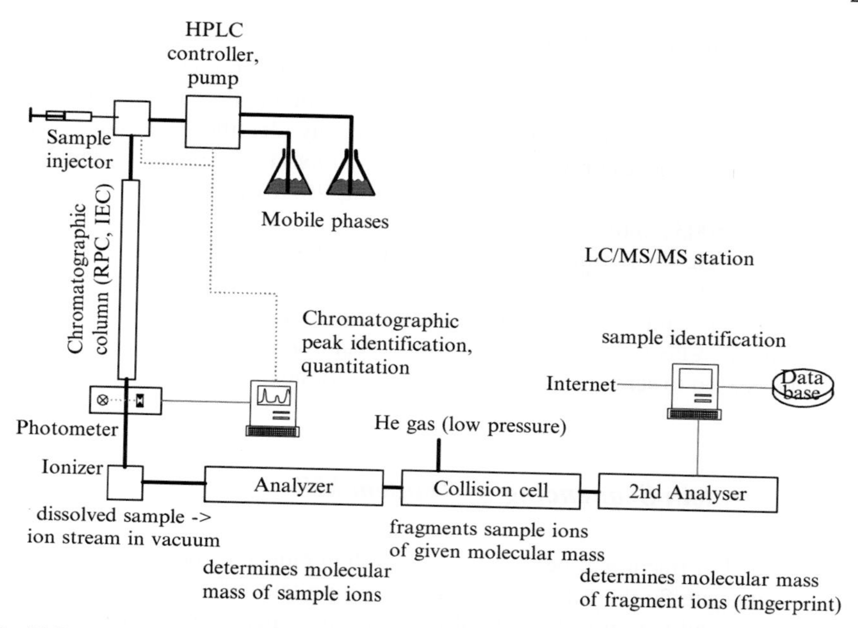

Fig. 30.7 Chromatography coupled to tandem mass spectrometry can be used to find and identify unusual compounds in biological fluids. Samples are subjected to either gas or liquid chromatography, any peaks are investigated by tandem mass spectrometry, giving the molecular mass and the fragment spectrum of the compound involved. This is usually sufficient to identify that compound, using in-house or internet databases. Applications include forensic laboratories and newborn screening for metabolic diseases

with M_r = molecular mass of the protein, n = number of charges added and X = mass of added group (usually proton, $X = 1$).

Each pair of peaks allows determination of M_r with a standard deviation of about 0.01 %, calculating the averages for all pairs increases precision further. This calculation is done automatically by the computer integrated into modern instruments.

30.2.4 Tandem Mass Spectrometry

In the simple mass spectrometer described above, ions are separated by *m*/*z* and then reach a detector. However, compounds with the same *m*/*z* ratio cannot be distinguished by simple mass spectrometry. Tandem mass spectrometry solves this problem.

All ions of a given *m*/*z* ratio are fed into a collision cell, where a "collision gas" (usually helium or argon under reduced pressure) collides with them. The collisions induce breakages in the sample molecules (collision activated dissociation), *i.e.*, each species present in the sample is disintegrated into a set of fragments which are fed into a second analyser. There the fragments are analysed for their m/z ratio. Each compound under these conditions will give a specific set of fragments (a fingerprint spectrum), as breakage occurs preferentially on certain bonds. The combination of molecular mass and fragmentation spectrum uniquely identifies a substance which can be used for forensic purposes. For recommendations on the use of tandem MS in newborn screening for metabolic diseases see http://www.cdc.gov/mmwr/preview/mmwrhtml/rr5003a1.htm (see also Fig. 30.7).

In a TOF-detector fragmentation may occur in the drift tube. The fragments of a particular molecule have the same velocity as the original ion, but a smaller kinetic energy . The reflectron may therefore be used as analyser, especially so-called "curved field" reflectrons. In this application beam blanking by an ion gate in the drift tube is used to select a particular m/z-range for analysis.

30.2.5 Protein Sequencing by Tandem MS

Proteins tend to fracture at peptide bonds, creating radical electrons on the carbon and the nitrogen. So the fragment spectrum of a protein will consist of two series of peaks (N- and C-terminal), each differing from its neighbour by the molecular weight of one amino acid. Leu/Ile/Hyp and Lys/Gln have the same molecular mass (isobaric amino acids), as have Asn and Gly–Gly. They can be distinguished by the high-energy ions d and w (see Fig. 30.8) where the R-group is fragmented.

Because the spectrum will be generated both by the N- and C-terminal fragments the sequence is determined twice and the results can be checked for agreement. Again the tedious task of converting a spectrum into a sequence is done automatically by computer, supported by database searches for known peptides.

The computer will also take care of the complications like chain breakage in other places than peptide bonds or uneven distribution of charges between the two fragments.

If a protein is too large for this approach, it can be digested with endoproteases and the resulting peptides analysed. This digestion can be performed directly in a protein spot isolated from a **2-D electrophoresis** gel. The peptides are then eluted and analysed by MS, an approach frequently used in **proteomics**. However, the resolution of 2-D electrophoresis is limited, it discriminates against proteins with extreme pI, M_r or hydrophobicity and this method is difficult to automate. Thus electrophoresis-free approaches have also been tried, for example **multi-dimensional protein identification technology (MudPIT)**: A mixture of proteins is digested by proteases and then subjected to anion exchange chromatography with a step gradient. The eluate of each step is subjected to reversed phase chromatography with an ethanol gradient, the second eluate is fed into a mass spectrometer (LC/MS/MS). In effect this results in 3-D separation of the peptides (charge, hydrophobicity, molecular mass), which are then identified by their fragment spectrum. Unfortunately, this method does not give quantitative information, only present/absent. In addition, running an electrospray ioniser continuously for many hours without clogging is a challenge. 2-D PAGE and MudPIT

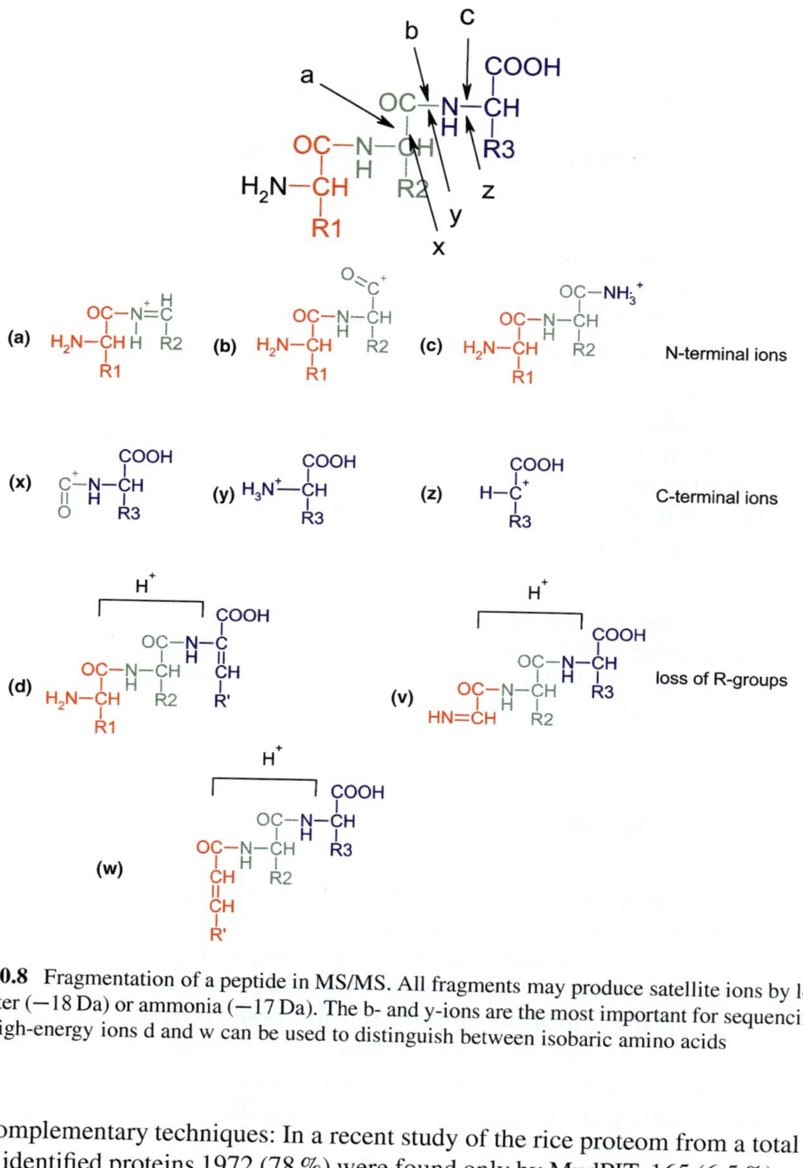

Fig. 30.8 Fragmentation of a peptide in MS/MS. All fragments may produce satellite ions by loss of water (−18 Da) or ammonia (−17 Da). The b- and y-ions are the most important for sequencing. The high-energy ions d and w can be used to distinguish between isobaric amino acids

are complementary techniques: In a recent study of the rice proteom from a total of 2528 identified proteins 1972 (78 %) were found only by MudPIT, 165 (6.5 %) only by 2-D PAGE and 391 (15.5 %) by both methods [206].

For a list of commercial and free software used in mass spectrometry of proteins see openwetware.org/wiki/Wikiomics:Protein_mass_spectrometry.

Collision with energetic electrons rather than gas is used to fragment ions in a FOURIER-transform ion cyclotron detector (**electron capture dissociation, ECD**), this method is very gentle and leaves post-translational modification (phosphoproteins, O-glycosides) in place. The method produces mostly c- and z-ions [391].

30.2.6 *Digestion of Proteins*

Fragmentation of protein spots in gels or on Western-blots followed by mass-spectrometric separation of the fragments has become a favourite tool for identification of proteins (**peptide mass fingerprinting**). The fragment spectrum obtained is compared to the spectrum of known proteins in a database ([321], www.matrixscience.com/, http://wolf.bms.umist.ac.uk/mapper/, http://prowl.rockefeller.edu/prowl/prowl-description.html). MALDI is usually used for ionisation since it produces mostly single-charged ions, resulting in simple spectrograms. Cleavage can be achieved by proteases (see Table 30.1) or by chemicals:

Cyanogen bromide (BrCN) in 70 % formic acid splits C-terminal of Met. Incubation is 24 h at room temperature in the dark. For the chemistry, see Fig. 21.8 on p. 217. **Careful**: Cyanogen bromide is highly toxic.

Controlled hydrolysis by 250 mM acetic acid 8 h at 110 °C splits proteins C-terminal of Asp.

Iodosobenzoic acid 0.5 % in 4 M guanidinium hydrochloride, 100 mM *p*-cresol, 80 % acetic acid splits C-terminal of Trp. Incubation 24 h at room temperature in the dark.

Note that if the hydrolysis of one sample is performed in normal water ($H_2{}^{16}O$), and that of another in $H_2{}^{18}O$, the mass difference may be used to compare relative concentrations of a particular protein in the samples in a similar way as DIGE (see Sect. 8.2.9.7 on p. 88, for a recent review see [279]).

30.2.7 *Ion–Ion Interactions*

For proteomics purposes, identifying proteins by proteolytic digestion and peptide MS/MS-sequencing is non-ideal, as variations in splicing, mRNA-editing, and post-translational modifications may be missed. Shotgun-sequencing increases the complexity of an already complex sample, which is perhaps not the optimal thing to do. Therefore the sequencing of whole proteins would offer distinct advantages. The FOURIER-transform ion cyclotron detector with its wider mass range is a step in this direction. In addition, the use of anions instead of nobel gasses for fragmentation has recently gained popularity (reviewed in [120]). There are three possible reactions:

Proton transfer $[MH_n]^{n+} + A^- \rightarrow [MH_{n-1}]^{(n-1)+} + HA$
Electron transfer $[MH_n]^{n+} + \dot{A}^- \rightarrow [\dot{M}H_n]^{(n-1)+} + A$
Anion attachment $[MH_n]^{n+} + A^- \rightarrow [MAH_{n-1}]^{(n-1)+}$

Of these, proton transfer from the protein cation to the anion is the most probable. With polyaromatic hydrocarbon radicals, one can obtain significant electron transfer; phosphor hexafluoride or iodide can form relatively long-lived complexes with protein cations as intermediates to proton transfer.

Table 30.1 Proteases used in proteomics

Name	EC-#	Source	Cleavage site	Size (kDa)	Type	*p*H-optimum
Arg-C-protease (clostripain)	3.4.22.8	*C. histolyticum*	Arg↓Xaa	50	Cys	8–8.5
Asp-N-protease	3.4.24.33	*Pseudomonas fragi*	$\text{Xaa}\downarrow^{\text{Asp}}_{\text{Glu}}$	27	Metal	7–8
Chymotrypsin (TLCK-treated)	3.4.21.1	bovine pancreas	aromatic ↓ Xaa	25	Ser	7.5–9
Cathepsin C	3.4.14.1	bovine phagosomes	N-terminal dipeptidase	21	Cys	4
Factor Xa	3.4.21.6	bovine plasma	$\text{Ile}\ ^{\text{Glu}}_{\text{Asp}}\ \text{Gly Arg}\downarrow$	43	Ser	7–8
Lys-C-protease	3.4.21.50	*Achromobacter lyticus*	Lys↓Xaa	30	Ser	8–8.5
Pepsin	3.4.23.1	bovine or pig stomach	between hydrophobic or aromatic aa	35	Asp	2–4
Submaxillaris-protease	3.4.21.?	mouse	$^{\text{Arg}}_{\text{Lys}}\downarrow\text{Xaa}$	21	Ser	7.5–9
Subtilisin	3.4.21.62	*B. subtilis*	large neutral ↓Xaa	27	Ser	7–8
Thermolysin	3.4.24.27	*B. thermoproteolyticus*	$\text{Xaa}\downarrow^{\text{Phe}}_{\text{Leu}}$	35	Metal	7–9
Trypsin (TPCK-treated)	3.4.21.4	bovine pancreas	$^{\text{Arg}}_{\text{Lys}}\downarrow\text{Xaa}$	23.8	Ser	7.5–9
V8-protease	3.4.21.19	*Staph. aureus*	$^{\text{Glu}}_{\text{Asp}}\downarrow\text{Xaa}$	27	Ser	4 and 8

As the rate of proton transfer increases with the square of the number of charges, this reaction can be used to simplify a spectrum (especially after ESI) by increasing the amount of lower charge states at the expense of the higher. This increases sensitivity and resolution; in addition, the fragment spectrum in tandem MS is simplified too.

Electron transfer with multivalent anions leads to **electron transfer dissociation** of the protein, producing mainly c^+ and z^+ ions. The reaction is independent of protein length and post-translational modification. It is possible to perform electron transfer after spectrum simplification by proton transfer. This allows protein sequencing up to about 70 kDa, compared to a limit of traditional collision activated dissociation of 1.5 kDa.

30.3 Special Uses of MS

Electrospray ionisation is gentle enough to leave bound substrates on the protein. Thus ESI-MS can be used to monitor substrate binding and turnover for rapid reaction kinetics [15, 205, 241, 355, 430].

Differential mass spectrometry may be used to monitor changes in protein association, for example after hormone stimulation of a cell. For this purpose cells are grown on ^{13}C-Arg, stimulated with the hormone and lysed. The lysate is mixed with a lysate from non-stimulated cells grown with ^{12}C-Arg. The receptor with any bound proteins is purified from the mixed lysate by affinity chromatography and subjected to mass spectrometry. Proteins from stimulated and non-stimulated cells can be distinguished by their mass difference. If the interaction of a protein with the bait is unaffected by hormone stimulation, the height of the corresponding protein peaks will be equal. If interaction increases after hormone stimulation, the ratio $^{13}C/^{12}C$ will be > 1. If interaction is reduced, it will be <1 [30].

Deuterium exchange can also be measured by MS. The protein is first extensively labelled with deuterium under physiological conditions. All labile hydrogen atoms, in particular amide hydrogens, are exchanged for deuterium. The protein is then bound to an affinity column, this allows the following reactions to be performed quickly and with a minimum volume of reagents. Incubation with a substrate or ligand traps some of the deuterium between ligand and protein, during the following off-exchange these deuterium atoms stay on the protein selectively. The exchange reaction is quenched at different time points by transfer to *p*H 2.7, the protein is denatured (usually with guanidinium hydrochloride) and reduced (with phosphine since SH-reagents don't work at low *p*H). Remaining deuterium atoms are detected by mass spectrometry after proteolytic digestion (ESI-MS/MS). For each peptide bond a protection factor (rate of exchange before addition of ligand divided by rate of exchange after addition) is determined [434].

30.3.1 Disease Markers

Finding differences in the proteome of normal people and patients suffering from diseases (say, cancer) has long been a goal of proteomics research. Tests based on such differences have considerable market value, thus industrial funding is available. However, the overall results have been largely disappointing. Although some such proteins have been found, these produce too many false positive and/or false negative results to be suitable for mass screening. Attention has shifted to analysis of **proteomics patterns** [326], that is, the relative abundance of various peaks in a sample. This approach is promising, but still in its infancy.

30.3.2 Shotgun Sequencing of Proteins

Shotgun approaches work on complex samples without prior separation. Proteins in a sample are digested with proteases and the resulting peptide mixture is separated by 2-D chromatography and sequenced by MS/MS [237]. Computer programs (see for example fields.scripps.edu/sequest/, http://fields.scripps.edu/DTASelect/, www.protein.osaka-u.ac.jp/rcsfp/profiling/Seqms/SeqMS.html) are then used to identify (and sometimes even quantify) the proteins present from partial sequence information.

30.4 Characterising Post-translational Modifications

Identification of phospho-proteins is discussed in Fig. 21.2 on p. 210.

30.4.1 Ubiquitinated Proteins

Trypsin digestion of ubiquitinated proteins results in peptides that have a Gly–Gly residue attached to the ϵ-amino group of Lys, resulting in an additional mass of 114.1 Da. The remaining amino acids of ubiquitin are removed. In addition, tryptic cleavage after the modified Lys is prevented [320].

30.4.2 Methylation, Acetylation and Oxidation

Result in a mass increase of 14, 42 or 16 Da, respectively.

30.4.3 Glycoproteins

Deglycosylation with base (O-linked carbohydrates) or N-glycanase before MS can be used to identify glycosylation sites in proteins and peptides. MS can also be used to characterise the sugar chains, but this is beyond the scope of this text. Interested readers are referred to recent reviews, *e.g.*, [7, 158, 425].

Chapter 31
Synthesis of Peptides

In the last chapter we have seen how to determine the sequence of a protein. For many purposes it would be useful to synthesise proteins of a given sequence. The solid-phase approach of MERRIFIELD (Fig. 31.1) serves this purpose.

Peptides of up to about 100 amino acids can be produced this way, addition of each amino acid takes about 30 min. Compare this with cellular biosynthesis, which produces a 1 000-amino acid protein in about 5 min!

The process can be automated, several companies offer peptide synthesisers that use MERRIFIELD's chemistry. Other companies do the synthesis as a service for a very reasonable price. Thus it is usually not cost effective to perform it in the laboratory, unless one has to do it routinely (*e.g.*, in a core facility).

Modern approaches of **combinatorial chemistry** use special paper as solid support instead of polystyrene beads. The reaction is performed on small spots of the paper, on each spot a slightly different peptide is synthesised. Thus several hundred or even several thousand peptides can be tested for a particular biological activity.

E. Buxbaum, *Biophysical Chemistry of Proteins: An Introduction to Laboratory Methods*, DOI 10.1007/978-1-4419-7251-4_31,

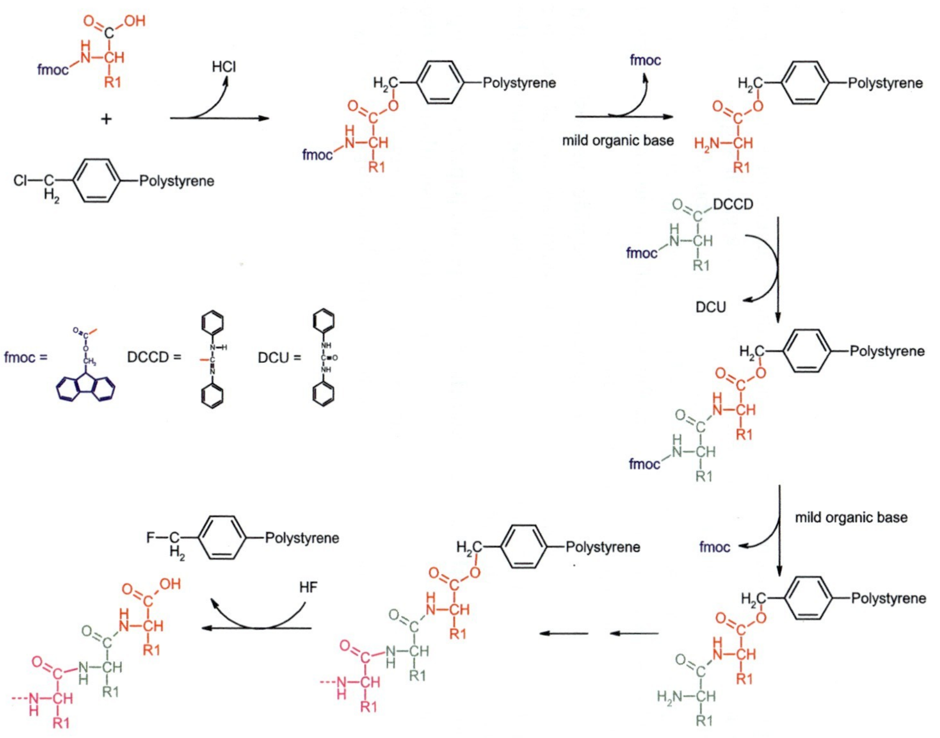

Fig. 31.1 Solid phase peptide synthesis by the method of MERRIFIELD. The amino-groups of the amino acids are protected with a FMOC-group. Then the carboxy-groups are activated with DCCD, so it may be coupled to a free amino group. After each coupling step the FMOC-protection group is removed with a mild organic base. Note that the synthesis goes from the C- to the N-terminus, the reverse order of biological protein synthesis. This should be kept in mind when talking to a peptide chemist! Finally the crude peptide is removed from the solid support with HF, it then needs to be purified, usually by reversed phase chromatography on a C_{18}-column

Chapter 32
Protein Secondary Structure

There are several methods available that can be used to gain information on secondary structure, apart from X-ray crystallography, which remains the 'gold standard'.

32.1 Circular Dichroism Spectroscopy

CD-spectroscopy measures the difference in absorbtion ($\mathcal{E}$) of right- and left-circular polarised light by optically active samples [193]. Circular polarised light can be described as light with an electric vector that is rotating around the direction of movement of the photon (see Fig. 32.1 and http://www.photophysics.com/opticalactivity.php for an animation). Thus the direction of the electric field vector changes, and its length stays constant. With linearly polarised light it is the other way around. If one looks toward the light source and the field vector is rotating clockwise, we speak of right circular polarised light, if rotation is anti-clockwise, it is left-polarised light. The field vector rotates once for each wavelength.

Linear polarised light (which is an equal mixture of right and left circular polarised light) when moving through a CD-active sample is converted to elliptically polarised light by the difference in absorbance $\mathcal{E}$ for right and left polarised light:

$$\mathcal{E} = -\log\left(\frac{I}{I_0}\right) = \epsilon * c * d, \tag{32.1}$$

$$\Delta\epsilon = \epsilon_\mathrm{l} - \epsilon_\mathrm{r} = (\mathcal{E}_\mathrm{l} - \mathcal{E}_\mathrm{r}) * c * d, \tag{32.2}$$

$$A_\lambda = \frac{\Delta\epsilon}{\epsilon} = \frac{2 * \epsilon_\mathrm{l} - \epsilon_\mathrm{r}}{\epsilon_\mathrm{l} + \epsilon_\mathrm{r}}, \tag{32.3}$$

$$\Theta_\lambda = 32.98 * \Delta\mathcal{E} \qquad \left[\frac{cm}{dmol}\right], \tag{32.4}$$

$$[\Theta]_\lambda = \frac{100 * \Theta_\lambda}{c * d} = 3298 * \Delta\epsilon \tag{32.5}$$

E. Buxbaum, *Biophysical Chemistry of Proteins: An Introduction to Laboratory Methods*, DOI 10.1007/978-1-4419-7251-4_32,

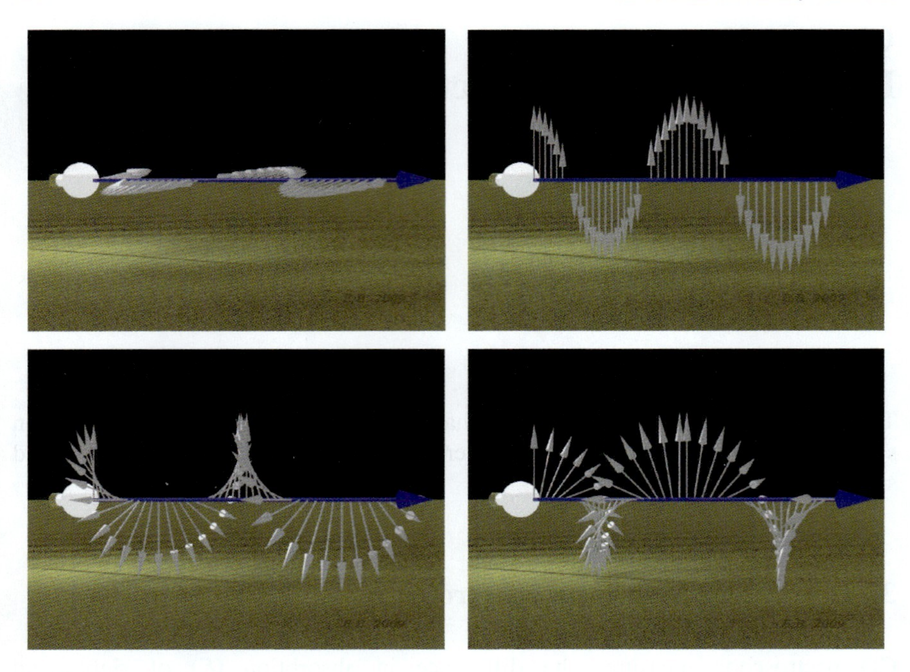

Fig. 32.1 *Top*: Horizontally and vertically linear polarised light. The plane of the electrical field vector is constant, but its length oscillates once every wavelength. *Bottom*: Left and right circular polarised light. The electrical field vector rotates once every wavelength around the propagation vector, but its length remains constant. Linear polarised light is a equal mixture of left and right circular polarised light

with the unit $^\circ \mathrm{l\,cm^{-1}\,mol^{-1}}$. Sometimes the anisotropy A_λ is also used to quantify CD-activity. Both Θ_λ and A_λ are redundant, as $\Delta\epsilon_\lambda$ is the most natural way to quantify CD-activity.

The rotational strength R is the imaginary part of the scalar product of the electric ($\boldsymbol{\mu}$) and magnetic ($\overrightarrow{\mathfrak{m}}$) transition moments of the electronic transition that causes the absorbance and is proportional to the integral under the peak in the CD-spectrum:

$$\begin{aligned} R &= im(\overrightarrow{\mu} \otimes \overrightarrow{\mathfrak{m}}) \\ &= \frac{hc}{32\pi^3 N_\mathrm{a}} * \int \frac{\Delta\epsilon}{\lambda} d\lambda \end{aligned} \tag{32.6}$$

Similar measurements can also be made using refractive index instead of absorbtion (**optical rotary dispersion (ORD)**), RAMAN and RAYLEIGH light scattering, and phosphorescence.

The light from a suitable source is passed through a monochromator and a linear polariser. Then it goes through a photoelastic modulator. This is a fused silica bar in contact with a piezoelectric element. The latter is excited at the fundamental frequency of the bar (usually about 50 kHz), causing strain in the silica and thus changing its birefringence. The bar thus switches rapidly between right and left circular polarized light. The light goes through the sample to a detector, either a photomultiplier or (in modern instruments) an avalanche photodiode. The latter has a higher quantum efficiency and can be used from the far UV to about 1 000 nm.

The average signal from the detector (DC-component) is proportional to $\mathcal{E}_\lambda$, if there is CD-activity in the sample the signal will be modulated with the frequency of the modulator (AC-component). The DC-component is kept constant by a feed-back loop that regulates the gain of the detector (*e.g.*, photomultiplier high voltage). Under these conditions the AC-component is directly proportional to $\Delta\epsilon_\lambda$. The instrument is calibrated with a reference compound of known CD-activity, most commonly used is (+)-10-camphorsulphonic acid with $\Delta\epsilon_{290.5} = 2.361\,\text{mol}^{-1}\,\text{cm}^{-1}$ and $\Delta\epsilon_{192.5} = -4.91\,\text{mol}^{-1}\,\text{cm}^{-1}$. Of particular importance is the selection of a sample cuvette with little strain in the material, as this would cause background.

The signal-to-noise ratio in CD-spectroscopy depends on the intensity of the light source I_0, the quantum efficiency of the detector Q, the measurement time τ, the absorbance $\mathcal{E}$ and on $\Delta\epsilon$:

$$S/N \propto \Delta\epsilon \times \sqrt{I_0 \;\; Q \;\; \tau \;\; 10^{-\mathcal{E}}} \tag{32.7}$$

Both I_0 and Q have been improved by instrument manufacturers in recent years, the selection of a proper $\mathcal{E}$ is in the hands of the investigator.

Because many commonly used buffer components react with far-UV light, experiments are usually performed in 10 mM phosphate or Tris buffer, with fluoride, perchlorate or sulphate added for ionic strength.

CD-spectra arise from the transition of molecular orbitals to excited states, $n \rightarrow \pi^*$ and $\pi \rightarrow \pi^*$.

- Several prosthetic groups have, when bound to protein, CD-activity in the visible region of the spectrum, this is called **induced CD**.
- Aromatic amino acids have CD-activity between 250–300 nm (a $\pi \rightarrow \pi^*$ transition), which is sensitive to tertiary structure. For example in haemoglobin the transition from the (oxygen-free) T to the (oxygen bound) R conformation is accompanied by a change in the CD-activity of Tyr-42 of the α-chain.
- The CD-activity of disulphide bonds at 260 nm ($n \rightarrow \pi^*$ transition) depends on the dihedral angle of the (S−S)-bond.
- The backbones of proteins (C=O bond) have considerable CD-activity ($n \rightarrow \pi^*$ transition) for wavelengths below about 250 nm. This transition is forbidden for symmetrical carbonyl groups (like acetone), but allowed for unsymmetrical. Thus the elipticity depends on the symmetry of the carbonyl-group, which is influenced by the hydrogen bonding this group participates in, and hence by secondary structure.

With usual laboratory instruments CD-spectra can be measured down to about 190 nm, below this value light sources are too weak and absorbtion by air and buffers is to high. However, recent work in vacuum with synchrotron radiation has revealed that wavelengths below 190 nm are even more informative, wavelengths down to 140 nm have been technically accessible [37, 263, 421].

α-Helix (λ_{min} 222, 208 and 160 nm, λ_{max} 190 nm, shoulder at 175 nm), β-sheet (λ_{min} 215 and 175 nm, λ_{max} 198 and 165 nm), collagen helices (λ_{min} 200 and 175 nm, λ_{max} 220 and 165 nm), turns (λ_{max} 200 nm) and coils (minimum 200 nm) have different CD-spectra, the CD-spectrum of a protein is the sum of the spectra of all its structural elements. With the given spectra of pure structures a computer can de-convolute the CD-spectrum of a protein and estimate the relative amount of the various structural elements.

Mathematically, this can be done by **singular value decomposition (SVD)**, a matrix algebra technique [337].

Alternatively, if the secondary structures to be investigated are saved in a row vector $\vec{\mathfrak{f}}$ (*e.g.*, % α-helix, β-sheet, γ-turn and coil) and the spectrum in a column vector $\vec{\mathfrak{c}}$ (*e.g.*, $\Delta\epsilon$ at 220, 219, ..., 180 nm), then there exists a transformation matrix $\mathcal{X}$ so that $\vec{\mathfrak{f}} = \mathcal{X}\vec{\mathfrak{c}}$. This transformation matrix $\mathcal{X}$ can be determined from n proteins with known secondary structure content (from X-ray diffraction or the like), which are saved in a structure matrix $\mathcal{F}$ (n rows) and CD-spectrum saved in a spectral matrix $\mathcal{C}$ (n columns). The system $\mathcal{F} = \mathcal{X}\mathcal{C}$ is then solved by **least square techniques**. Verified CD-spectra may be found on pcddb.cryst.bbk.ac.uk/.

Structure content determined this way is fairly reliable for α-helix and coil, β-sheet and γ-turn determinations have a higher error margin.

Sample requirements are about 10–30 μg of protein, the path length (1 or 0.1 mm) and sample concentration should be chosen for an absorbtion of less than 1.0 in the entire wavelength range to be measured.

A specialised technique for studying small molecules with conjugated π-systems (*e.g.*, porphyrins) is **magnetic CD**, where the CD-spectrum is determined with a magnetic field parallel or anti-parallel to the light beam.

32.2 Infrared Spectroscopy

While in UVIS-spectroscopy the absorbtion $\mathcal{E} = -\log(I/I_0)$ is plotted against the wavelength λ, in an infrared (IR)-spectrum the relative transmission ($100 * I/I_0$) is plotted against the wave-number $\tilde{\nu}$ (number of waves per centimeter). The wave-number is directly proportional to the energy in a particular oscillation. Light source is usually a pin of a high melting compound (zirconium oxide or silicon carbide) which becomes conducting at high temperature. The electrical current then keeps the pin at that temperature. All optical components need to be manufactured from IR-transmitting materials, *e.g.*, calcium fluoride. Modern instruments usually have photodiodes as detectors. In chemistry, samples (about 1 mg) are mixed with KBr. Application of high pressure turns the KBr into a pellet, which is mounted in the IR-spectrophotometer (spectroscopists don't like it when you say "pill" to that pellet, even though that's exactly what it looks like). Thorough desiccation of sample and KBr is essential, as water strongly absorbs in many interesting regions of the IR-spectrum.

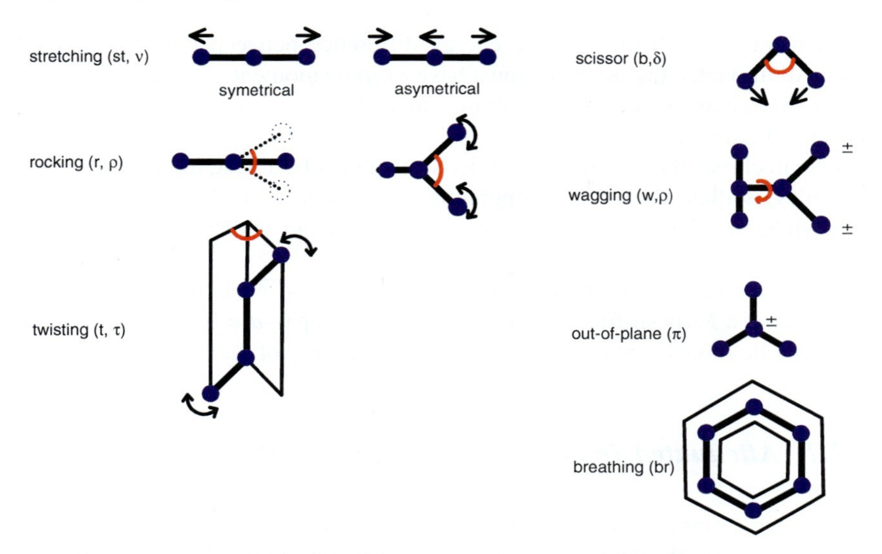

Fig. 32.2 Various vibrations in a molecule. In stretching vibrations (abbreviated st or ν) atoms move along the bond. They can be symmetrical (ν_1) or asymmetrical (ν_2). In bending vibrations atoms move orthogonal to bonds: scissor (s, δ), wagging (w, ρ_w), rocking (r, ρ_r), and twisting (t, τ). In addition, breathing (br) vibrations change the size of molecules and in out-of-plane vibrations (π) one atom moves below and above the plane formed by other atoms. Movement in the paper plane is indicated by *arrows*, movement perpendicular to the paper plane by $\pm$

If infrared light is absorbed by a molecule, its nucleï start vibrating (see Fig. 32.2)[1]. If the bonds between the atoms are assumed to act like springs, the force required for a certain extension or compression of the spring is described by HOOK's law ($\vec{\mathfrak{F}} = -k * (\vec{r} - \vec{r}))$ and the potential energy of the system by $E = 1/2k(\vec{r} - \vec{r})^2$. This describes a symmetrical parabola with a minimum at r_0. The fundamental frequency of this system is

$$\nu = \frac{1}{2 * \pi} \sqrt{\frac{k}{\frac{m_1 * m_2}{m_1 + m_2}}} \tag{32.8}$$

Multiplication of the frequency ν with the speed of light gives the wave-number $\tilde{\nu}$. This description is derived from classical mechanics. In nuclear dimensions however quantum mechanics applies and only discrete energy levels are allowed, which can be described by integer quantum numbers. Absorbtion of a light quantum lifts the molecule to the next higher energy level, and the energy contained in the light

[1] A very instructive demonstration of vibration modes of acetophenone at various wavenumbers by E. MOTYKA, P.M. LAHTI AND R. LANCASHIRE is available for the protein explorer at www.umass.edu/microbio/chime/.

quantum must be identical to the energy difference between the two molecular states. Additionally, the molecule must have a dipole moment aligned with the electrical field vector of the light quantum. Thus IR-active functional groups usually have asymmetric bonds.

Also, in classical mechanics the entire system may be at rest, in quantum physics molecules oscillate unless the temperature were lowered to 0 K, which is not possible. Molecules have a certain basic energy E_0, from which they are exited by IR-light.

Thus the energy as function of the extension ($\vec{r}$ - $\vec{r}$) is not described by a parabola, but a MORSE-function ($E_r = D_e * [1 - \exp(-a * (r - r_0))]^2$, with a = bond-specific force constant and D_e dissociation energy of the bond).

32.2.1 *Attenuated Total Internal Reflection IR-Spectroscopy*

Since water strongly absorbs IR-radiation, but proteins loose their conformation if water is removed, methods of reducing water contribution to the spectrum are valuable. For ATIR-spectroscopy the surface of a high optical density crystal (*e.g.*, germanium, ZnSe or diamond) is covered with the protein molecules. The IR-light shines at the interface between crystal and buffer at such an angle that repeated total reflection occurs, until the light hits the detector (see Fig. 4.3 on p. 30). The evanescent wave penetrates a short distance ($\approx \lambda$) into the buffer. Because of the low penetration depth the sample can be hydrated, the relatively few water molecules which can interact with the IR-light do not interfere with measurements. This technique is also useful for membrane proteins which can be investigated inside the lipid bilayer.

Adsorption of proteins to the crystal forms a monolayer. This is a saturatable process essentially complete at about 0.5–1.0 mg/ml. Thus the protein concentration at the crystal is not identical with that in solution, in addition protein binding to the surface may lead to conformational changes. The effects of these processes on the IR-spectrum can be eliminated by subtracting the spectrum measured at low concentration (say, 1.0 mg/ml) from that measured at high concentration (say, 10 mg/ml). The difference spectrum is identical to that of the dissolved protein at, in this example, 9 mg/ml. Warning: changes in protein concentration result in changes of the refractive index of the solution and this in turn changes the penetration depth of the evanescent wave. If the IR-spectra of the protein at different concentrations need to be compared, this needs to be taken into account by subtracting weighted buffer-spectra [118].

32.2.2 *Fourier-Transform IR-Spectroscopy*

In a conventional IR-spectrophotometer the light beam coming from the lamp is passed through a monochromator, and the resulting monochromatic light passes the sample before reaching the detector. However, mechanically changing the wavelength of the beam to get a spectrum is a time consuming process (about 10 min). Additionally, monochromators are expensive, especially if they have to be made from IR-transparent material. Slit width in the monochromator is always a balance between wavelength-resolution (which requires a narrow slit) and sensitivity (which requires high light intensity, hence wide slits).

A Fourier transformed infrared (FTIR)-spectroscope is essentially a MICHELSON-interferometer, (see Fig. 32.3). One of the mirrors is moved into the light beam during measurement. The detector measures the light resulting from the interference of the measuring and reference beam. Because of the movement of the mirror, different wavelengths will undergo constructive and destructive interference at different times. The inverse FOURIER-transformation of that signal gives the spectrum of the light source if no sample is present. If some wavelength in that spectrum is attenuated by the sample, this will be missing in the spectrum. Light from a He/Ne-laser with known wavelength serves as a ruler. Measurement requires only a few seconds (so called **multiplex advantage**: the time required for a spectrum with n spectral elements is reduced n-fold by using FT). Because FT requires a computer anyway, operations like subtraction of a solvent spectrum can be handled automatically. Also, FTIR spectra are less affected by noise than conventional spectra and resolution is higher (compare the conventional and FTIR spectra in Fig. 32.4). The spectrum can be further improved by repeating the measurement several times and averaging the results.

The main disadvantage of FTIR-spectroscopy is the difficulty of time-resolved measurements.

32.2.3 IR-Spectroscopy of Proteins

In protein biochemistry, IR-spectroscopy gives us information on the secondary structure of a protein (and its changes, for example during enzymatic turnover). The **amide A-band** is caused by stretching of N−H-bonds, **amide-I** by stretching of the C=O-bond, and **amide-II**, and **amide-III** mostly by in-plane bending of N−H-bonds (with some contributions by C−N and $C^{\alpha}-C'$ stretching, and C=O in plane bending), their maximum depends on secondary structure (see Table 32.1).

Cys −SH stretching results in absorbtion at 2 290–2 700 cm^{-1} depending on hydrogen bonding strength. This region is not subject to interference by other groups and can be used to follow disulphide bond formation and solvent accessibility of the Cys side-chain.

About 100 μg of protein are required for such an experiment, which can be performed even in turbid solutions as scattering is low for the long wavelengths involved.

The **amide-I** region spectrum depends on the hydrogen bonding of the C=O to NH groups and on transition moment coupling of the C=O-bonds. In repetitive structures, coupling of oscillators leads to delocalised states involving large regions

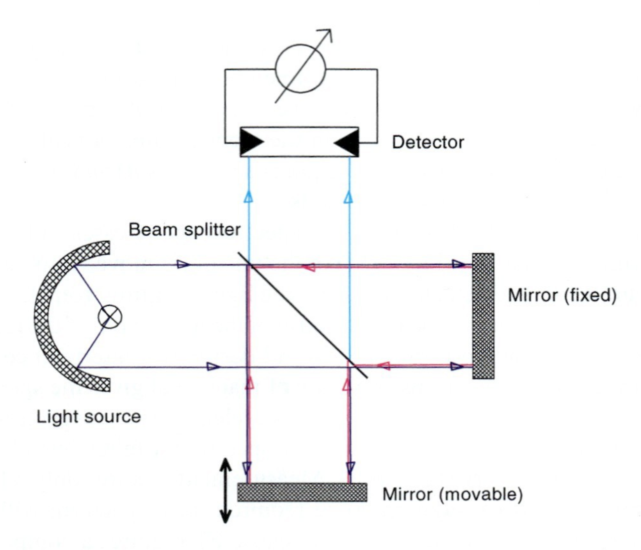

Fig. 32.3 The MICHELSON-interferometer. The light beam (*blue*) coming from the source is split by a semi-transparent mirror into two beams, which are reflected back by two mirrors. The returning beams (*purple*) interfere at the beam splitter, the resulting light beam (*cyan*) is detected. Changes in the distance between the mirrors and the beam splitter can be detected with a precision of less than one wavelength of the light used, also any other delay of one of the light beams

of the protein. Thus the IR-spectrum is sensitive to size and conformational changes. In β-**sheets**, there are two vibration modes: $\tilde{\nu}_{\perp}$ at about 1 620 cm^{-1}, where the oscillators are in phase perpendicular to the strand but out of phase with their bonded neighbours and $\tilde{\nu}_{\parallel}$ at 1 680 cm^{-1} where this is reversed. Intermolecular β-sheets in **amyloid** gives peaks at 1 610–1 624, 1 654 and 1 684–1 700 cm^{-1}. In α-**helices** the $\tilde{\nu}_A$ band at 1 639 cm^{-1} with all oscillators in phase is much stronger than the $\tilde{\nu}_{E1}$-band at 1 652 cm^{-1}, where the phase of the oscillators varies with a period of 3.6 residues. Note that Asn (1 622 and 1 678 cm^{-1}) and Gln (1 610 and 1 670 cm^{-1}) produce vibrations in the amide-I region, whose exact frequency depends on solvent exposure. This can become a problem in proteins containing a large amount of these amino acids.

The amid-I region is CD-active, however, the effect is not very strong and CD–IR-spectroscopy (vibrational CD) is a rather unusual technique.

De-convolution of the IR-spectrum of a protein in the amide region can be used to determine the percentages of α-helix, β-sheet and coil in a similar way as discussed for CD-spectra (see p. 294). The process starts with the removal of water (solute and vapour) contributions to the spectrum. The simplest way to analyse the resulting spectrum is to calculate its second derivative which has downward peaks at each shoulder or inflection point, their intensity is proportional to the amount of the corresponding secondary structure. **Band fitting** can be used to determine relative secondary structure content, the number and position of the components may be obtained from second derivative spectrums. **Resolution enhancement data analysis**

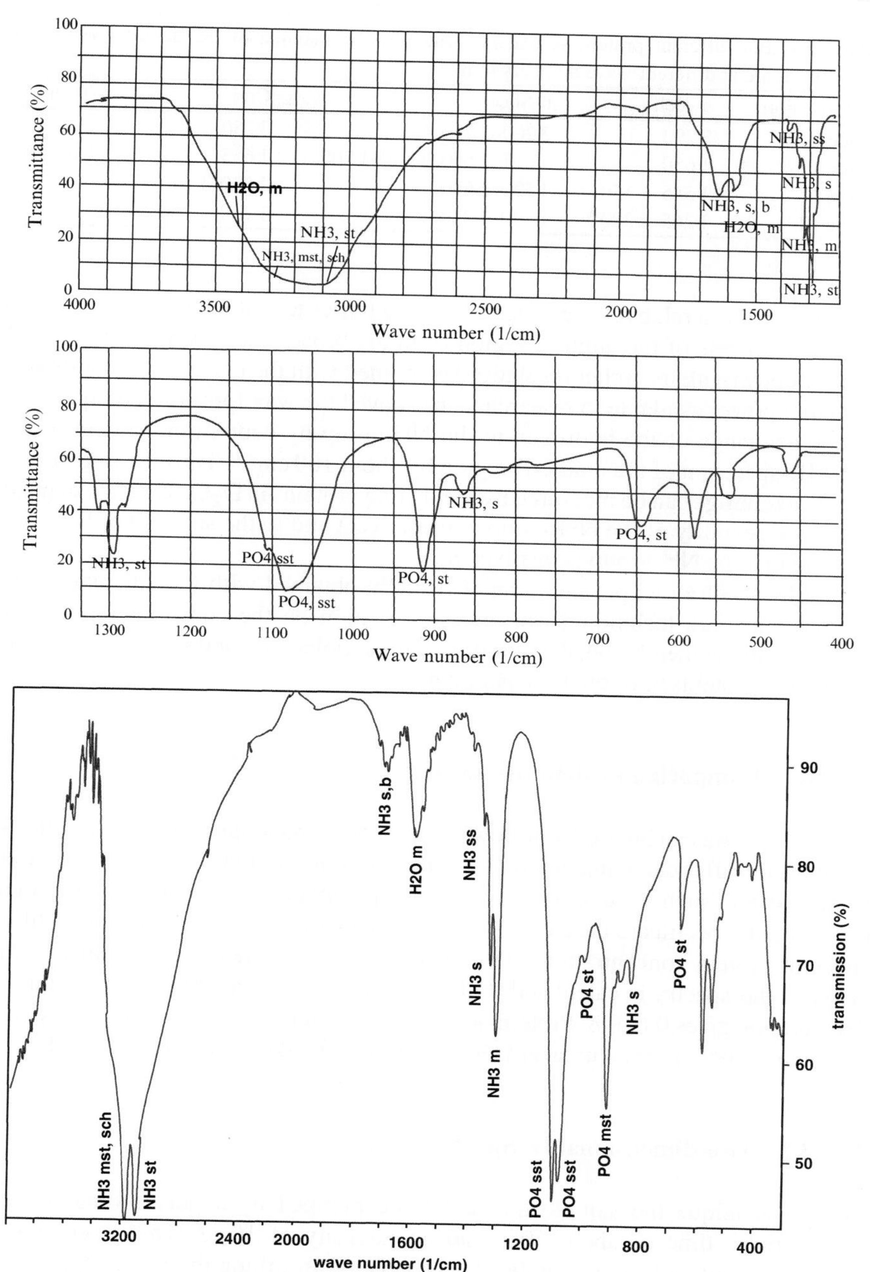

Fig. 32.4 Infrared spectra of $[Co(NH_3)_4PO_4]$. This compound can be used as an analogue of the Mg–PO_4 complex to probe the active site of enzymes. Each of the peaks is caused by specific movements of a specific functional group. *Top*: Spectrum of a conventional IR-spectrophotometer. *Bottom*: Spectrum from a FTIR-spectrometer, note the much higher resolution

Table 32.1 For different protein secondary structures the maxima of the peptide bond in an IR-spectrum are at different wave numbers (cm^{-1})

Conformation	α-Helix	β-Sheet	Coil	β-Turn
Amide A	3 290	3 260–3 290	3 250	
Amide I	1 650	1 615–1 650, 1 680–1 700	1 655	1 660–1 680
Amide II	1 545–1 550	1 520–1 530	1 520–1 545	
Amide III	1 265–1 300	1 230–1 240	1 240–1 260	

[95, chap. 10] is a relatively new technique which tries to mathematically extract the spectral elements of the amide-I region from the broad peak, the advent of FTIR-spectroscopy (with its higher resolution) combined with the phenomenal increase in computer power available to researchers has paved the way for this technique.

The exchange of the hydrogen in the NH-group by deuterium leads to a shift in the IR-spectrum of the amide-II region by about 100 cm^{-1}. Thus the rate of **deuterium exchange** can be measured by incubating proteins in D_2O. This rate depends on the accessibility of the NH-group to the solvent and to the strength of hydrogen bonding that the NH-group is involved in.

The IR-spectra of lipid membranes (usually obtained with the attenuated total internal reflection technique) can give information on the conformation of lipid molecules in the membrane, the mobility of lipid molecules, and the influence transmembrane proteins have on these parameters.

32.2.3.1 Comparing Protein IR-Spectra

Amide-I spectra can be used to follow protein conformational changes, unfolding or denaturation after changing environmental conditions or after ligand binding. Apart from research, such studies are also used for quality control in commercial protein purification. Spectra are baseline-corrected and area-normalised to eliminate differences in protein concentration and cuvette pathlength. Then the **area of overlap** between the spectra is calculated. Full identity results in an AO of 1.0, complete denaturation gives 0.6–0.8. Protein denaturation by urea can be measured only with ^{13}C-urea, as both normal urea and GuHCl produce peaks in the amide-I band.

32.2.3.2 Two-dimensional IR Spectroscopy

In this technique the sample molecules are pumped by a pulsed tunable laser with a pulse time of about 50 fs and an intensity of 10 μJ. This is enough to bring essentially all sample molecules capable of absorbing the wavelength of the laser to a higher vibrational level. Because of the low energy of IR photons, this will not induce any chemical reactions. Rather, molecules are "labelled" by the change of vibrational state. After a waiting time in the ps range, a probe pulse of broad-band IR-radiation is used to measure a FTIR-spectrum. This is repeated for different pump wavelengths, resulting in a two-dimensional spectrum with cross-peaks originating from induced absorbtion or stimulated emission. This method is used to follow protein structural changes (folding, binding) on a very fast time scale [110, 256]. Alternatively, multi-pulse regimes are also used [447].

32.2.4 *Measuring Electrical Fields in Enzymes: The* STARK*-effect*

Both the pK_a and the standard redox potential of residues in the catalytic centre of an enzyme can vary significantly from the values of the corresponding free residues in aqueous solution (see [132] for a recent review). The main reason for discrepancies in pK_a-values are:

Charge–charge interactions positively charged environment lowers, negatively charged environment increases pK_a.

Charge–dipole interactions hydrogen bonding is stronger with a charged than with a a polar group, hence the charged form is energetically favoured.

BORN-effect inside the protein the dielectric constant is usually lower than in water, this increases the pK_a of acidic, and decreases the pK_a of basic groups.

This has considerable mechanistic implications. For example, in the xylanase from *Bacillus circulans* [268] Glu-172 acts as a proton donor and has a pK_a of 6.7, while Glu-78 with $pK_a = 4.6$ is ionised and stabilises the positively charged intermediate. The redox potential of Fe^{3+}/Fe^{2+} is 771 mV in aqueous solution, for protein bound iron the potential can reach from +365 mV (cytochrome f) down to −432 mV (ferredoxin). Attempts to calculate such effects have been made [16, 387].

Electrical fields between charged residues and ligands affect pK_a-values. Although the potential differences are small, they act over atomic distances, hence the field strength can be very high in the order of 10^7 V/cm. A dipolar transition state that separates a unit charge over a distance of 1 Å parallel to that field would result in a lowering of the activation energy by 9.6 kJ/mol [394]. Thus electrical fields are of considerable importance for our understanding of enzyme mechanism.

The stretching frequency of a polar bond changes with the electrical field strength (**electrochromic band shift**, STARK-effect) [38], that change can be calibrated by applied external fields. The change in wave number $\Delta\tilde{\nu}$ as a function of the change in electrical field strength $\Delta\vec{\mathfrak{E}}$, for example as a result of a conformational change in an enzyme, can be calculated to

$$hc\Delta\tilde{\nu} = -\Delta\vec{\mu} \times \Delta\vec{\mathfrak{E}} \tag{32.9}$$

with h PLANK'S quantum, c the velocity of light and $\Delta\vec{\mu}$ the projection of the change in dipole moment along the electric field vector.

Carbon monoxide and nitriles (−C≡N) make good probes for such studies as they have intense vibration lines different from those found in protein, and the change of dipole moment is large and oriented along the probe axis. The experiment is performed in a frozen glass so that the proteins have no rotational or translational freedom.

32.3 Raman-Spectroscopy

Scattering of light on molecules depends on the polarisability of the molecule's electrons. Molecular vibrations which change the polarisability of bond electrons will change scattering. Thus the incident light will be amplitude-modulated with the frequency of the vibration. This beating of ν_{vib} against ν_0 generates two additional light-beams, with frequencies $\nu_0 \pm \nu_{vib}$ (ν_0 = frequency of the incident light, ν_{vib} = frequency of the vibration). Thus scattered light contains on both sides of the RAYLEIGH-band additional signals, which were discovered by CH. RAMAN. Bands with longer wavelength are called STOKES-, those with shorter wavelength anti-STOKES signals.

Resonance RAMAN-spectra are obtained when the wavelength of the incident beam is close to an absorbance maximum of the sample molecules. The photons are absorbed and almost immediately (within 10 ps) emitted again. In the absence of molecular vibration, the emitted photon would have the same wavelength as the absorbed, molecular vibration leads to the RAMAN-effect. The process is different from fluorescence, where the return of the electrons to the ground state is slower and the emitted light has a different (usually longer) wavelength than the exciting photon. However, if the sample molecule shows fluorescence, measuring resonance RAMAN spectra becomes more complicated. Because absorbtion of the light leads to a magnification of the RAMAN-signal for a specific chromophore, this method can be used to measure a particular molecule in a mixed sample.

IR and RAMAN-spectroscopy both measure molecular vibrations. For a vibration to be IR-active requires a change in dipole moment of the molecule during vibration (usually in unsymmetrical groups like −CO−), RAMAN-activity requires a change in polarisability, usually observed in symmetrical groups like > C=C <. Some vibrations however have both effects and can be measured by both methods. One of the advantages of RAMAN-spectroscopy is that because water causes a much weaker signal, RAMAN-spectra of proteins can be measured in solution.

RAMAN-spectrometer use lasers as monochromatic, high intensity light sources. Since scattering increases with ν^4, short wavelengths are preferred. Ideal are tunable lasers (*e.g.*, dye lasers). The scattered light is picked up under 90° and passed through a monochromator, which has to separate the RAMAN-bands from the 10 000-fold stronger RAYLEIGH-band. This requires long distances between grid and slit, which determine the physical size of the instrument. The light is detected by photomultipliers.

Chapter 33
Structure of Protein–Ligand Complexes

33.1 Electron-Spin Resonance

A more detailed description of electron spin resonance (ESR) than possible here may be found in [334].

The spin of unpaired electrons causes a magnetic moment μ_e. Similar to the nuclear spin, this magnetic moment can be oriented parallel or anti-parallel to an external magnetic field (of strength B_0), that is, an outer magnetic field leads to a splitting of the energy level of the electron (ZEEMAN-effect). The energy difference between these states can be covered by electromagnetic radiation of suitable frequency ν:

$$\Delta E = h\nu = \mathsf{g}\mu_0 B_0 \tag{33.1}$$

$$\nu = \frac{\mathsf{g}\mu_0}{h} B_0 \tag{33.2}$$

$$\mu_0 = \frac{eh}{4\pi m_e c} \tag{33.3}$$

with $\mu_0 =$ BOHR's magneton. Similar to NMR, nearby particles create their own magnetic fields, which influences the energy difference ΔE, this is expressed in LANDÉ's **g**-factor $2 * \frac{\mu_e}{\mu_0}$. For free electrons $\mathbf{g} = 2.00232$. Typesetting of **g** indicates that it is actually a tensor[1], and hence dependent of the orientation of the label with respect to the outer magnetic field.

In **electron spin resonance (ESR)**, also called **electron paramagnetic resonance (EPR)**, **g** is determined as a measure of the environment of an electron, for unpaired electrons in transition metals $1 < \mathbf{g} < 10$. Further information

[1] Tensors are a mathematical concept introduced by WOLDEMAR VOIGT in 1898. Tensors of order 0 are scalars, of order 1 vectors, of order 2 square matrices, of order 3 cubic matrices and so on. The word tensor in physics is often used as abbreviation for "tensor field", a structure that connects each point in n-dimensional space with a vector of length n. An example would be a weather map, where each point has a vector indicating wind-speed and -direction associated with it.

E. Buxbaum, *Biophysical Chemistry of Proteins: An Introduction to Laboratory Methods*, DOI 10.1007/978-1-4419-7251-4_33,

may be obtained from the hyper-fine splitting that results from the interaction of the magnetic moment of the electron with that of the atoms in its vicinity. Line width of EPR spectra increases as mobility of the probe is restricted.

Variation of the frequency of about 9 GHz (X-band) is technically difficult, hence ESR-experiments are performed with a constant ν (emitted from a klystron) and varying magnetic flux density (0–1 Tesla). A second electromagnet supplied with an AC-voltage of about 100 kHz amplitude-modulates that field. Resonance leads to absorbtion of microwave radiation by the sample which reduces the microwave energy in the sample chamber. This reduction is picked up by a crystal detector which acts as a rectifier. The amplitude of the signal at the rectifier (which is of course amplitude modulated with 100 kHz and hence can be distinguished from background noise) is proportional to the 1st derivative of the ESR-signal. Sample concentrations of about 10 μM are required for a decent signal. The sample needs to be arranged in the node of the electric field to minimise heating of sample water, at the same time this position has the maximal magnetic field and hence maximal sensitivity.

The population difference in ESR is only 0.1 % at room temperature, but 10 % at the temperature of liquid nitrogen, additionally the averaging out of the anisotropy by BROWNIAN motion is reduced at lower temperature. At the same time microwave absorption by water is lower in the frozen state. Thus many ESR experiments are performed at **cryogenic temperatures**. Addition of glycerol, sucrose or ethylene glycol to the sample results in a glass-like state after freezing and prevents concentration changes by freezing out during water crystal formation. Such freezing out would lead to increased spin-spin coupling and to altered ligand concentration.

33.1.1 Factors to Be Aware Of

The high sensitivity of ESR requires pure reagents, from which all ESR-active metals (Mn, Fe, Cu...) must be excluded. Oxygen must also be removed by careful de-aeration of samples. Fused silica tubes are used as sample containers as other glass sorts frequently contain contaminants. Finger prints and smoke are a frequent source of contamination (fortunately the filthy habit of smoking in the lab has finally been eliminated virtually everywhere).

Quantitative measurements need to be performed at a microwave power which is not saturating. As saturation is a phenomenon independent of the probe used, the effect of microwave power on ESR signal strength can be determined with a convenient standard ($CuSO_4$ or TEMPO), whose concentration can be determined by optical absorption. Even under optimal conditions however quantitation by ESR is subject to an error of $\approx$10 %.

Side reactions may cause the appearance of small amounts of ESR-active compounds, which may be taken as evidence for their participation in the main reaction because of the high sensitivity of ESR.

33.1.2 Natural ESR Probes with Single Electrons

ESR can be used to follow the creation of single electrons or a change in their environment during chemical reactions. **Redox-reactions** can be followed by ESR, but relaxation is often so fast that measurements need to be performed at low temperatures in liquid nitrogen. However, some very delocalised radicals like chinons and flavines can also be followed at room temperature.

Cu^{+} has 10 d-electrons (*i.e.*, all paired) and is not ESR-active, Cu^{2+} has a single unpaired electron and is ESR-active. Fe^{3+} has five d-electrons and can occur in a high-spin (five unpaired electrons) or low-spin state (one unpaired electron). In Fe^{2+}, with six electrons, only the high-spin state has ESR-activity, the low-spin state is undetectable. Whether iron is in the high- or low-spin state depends on the ligands bound to it.

Many biological redox-reactions have free radicals as intermediate products, which can be detected by ESR, for example those involving FAD. Radiation damage to DNA involves thymidine-radicals.

The orientation of a molecule with respect to the magnetic field determines **g**, so if a protein crystal is turned in a magnetic field the orientation of an ESR-active functional group (for example haem) with respect to the crystallographic axes of a protein can be determined.

33.1.3 Stable Free Radical Spin Probes

Stable **free radicals** can be used as **spin-probes** to analyse the environment of a protein (see Fig. 33.1).

With spin labels the speed of rotational tumbling can be determined in the form of the **rotational correlation time** $\tau_R = V\eta \,/\, kT$, with V = molecular volume, η = viscosity of the medium, k = BOLTZMANN-constant, T = absolute temperature.

ESR with spin-labelled proteins is suitable to measure conformational changes in proteins on a picosecond time scale (that is, statistical sub-states, see Chap. 35 on p. 343). To distinguish between motion of the protein and the label (*e.g.*, 'R1' on Fig. 33.1), one can stabilise (or destabilise) protein structure with osmolytes like sucrose (see Sect. 34.3.1.1 on p. 325). This shifts the equilibrium between protein sub-states, which can be detected in the ESR-spectrum [244]. Effects of viscosity are controlled for with Ficoll 70, a high molecular mass synthetic sucrose polymer that increases viscosity with little effect on osmolarity.

Labelling lipids with oxazolidine-spin labels allows the degree of order in a membrane to be determined. Binding of lipid to transmembrane proteins increases order, if the ESR-signal is determined at different lipid:protein ratios, the number of lipid molecules associated with the protein (**annular lipids**) can be determined. In addition, it is possible to determine whether labelled lipids are in the extracellular or cytoplasmic leaflet of the plasma membrane, or in intracellular membranes [369].

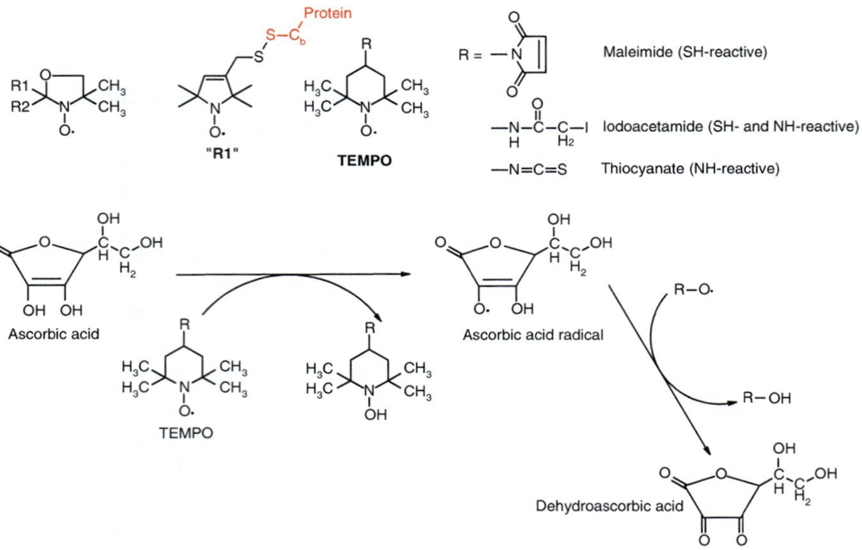

Fig. 33.1 Nitroxide-radicals like TEMPO are stable for several days in aqueous solution. In polar solvent the unpaired electron is located on the nitrogen, in apolar solvents on the oxygen. With suitable reactive groups they can be used to label proteins or other molecules of biological significance for ESR-studies. Reduction of the nitroxyl-radical with ascorbic acid destroys ESR-activity, but only if the label is exposed to the aqueous environment. In the oxazolidine spin probe (*top of diagram*) the orientation of the probe with respect to the rest of the molecule is fixed

Spacer length between probe and labelled molecule can be varied to determine the depth of a binding site. Once the spacer is long enough for the spin probe to be outside the binding side, its mobility increases, which can be detected with ESR.

With **spin label relaximetry** it is possible to determine the structure of membrane proteins, or an interesting region thereof. The method is based on spin relaxing agents increasing the relaxation rate, which influences ESR signal strength as function of microwave power. In the protein the amino acids are mutated to Cys one by one (ensuring that the resulting proteins still have proper folding and function), these Cys-residues are reacted with a SH-reactive spin probe ("R1", see Fig. 33.1). If the labelled amino acid is exposed to the aqueous environment, it will be sensitive to both oxygen and to chromium oxalate (CROX). If it is inside the lipid bilayer, it will still be sensitive to oxygen (which as a small gas is lipid soluble), but not to CROX. If it is buried inside the protein matrix, it will be sensitive to neither oxygen nor CROX.

The rate of flipping of lipids between the leaflets of a membrane can be determined if the spin-probes on the head-groups of the lipids in the outer leaflet of a vesicle are destroyed with ascorbic acid. Because ascorbic acid is very hydrophilic, it can not reach the inner leaflet. Flipping moves spin label from the inner to the outer leaflet in a time- and temperature dependent manner. This label can then be destroyed with ascorbic acid, leading to a reduction in ESR-signal strength.

33.1.4 *Hyperfine Splitting: ENDOR-Spectroscopy*

In electron nuclear double resonance (ENDOR) the ESR signal measured is influenced by hyperfine splitting due to the nuclear spins in the vicinity of the unpaired electron (up to two bonds away). The populations of these nuclear spins however can be influenced by radio frequency radiation (see the section on NMR on p. 309). Thus if such a radio frequency field is introduced into the cavity of an ESR-spectrometer, one can use the ESR spectrum as a method to measure NMR, combining the high sensitivity of ESR and the high spectral resolution of NMR. Additionally, the nucleï responsible for hyperfine splitting can be identified.

In the case of ENDOR-spectroscopy, the microwave power is increased to partially saturating; if the radio-frequency introduced matches the NMR-frequency of one of the coupled nucleï, the alternative pathway leads to desaturation and an increase in the ESR signal.

This technique has also been used to measure the binding of substrates or inhibitors to metalloenzymes.

33.1.5 *ESR of Triplet States*

Triplet states are ESR-active, the ESR transition influences the absorbance and phosphorescence (**optically detected magnetic resonance (ODMR)**). The main advantage of this method, useful for aromatic amino acids, purines, pyrimidines, photosynthetic pigments, and flavines, is the extremely high sensitivity, allowing single molecule detection under ideal circumstances.

33.2 X-ray Absorbtion Spectroscopy

33.2.1 *Production of X-rays*

Electrons from a heated filament, accelerated by an electric field, can kick other electrons from the inner orbitals of a target metal (usually Cu, Cr or Mo) into the continuum. Electrons from higher orbitals will fall down into these holes, emitting X-rays of a specific wavelength. The X-ray energy where this occurs (edge) depends on the element and the orbital of the electron involved. If the electron comes from the L-shell and drops to the K-shell (L→K), we call the edge K_α, for M→L K_β and so on. The wavelengths are longer for elements with lower Z, as there are fewer protons in the nucleus to attract the electron, thus we get less energy from the process.

33.2.2 Absorbtion of X-rays

All elements of the periodic system, even those which are silent in optical and magnetic spectroscopy techniques (Zn, S) absorb X-rays. The absorbtion coefficient increases with wavelength, but drops sharply just below K_β, to increase again at even longer wavelengths. This results in an absorbtion edge, the wavelength of which is characteristic for each element, and again becomes longer with decreasing Z.

Chapter 34
3-D Structures

34.1 Nuclear Magnetic Resonance

34.1.1 Theory of 1-D NMR

This introduction to NMR must be necessarily brief. Further information can be found in [344, 436] and references therein.

Particles that rotate have an angular momentum $I * \overrightarrow{\omega}$ (I = moment of inertia, $\overrightarrow{\omega}$ = angular velocity), which can be represented by a vector in the direction of the axis of rotation. In classical physics, this axis can have any angle in space. In quantum physics, both direction and magnitude of the angular momentum can take on only limited values.

The magnitude is restricted to the **angular momentum quantum number** j, which can take on only half-integer values (0/2, 1/2, 2/2, 3/2...).

The angle, specified relative to a standard direction, is limited to values of $\frac{nh}{2\pi}$, where n can take on values of j, j-1,... 0, ..., -j. Thus there are 2*j+1 possible directions for the angular momentum. In NMR experiments, the standard direction is defined by the field vector of the external magnetic field.

NMR spectroscopy is based on the fact that atoms with an odd number of nucleons (^{1}H, ^{13}C, ^{15}N,^{19}F, ^{31}P, spin quantum number $j = 1/2$, 2 possible directions relative to an external magnetic field) or with an odd number of both protons and neutrons (^{2}H, ^{14}N, $j = 1, 3$ possible directions relative to an external magnetic field) have a spin, that is an angular momentum, different from zero. Atoms with an even number of both protons and neutrons have $j = 0$ and cannot be detected by NMR.

Because of their positive charge and the spin, nucleï with $j \neq 0$ have a magnetic moment μ_N and can orient themselves parallel or antiparallel to an external magnetic field (with a strength B_0 of several Tesla), similar to a compass needle in the magnetic field of the earth. External magnetic fields are generated in superconductive coils (Nb-alloys), cooled by liquid helium. The sample, however, is insulated from the cryogenic fluid and usually kept at room temperature.

The energy difference between these orientations is calculated to:

$$\Delta E = 2\mu_N B_0 = \hbar\omega_L \tag{34.1}$$

E. Buxbaum, *Biophysical Chemistry of Proteins: An Introduction to Laboratory Methods*, DOI 10.1007/978-1-4419-7251-4_34,

with ω_L being the frequency of an electromagnetic wave that has an energy equivalent to that required for flipping. With the magnetic field strength currently available (several Tesla), this LARMOR-frequency is in the order of several 100 MHz (500 MHz at 12 T is typical for current instruments).

34.1.2 BOLTZMANN-*Distribution of Spins*

The energy difference between parallel and antiparallel orientation is small, even BROWNIAN movement can lead to flipping of these magnets. The BOLTZMANN-distribution can be used to calculate the number of atoms in either orientation:

$$\frac{\eta_1}{\eta_2} = \exp\left(\frac{-\Delta E}{kT}\right) = \exp\left(\frac{-2\mu_N B_0}{kT}\right). \tag{34.2}$$

The product kT is the energy of thermal motion at the given temperature. So in a field of 2 T out of 2×10^6 hydrogen atoms 10 00 005 will be parallel and 9 99 995 antiparallel to that field, a difference of 1/2 00 000. For other nucleï, this ratio may be even smaller (or the NMR-active isotope may be less abundant than ^{1}H, *e.g.*, ^{13}C, see Table 34.1), hence their signals in NMR are weaker than for ^{1}H. This small difference in occupation of high and low energy states is responsible for the low sensitivity of NMR compared to other spectroscopic techniques. The ratio can be improved by using a stronger magnetic field B_0, which is expensive, or by decreasing the temperature, which is often inappropriate, especially with biological samples.

If energy in the form of a brief (μs) radio pulse of frequency ω_L is applied, some of the atoms in parallel orientation switch to the antiparallel, until (if the pulse is long and strong enough) $\eta_1 = \eta_2$. The magnetic moments of all the atoms in a sample results in a magnetisation in the direction of the outer magnetic field

Table 34.1 Elements important in biological NMR. The signal strength depends, among others, on the abundance of the active isotope. Rare isotopes like ^{13}C are useful only in isotopic labelling experiments. Fluorine is rare in biological compounds, but can easily be introduced as a label. Relevant data for other elements may be found at www.pascal-man.com/periodic-table/periodictable.html

Element	Biol. abund. (%)	NMR-act. isotope	Isotope abund. (%)	Unpaired protons	Unpaired neutrons	Spin	γ (MHz/T)
Hydrogen	63.0	^{1}H	99.985	1	0	1/2	42.58
		^{2}H	0.015	1	1	1	6.54
Carbon	9.40	^{13}C	1.11	0	1	1/2	10.71
Nitrogen	1.5	^{14}N	99.63	1	1	1	3.08
		^{15}N	0.37	1	0	1/2	−4.32
Phosphorus	0.24	^{31}P	100	1	0	1/2	17.25
Fluorine	trace	^{19}F	100	1	0	1/2	40.08
Xenon	0	^{129}Xe	26.4	0	1	1/2	−11.84

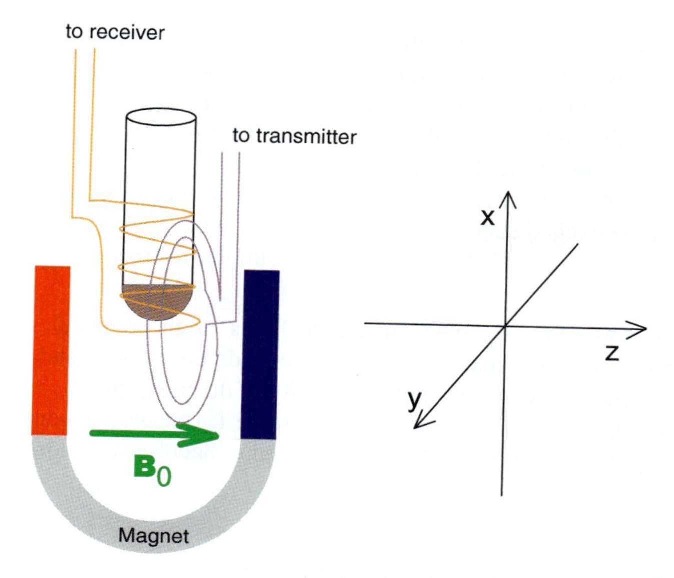

Fig. 34.1 Principle components of a NMR-spectrometer. The sample is placed into a strong magnetic field, transmitter- and receiver antennas are placed close to it. By convention, the direction of the magnetic field defines the z-axis of the system, the sample and receiver antenna the x-axis and the transmitter antenna the y-axis

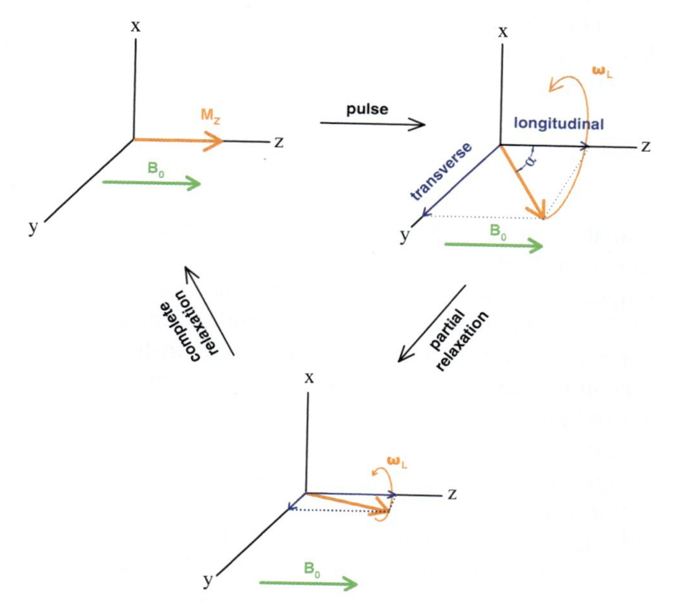

Fig. 34.2 The total magnetisation of the sample can be represented by a vector (*orange*), which in equilibrium will point into the direction of the external magnetic field B_0 (*green*). A brief radio-frequency pulse causes the magnetisation vector to flip by a certain angle α, the tip of the vector will precess around the z-axis with the LARMOR-frequency ω_L. Once the external pulse stops the flip angle will decrease exponentially over time

(see Figs. 34.1 and 34.2), and the flipping of atoms by the radio pulse changes this magnetisation by a certain **flip angle** α, which depends on strength and length of the radio pulse. Thus one speaks of a 30° pulse, a 60° pulse and so on.

Once the radio pulse stops, these atoms flip back over a period of time, this is called **relaxation**. If the flip angle is plotted over time, an exponential curve is obtained with a relaxation time in the order of milliseconds to minutes. The energy contained in the excited state is released in the form of radio waves which can be detected by a receiving coil placed close to the sample. There are two causes of this back-transition:

- Magnetic noise caused by the magnetic fields of the molecules in the system, which are in thermal motion. This depends on the size and shape of the molecules, the temperature, and the viscosity of the medium.
- Spin exchange between molecules of the same LARMOR-frequency that come close to each other. This effect increases when molecules stay close together for longer time.

34.1.3 Parameters Detected by 1-D NMR

The key point of NMR is that part of the external magnetic field is shielded from the nucleï by the electrons (diamagnetic effect) and nucleï surrounding it (within a maximal distance of ≈ 8 Å), hence $B_{\text{eff}} < B_0$. Thus the energy difference between the antiparallel and parallel orientation, and hence ω_L of the radio pulse emitted, depends on the kind of bonds that the atom is involved in. The **chemical shift**,

$$\delta = \frac{B_0 - B_{\text{eff}}}{B_0} = \frac{\omega_L - \omega'_L}{\omega_L} \tag{34.3}$$

is quite small, in the order of 1–50 ppm. Beating the signal picked up by the antenna against the stimulating radio wave results in a difference signal with frequencies in the audible range, FOURIER analysis of this difference signal gives the NMR spectrum (see Fig. 34.3). This is performed after digitalisation of the measured signal by a computer algorithm known as **fast FOURIER transform (FFT)** [337]. It is somewhat difficult to measure the absolute value of chemical shifts (in Hz), therefore **tetramethylsilane** is added as reference substance. The proton resonance peak of this compound is given arbitrarily the value 0 ppm, the other peak positions are scaled accordingly. This allows shift values from different laboratories and instruments to be compared.

If the NMR-signal were constant over time, the FOURIER-transform would consist of sharp lines. The exponential decay of the NMR-signal results in peaks that have a certain width, the shape of the peak is a LORENTZIAN curve[1] : The faster the decay time, the wider the peak. Since the decay time depends on molecular

[1] $f(x) = \frac{a}{1+(\frac{x-x_0}{b})^2}$.

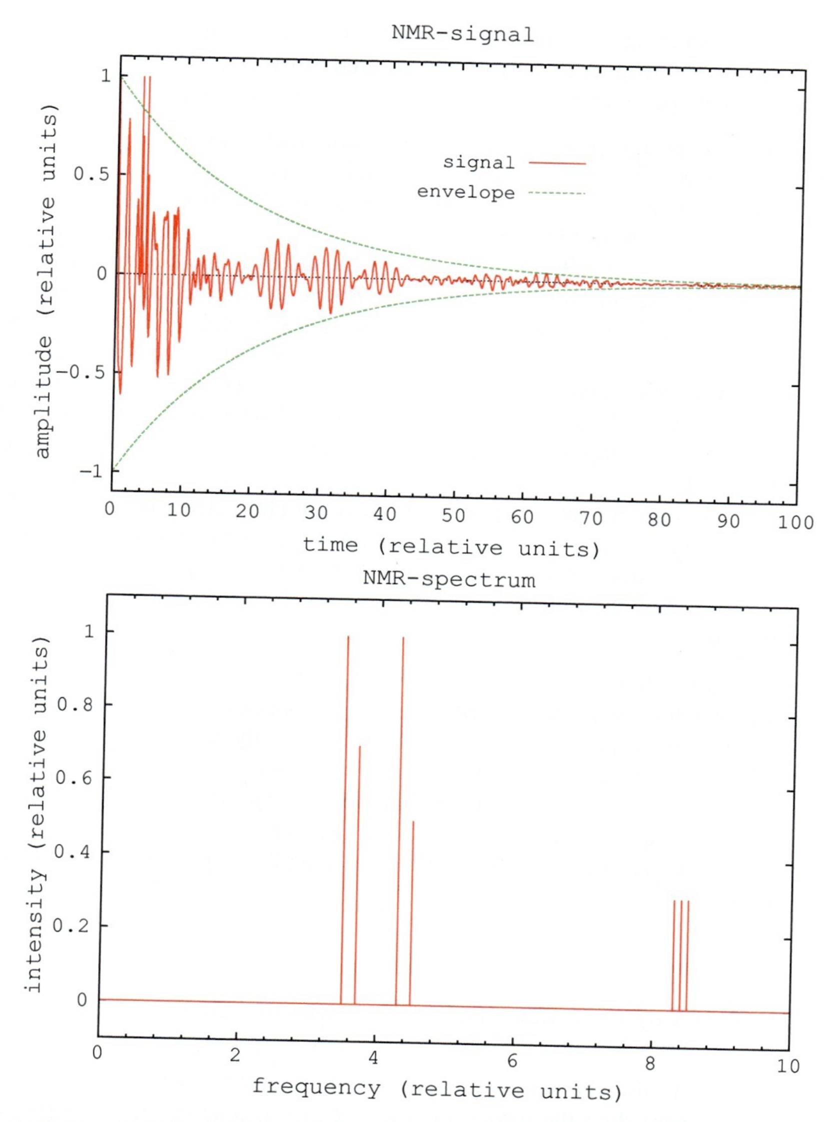

Fig. 34.3 *Top*: A NMR spectrum is the sum of several frequencies, the signal strength decays exponentially over time. Because the spectrometer uses a superheterodyne amplifier, the signal is in the audible (kHz) range. *Bottom*: FOURIER-transformation isolates the component frequencies from the mixture

motion, and this in turn on **molecular interactions**, such interactions can be probed by NMR.

This line-widening is also one of the reasons that excitation in NMR is performed by radio pulses, the resulting line-widening allows nucleï to become excited despite their chemical shift. This means that for nucleï with large shifts, pulses need to be shorter, and hence stronger to achieve the required flip angle.

The following parameters are of interest in an NMR-spectrum:

Chemical shift characterises the chemical environment of a nucleus.

Peak area is proportional to the number of nucleï that contribute to the signal.

Peak width (measured at half the hight of the peak) is proportional to the relaxation time and hence to molecular motion.

Multiplicity is caused by closely spaced identical nucleï. If nucleus A is in a particular orientation, it will influence the atoms around it, and the shift of nucleus B. Hence the peak for nucleus B will split into two sub-peaks. Of course, B will split the spectrum of A in a similar manner. Several such coupled nucleï will cause splitting into more sub-peaks. For example in ATP the β-phosphate peak is split into three sub-peaks, by the α and γ-phosphate groups. The peaks for α and γ-phosphate groups will be split into two sub-peaks each by the β-phosphate (see Fig. 34.4).

$J_{\mathbf{AB}}$**-value** (or coupling constant) is the distance (in Hz) between multiplets. This depends on the number of bonds between the nucleï and on the geometry of the bond, hence J-values of coupling between protons in the peptide bond can be used to determine the dihedral angles in this bond, *i.e.*, the secondary structure of a protein.

It is usually not sufficient to collect one NMR spectrum, as the noise would make it impossible to detect the peaks. For this reason many spectra are taken and averaged. Thus although the acquisition of an NMR-spectrum only requires a few seconds, the time for each experiment is in the order of hours or days. The lower the amount of sample, the noisier the spectrum. For reasonable spectra in a affordable time 0.1–1 μmol of sample are required. 1 μmol of a protein of 50 kDa would be 50 mg. These have to be dissolved in a small volume of buffer (typically 0.5 ml) without aggregation.

34.1.4 NMR of Proteins, Multi-dimensional NMR

If all the lines of this page were printed on top of each other, they would be unreadable. Only by spreading the information into 2 dimensions (by printing the lines beneath each other) can we make use of the information contained.

In small molecules (M_r in the order 10^2) the number of interacting nucleï, and hence peaks in the NMR-spectrum, is small and can be fully resolved by 1-D NMR. However, proteins have a molecular mass of 10^4–10^5, peaks in the spectrum of such molecules in 1-D NMR overlap and can not be resolved.

Assume two nucleï in close proximity with their magnetic moments parallel to the external field B_0 (↗↗). Then an external radio signal may flip one of them into an anti-parallel orientation (↗↙). Cross-relaxation allows the magnetic state to be exchanged between the nucleï, if they are close to each other (less than 5 Å apart):

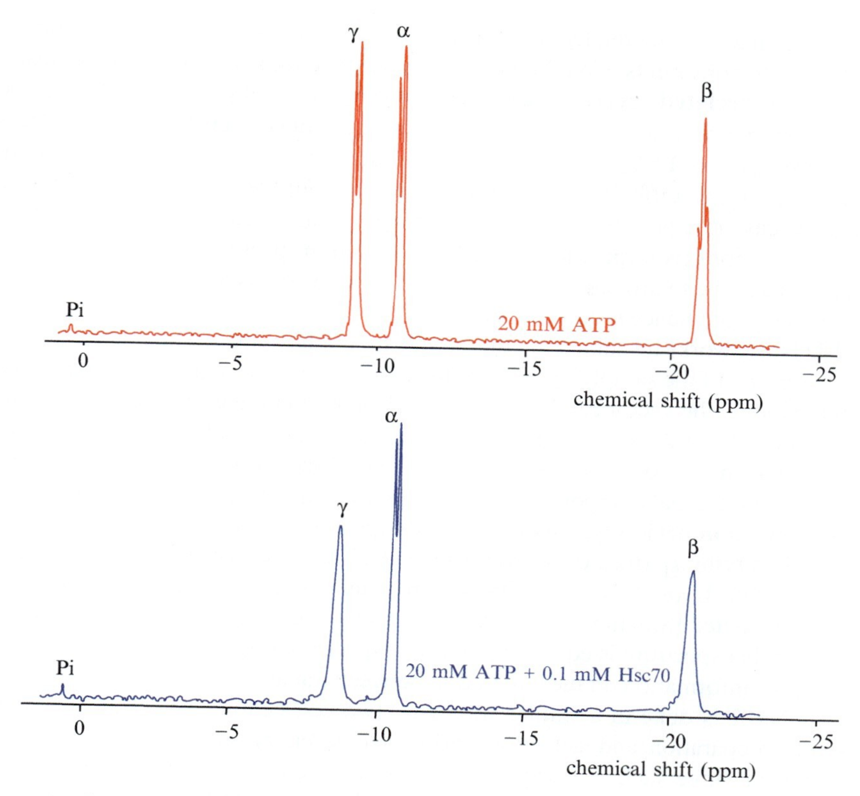

Fig. 34.4 ^{31}P-NMR spectrum of ATP. *Top*: Free ATP. Interaction between the three phosphate groups results in peak-splitting. α- and γ-phosphate groups have one neighbour each, hence their peaks are split into two sub-peaks. The β-phosphate group has two neighbours, hence three sub-peaks are visible. *Bottom*: ATP bound to Hsc70. Binding of the nucleotide to the much bigger enzyme molecule reduces its mobility, even though the enzyme concentration is 2 orders of magnitude lower than that of the nucleotide. The peaks become wider and the peak-splitting is less well resolved

$\nearrow\swarrow \rightleftharpoons \swarrow\nearrow$. The cross-relaxation rate σ depends on the distance r and the **effective rotational correlation time** τ_{eff}:

$$\sigma = \tau_{\text{eff}}/r^6. \tag{34.4}$$

In equilibrium the rate for the forward and backward reactions are the same, but if we perturb one nucleus, then the magnetisation of the second will change with a rate σ, that depends on r^6. σ will fluctuate because of internal motion within the protein, thus only an upper limit for the distance can be obtained. The sum of the VAN DER WAALS radii of the atoms involved (2 Å for hydrogens) is used as lower limit.

Since both nucleï individually also correlate with other nucleï, we get a network of 3-D distance constraints. Coupling can be by the electrons of shared bonds (scalar coupling, **correlated spectroscopy (COSY)** or by physical proximity of nucleï in the 3-D structure of a molecule (bipolar coupling, **nuclear OVERHAUSER effect spectroscopy (NOESY)**. Note that coupled nucleï can be of the same (homonuclear coupling) or different species (heteronuclear coupling, *e.g.*, ^{1}H with ^{13}C). In the latter case it is possible to control coupling by introducing isotopes like ^{13}C at specific positions (**isotopic labelling**). The correlation spectrum between the amide hydrogen and the hydrogen on C^{α} of the same amino acid is a measure of the dihedral angle ϕ, and hence secondary structure.

2-D NMR works by subjecting the sample to 3 radio-pulses: The first rotates the magnetisation of the sample by 90° into the x, y-plane. During the following **evolution period** t_1, the vector can be measured. Then the sample is exposed to a second pulse which flips the magnetisation by 90° into the x, z-plane. During the following **mixing period** τ_m cross-correlation between nucleï occurs. Then a third radio-pulse flips the longitudinal component of M (along the z-axis) by 90°. This is followed by the **detection period** t_2, where the NMR-signal is measured. This experiment is repeated, varying t_1 from 0 to several seconds. The LARMOR-frequency of the z-component of M (ω_1) changes with t_1 ("frequency labelling"). The final spectrum is then calculated from the sets of NMR-signals by 2-D FOURIER-analysis . The diagonal of that spectrum is equivalent to a 1-D spectrum ($\omega_1 = \omega_2$), the off-diagonal points give information on the interaction between nucleï.

3-D NMR is done in a similar way, with two consecutive 2-D sequences with a single preparation and detection period. However, t_2 and τ_m are varied. For 4-D spectra, three such 2-D sequences are used. Such experiments use uniformly or specifically ^{15}N and ^{13}C labelled proteins. While 2-D NMR can be used for proteins up to about 15 kDa, higher dimensions can work up to twice that size. Proteins need to be soluble in high concentrations (up to 2 mM) under slightly acidic conditions (to ensure that they are fully protonated and prevent proton exchange with the solvent). To obtain a high S/N ratio, temperatures need to be >15°C.

Each peak in a 2-D NMR spectrum is one constraint on the proposed structure of a molecule, about seven peaks per residue are sufficient to build a computer model of a protein backbone (taking also into account other constraints, *e.g.*, RAMACHANDRAN-plots). With 15–20 peaks per residue one can build a full 3-D structure of a protein molecule. Usually the structure inside a protein is better resolved than that of the surface, because amino acids inside have more neighbours to interact with and because their lower mobility increases the efficiency of the NOE.

Measuring 2-D spectra itself is only the easiest part of structure determination, it results in a pack of A1- or A0-sized plots in which the peaks need to be assigned. This can not be done automatically, computer programs for multi-dimensional NMR simply keep track of assignments.

A computer then tries to determine a model structure that is consistent with the distances of nucleï thus determined. This can be done *ab initio*, or one can try to fit the data to structures from a protein structure database (**knowledge-based** structure prediction).

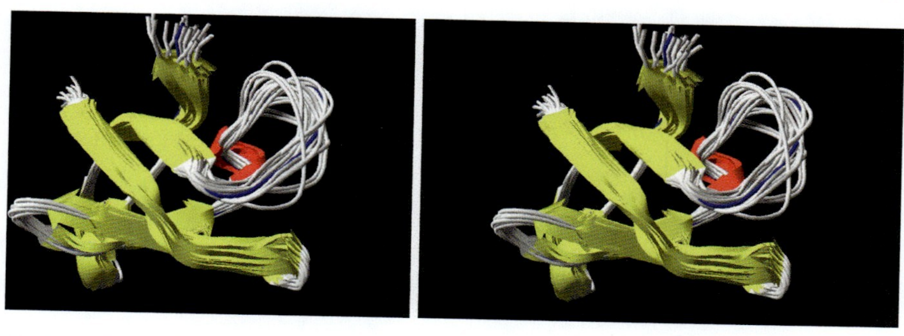

Fig. 34.5 Result of protein structure determination by NMR, here major cold shock protein. The constraints determined by NMR are consistent with a number of models (here 20, PDB-code 1nmf). From these, an average structure can be calculated, this is then run through energy minimisation. The resulting structure (PDB-code 1nmg) is shown in blue. Note that a large part of the structure is well defined, except for the N- and C-termini and for an exposed loop to the right. Important: The variability in the ensemble does *not* represent the flexibility of a protein molecule, but different possible solutions for the given data set. This is a stereo image: If you look at it cross-eyed, you should see three images, the centre one appears 3-D. Stereo viewer lenses, which make viewing of such images easier, are available from many internet sources and included in [44]

The result is a set of possible structures that are all consistent with the data (see Fig. 34.5). Average positions for all atoms in the protein are then calculated and the result run through energy minimisation (*vide infra*). This results in an average, energy minimised model for the protein under investigation. Quality of the model can be expressed as RMS deviation of the individual structures from the minimised average, which should be in the column called "B-value" in the PDB-file (PDB-files were originally designed for models from X-ray crystallography, where the B-factor has a different, but related meaning, see below. For a detailed description of the PDB-format see www.wwpdb.org/docs.html). Many computer graphics packages can colour models by B-factor. In a good model, the RMS deviation is about 0.5 Å.

34.1.4.1 ^{129}Xe NMR

The noble gas Xenon is not normally part of biologically relevant compounds. However, it dissolves into hydrophobic pockets of proteins. Since its NMR signal is then modified by the protein surrounding, Xe-NMR can be used to measure conformational changes in proteins, for example after ligand binding. The signal/noise ratio can be increased by four orders of magnitude by optical pumping with a laser [122], thereby temporarily (for minutes to hours) creating much higher spin polarisation than would be possible with a magnetic field. Since Xe does not react chemically with the protein, this method is non-destructive and the protein can be recovered later. Optically pumped Xe NMR can also be used for magnetic resonance imaging in living organisms as Xe is non-toxic, although it has an anaesthetic effect in high concentrations.

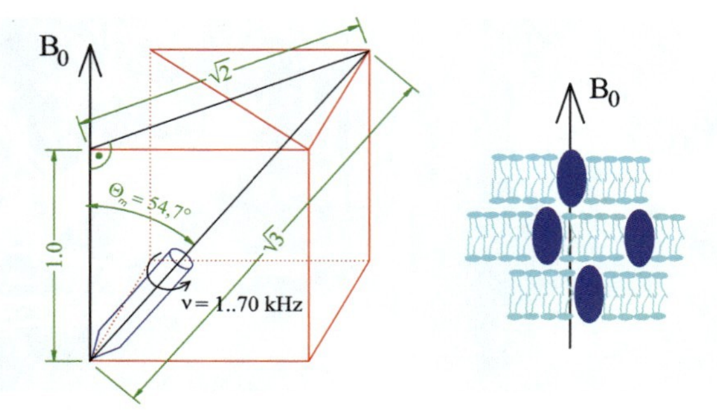

Fig. 34.6 Solid state NMR. *Left*: In magic angle spinning NMR the sample is spun rapidly at $\theta_m = \arctan(\sqrt{2}) = \arccos(1/\sqrt{3}) = 54.7°$ with respect to the magnetic field. θ_m is the angle between the space diagonal of cube and its connecting edge. *Right*: In oriented solid state NMR the sample (*e.g.*, biomembranes) are oriented perpendicular to the magnetic field. For details see text

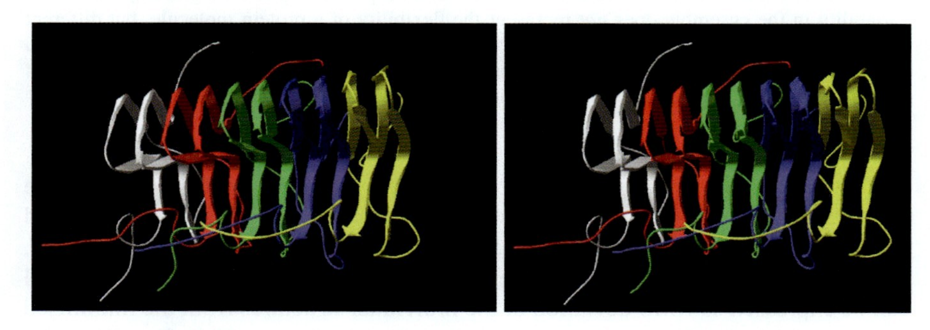

Fig. 34.7 Amyloid fibril formed by HET-s(218-289), fragment of a protein from the yeast *Podospora anserina*, PDB-code 2rnm. Each protein is represented by a different colour. Magic angle spinning NMR – unlike classical NMR and X-ray crystallography – is able to determine the structure of such insoluble, non-crystalline material

34.1.5 Solid State NMR

NMR is usually used for freely mobile molecules, Fig. 34.4 shows how spectral resolution declines as rotational correlation times become longer than the heteronuclear dipole-dipole and spin interaction times (milliseconds). Such samples are called "solid" by NMR-spectroscopists, irrespective of their physical state. If they were investigated by conventional NMR, line-broadening by anisotropic spin-interactions would degrade the signal to the point of uselessness. For small molecules in solution, BROWNIAN motion averages these interactions out so they can be neglected. Increasing the temperature reduces sample viscosity and increases molecular velocity, but for proteins that is often not an option.

There are proteins that are difficult to crystallise – and hence not amendable to X-ray crystallography – but that at the same time can not be kept in solution for conventional NMR: fibrous, amyloid and transmembrane proteins, and those fixed in large supramolecular structures. Two methods have been developed to obtain useful NMR-spectra for such samples:

Magic angle spinning (MAS-) NMR If the sample is tilted at the magic angle ($\theta_m = 54.7°$) and rotated rapidly ($1 - 70\,\text{kHz}$), line broadening that can be reduced as anisotropic interactions are averaged during rotation (see Fig. 34.6 and 34.7). Further reduction of line width can be achieved by replacing non-exchangeable hydrogens with deuterium. Solving structures is then accomplished by multi-dimensional NMR with ^{15}N- and ^{13}C-labelled proteins. ^{31}P-NMR can be used to study protein-lipid interaction.

Limiting the use of this method is the sample requirement (1 μmol) of protein in the maximal rotor volume, which is smaller for rotors operating at higher speeds (down to a few μl). Thus there is a target conflict between narrower lines (which require high rotor speeds) and high sensitivity (which requires larger rotor volumes). For measurements with proteoliposomes, the protein/lipid ratio during reconstitution needs to be 1 in a few hundred.

Oriented ^{2}H and ^{15}N solid state NMR This method is used to investigate membrane proteins [5]. Membranes are oriented perpendicular to the magnetic field, for example between two glass plates. Under these circumstances ^{15}N NMR is used to determine the tilt angle of helices in the membrane, NMR of 3,3,3-^{2}H-Ala labelled proteins are used to complement this information.

[301] provides an extensive bibliography on these methods.

34.2 Computerised Structure Refinement

If the structural information obtained by experimental methods is insufficient, computer modelling can sometimes improve it. There are three main methods: Energy minimisation, molecular dynamics and Monte Carlo simulations. For the interested reader [161, 165, 228] give more details on this subject.

34.2.1 Energy Minimisation

Protein structures are refined by energy minimisation. This is done by adding the energies resulting from stretching or compression of bond lengths($\Delta E_l = 1/2\ K_b{*}(b\text{-}b_0)$, $K = 800\text{–}4\,000\,\text{kJ mol}^{-1}\,\text{Å}^{-2}$), bond angles ($\Delta E_\angle = 1/2\ K_\theta{*}(\theta - \theta_0)$, $K_\theta = 200\text{–}500\,\text{kJ mol}^{-1}\text{rad}^{-1}$) and dihedral angles ($\Delta E_d = 1/2\ K_\phi{*}(1{+}\cos(N{*}\phi\text{-}\delta))$, $K_\phi = 5\text{–}25\,\text{kJ mol}^{-1}\,\text{rad}^{-1}$) from their normal values.

To this, the energies from non-covalent interactions have to be added: electrostatic ($\Delta E_{\text{es}} = q_1 q_2 / D r^2$, with D depending on the dielectric constant of the medium between the charges), VAN DER WAALS ($\Delta E_{\text{vw}} = A/r^{12} - B/r^6$, with A and B representing the short-range attractive and shortest-range repulsive forces, respectively) and hydrogen bonds ($\Delta E_{\text{hb}} = A/r^{12} - B/r^{10} * \text{f}(\theta', \theta'')$, with A and B being constants and f being a function of the bond angle).

The potential energy of each atom is the sum of all the above energies, plus a penalty if its distance from other atoms deviates from the NMR-determined distances. The energy of a molecule is the sum of the energies of all its atoms and is minimised by slowly changing the conformation. The NMR-penalty is initially large, with each round of optimisation it decreases and becomes 0 as the distances settle toward a final structure.

The position of each atom is characterised by three parameters (x, y and z coordinates), hence the structure of a molecule of n atoms has $3n$ parameters. Energy minimisation in a $3n$-dimensional space is laborious and complicated by the fact that there are many local minima. This is the reason why methods like NMR are needed to narrow the search space to something that our computers can handle, *ab initio* calculations of protein structures are not possible with currently available technology. Energy minimisation can be either by search- (Simplex) or gradient-methods (steepest descent) or using the NEWTON–RAPHSON-algorithm (see [337] for explanation of these techniques).

34.2.2 Molecular Dynamics

All atoms, even inside molecules, are in constant BROWNIAN motion, the mobility increases with temperature. Thus the atoms of a molecule sample occupy a considerable structural space. In molecular dynamics one starts with a computer model of a molecule and allows the atoms to move around as they would do in a molecule at high temperature. This requires the calculation of the first derivative of the energy function for each atom which determines the force acting on the atom under NEWTON's law of motion. If the temperature (that is the freedom of movement of all atoms) is slowly lowered, structure should automatically approach an energy minimum. This is called **simulated annealing**. Calculation time for such simulations are very high, a few picoseconds of simulated time may require several hours of computer time.

34.2.3 Monte Carlo Simulations

These are named after the famous casino, where money is redistributed from customer to bank according to the laws of chance. Monte Carlo simulations also use chance: From one simulation step to the next the system changes its state. There are several options for the new state, each associated with a certain probability P. Say, we have 3 options A, B and C, with associated probabilities 50, 30 and 20 %. To carry out a simulation step, the computer calculates a (pseudo)-random number i, these are in the range $0 \leq i \leq 1$. This random number is then used to determine which of the paths the system goes next. In our example

$0 \le i \le 0.5$ the computer chooses path A
$0.5 < i \le 0.8$ (0.5+0.3) the computer chooses path B
$0.8 < i$ the computer chooses path C

By repeating this basic operation many times, one can predict the behaviour of complex systems.

For the simulation of protein folding, each atom can move independently in all directions of space. The change in energy resulting from such moves is calculated as discussed for energy minimisation. Some of these movements will reduce the energy of the protein, these moves are favourable and have a high probability. Other moves are unfavourable and increase the energy of the protein. These moves have a low probability, but it is essential that the probability stays larger than zero even for very unfavourable moves. If the simulated protein already has achieved a near-optimal structure, such an unfavourable move probably will be reversed during the next couple of steps. However, if the protein was trapped in a local minimum, such a seemingly unfavourable move might get it out, and the next couple of moves will then most likely go downhill towards the global minimum.

34.2.4 Future Directions

Instrument manufacturers are constantly working to build devices with stronger magnetic fields, so that the ratio of atoms in parallel to those in antiparallel orientation, and hence the signal strength, is higher. Because of the higher field strength, more energy is required to flip the atoms between orientations, hence such machines also operate at higher frequencies. Operating frequency thus is a way to characterise the sensitivity of an instrument. While simple first generation NMR spectrometers operated with radio pulses of 100 MHz or less, the most modern work at 1 200 MHz. Such equipment is very expensive, and the operating costs are also high due to the need for liquid helium and skilled manpower.

In addition, progress in bio-computing will certainly drive this field. "**Threading**" of proteins with unknown structure onto related proteins has become routine, and there is now a ≈98 % chance to find a related protein with solved structure in the data base. **Distributed computing** (*e.g.*, folding@home at folding.stanford.edu/) has increased the computing power available to scientists by allowing the users of internet-linked computers to donate computing time otherwise not required.

34.3 X-ray Crystallography of Proteins

This method is still the "gold-standard" for protein structure determination. Astounding progress has been made since BERNAL AND CROWFOOT published the first high resolution diffraction patterns of an enzyme (pepsin) in 1934 [26]. This

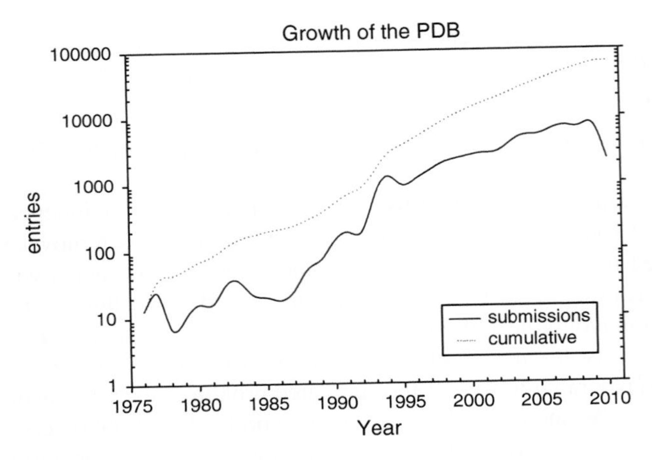

Fig. 34.8 From humble beginnings: Growth of the PDB-database over time. Data from www.rcsb.org/pdb/static.do?p=general_information/pdb_statistics/index.html

is especially true for soluble proteins whose crystallisation is now almost routine. When the Brookhaven Protein Database (PDB, www.rcsb.org/pdb/home/home.do) started its operation in 1971, it contained the structures of seven proteins. In April 2010 there were 64 623 structures available (see Fig. 34.8), of those 3873 were representative (non-redundant). The crystallisation of membrane proteins is still very difficult, structures of only 234 of them are available[2], of those 36 were solved by NMR, the reminder by crystallography. However, that number is also growing exponentially.

In the ideal situation, a protein is crystallised with and without bound substrate(s) or inhibitor(s), so that snapshots of the protein during its enzymatic cycle can be taken. This allows the reaction mechanism of the enzyme to be elucidated, and the role of individual amino acids to be determined. bio5.chemie.uni-freiburg.de/ak_movie.html. has an instructive example.

34.3.1 Crystallisation of Proteins

Proteins are usually crystallised by slowly lowering their solubility by the addition of precipitating agents (see [50, 269] for a more detailed discussion). Aim in such experiments is to bring about 40 protein molecules together into a stable nucleus, which can then grow to a macroscopic crystal, containing about 10^{15} molecules. A solution can be unsaturated, meta-stable supersaturated (only crystal growth is possible) and labile supersaturated (both crystal growth and nucleation are possible). This is determined by the forces between protein molecules: If they are repulsive, no crystal formation is possible, if they are strongly attractive, precipi-

[2] A current list of membrane proteins with solved structure is kept on blanco.biomol.uci.edu/Membrane_Proteins_xtal.html

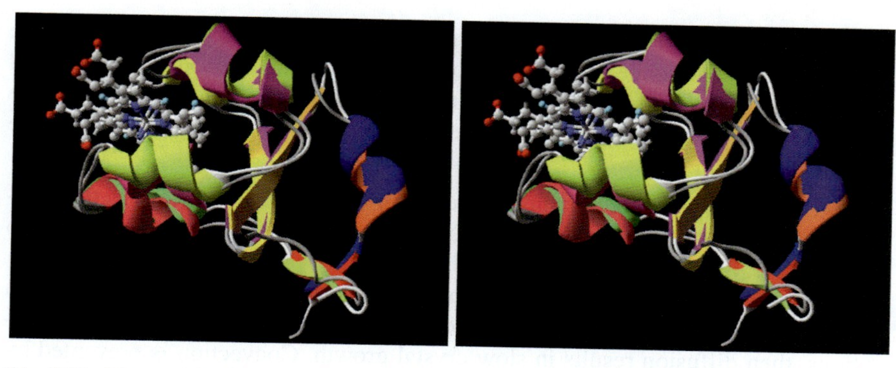

Fig. 34.9 The models of cytochrome B_5 determined by X-ray crystallography (PDB-code 1ehb, *red→yellow→green*) and NMR in solution (PDB-code 1aqa, *blue→purple→red*) are largely identical. This appears to be the general finding when crystal and solution structures of the same protein are compared. The reason probably is that protein crystals contain a lot of bound water, so the protein is actually in a fully hydrated state, just as in solution. If protein crystals are dried, they lose the ability to diffract X-rays

tation will occur. The second virial coefficient (see footnote 1 on p. 251) determines the inter-molecular forces, slight attraction between molecules correspond to a small, negative B_2 [239]. The crystallisation slot is -2 to -8×10^{-4} mol ml g^{-2} [113]. Apart from measuring osmotic pressure and light scattering, B_2 can also be determined by sedimentation velocity centrifugation [384], self-association chromatography [410], fluorescence correlation spectroscopy and small angle X-ray scattering.

During crystallisation, proteins loose freedom of movement (entropy), however, at the same time some water molecules bound to the protein's surface are released, increasing the entropy of the system. The movement of large, hydrophilic surface-exposed side chains can increase the conformational entropy of the system to a point were crystallisation is inhibited. Replacing them by small, non-polar residues – usually Ala – can ease crystallisation (**surface entropy reduction (SER)**). Critical residues can be identified by a computer algorithm [119]. Obviously, the effect of such mutations on protein structure and stability needs to be evaluated in each instance.

Electrostatic repulsion between equally charged protein molecules and hard-sphere repulsion (caused by the fact that no two objects can occupy the same space at the same time) are the strongest repulsive forces, electrostatic forces between oppositely charged molecules and VAN DER WAALS-bonds are attractive forces. Thus crystallisation can be eased by:

- Working near the *p*I of the protein to minimise electrostatic repulsion. However, protein stability is lowest at the *p*I.
- Reducing electrostatic repulsion with salts.
- Increase attraction with osmolytes.

For crystallisation to occur, the solution needs to be brought briefly into the labile supersaturated state; once nucleï have formed, concentration needs to be reduced into the meta-stable supersaturated region to prevent the formation of a precipitate and allow the slow growth of a well formed crystal. Several techniques are available to achieve this aim:

Batch crystallisation Precipitant is added until a slight haze develops in the solution (TYNDALL-effect, sign of nucleus formation). Then the solution is left to crystallise. Seed crystals can be added.

Free interface diffusion Protein solution and precipitating agent are layered on top of each other. Nucleus formation occurs at the interface between both solutions, then diffusion results in slow crystal growth. Convection is prevented by adding gelling agents like 0.15 % low-melting agarose. This type of experiment is often performed in capillaries.

Locally induced nucleation Conditions are selectively changed in a particular spot of the sample to induce nucleation, for example by lowering the temperature with a cold needle.

Dialysis The protein solution is dialysed against a solution of the precipitant, resulting in a slow increase of precipitant concentration.

Vapour diffusion A few micro-liters of protein solution (say, 10 mg/ml) are mixed with an equal volume of the precipitating solution. The resulting mixture is placed next to a much larger reservoir of precipitating solution in an enclosed space. This assembly is placed into a thermostated box. The water vapour pressure of the protein solution is, because of its lower concentration of dissolved matter, higher than that of the precipitant solution. This results in a net transport of water from the drop with protein solution to the reservoir (isothermal distillation). The increasing concentration of precipitant reduces the solubility of the protein and (hopefully) leads to its crystallisation. Two methods are in common use:

- **Hanging drop:** the drop with protein precipitant mixture hangs from a silanised cover glass placed over the reservoir with the precipitating solution.
- **Sitting drop:** the drop is placed into a separate well next to the well with precipitating solution.

For both methods, special plastic plates are available which allow a large number of crystallisation conditions to be checked in parallel.

Ultrafiltration of the mixture of protein and precipitant increases protein concentration until crystallisation occurs (*e.g.*, using centrifugal concentrators [329]).

It is interesting to note that although these techniques seem to produce crystals with similar efficiency, results with a particular protein vary depending on the method used.

Commonly used precipitants are ammonium sulphate (1–2 M), sodium chloride (2–4 M), sodium malonate (up to 14 M), K-citrate (1–1.6 M), K-phosphate (400–750 mM), K-acetate (0.5–1.5 M), $MgSO_4$ (0.5–1.6 M), Li_2SO_4 (1–2 M), ethylene glycol (20–30 %) and polyethylene glycol (of different molecular mass, 2–30 %), ethanol (5–30 %), i-propanol (5–30 %), acetone (5–30 %), or 2-methylpentane-2,4-diol (MPD) either alone or in combination of two precipitating compounds (using about half the normal concentration each). Salt ions can neutralise ionic groups on the protein and thus reduce electrostatic repulsion between protein molecules. They also create a short range (3 Å) attraction between protein molecules which increases with decreasing temperature. If the *p*H is higher (lower) than the *p*I the strength of interaction follows the (non-)inverted HOFMEISTER series (see p. 145) and is determined mostly by the anions. No physical theory for this force has been found yet, it is relevant mostly for small proteins. In addition salts, solvents, and small molecular mass PEGs seem to compete with the proteins for the available water, while large molecular mass PEGs seem to reduce the space available for proteins (**crowding**). Since PEGs and proteins are mutually impenetrable, the PEGs are excluded from the vicinity of protein molecules. This creates a zone of lower osmotic pressure around the protein molecules (**depletion zone**). If protein molecules get close enough together for their depletion zones to overlap, additional space is recovered for the PEG molecules to move in, resulting in an increase of entropy. This depletion effect depends on PEG size and concentration, but is largely independent of temperature. Interestingly, large molecular mass PEGs are more successful in the crystallisation of soluble proteins, while small molecular mass PEGs are preferable for membrane proteins [291]. The higher the molecular mass, the lower the concentration required for crystallisation. For a fuller treatment of the physical chemistry of crystal growth induction see [100].

Suitable buffers are Tris-HCl *p*H 8.5, HEPES-KOH *p*H 7.5, MES-KOH (or cacodylate) *p*H 6.5, K-citrate *p*H 5.6 and K-acetate *p*H 4.6, each at 100 mM. A database of successful crystallisation conditions is available under www.bmcd.nist.gov:8080/bmcd/bmcd.html. A recent study has shown that successful crystallisation of the majority of proteins is possible under a small number of carefully selected conditions [198].

34.3.1.1 Protein Interactions with Osmolytes

Stabilisers like polyols (*e.g.*, glycerol, 1,2,3-heptanetriole [111], sucrose, trehalose (α-glucopyranosyl-1,1-α-glucopyranoside)) may be added (see Fig. 34.10). They tend to reduce the available water concentration, limiting the flexibility of the protein molecules. In addition, polyols may replace water in forming hydrogen bonds to the protein, further stabilising the protein. This reduced conformational freedom may aid crystallisation, increase stability to thermal and cold denaturation and reduce enzymatic activity. For a recent review on trehalose–protein interaction see [175]. Another way to reduce conformational freedom is to crystallise the protein as complex with antibodies or, even better, their F_v-fragments [306].

According to [33, 34] protein solute interactions are determined by the following effects:

Solvophobicity free energy change for protein-water interaction is more favourable than the free energy change for solute interaction with protein. The opposite situ-

trimethylamine-N-oxide (TMAO)

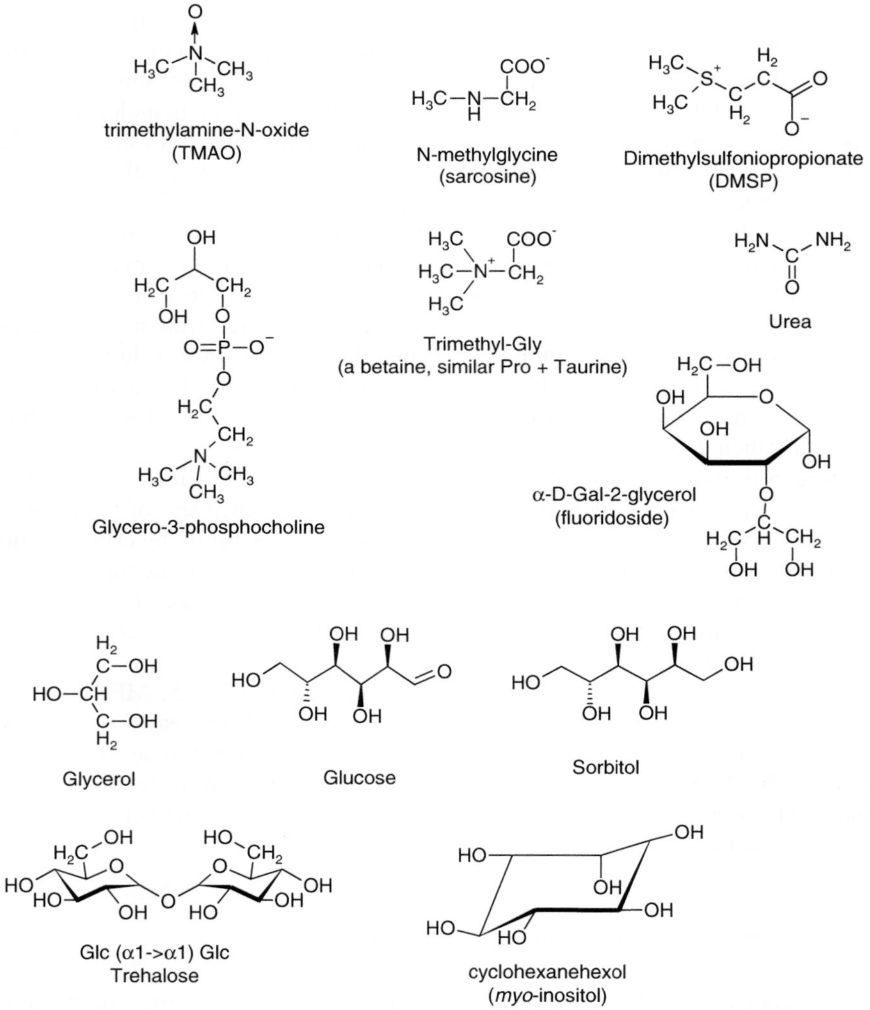

N-methylglycine (sarcosine)

Dimethylsulfoniopropionate (DMSP)

Glycero-3-phosphocholine

Trimethyl-Gly (a betaine, similar Pro + Taurine)

Urea

α-D-Gal-2-glycerol (fluoridoside)

Glycerol

Glucose

Sorbitol

Glc (α1->α1) Glc Trehalose

cyclohexanehexol (*myo*-inositol)

Fig. 34.10 Natural osmolytes that are used to stabilise cellular proteins against environmental stress

ation, where the proteins prefer solute over water, would be solvophilicity. Since the exposure to solvent increases upon denaturation, solvophobic substances shift the equilibrium to the native, solvophilic substances to the unfolded state.

Excluded volume The distance of closest approach of a small molecule (of radius r_s) to a protein (of radius r_p) depends on r_s. If the small molecule is rolled over the surface of the protein, we get a volume $4/3\pi(r_p + r_s)^3$, from which the small molecule is excluded. Water has a smaller radius than osmolytes, and hence a smaller excluded volume, and the effect is significant because the radius occurs in the third power. The effect will be stronger for large (*e.g.*, PEG) than for small solutes (*e.g.*, urea).

Surface tension A protein molecule is a "bubble" in the water and is surrounded by water molecules that have surface tension (just like an air bubble would be). For a solute that increases the surface tension of water the concentration at the interface should be lower than in the bulk solution, and for a substance that decreases surface tension the concentration at the interface should be higher. A denatured protein exposes more surface area, so denaturation is opposed by molecules that increase surface tension.

The actual effect of a solute on a protein is the sum of all these effects. For example, urea increases surface tension and has a larger excluded volume than water. Both effects should result in a stabilisation of protein structure. However, they are swamped by the solvophilicity of urea, which makes this reagent strongly caotropic.

The **transfer free energy** $\Delta G_{\mathrm{sol}}^{\mathrm{aa}}$ can be used to quantify this. Solubility of an amino acid is determined both in water s_{w} and 1 M solute s_{s}. At maximum soluble concentration the free energy of the amino acid in solution is equal to that in crystals. Since the free energy in the crystals is independent of solute concentration, the transfer free energy becomes $\Delta G_{\mathrm{s}}^{\mathrm{aa}} = -RT \ln(s_s/s_{\mathrm{w}})$. The difference in $\Delta G_{\mathrm{s}}^{\mathrm{aa}} - \Delta G_{\mathrm{sol}}^{\mathrm{Gly}}$ is the transfer energy of the amino acid R-group. With poly-Gly one can determine the transfer energy of the peptide backbone. The difference of the unfolding free energies in water and solvent $\Delta G_{\mathrm{s}}^{\mathrm{u}} - \Delta G_{\mathrm{w}}^{\mathrm{u}}$ must be equal to the difference of transfer energies of unfolded and folded state from water to osmolyte.

34.3.1.2 Evaluation of Crystallisation Experiments

Some of the tested conditions will lead to rapid precipitation of the protein, under other conditions the protein remains soluble. The trick is to find the combination of buffer, precipitating agent and substrates or inhibitors, which leads to the slow formation of big (that means about 0.1–1 mm!), regular, clear crystals with smooth faces. This is best checked under a polarisation microscope. Once such conditions have been identified in a search experiment, they need to be optimised, until crystals have been obtained which refract well at high resolution.

Crystallisation takes time, so all solutions should be sterile-filtered and handled under aseptic conditions. 0.02 % sodium azide may be added to prevent bacterial growth (provided it does not interact with the protein). All handling should be done under a laminar flow hood, to avoid not only microbial contamination but also dust. Protein crystals formed around a dust nucleus may give strange diffraction patterns!

All reagents should be of highest purity; especially detergents and PEG can contain peroxides and other contaminants that inactivate proteins. Recipes for re-purification are available in the literature, but are laborious and often involve the use of noxic solvents. Buying certified chemicals is probably the easier and, in the end, cheaper option.

Plates can be scored after 1 and 3 days, one week and one month (possible outcome: clear, precipitate, crystals, phase separation). If crystals are found (they are easiest to spot between crossed polarisers), it needs to be verified that they are made of protein rather than precipitant. There are several ways to do this. One indicator is crystal shape: Cubic crystals in a sodium chloride solution can most likely be discarded. Protein crystals have an open structure, they consist mostly of water (about one molecule of ordered water per amino acid bound to the protein, plus disordered water between the protein molecules). If a dye like methylene blue is added to the mother liquor, it will diffuse into protein, but not the denser salt crystals. The third option is to press down onto the crystal with a fine needle. Salt crystals crack with a breaking sound. If the crystal gets squished instead, you have destroyed a perfectly good protein crystal!

Sometimes crystallisation is impeded by flexible coil regions in the protein. These can be identified by deuterium exchange mass spectrometry (DXMS), engineering proteins without those loops can lead to crystallisable samples [311]. However, the extent to which such samples still reflect the structure of the original material has to be critically evaluated case-by-case. Demonstration of remaining enzymatic activity is a good indicator of an essentially unchanged structure.

34.3.1.3 Membrane Proteins

About a quarter of a proteome are transmembrane proteins [213], but about 70 % of the ≈500 are known drug targets. About 1.3×10^6 protein sequences had been deposited in public databases, 200 genomes have been sequenced completely. Of the 64 623 protein structures (3873 non-redundant) available from PDB in April 2010, only 234 were transmembrane proteins. Of the latter many are β-barrel proteins, which are easier to crystallise than the α-helical proteins which presumably form the vast majority of naturally occurring membrane proteins. It is estimated that we have structural knowledge of less than 10 % of all membrane proteins. About 50 % of all drugs bind to G-protein coupled receptors (including histamine, 5-HT, angiotensin, acetylcholine, dopamine, prostaglandin, leucotriene, and ADP receptors), of which only few have been crystallised. This is unfortunate because the development of new drugs is seriously impeded by the lack of structural information on membrane proteins. Surveys of methods used for successful crystallisation of membrane proteins were published recently [208, 291, 385], together with suggestions for sparse matrices of such conditions.

The following types of crystals from membrane proteins are distinguished:

Type I stacks of 2-D sheets of proteins embedded in lipid, the transmembrane segments are responsible for crystallisation in the membrane plane, the hydrophilic segments form the contacts between the sheets.

Type II crystals are formed from detergent-solubilised proteins, only the hydrophilic segments are involved in crystal formation. The hydrophobic parts are covered by detergent micelles, this requires the micelles to be small enough not to interfere with protein–protein contacts. Monoclonal antibody F_v-fragments against the protein of interest potentially increase hydrophilic contacts and may aid in crystallisation. The same effect may be achieved by genetically engineering soluble protein segments (*e.g.*, lysozyme) between transmembrane loops of the protein of interest.

Membrane proteins can sometimes be crystallised after solubilisation with detergent. As far as has become clear from the successful cases, the detergent should form monodisperse micelles just large enough to cover the hydrophobic parts of the protein. If the micelles are too small, protein precipitation will occur, if they are too big, detergent will interfere with crystallisation. A simple test for the former involves ultracentrifugation of the sample at 350 000 g for 45 min and analysing the supernatant for the protein (*e.g.*, by SDS–PAGE). Loss of protein compared to samples taken before centrifugation indicates the formation of aggregates [134]. In modern desktop-ultracentrifuges only a few micro-liters of sample are required for this test, and several different samples can be examined in parallel. Head-groups of the detergent should be small, the tail should be more broad than long (some newer detergents have phenyl- rather than alkyl-tails). The correct choice of detergent, however, is more art than science, for example cytochrome c-oxidase can be crystallised from C_{11}-maltoside to form crystals which refract to 2.6 Å, crystallised from C_{12}-maltoside the crystals refract only to 8 Å, from C_{10}-maltoside no crystals were obtained [306]. In some cases phospholipids with short fatty acids (C6-C8) have been used instead of detergents [85].

The concentration of detergent micelles should be such that each micelle contains at most 1 protein molecule, but a large excess of "empty" micelles interferes with crystallisation. Small amphophilic molecules (polyols like 1,2,3-heptanetriole, benzamidine) can decrease micellar size and increase micellar curvature by partitioning into the micelle and displacing some of the detergent. They also stabilise protein structure and increase protein-protein contacts [111]. A second detergent, below its cmc, can be used to achieve the same effect. Protein–protein contacts can be increased by crystallisation in the presence of the F_v-fragment of an antibody directed against the membrane protein.

The concentration of precipitating salts should be just below that which results in phase separation of the detergent, that is at the consolute boundary above which salt-rich and detergent-rich phases form. The second virial coefficient of the solution should be small and negative, indicating weakly attractive solute interactions. B_2 is determined by the detergent and the temperature, the protein has little influence (see [208] for phase boundaries of commonly used detergent/PEG/salt combinations).

34.3.1.4 Related Techniques

For crystallisation of membrane proteins from lipids in liquid-crystal "meso"-phases see [48], the use of **bicells** (small lipid disks stabilised by amphophilic caps) is described in [93, 361].

34.3.2 *Sparse Matrix Approaches to Experimental Design: The* TAGUCHI*-method*

If the result of an experiment (say, the size of protein crystals or their resolution in X-ray diffraction) depends on k factors (concentrations of protein, precipitants and modifiers, pH, temperature...), and each factor is to be tested in n different values, then a test of all possible combinations would require $E = n^k$ different experiments to find the optimal combination. Thus even moderate increases in n or k result in a very large number of experiments, which may not be economically feasible.

TAGUCHI [398] has shown that the number of required experiments can be reduced to $E = 2k + 1$ (or the next higher number dividable by n) without significant loss of information. So in the case of three factors with three values each, only nine experiments are required instead of 27. The prerequisite is that the various factors are independent of each other, *i.e.*, that changes in factor 1 do not change the response of the system to factor 2. In practice, this requirement is often fulfilled.

In the above example the $n =$ three different values (designated a, b and c) for each factor only need to fulfill the inequality $a > b > c$. The values need not be equidistant, nor is it required that their numerical values be identical for all factors. The experimental design is then described by a **balanced (orthogonal) matrix**, in which:

- All n values occur with equal frequency for all k factors.
- All combinations of the n values occur with equal frequency for all pairs of factors.

The design of such matrices is described in [399], a computer program for the design of such matrices is described in [14] and available under igs-server.cnrs-mrs.fr/samba. NIST has a table of frequently used designs under www.itl.nist.gov/div898/software/dataplot/designs.htm#taguchi. For our example, the matrix is:

Exp. No	Factor 1	Factor 2	Factor 3	Outcome
1	a	a	a	
2	a	b	b	
3	a	c	c	
4	b	a	c	
5	b	b	a	
6	b	c	b	
7	c	a	b	
8	c	b	c	
9	c	c	a	

You can easily verify that both condition for an orthogonal matrix are fulfilled: Each value occurs exactly three times for all factors and each pair of values occurs exactly once for each pair of factors.

The experiments are performed and the outcomes entered into the table above. Then the outcomes are also tabulated against the value of each factor:

	a			b			c		
Factor 1	O_1	O_2	O_3	O_4	O_5	O_6	O_7	O_8	O_9
Factor 2	O_1	O_4	O_7	O_2	O_5	O_8	O_3	O_6	O_9
Factor 3	O_1	O_5	O_9	O_2	O_6	O_7	O_3	O_4	O_8

From these data the signal-to-noise ratio can be calculated for each factor by the following formula:

$$\mathrm{SNL} = -10 * \log\left(\frac{1}{n} * \sum_{i=1}^{n} O_i^2\right), \tag{34.5}$$

where the O_i are the outcomes for each of the n values of a factor. The signal-to-noise levels (SNL) are again tabulated:

	a	b	c
Factor 1	SNL_{1a}	SNL_{1b}	SNL_{1c}
Factor 2	SNL_{2a}	SNL_{2b}	SNL_{2c}
Factor 1	SNL_{3a}	SNL_{3b}	SNL_{3c}

For each factor the value with highest signal-to-noise is identified and used in the future. For even higher precision the SNL may be maximised by regression analysis ($f(x) = ax^2 + bx + c$ is a useful function for this purpose).

In such optimisation studies it is not necessary to wait for crystallisation, because crystallisation is preceded by the formation of small clusters of protein molecules which serve as nucleï. Such cluster-formation can be detected, for example, by laser light scattering, giving not only faster but also more quantitative results.

34.3.3 X-Ray Structure Determination

Textbooks on physical biochemistry [49, 117, 167] give a detailed mathematical treatment of this subject, [350] deals specifically with X-ray crystallography. www.msm.cam.ac.uk/doitpoms/tlplib/xray-diffraction/index.php has a nice animated lecture unit, www-structmed.cimr.cam.ac.uk/Course/Overview/Overview.html is also nice. For teaching purposes the use of laser light and a repetitive photographic image as sample may be used as originally described in [409], a modern version of this experiment is available online at ibmc6187.u-strasbg.fr:8080/webMathematica/bioCrystallographica/index.html. If only qualitative results are required, diffraction of a laser beam by a dried blood smear (see Fig. 34.11) results in the diffraction pattern of packed erythrocytes: concentric rings.

Once you have grown your crystal, it is placed into **harvest buffer**, that is, mother liquor with cryoprotectant, and taken up into a small loop (≈1 mm Ø) with that buffer. The loop is then dipped into propane cooled by liquid nitrogen (−196°C) to snap-freeze (vitrify) the sample without ice-crystal formation. Even better results are obtained when the liquid nitrogen is cooled to its melting point (−210°C) by evacuation (**warning**: do not allow prolonged exposure of this liquid to air, as

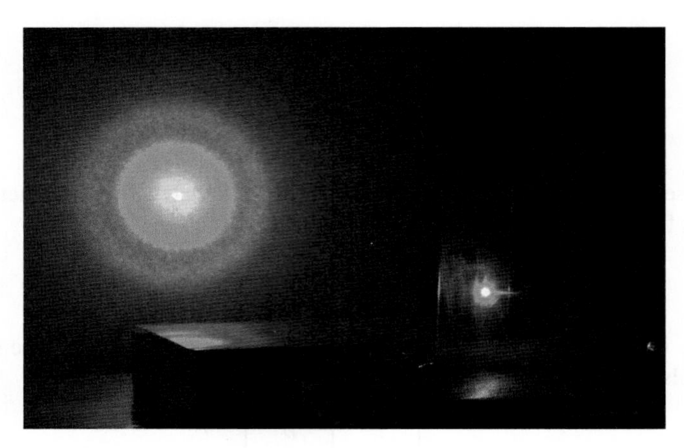

Fig. 34.11 Model experiment to show diffraction. A slide with a blood smear is illuminated with a laser pointer (from the right). The regular pattern of closely packed erythrocytes diffracts the light, resulting in a primary diffraction pattern: concentric rings. This works best when the erythrocytes are densely packed but in a single layer

otherwise oxygen would condense and form an explosion risk). These procedures avoid the formation of insulating gas bubbles around the sample, which would slow down cooling (LEIDENFROST effect). The water in the loop should appear clear and glass-like, rather than frosty. In this state the crystal can be shipped in liquid nitrogen to the X-ray source. X-rays can be produced in laboratory-based generators, but the high-intensity, monochromatic, and tunable beams generated by an electron synchrotron offer considerable advantages (see later). There the crystal is carefully centred onto a goniostat, which allows the crystal to be turned in the X-ray beam around all three axes of space with great precision (fraction of an arc second). It is constantly cooled with a stream of cold nitrogen, to maintain a temperature of −100°C during data collection. At this low temperature radiation damage is reduced so that more data can be generated per crystal. In addition, since lower temperatures reduce BROWNIAN motion, sharper diffraction peaks are obtained.

The X-ray photons have an energy of about 8 keV, their electric vector can interact with the electrons of the molecules in a crystal. This induces a **forced oscillation** (no resonance) of the electrons, until the energy is released at a wavelength identical to that of the exciting photon, but with a phase delay of 180° (**coherent scattering**). Scattering intensity depends on the angle of incident, the wavelength and the atomic number Z of the element. Hydrogen is only a very weak scatterer of electrons and hence invisible in X-ray crystallography, while heavy metals scatter electrons very effectively. Since C, N and O have similar Z, they are difficult to distinguish by X-ray crystallography. Therefore, the amino acid sequence of the protein under study needs to be known.

If lenses for X-rays were available, the scattered rays could be used to generate a magnified picture of the protein molecules in the same way as an intermediate picture is created in a microscope (see p. 5). Unfortunately, such lenses are not available and have to be replaced by mathematical procedures. The diffraction pattern (position and intensity) is recorded either on photographic film or on solid state detectors (a phosphor screen that produces light flashes where hit by X-rays, the flashes are collected by tapered optical fibres and directed onto a 10 Megapixel CCD-element, similar to those used in digital cameras). Their output is fed directly into a computer.

A crystal is made up of **unit cells** (see Fig. 34.12), the simplest repeating unit in the crystal. Each unit cell may contain several protein molecules. The crystal is generated from repeating unit cells by translation along the axes of space.

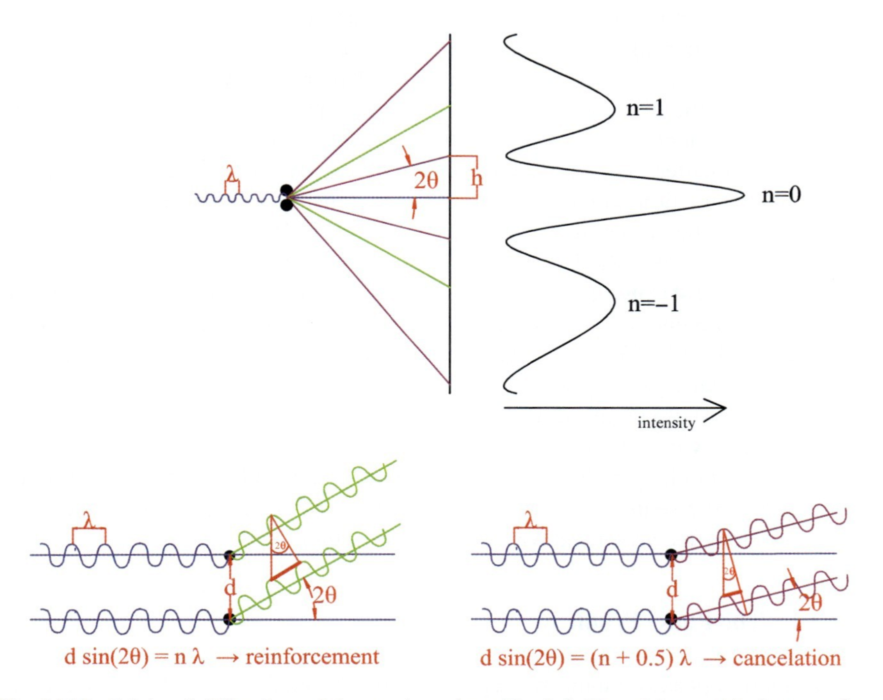

Fig. 34.12 Origin of diffraction minima and maxima: *Top left*: Two objects with distance d are irradiated with a beam of wavelength λ. The scattered waves are collected on a screen placed at the distance b, where, depending on the scattering angle 2Θ they hit a point at a height h. d and λ have nuclear dimensions (Å), while b and h have macroscopic dimensions (cm). For this reason the rays scattered by either of the two objects onto any point of the screen will appear parallel, just as the light beams coming from the sun appear parallel to us. *Top right*: Reinforcing and destructive interference cause regularly spaced maxima and minima in beam intensity. *Bottom left*: If $d \sin 2\Theta = n\lambda$ (BRAGG's law) the beams coming from both objects are in phase, they reinforce each other and create a diffraction maximum (*green* in top figure). *Bottom right*: If, however $d \sin 2\Theta = (n + 0.5)\lambda$ the beams are out of phase, they will cancel and result in a diffraction minimum (*purple* in top figure). *Note*: (**a**) The angles where maxima occur ($2\Theta = \frac{n\lambda}{d}$ is reciprocally related to the distance of the scattering points d, smaller d results in wider spacing of the scattering maxima. (**b**) The diffraction pattern of a protein is the convolution of the contributions by all the atoms within it. An X-ray lens could create an image of the protein from the diffraction pattern as discussed for light microscopy (see p. 5), unfortunately such lenses are not available

Each scattering point in a unit cell contributes to all diffraction peaks, and each peak contains contributions from all scattering points. If one were to walk through the crystal along any straight path, the electron density would be a periodic function of the distance covered with one repeat per unit cell. We recall from physics that all periodic waves can be generated by summing of sine or cosine waves (ground frequency and harmonics). This is called FOURIER's theorem.

Interaction of the scattered rays causes amplification in some spots and cancellation in others, depending on the phases of the rays. This is called **diffraction**. If there are only 2 scattering points, waves would cancel only if their phase difference were exactly 180°, resulting in fairly broad peaks. In a crystal with many scatterers the summing up leads to cancellation even when the phase differences are somewhat different from 180°. Hence the larger and more ordered the crystal, the sharper the diffraction peaks will be and the higher the signal/noise ratio during measurement. The sum of all the scattered X-rays at any point (h,k,l) of the diffraction pattern is called its **structure factor**. We therefore have to find the mathematical function which links the electron density at the coordinates (x,y,z) of a unit cell to the structure factors in position (h,k,l) of the diffraction pattern. A diffraction pattern is the FOURIER-**transform** of the crystal structure, hence an **inverse** FOURIER-**transform** can be used to generate the structure from the diffraction pattern. The details are beyond the scope of this book, but may be found for example in [49, 117, 167, 350]. If you get to this stage, you will have to collaborate with professional crystallographers who not only have access to the X-ray sources and know how to use them safely, but who can also help you with the maths.

Once a density map has been obtained, a molecular model is built that fits into the map. Such a model, apart from crystallographic data, also has to take into account knowledge of the sequence of the protein and the rules of stereochemistry (bond angles and -lengths, VAN DER WAALS-radii of atoms).

34.3.3.1 The Phase Problem

The diffraction peaks contain two types of information, the intensity (related to the number of scattering points that have a given distance, say, 1 Å from each other) and the phase (related to the exact locations of these scattering points). Phase information is the more desirable of the two. Unfortunately, the phase is not recorded in scattering experiments. For molecules smaller than 500 atoms the phase information can be added back by trial and error ("direct method") in a computer. However, computation time becomes excessive for larger molecules. Thus the missing phase information has to be obtained by other experiments, before computer models of protein structure can be built. A special case are centro-symmetric crystals, as phase angles are limited to certain values. These crystals are derived from centro-symmetric molecules, or from racemates that contain equal amounts of a molecule and its mirror image. Plectasin (40 amino acids) has recently been crystallised together with its chemically synthesised *d*-variety [255] to solve the crystal structure directly. Obviously, this approach is possible only for very small proteins.

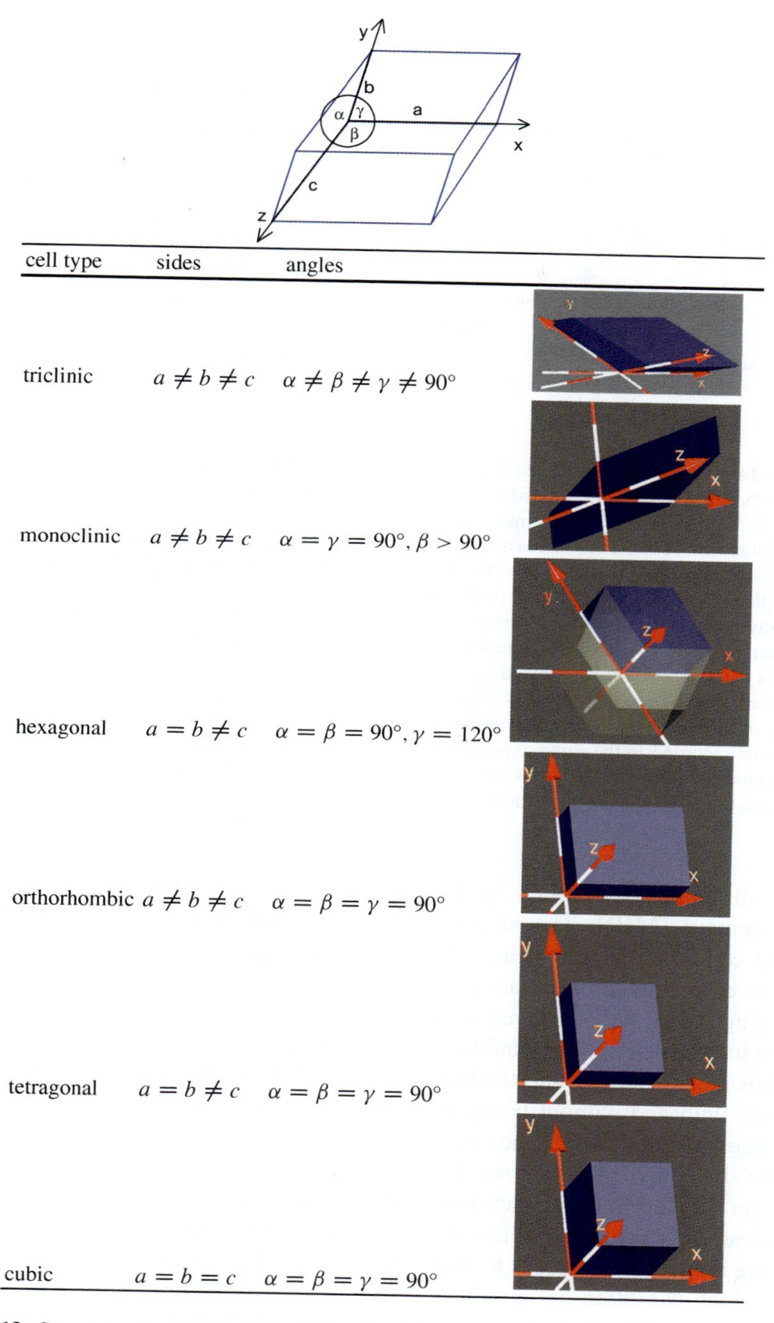

cell type	sides	angles
triclinic	$a \neq b \neq c$	$\alpha \neq \beta \neq \gamma \neq 90°$
monoclinic	$a \neq b \neq c$	$\alpha = \gamma = 90°, \beta > 90°$
hexagonal	$a = b \neq c$	$\alpha = \beta = 90°, \gamma = 120°$
orthorhombic	$a \neq b \neq c$	$\alpha = \beta = \gamma = 90°$
tetragonal	$a = b \neq c$	$\alpha = \beta = \gamma = 90°$
cubic	$a = b = c$	$\alpha = \beta = \gamma = 90°$

Fig. 34.13 Space groups. A unit cell is the smallest group in a crystal that repeats by translation only. A unit cell can be described by three axis x, y and z which form angles α (between y and z), β (between x and z) and γ (between x and y). The lengths of the cell along these axes are called a, b and c

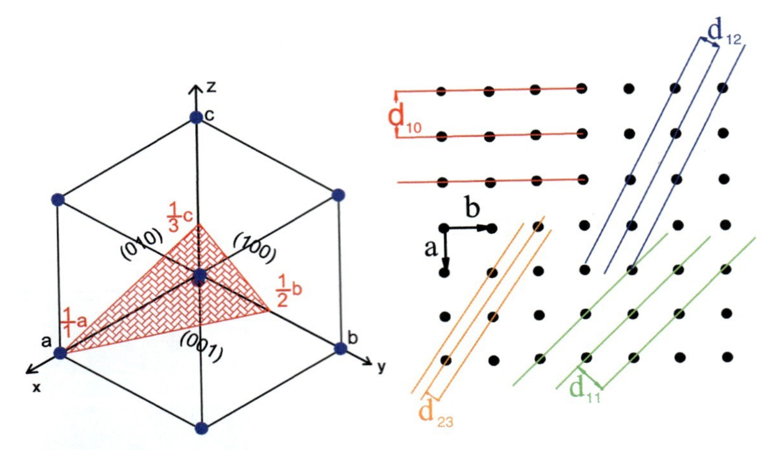

Fig. 34.14 *Left*: MILLER-indices. The cubic unit cell is cut by the red plane at the full length of a, half the length of b, and a third of the length of c (seen from the origin). An equivalent way of expressing this is that a is cut into one, b into two and c into three pieces. The MILLER-indices of this plane are **(123)**. The indices are natural figures and have 1 as greatest common denominator. If the plane is parallel to one of the axes then this axis is not cut at all and the MILLER-index is 0. If the plane intersects the unit cell in the origin, another corner is selected as origin, resulting in a negative MILLER-index. This is marked by a bar, *e.g.* **(1$\bar{2}$3)**. Multiplying all three values with a constant multiplier gives a plane parallel to the original one. Note the MILLER-indices of the crystal faces in the diagram. *Right*: Depending on the direction of the X-rays relative to the crystal lattice the distance of the refracting planes changes (here for a 2-D crystal, to make drawing easier). This results in different refraction patterns for different MILLER-indices. By rotating the crystal these can all be recorded in the same experiment. Atoms contribute most strongly to the reflection that correspond to lattice planes intersecting with that atom, and much weaker to all others. From this data set a model of the crystal structure is generated by a computer

Traditionally, the phase problem was solved by **multiple isomorphous replacement (MIR)**: The protein crystals are soaked in a solution of a compound containing a very heavy element like mercury, uranium, platinum metals or rare earths, which bind at specific residues of the protein (His, Cys) without changing the crystal structure (isomorphous from Gr. equal shape). If the structure were changed, the crystal would shatter, thus isomorphy is easy to verify. Then the crystals are taken back into the beam line. The difference in diffraction patterns of crystals with and without heavy elements is equivalent to a crystal that contains only the heavy element. Since the number of substitutions is small, this structure can be solved by trial and error. The heavier the atoms used for this technique, the stronger the additional refraction patterns; because the refraction signals from all electrons are coherent and have to be added, the intensity is proportional to Z^2. A more modern technique is to metabolically label the proteins with selenomethionine which can often be introduced instead of Met without changing the protein conformation. If only a single such experiment is performed, the method is called **single isomorphous replacement**. Several different derivatives, with heavy atoms in different positions in the unit cell, are used in MIR, leading to better solutions of the phase problem.

Alternatively, **multiple anomalous dispersion (MAD)** can be used. This method relies on the ability of synchrotron X-ray sources to produce beams of different, but exactly defined, wavelengths. Atoms absorb specific wavelengths and re-emit it again with a phase difference that is not 180°, as in coherent scattering (see X-ray spectroscopy on p. 307). As a result, FRIEDEL**'s law**, that the intensity of the reflection (h, k, l) should be equal to the intensity of the reflection at $(\bar{h}, \bar{k}, \bar{l})$, and their phases should be opposite, no longer holds. The X-ray wavelengths used in crystallography are near the K_β-absorption edges of heavy metals. Thus any heavy element (belonging either to the native structure of the protein like Fe in haemoglobin, selenomethionine introduced by genetic engineering, or a metal added by isomorphous replacement) can be used to induce a wavelength-dependent phase and intensity difference in FRIEDEL-pairs, from which phases can be reconstructed. To ensure that the crystal is not degraded during measurements at different wavelengths, it is essential to work at cryogenic temperatures.

Molecular replacement uses the fact that most proteins belong in a few large families. For each new protein whose structure is solved nowadays there is a better than 98 % chance that the structure of another member of the same family is known. In this case one can use the already determined phases of the known protein together with the freshly determined intensities of the unknown to get an initial estimate of the latter's structure. This method may also be used to determine crystal structures of a protein after minor changes, *e.g.*, addition of a ligand or introduction of a mutation. The greatest computational challenge in this method is to get the relative orientation of known and unknown protein to superimpose.

34.3.3.2 Measuring and Improving the Quality of the Result

The information available from the diffraction pattern varies with the size and quality of the crystals. Larger, more ordered crystals produce diffraction peaks at higher **resolution**:

10–20 Å The outer shape of the molecule becomes visible.
3.5 Å Secondary structure elements can be resolved partially.
3.0 Å Secondary structure elements can be resolved completely.
2.5 Å The electron density map is so detailed that individual amino acids can be located if the sequence of the protein is known.
2.0 Å Amino acids can be placed into the electron density map without reference to the protein sequence. Position of individual atoms can be determined by using the amino acids as building blocks.
<1.5 Å Individual atoms can be located in the electron density map.

In this context "resolution of 2 Å" means that diffraction peaks up to 1/2 Å have been included in the calculation of the model. The root mean square (RMS) error for the position of an atom is usually a few tenths of an Å (compare to a C–C bond length of 1.54 Å).

The quality of a molecular model can be expressed in terms of R-**factor** by comparing the observed amplitudes of all diffraction spots with those calculated from the model:

$$R = \frac{\sum ||F_{\text{obs}}| - |F_{\text{calc}}||}{\sum |F_{\text{obs}}|}. \tag{34.6}$$

Since all methods of phasing diffraction data give only first estimates, the first model calculated will probably be of poor quality. However, additional phasing data can be extracted from it, which lead to a second model with, hopefully, a better R-factor. From this additional phasing data are extracted and the process is iterated until no further reduction in R is seen, *i.e.*, all information present in the diffraction pattern has been extracted into the model. This process is called **bootstrapping**. A random distribution would yield $R = 0.65$; 0.4 would be a decent starting point for refinement of protein structure, which should yield a final $R \approx 0.1$. For small molecules, $R \approx 0.01$ is achievable. If the resolution is still insufficient, molecular dynamics and energy minimisation can be used to get higher resolution models, just as was described in the section on NMR (see p. 319).

Another important way to determine the quality of models is the **temperature factor** (also known as **displacement parameter**) B_j of the jth atom,

$$B_j = 8\pi^2\{u^2\} \tag{34.7}$$

with $\{u^2\}$ the mean square displacement of the atom. $B_j < 20\,\text{Å}^2$ indicates a well-resolved structure, $B_j > 30\,\text{Å}^2$ high disorder. The charm of B_j is of course that it provides information for individual residues, rather than the protein as a whole. A useful tool for assessing model quality is the RAMACHANDRAN-plot of the torsion angles ϕ, ψ. If a non-Gly residue lies in the forbidden area, then there better be a convincing explanation!

All these sanity checks have not prevented structures with errors from being submitted to PDB, where they sometimes survive without correction for years. For methods to detect such cases see [414]. The following web sites are relevant:

EDS-server eds.bmc.uu.se/eds/ at Uppsala University contains the electron density maps you need for checking.

MolProbity molprobity.biochem.duke.edu/ at Duke university allows the search for collisions between atoms in a structure and other geometric problems.

Another point of importance is that the model shows the **asymmetric unit** of a crystal, not the **functional assembly** of the protein. The asymmetric unit is the smallest part of a unit cell without symmetry. Unit cells are assembled from asymmetric units by the fundamental operations of rotation, mirroring on a plane, inversion on a point and translation (either alone or in combination). Remember: Several asymmetric units may be required for a functional protein, or an asymmetric unit may contain several functional proteins.

Once model building is finished, the final model should be made available to the scientific public, this is done by submitting it to the **protein database (PDB)** at http://www.rcsb.org/pdb/home/home.do. From there, anybody with internet access can retrieve it for further studies. Note however, that the maintainers of the database do not vouch for the quality of the entries, that remains the task of both submitter and users.

34.3.3.3 Radiation Damage and X-Ray Flashes: A Glimpse of the Future?

In common experimental setups only about 10% of all X-rays entering a crystal are diffracted, some others are absorbed. Most of the latter kick out electrons from lower orbitals of atoms (photoelectric effect). As a consequence, in high Z atoms, electrons drop from higher orbitals into the vacant places, causing X-ray fluorescence. In the low Z atoms of biological samples the energy difference is large enough to cause the emission of additional electrons (known as AUGER-**effect**, even though it was first discovered by LISE MEITNER). In addition, the sudden change in the atoms electronic state can result in quantum-mechanical instability of bound electrons, which dissociate in a "nuclear shake up" event. All this increased ionisation leads to **sample heating** and **radiation damage**. Thus usually the radiation dose has to be limited to 200 photons/Å^2 (at 1 Å wavelength) to prevent artifacts. Only by averaging the signal from many molecules is it possible to get useful structures; this is one of the reasons why crystals are required.

However, the events leading to radiation damage take time, in the order of 10^{-14} s. If very brief, very bright X-ray flashes were used, "quick pictures of slowly exploding samples" [137] could be taken. Such flashes could possibly be produced by free electron X-ray lasers, currently under construction at DESY, Hamburg (expected to come online in 2014, www.xfel.eu). According to computer simulations, radiation dose could be increased by 2–5 orders of magnitude, this could be enough to take X-ray pictures of single molecules. It is envisaged to bring these molecules into the beam by electrospray techniques similar to those used in mass spectrometry.

RAINES et al. have recently shown that it is possible to reconstruct 3-D structures from a single exposure made with soft X-ray lasers [342].

34.3.4 *Other Diffraction Techniques*

Not all biologically interesting materials can be crystallised, so that their 3-D structure can be determined by X-ray crystallography. However, it is possible to determine structures in 2-D by fibre X-ray diffraction and 1-D by powder X-ray diffraction.

34.3.4.1 Fibre Diffraction

Some biological materials come in the form of fibres, where molecules are naturally aligned along an axis (for example DNA , silk, collagen), in essence forming a 1-D crystal. Some fibres occur naturally (*e.g.*, silk), in other cases (*e.g.*, DNA) the fibre is pulled form a sample with tweezers. If these samples are irradiated with X-rays perpendicular to their long axis, the resulting diffraction pattern is averaged over all possible rotations of the molecules around the long (y-) axis, but the remaining two dimensions can be studied.

JAMES WATSON AND FRANCIS CRICK deduced the structure of B-DNA from fibre diffraction data obtained by ROSALIND FRANKLIN AND RAYMOND GOSLING. The clearly visible "X" on "Photo 51" (www.pbs.org/wgbh/nova/photo51/anat-flash.html told the researchers that DNA had to be helical. From the diffraction image CRICK could work out the pitch of the helix, its radius and the number of residues per turn. ERWIN CHARGAFF had already shown that all DNA contains as much A as T and as much G as C. From these information WATSON could form a credible structure of DNA, using precise molecular models of nucleotides. This shows how fibre diffraction, together with other techniques, can be used to derive molecular models. But now back to proteins!

34.3.4.2 Powder Diffraction

In an amorphous powder, the diffraction pattern is averaged in all three axis of space. However, if a protein contains strongly scattering elements like heavy metals, their distance can be determined from powder diffraction. At very small angles, X-ray scattering is a function of the radius of gyration of a molecule, which may be determined this way. Methods for size and shape determination have been discussed already in Sect. 29 starting on p. 261.

34.3.4.3 LAUE Diffraction

The crystal is exposed to a polychromatic X-ray beam, containing wavelengths between, say, 0.5–2 Å. Each wavelength creates its own diffraction pattern according to BRAGG's law. A single exposure therefore results in many more peaks than with a monochromatic X-ray beam. Thus a structure may be calculated, under ideal circumstances, from a single, brief exposure of the crystal to the beam. This allows **time resolved X-ray crystallography**, where structural changes during the reaction cycle of a protein are followed, with shots taken every 200 ps or so (in a synchrotron source). For this to work, essentially all molecules of a crystal have to react simultaneously. Caged substrates (see p. 374) triggered by a fast, powerful laser are one option.

34.3.4.4 Neutron Diffraction

Unlike X-rays, neutrons are scattered by interaction with nucleï, not electrons. **Hydrogen** does scatter neutrons very well, but is nearly invisible in X-ray diffraction studies. This makes neutron diffraction a valuable complementary technique, for example to distinguish amide nitrogen and oxygen in Gln and Asn. In addition, H and D have very different scattering lengths, **exchangeable hydrogens** in a protein can therefore be labelled. In methyl-groups that are unable to rotate, the 3 hydrogens appear as distinct scatterers, in a group that can rotate freely, they appear as confluent torus. **Hydrogen bonding** can also be detected by neutron scattering (including indirect hydrogen bonds via water), as can the **ionisation** state of basic and acidic groups.

For neutron diffraction the brightness of nuclear reactors is insufficient, neutrons are therefore produced in **spallation sources** (proton synchrotrons that smash fast protons onto tungsten targets to kick out neutrons). The neutrons are then decelerated in heavy water, until their DE BROGLIE wavelength is in the 1–2 Å range, these are called **thermal neutrons** as their velocity is in the same range as that of molecules at room temperature (about 2.2 km/s). Note that the speed, and hence wavelength, of neutrons can be adjusted simply by changing the temperature of the water bath! Longer wavelength "cold" neutrons (6 Å) for SANS studies are passed through liquid hydrogen. Neutron are then collimated and passed through a monochromator (a single crystal whose molecular layers act as grating). After interaction with the sample, the neutrons are detected by a Gd_2O_3 screen, which produces γ-rays when struck by neutrons. These can be detected by phosphorescence as described above for X-rays.

Since there are no elements that scatter neutrons much better than others (like heavy elements do for X-rays), **phasing** of neutron diffraction patterns has to rely on a previously existing X-ray structure.

34.4 Electron Microscopy of 2-D Crystals

As mentioned, obtaining crystals from solubilised membrane proteins is far from trivial. However, if the protein concentration in the membrane is high enough and little or no interfering proteins are present, some membrane proteins form ordered 2-D arrays in the membrane plane. Indeed, the photoreaction centre of certain photosynthetic bacteria form such arrays even *in vivo*, this was the first membrane protein whose structure could be solved [155]. In other cases the protein in question is abundant enough that removal of lipid and other proteins by detergent extraction leads to crystallisation (*e.g.*, acetylcholine receptor in the electric organ of *Torpedo californica*) . In most cases however, 2-D crystals are grown from detergent solubilised and purified proteins by reconstitution into membranes. From the reconstitution for biochemical purposes (see Sect. 15.3 at p. 174) the 2-D crystallisation is distinguished only by the low lipid/protein ratio (10–50 rather than 2 000–10 000 lipid molecules per protein). While in 3-D crystals of membrane proteins protein–protein contacts occur mostly in the hydrophilic parts outside the membrane, 2-D crystals are formed by interactions between the transmembrane parts of the proteins.

Originally it was thought that the growth of 2-D crystals was a slow process and consequently low-cmc detergents were used for solubilisation, these were then removed by micro-dialysis (which is a slow process due to the minute concentration of monomeric detergent molecules). Today we know that even the comparatively rapid removal of detergent by absorption to hydrophobic beads provides ample time for crystallisation. So the only requirement for the detergent is that it can maintain the structure and function of the protein in the solubilised state. Neither its cmc nor its chemical structure are directly relevant to the results of 2-D crystallisation studies. The latter is not surprising, since the detergent is not present in the final 2-D crystal (unlike the 3-D ones). Since reconstitution is fast and occurs in the presence of lipid even detergents to harsh to be used for purification may be tried for crystallisation after detergent exchange. The growth of 2-D crystals was reviewed recently by RIGAUD et al. [352].

Such 2-D crystals can be examined in an electron microscope. The same crystal is examined several times, each time tilted further by a few degrees. This series of photos can be used to reconstruct the structure of the embedded protein molecules. Resolution is often limited to the 10–20 Å range, however, in exceptional cases 3.5 Å has been achieved. Since an electron microscopic picture is obtained using both phase and intensity, separate phasing is not required.

In order to prevent damage to the proteins from the high energy electron beam, low temperatures are used (cryoelectron microscopy). The radiation dose needs to be as low as possible; this results in pictures where the structure is overlayed with a high amount of noise which needs to be removed. This is done in a computer, using 2-dimensional FOURIER-filtering. This technique effectively separates repetitive structures (low frequency signal) from non-repetitive noise (high frequency).

While X-rays interact with the electrons of a molecule and neutrons with its nucleï, electrons interact with the electrostatic potential, and thus allows a view of a molecule that is complementary to the other techniques.

Not only crystals can be investigated by this technique. Medium-resolution images have been obtained from ribosomes by photographing many of them [114,420]. All these photos were fed into a computer that sorted them according to orientation using the technique of cluster-analysis [383]. Once all photos had been turned into the same orientation, the FOURIER-transforms can be added and back-transformed, in effect calculating an average image with the noise removed.

Results of protein studies by electron microscopy are collected at the **electron microscopy database (EMDb)** at www.ebi.ac.uk/pdbe/emdb/.

Chapter 35
Folding and Unfolding of Proteins

Proteins are stabilised by the difference between the bond energies between interacting amino acids and the energy of interaction between these amino acids and water (several MJ/mol each). Protein folding decreases not only the enthalpy, but also the entropy of the molecule. Hence the total stabilisation energy is $\approx$20–60 kJ/mol which should be compared with thermal energy kT ($\approx$2 kJ/mol at room temperature).

Because of the **marginal stability** of proteins, their molecules undergo constant motion between conformational sub-states. We distinguish

Taxonomic sub-states with lifetimes in the μs to ms range that can be characterised in molecular detail. These may correspond to specific conformations along the catalytic pathway of an enzyme, T- and R-state in co-operative proteins, open and closed channels and the like.

Statistical sub-states with lifetimes in the ps to ns range. These sub-states are along the pathway between the taxonomic ones, for example during binding of a ligand.

Intrinsically disordered proteins can assume a particularly large number of taxonomic sub-states, allowing for their catholic binding properties.

35.1 Inserting Proteins into a Membrane

When protein helices are inserted into a membrane, the flexible lipid molecules adapt to the fairly rigid protein, losing entropy in the process. There is little ΔH associated with dehydration, as hydrogen bonds between water and amino acids are replaced by bonds between water molecules. Moving a hydrophobic amino acid from cytosol into lipid results in a hydrophobic effect, giving a ΔG of about −6.3 kJ/mol. Moving the polar peptide bond into a hydrophobic environment costs about +9 kJ/mol, partitioning of C^{α} into lipid brings about −4 kJ/mol. Thus the overall gain of moving a transmembrane α-helix into the lipid environment is −1.3 kJ/mol per amino acid, or −26 kJ/mol for a helix of 20 amino acids.

E. Buxbaum, *Biophysical Chemistry of Proteins: An Introduction to Laboratory Methods*, DOI 10.1007/978-1-4419-7251-4_35,

35.2 Change of Environment

If proteins are exposed to a change in environmental conditions (pH, temperature, concentration of urea or guanidinium hydrochloride) bonds involved in the stabilisation of secondary, tertiary, and quaternary structure may be broken, and a different conformation may become the energetically favoured one. Provided that these changes are fully reversible they may be investigated by kinetic methods. This type of investigations should further our understanding of the folding pathway of proteins and why a protein with a given sequence adopts a particular 3D structure. For a review see for example [174].

Most of this work has been done with small, single-domain proteins. These often fold with a two-state, reversible kinetics $\mathsf{F} \rightleftharpoons \mathsf{U}$ (see Fig. 35.1). Such proteins are called **class II**, they fold rapidly and their structure is stabilised mostly by local interactions. **Class I** proteins fold with an intermediate $\mathsf{F} \rightleftharpoons \mathsf{I} \rightleftharpoons \mathsf{U}$, often a molten globule (see Fig. 35.2). In these proteins tertiary structure is stabilised by interactions between amino acids that are far apart in primary structure, leading to hierarchical folding.

It has long been believed that proteins in solutions with high denaturant concentrations (for example 6 M guanidinium chloride or 10 M urea) are completely unfolded. This view has been challenged recently (review in [102]). Today we view denatured proteins as stiff segments of secondary structure connected by flexible

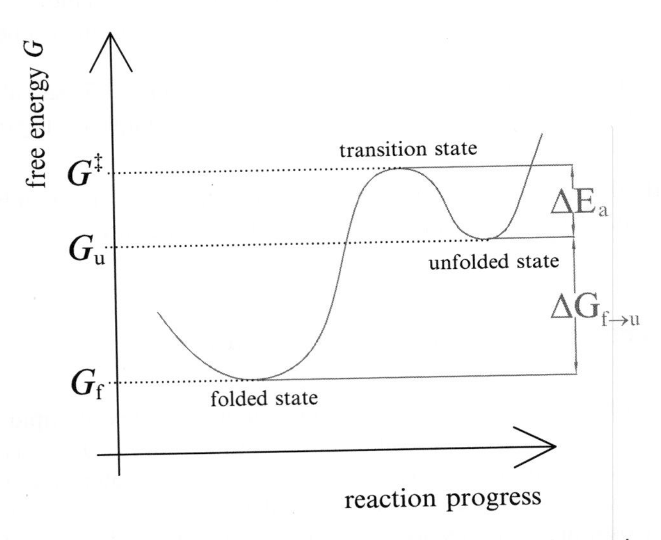

Fig. 35.1 Energy diagram of a two-state folding/unfolding reaction. The protein occurs only in the folded or unfolded state, no folding intermediates exist. Note that the transition state ‡ does not count as intermediate, in a three-state folding process there would be two transition states, one between folded and intermediate, and one between the intermediate and unfolded state. For the following deliberations, we assume that the folding/unfolding reaction is fully reversible, which is often the case if no covalent bonds are affected (disulphides, Pro *cis/trans*-isomerisation)

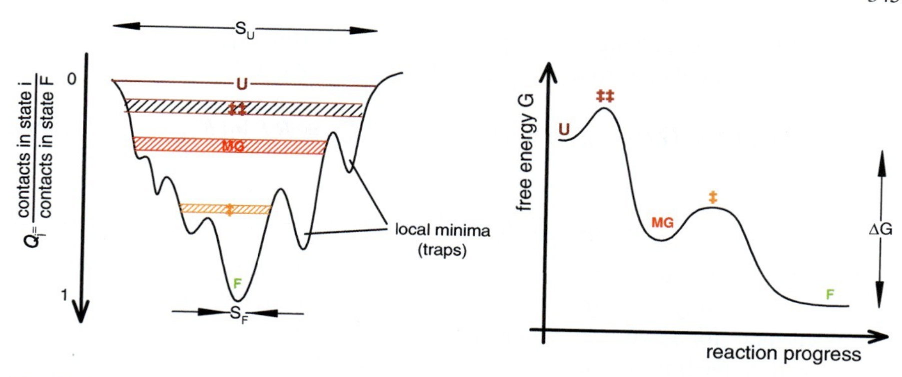

Fig. 35.2 *Left*: Folding funnel of proteins. In the totally unfolded state U the amino acids in the protein have no native contacts ($Q_i = 0$). However, their freedom of movement, and hence entropy S, is maximal. In the folded state F, their entropy is minimal, but they have all native contacts between amino acids ($Q_i = 1$). All other folding states can be interpreted as points in the coordinate system S_i/Q_i. Local minima act as traps where protein molecules can get stuck for a while. In the cell, molecular chaperons and chaperonins help proteins out of such traps. *Right*: Free energy of folding for a class I protein. The molten globule MG is a local minimum, that is an obligatory station along the folding pathway, while the local minima in the energy landscape on the left are not. From there the folded state is reached via a transition state ‡

linker regions. However, since intra-molecular hydrogen bonds within the folded protein molecules are replaced by inter-molecular bonds with water, secondary structure may be significantly different from what is observed in the folded state. This model still fits the EINSTEIN–SMOLUCHOWSKI theory of BROWNIAN motion within the chain, which states that the root-mean end-to-end distance in a flexible chain is proportional to the square root of the number of monomers. Especially the polypeptide II-helix ($\phi, \psi \approx -65°, 140°$, also known as P_{II} or poly-proline helix since it occurs in the proline-rich collagen) is favoured in the denatured state since its formation requires neither intra-segment (like α-helix) nor inter-segment (like β-sheets) hydrogen bonds. Instead the P_{II}-helix is stabilised by entropic forces: it allows its amino acids maximum "wiggle"-space before encountering steric constraints and minimises ordering of solvent.

Measurements are often required over a time scale of μs to ks that is possible with stopped flow, continuous flow, or temperature or pressure jump setups (see Chap. 38 on p. 369 for a discussion of these techniques). Ligand binding, enzymatic turnover, absorbance, fluorescence (unfolding brings aromatic amino acids into a more polar environment, their fluorescence quantum yield is reduced, and the emission maximum red-shifted), long-wavelength CD (aromatic amino acids 250–300 nm), viscosity, or deuterium-exchange coupled to ESI-MS or NMR can be used as reporter for tertiary, short-wavelength CD and IR spectroscopy for secondary structure (*vide supra*).

As the denaturant concentration is increased, the rate for unfolding increases and the rate for folding decreases. Thus the equilibrium constant between folded and unfolded protein changes with [denaturant], and from K_{FU} the change in free energy can be calculated from the GIBBS–HELMHOLTZ equation as

$$\Delta G(T) = G_U - G_F = \Delta H(T) - T\Delta S(T) = RT \ln(K_{FU}) \quad (35.1)$$

with

$$K_{FU} = f_F / f_U = (y - y_F)/(y_U - y) \quad (35.2)$$

ΔH, ΔS and ΔG are the heat of reaction and the changes in entropy and free energy, respectively. f_F and f_U are the mol fractions of protein in the folded and unfolded state (for a two-state process, $f_F + f_U = 1$), y is the measured parameter and y_F and y_U the limiting values for fully folded and unfolded protein, respectively. For approaches with more than two states, see [310].

Both ΔH and ΔS of unfolding increase strongly with temperature, thus their effect on ΔG cancel to a large extent. As a result the function of ΔG versus temperature is an inverted parabola, proteins are most stable at an optimum temperature. Both lower and higher temperatures reduce their stability (cold and heat denaturation, respectively). Optimum temperatures of proteins vary with the environment that an organism lives in (higher for thermophiles and lower for mesophiles and psychrophiles). The melting temperature (where $K_{eq} = 1$) is usually between 20 and 50°C for mesophiles and 60 and 90°C for thermophiles.

From the GIBBS–HELMHOLTZ equation (35.1) the VAN'T HOFF-equation can be derived:

$$\partial \ln(K_{FU}) = \frac{\Delta H^0}{R} * \partial(1/T). \quad (35.3)$$

Thus a plot of $\ln(K_{FU})$ as a function of reciprocal temperature can be used to determine ΔH^0 from the slope of the regression line.

35.2.1 *Standard Conditions for Experiments*

Unless dictated otherwise by the properties of protein at hand, folding experiments should be performed under the following conditions: 25°C (298 K), *p*H 7.0 with phosphate as buffer, and at the minimum ionic strength compatible with the protein. Urea is the preferred denaturant, if saturated urea solutions do not lead to unfolding GuHCl or GuSCN may be used. Following these guidelines ensures that results can be easily compared between laboratories.

35.3 The Chevron-Plot

The relaxation times of unfolding by mixing with denaturant and the relaxation times of refolding by diluting protein/denaturant mixtures are determined and their logarithm plotted against the denaturant concentration (see Fig. 35.3). If the free

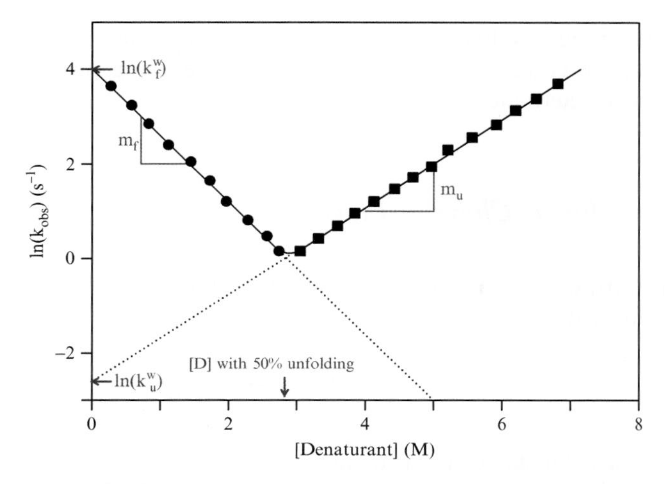

Fig. 35.3 Simulated chevron-plot. Data are obtained by mixing native (*circles*) or fully denatured protein (*squares*) with denaturant. At least ten data points should be obtained for each arm of the chevron. For details see text

energy difference between the folded and unfolded state depends linearly on the denaturant concentration the resulting straight lines will form an inverted 'V' with smooth connection between them (or a upright 'V' if rate constants are used). These figures are called chevrons [99, 261, 445]. The slope of the lines m depends on the change of surface area during the folding/unfolding reaction, small conformational changes or proline *cis/trans*-isomerisation result in flat chevron-plots. If the relaxation time decreases with decreasing urea concentration, the surface area of the protein is reduced during the reaction. Data are fitted to the following equation:

$$\ln(k_{\mathrm{obs}}) = \ln(k_{\mathrm{f}}^{\mathrm{w}}) - m_{\mathrm{f}} \times [D] + \ln(k_{\mathrm{u}}^{\mathrm{w}}) + m_{\mathrm{u}} \times [D] \qquad (35.4)$$

where m_{u} and m_{f} are the slopes for unfolding and folding, $k_{\mathrm{f}}^{\mathrm{w}}$ and $k_{\mathrm{u}}^{\mathrm{w}}$ the folding and unfolding rate constants in water and $[D]$ the denaturant concentration. Note that because of data weighing this logarithmic form of the equation is preferred over the non-transformed sum of exponentials.

35.3.1 Unfolding by Pulse Proteolysis and Western-Blot

For large-scale screening the need to purify a protein in sufficient quantities for chevron-plot analysis is a considerable obstacle. A quicker approach was suggested in [197]: Cell lysates are incubated with proteases (*e.g.*, thermolysin) in the presence of varying concentrations of urea. Incubation time is selected so that unfolded but not folded protein is digested. Then proteins are separated by SDS–PAGE and subjected to quantitative Western-blotting. From the resulting curves the transition midpoint can be determined.

35.3.2 Non-linear Chevron-Plots

Non-linear arms, usually seen in the folding arm, in the chevron-plot ("rollover") result if the mechanism is not 2-state:

$$\mathrm{F} \underset{k_{if}}{\overset{k_{fi}}{\rightleftharpoons}} \mathrm{I} \underset{k_{ui}}{\overset{k_{iu}}{\rightleftharpoons}} \mathrm{U}$$

Note that the meta-stable intermediate I is different from the mere energy barrier transition state ‡, which also occurs in two-state reactions. In this analysis, k_{if} is assumed to be the rate limiting step of the folding reaction. Given K_{iu} the equilibrium constant between unfolded and intermediate state, the rate of folding becomes:

$$k_f = \frac{K_{ui}}{1 + K_{ui}} \times k_{if} \tag{35.5}$$

Then the chevron can be fitted to [445]:

$$\ln(k_{obs}) = \ln\left(\frac{K_{ui}\exp\left(\frac{-m_{u\to i}}{RT}\right)[D]}{1 + K_{ui}\exp\left(\frac{-m_{u\to i}}{RT}\right)[D]}\right) \times k_{if}\exp\left(\frac{-m_{i\to\ddagger}}{RT}\right)[D] + k_{fi}\exp\left(\frac{m_{f\to\ddagger}}{RT}\right)[D] \tag{35.6}$$

35.3.3 Unfolding During Electrophoresis

A very neat way of analysing protein unfolding is electrophoresis in a transversal urea gradient [71], that is, a gradient at right angle to the direction of electrophoresis. Since unfolding changes the molecular volume and charge, the electrophoretic mobility of a protein depends on the urea concentration. If the native to unfolded transition is fast compared to the time required for electrophoresis, the mobility of the protein in an urea gradient from 0 to 8 M is the weight average of the mobilities of the native and unfolded protein. A clear transition will be visible at the urea concentration where $K_{fu} \approx 1$. Because urea affects the sieving property of the polyacrylamide gel, the concentration gradient of urea needs to be accompanied by an inverse concentration gradient of acrylamide (15 to 11 %).

35.3.4 *Membrane Proteins*

Membrane proteins often have several domains or even several subunits. They are contained in a lipid environment, the native lipid composition not only changes from the luminal to the cytosolic leaflet, but also laterally (*e.g.*, in lipid rafts). Membrane proteins are often very sensitive to these environmental factors. Therefore, protein unfolding is often not reversible; unfolded membrane proteins usually aggregate. In addition, the expression, purification and stabilisation of membrane proteins for investigation is still more art than science.

However, there have been examples in the literature [35] of proteins that can be reversibly unfolded in micelles of a mild detergent (*e.g.*, CHAPS) or detergent/lipid bicells, by addition of denaturing detergents like SDS or trifluoroethanol. Especially with SDS the usable concentration range is quite narrow, limiting the number of data points that can be collected for the chevron-plot. For some micelle-solubilised proteins, *e.g.*, GalP, urea can be used for denaturation, alleviating at least this problem.

One of the more interesting outcomes of such studies is that the per-residue stabilisation energy of water-soluble and membrane proteins is quite similar, even though the forces involved in maintaining the native structure are probably quite different. Another useful result is that protein activity depends on membrane **curvature** (measured as reciprocal of the curve radius). This may well be important for studies on membrane protein reconstitution and crystallisation.

35.4 The Double-Jump Test

Immediately after unfolding the peptide-bonds are still in their native form. The purpose of double-jump experiments is to measure the rate of proline *cis/trans*-isomerisation, the required time can be between several seconds and about 20 min. This reaction is usually much slower than other changes of protein structure.

The protein is mixed with the denaturant and after various delay times aliquots are mixed with water to induce refolding. The relaxation time and the amplitude for refolding is then plotted as a function of the delay time.

This type of experiment can also be performed in the reverse direction, unfolded protein is mixed with water to induce refolding, after various delay times samples are mixed with denaturant and the rate of unfolding is measured.

35.5 Hydrogen Exchange

The fully deuterated, unfolded protein is pulse-labelled for 10–20 ms, with ^1H_2O at alkaline *p*H at various times during the refolding reaction, then the exchange reaction is stopped by decreasing the *p*H. The amount of exchange between amide deuterium and hydrogen is assessed either by ESI-MS or by NMR. Deuterium atoms involved in hydrogen bonds (that is in secondary structure formation) exchange slower than free ones. The position of exchanged hydrogens can be determined by multi-dimensional NMR or by MS/MS.

35.6 Differential Scanning Calorimetry

Differential scanning calorimetry (DSC) measures the change in heat capacity $\Delta C_\mathrm{p} = \mathrm{d}H/\mathrm{d}T$ of a protein solution as a function of temperature. Two vials, one for protein solution and one for buffer, are heated by small electrical heating elements, maintaining them at identical temperature to within 10^{-3} K. The current required is measured and plotted as a function of temperature. The sample will have a slightly higher heat capacity than the reference solution as increased temperature is accompanied by increased intramolecular motion of the protein. At a characteristic temperature (inappropriately called "**melting temperature**") this difference becomes suddenly larger, because energy is required to break the secondary, tertiary, and quaternary structure of the protein during unfolding. Once the protein is completely denatured, the difference in heat capacity becomes smaller again.

$$< H^0 > \quad \propto \int_{T_1}^{T_2} c_\mathrm{p} \mathrm{d}T \tag{35.7}$$

is the average calorimetric enthalpy. If $< H^0 > \neq \Delta H_\mathrm{vH}$ (the ΔH from a VAN'T HOFF-plot) the process is probably not two-state (only native and denatured protein involved) but involves one or more intermediates. This may also be indicated by broadening or skewing of the DSC curve. This is a very simple way to measure temperature stability of the protein as a function of ligand concentration or similar parameters. Unfortunately, the instruments are expensive and large (milligrams) amounts of protein are required for each test.

35.7 The Protein Engineering Method

In this method, various mutants of the protein in question are created, each with a single amino acid mutation [99, 261, 265, 445]. Protein folding kinetics and thermodynamics are then analysed by the methods described above. The change in the free energy of unfolding $\Delta\Delta G_{\mathrm{f}\to\mathrm{u}}$ is calculated by

$$\Delta\Delta G_{\mathrm{f}\to\mathrm{u}} = -RT \ln\left(\frac{k_\mathrm{f}^\mathrm{mut}}{k_\mathrm{f}^\mathrm{wt}}\right) - RT \ln\left(\frac{k_\mathrm{u}^\mathrm{wt}}{k_\mathrm{u}^\mathrm{mut}}\right) \tag{35.8}$$

and the slopes

$$m_{f\to u} = -RT(m_f + m_u) = m_{\ddagger\to u} + m_{f\to\ddagger} \tag{35.9}$$

where ‡ refers to the transition state. Correlation between $\Delta\Delta G_{f\to u}$ and $m_{f\to u}$ values determined by kinetic and equilibrium methods should be high, to confirm that all mutants have two-state folding behaviour, with the folded and unfolded state separated by a high-energy transition state.

$$\Delta\Delta G_{\ddagger\to u} = -RT \ln\left(\frac{k_f^{mut}}{k_f^{wt}}\right) \tag{35.10}$$

$$\Delta\Delta G_{f\to\ddagger} = -RT \ln\left(\frac{k_u^{wt}}{k_u^{mut}}\right) \tag{35.11}$$

$$\phi_f = \frac{\Delta\Delta G_{\ddagger\to u}}{\Delta\Delta G_{f\to\ddagger}}, \tag{35.12}$$

ϕ_f is between 0.0 (all destabilisation is caused by increasing k_u, with no effect on k_f) and 1.0 (all destabilisation is caused by slowing k_f, with no effect on k_u). Simply put, in a mutation with $\phi_f = 1$ the amino acid has native-like stabilising interactions already in the transition state, at $\phi_f = 0$ these interactions do not exist in the transition state.

Negative ϕ_f-values result from opposite effects of a mutation on folded and transition state (*e.g.*, stabilise the transition and destabilise the native state). $\phi_f > 1$ results from substitutions, that either stabilise or destabilise both the transition and folded state. About 10–15 % of all substitutions result in such abnormal ϕ_f-values.

Condition for ϕ_f-analysis is that the mutation does not change the overall structure of the protein (either in the folded or unfolded state) or the folding pathway; hence conservative mutations are preferred. However, it has become customary to change all amino acids simply to Ala, because this is convenient and Ala is least likely to introduce new non-native interactions. Too conservative a substitution may produce a very small $\Delta\Delta G_{f\to u}$, which can not be determined reliably.

To further evaluate the results of protein engineering experiments, the $\ln(k_f)$ or $\ln(k_u)$ values for all substitutions are plotted as function of their respective $\Delta\Delta G_{f\to u}$ (BRØNSTED-plot). This may result in different shapes:

A single straight line with a slope β_f equal to the average ϕ_f. The transition state of the protein is uniformly expanded compared to the native state, with no single region being more (un)folded than any other.

Two distinct lines A region of the protein is more structured in the transition state than the rest. Mutants with high ϕ_f lie on the line with the steeper slope.

A curve would mean that there are two pathways, where folding starts in different regions of the protein. Such plots, although predicted in theory, have not been observed in practice.

The m-values are proportional to the difference in solvent exposure of the corresponding states. Thus, if the m-values of mutants do not vary significantly from wild type, then mutation has not affected transition state or unfolded state structure. β_t is the solvent exposure of the transition relative to the native state:

$$\beta_t = \frac{m_{\ddagger\to u}}{m_{f\to u}} \propto \frac{m_f}{m_u + m_f} = \alpha \tag{35.13}$$

Thus a high β_t corresponds to a compact transition state. β_t can be compared to β_f (relative energy between states). A diagram of β_t against $\Delta\Delta G_{f\to u}$ is called a HAMMOND-plot.

Part VI
Enzyme Kinetics

Chapter 36
Steady-State Kinetics

The theoretical aspects of enzyme activity, inhibition, and inactivation have been discussed in a separate volume [44]. More extensive treatments of both theory and practice of enzyme kinetics may be found in [27, 68, 69]. Ready-to use rate equations for most mechanisms may be found in the otherwise dated monography by SEGEL [372], rate equations may also be obtained by a computer program described in [438].

The purpose of enzyme kinetics is to elucidate the mechanism of enzymatic reactions. To this end, the reaction velocity is determined at different concentrations of substrates and inhibitors, and under different environmental conditions (*p*H, temperature, pressure,...). Enzyme kinetics are not directly interested in the physiology of the enzyme or its substrate(s), hence the reaction conditions chosen may well be unphysiological as long as they yield information on mechanism. Many reviewers of scientific journals fail to understand that difference, no doubt an indication for the deplorable state of undergraduate teaching in this field. This is also reflected by the fact that treatment of enzymology in most standard textbooks of physiology, biochemistry and molecular biology is dated, and in many cases wrong.

Enzymological parameters are subject to evolutionary pressures. The volume of a cell limits the number of protein molecules it may contain. In order to make space for new functions during evolution, an organism can

- Loose a function no longer required in its environment. Nutritional requirements for essential amino acids and vitamins are the result of this process.
- Ensure that K_m is in the range of $0.1[S] - [S]$ so that v is in the range of $0.5 - 0.9$ V_{max} and controlled by the metabolic situation of the cell.
- Increase k_{cat} of existing enzymes so that the same turnover is achieved at lower enzyme concentrations. The limiting factor for the efficiency constant k_{cat}/K_m is the maximal rate for the formation of the ES-complex in aqueous solution, $10^9\ M^{-1}s^{-1}$. Once an enzyme can produce product as fast as it can bind substrate (catalytic perfection), further improvements in k_{cat} have no beneficial effect.
- Formation of multi-enzyme complexes, where the product of one enzyme is delivered directly to the next enzyme as substrate, without time-consuming release, diffusion, and rebinding steps. This can overcome the above limitation of k_{cat}.

E. Buxbaum, *Biophysical Chemistry of Proteins: An Introduction to Laboratory Methods*, DOI 10.1007/978-1-4419-7251-4_36,

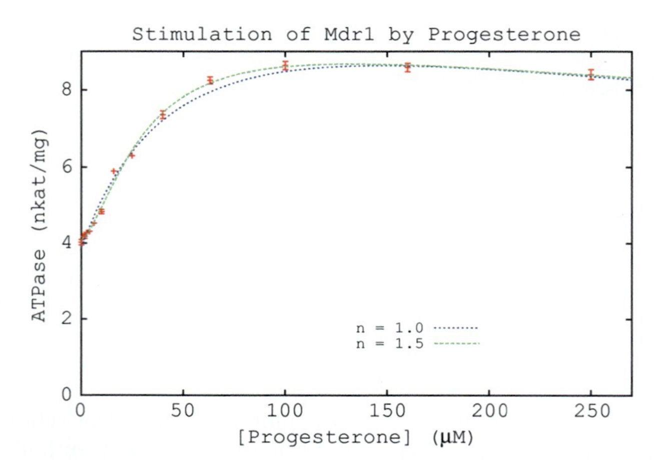

Fig. 36.1 Hyperbolic or sigmoidal? The ATPase-activity of Mdr1 (p-glycoprotein, an ABC-type transport ATPase) is stimulated by progesterone (one of its transported substrates) in a co-operative manner, with a HILL-coefficient of 1.5 which can be increased by the presence of other substrates (from [42]). A considerable number of good-quality data points are required to reliably detect such small deviations from standard hyperbolic behaviour. Other interesting features of this experiment include substrate inhibition and basal activity in the absence of progesterone

These processes also reduce the energy costs of producing the required enzymes, giving an organism an evolutionary advantage. Optimisation of K_m and k_{cat} has to take the average body temperature of the organism into account.

36.1 Assays of Enzyme Activity

Because so many repetitive measurements are required in enzyme kinetics a good assay is even more important than in protein purification. The assay should be simple, quick, cheap, well characterised, robust, and precise (see Sect. 47.1 on p. 428 for a definition of these terms). We distinguish

Real time assays produce a signal while the assay is being performed. They usually measure changes in absorbance, but changes in optical rotation, fluorescence, conductivity, *p*H and other physical properties have been used. Clearly, this type of assay is preferable, as any change in enzyme activity over time is immediately apparent.

End-point assays during the reaction aliquots are withdrawn at regular intervals, these are pipetted into a stopping-reagent which kills the enzyme without breaking down either educt or product (SDS, TCA, NaOH,...). Once all samples have been collected the concentration of either product or educt is determined by suitable means.

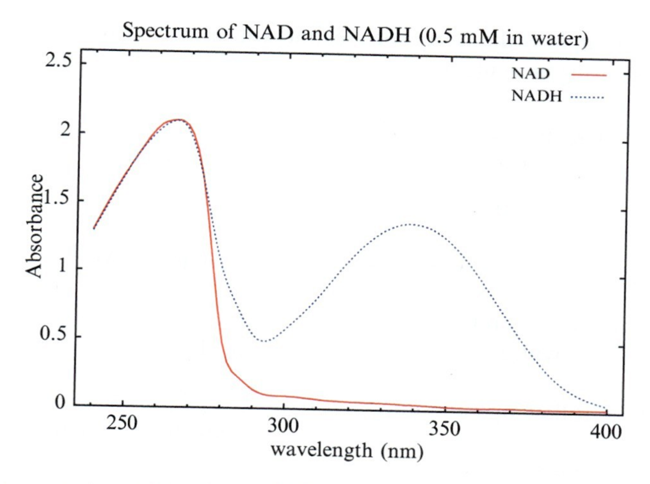

Fig. 36.2 Spectrum of NAD(P) and NAD(P)H in water. Both compounds show a maximum at 260 nm which is caused by the adenine ring. The reduced nicotinamide ring has an additional absorbtion peak at 340 nm, which allows quantitation in the presence of the oxidised form

In either case it is essential that control reactions without enzyme are run in parallel to account for reagent instability. The enzyme activity is then calculated from the concentration change over time, for this calculation of course the data have to be on a straight line. Failure to verify linearity in each and every experiment is a certain recipe for disaster! Samples must be free of dust and air bubbles (vacuum filtration over 0.22 μm low protein binding membrane filters) and thoroughly mixed.

36.1.1 The Coupled Spectrophotometric Assay of WARBURG

The most commonly used real time assay is based on the change of absorbance at 340 nm of NAD(P) upon reduction to NAD(P)H (see Fig. 36.2). Absorbtion of light at this wavelength by NAD(P)H (6220 M^{-1} cm^{-1} is associated with fluorescence at 465 nm, which can be used for even more sensitive (but less robust) assays. The sensitivity is high enough to determine metabolite and enzyme concentrations in a single cell [308, 309]!

Classically, assays were performed with a photometer equipped with a cuvette changer: The cuvette holder could accept several (four or six) cuvettes, a motor brought them into the beam for a few seconds each. The absorbtion was continuously recorded on a x/t chart recorder. The record was a pattern of dotted lines, whose slope was determined with a ruler. Nowadays 96-well photometers with sample injector and computerised data handling offer a much larger number of parallel samples, but the principle is the same.

Of course, not all enzymes use pyridine nucleotides. However, in many cases it is possible to **couple** such reactions with a reaction that does use pyridine nucleotides.

A classical example is the assay for creatine kinase, which is released from heart muscle cells during an acute myocardial infarct:

$$\text{Creatine} + \text{ATP} \xrightleftharpoons{\text{CK}} \text{Creatine phosphate} + \text{ADP}$$

$$\text{ADP} + \text{PEP} \xrightleftharpoons{\text{PK}} \text{ATP} + \text{Pyruvate}$$

$$\text{Pyruvate} + \text{NADH} \xrightleftharpoons{\text{LDH}} \text{Lactate} + \text{NAD}^+$$

Each molecule of ADP formed by CK leads to the conversion of one molecule of NADH to NAD^+, with a concomitant reduction in A_{340}.

Assume the following, simplified reaction: $\mathsf{A} \underset{\text{enzyme 1}}{\overset{\text{v}}{\rightleftharpoons}} \mathsf{B} \underset{\text{enzyme 2}}{\xrightarrow{V/K_\mathrm{B}}} \mathsf{C}$. If the following conditions are met:

- $[\mathsf{B}] < K_\mathrm{b}$, the MICHAELIS-constant of enzyme 2 for B
- $[\mathsf{B}] < K_\mathrm{i}$, the product inhibition constant of enzyme 1 for B

then

$$\frac{\mathrm{d}[C]}{\mathrm{d}t} = \frac{V}{K_\mathrm{B}} \times [B] = v\left(1 - \mathrm{e}^{-(V/K_\mathrm{B})t}\right). \tag{36.1}$$

Thus the measured rate increases from 0 at time 0 to a constant rate $\approx v$ after some lag time. This lag is the time required for [B] to become large enough that $\frac{V}{K_\mathrm{B}} \times [B] = v$. With increasing number of coupling steps n, the lag time τ increases:

$$\tau = \sum_{j=1}^{n} \left(\frac{K_j}{V_j}\right). \tag{36.2}$$

After 3τ, $\frac{\mathrm{d}[C]}{\mathrm{d}t} = 0.95v$, after $\approx 5\tau$, $\frac{\mathrm{d}[C]}{\mathrm{d}t} = 0.99v$. τ can be adjusted by the activities of the coupling enzymes V_j. For the large number of assays required, it is worthwhile to minimise the costs of the assay [68, 417]:

$$V_j = \left(\frac{K_j}{\tau}\right)\left(1 + \sum_{k(k \neq j)} \left[\frac{x_k K_k}{x_j K_j}\right]^{1/2}\right) \tag{36.3}$$

with x_j the cost for enzyme j (in, say, €/kat). The concentrations of the intermediates become $v\frac{V_j}{K_j}$, this should be less than 0.02 K_j. This limits the measured velocity v to $0.02V$, with $V = \min_{j=1}^{n}(V_j)$.

The concentrations of substrates for the coupling enzymes should be near-saturating. Note, however, that the concentration of NAD(P)H is limited to about 350 μM by the resulting absorbance, which is at the limit of the linear range of most photometers.

36.2 Environmental Influences on Enzymes

K_m and k_{cat} change with environmental conditions (*p*H, temperature, ionic strength, *etc.*), this can be used to help elucidate enzyme reaction mechanism. On the other hand, these factors need to be tightly controlled during investigations to avoid erroneous conclusions. Sometimes the attempt to control for such factors can, however, aggravate the problem; for example, Tris can deacetylate the chymotrypsin transition state (and presumably that of similar enzymes) [190].

36.2.1 pH

Chymotrypsin is a Ser-protease that utilizes a classic catalytic triad (Ser as nucleophile, His as proton-acceptor, and Asp to stabilise the charged state of His, see [44, chapter 6.6.1] for details). If the *p*H is low enough to protonate the catalytic His-57, k_{cat} becomes zero. The apparent k_{cat} is therefore the intrinsic k^0_{cat} multiplied with the fractional dissociation of His-57 α, which depends on the *p*H and the pK_a of His-57 ($\approx$7):

$$k_{cat} = \alpha k^0_{cat} = \frac{k^0_{cat}}{1 + \frac{[H+]}{K_1}} \tag{36.4}$$

On the other hand, an amino-group with $pK_a \approx 9$ must be protonated for optimal substrate binding. If this amino-group is deprotonated, the apparent K_m increases:

$$K_m = \frac{K'_m}{1 + \frac{K_2}{[H+]}} + \frac{K''_m}{1 + \frac{[H+]}{K_2}} \tag{36.5}$$

Thus the enzymatic rate becomes

$$v = \frac{k_{cat}E[S]}{K_m + [S]} = \frac{\frac{k^0_{cat}}{1+\frac{[H+]}{K_1}}E[S]}{\frac{K'_m}{1+\frac{K_2}{[H+]}} + \frac{K''_m}{1+\frac{[H+]}{K_2}} + [S]}. \tag{36.6}$$

If the rate is plotted as function of *p*H, a optimum-function is obtained, from which K_1 and K_2 can be determined by curve-fitting (see Fig. 36.3).

36.2.2 Ionic Strength

The ionic strength I_c is defined as:

$$I_c = \frac{1}{2}\sum_{b=1}^{s} z_b^2 c_b \tag{36.7}$$

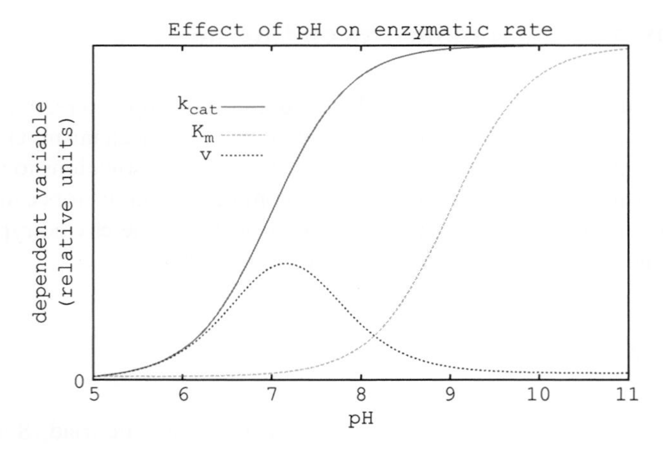

Fig. 36.3 Simulation of an enzyme that requires an unprotonated His ($pK_a \approx 7$) for catalysis and a protonated amino-group ($pK_a \approx 9$) for substrate binding. The curves were scaled to fit into a single diagram. For details see text

with s the number of ions present, c_b the concentration of the bth species, and z_b its proton charge. The ionic strength influences the chemical activity a of charged substrates according to the expanded DEBYE–HÜCKEL-theory:

$$a_S = [S]\gamma = [S]\exp\left(\frac{z_S^2 A\sqrt{I_c}}{1 + B\sqrt{I_c}}\right) \tag{36.8}$$

with $A \approx 0.5$ and $B \approx 1$ fitting constants.

γ is called the activity coefficient. This equation is valid for concentrations up to about 100 mM. The adjusted K_m and k_cat become:

$$k'_\text{cat} = k^0_\text{cat}\exp\left(\frac{z_\text{s} z_\text{e}\sqrt{I_c}}{1 + \sqrt{I_c}}\right) \tag{36.9}$$

$$K'_\text{m} = K^0_\text{m}\exp -\left(\frac{z_\text{s} z_\text{e}\sqrt{I_c}}{1 + \sqrt{I_c}}\right) \tag{36.10}$$

where z_e is the charge of the enzyme's active site. This correction can be neglected at low to medium substrate concentrations <100 mM.

36.2.3 Temperature

Binding of substrate to an enzyme is associated with a change in free energy, ΔG^s, this can be positive or negative. From there the reaction proceeds to the transition state, the difference in free energy between bound substrate and transition state is $\Delta G^\ddagger$, which is always positive. Thus k_{cat} always increases with temperature, whilst K_{m} may increase or decrease. Since usually $|\Delta G^\ddagger| \gg |\Delta G^s|$ (induced fit hypothesis!) the enzymatic rate will increase with temperature, however, above an optimum temperature enzymes unfold which have to be modelled separately. The temperature-effect on the reaction rate before unfolding can be modelled by

$$v = \frac{A[E][S]\exp\left(-\frac{\Delta G^\ddagger}{RT}\right)}{B\exp\left(-\frac{\Delta G^s}{RT}\right) + [S]}, \tag{36.11}$$

where A and B are fitting constants.

36.3 Synergistic and Antagonistic Interactions

In real life, a system is often exposed to not only one active agent, but to several at the same time. Examples include patients being treated with several different drugs, organisms exposed to several environmental pollutants, or enzymes in the presence of several inhibitors. It would be naïve to expect the total effect to be the sum of the individual effects of the agents (even though that is all too often the rational of treating patients with several drugs!). The detection of interactions between the agents has been reviewed in [24, 127].

36.3.1 *Nomenclature*

If the effects of n agents are investigated, each at a dose $d_1, d_2, \ldots, d_n$ or concentration $c_1, c_2, \ldots, c_n$, then the effect is described by $E(c_1, c_2, \ldots, c_n)$ or by the surviving activity $S(c_1, c_2, \ldots, c_n) = 1 - E(c_1, c_2, \ldots, c_n)$. The concentration of a single compound i that has the same effect (is **isoeffective** with) the combination $c_1, c_2, \ldots, c_n$ is called C_i. If an agent alone, no matter what its concentration, can not achieve the effect $E(c_1, c_2, \ldots, c_n)$ of the combination, we call the interaction **heterergic**. If none of the agents can individually achieve that effect, **coalitive**. The sum $\gamma = \sum_{i=1}^{n} c_i / C_i$ is called the **interaction index**.

36.3.2 *The Isobologram*

If in a plot of c_1 versus c_2 all points of equal effect $E(c_1, c_2)$ or surviving activity $S(c_1, c_2)$ (isoeffect curve or isobole, see Fig. 36.4) are joined by a line, we get three possible results:

No interaction: $\gamma = 1$ resulting in straight lines.
Synergy: $\gamma < 1$ resulting in upwardly concave curves below the line for no interaction.

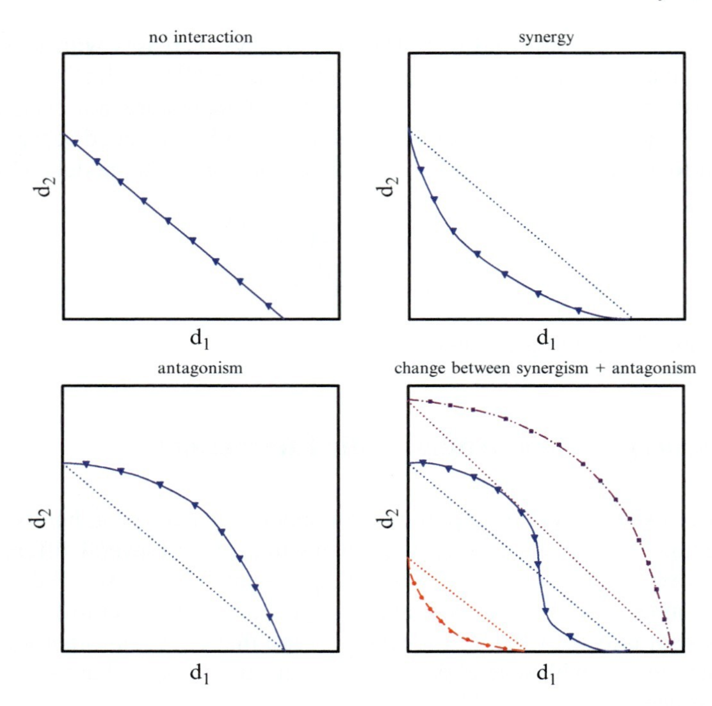

Fig. 36.4 Isobolograms. *Top left*: If all combinations of agents which result in the same effect (say, $S(c_1, c_2) = 0.5$) are connected, a straight line indicates no interaction between the agents. *Top right*: Upwardly concave curves below the line for no interaction (*dotted*) indicate synergism. *Bottom left*: Downwardly concave curves above the line for no interaction indicate antagonism. *Bottom right*: The isoboles for different effects (say, $S(c_1, c_2) = 0.3$, 0.5, and 0.7, *red*, *blue* and *maroon*, respectively) may indicate different interactions, this is even possible for a single isobole in different regions (here, the *blue* one)

Antagonism: $\gamma > 1$ resulting in downwardly concave curves above the line for no interactions.

It is, however, possible that for the same agent pair some isoboles show synergy, others antagonism. There are even situations where one and the same isobole has antagonistic and synergistic regions, crossing the line for no interactions. Results from kinetic studies may not always contain data sets with identical effects. In such cases, interpolation may be used (with due care!).

36.3.3 Predicting the Effect for Combinations of Independently Acting Agents

It is useful to predict the effect of a combination of agents, assuming that they do not interact. If the measured result deviates significantly from the predicted value, the agents must interact. Such prediction is simple if the dose-effect curves for all agents are known, and are of the same general shape for all agents:

Linear The effects are additive: $E(c_1, c_2, \dots, c_n) = \sum_{i=1}^{n} E(c_i)$

Exponential the surviving activity $S(c_1, c_2, \dots, c_n)$ is the product of the surviving activities in the presence of each agent individually: $S(c_1, c_2, \dots, c_n) = \prod_{i=1}^{n} S(c_i)$. Since $S(c_i) < 1$ multiplication leads to a reduction of $S(c_1, c_2, \dots, c_n)$.

Log–lin and log–log With either $E(c_i) = \alpha_i + \beta_i \times \log(c_i)$ or $\log[E(c_i)] = \alpha_i + \beta_i \times \log(c_i)$. Then $C_1[1 - C_1^{-\beta_1/\beta_2} \times 10^{(\alpha_1-\alpha_2)/\beta_2} \times c_2] - c_1 = 0$ and C_1 (and similarly, C_2) can be determined by iteration.

Sigmoidal The dose-effect curve for a single agent is $\frac{E(c_i)}{1-E(c_i)} = \left(\frac{c_i}{M_i}\right)^{m_i}$, with m_i the HILL-coefficient and M_i its $K_{0.5}$. This includes the case of $m_i = 1$, that is, HENRI–MICHAELIS–MENTEN (hyperbolic) behaviour. Then we distinguish

Competitive agents, which cannot bind to the target at the same time. The effects of the agents should be zero-interactive. In this case the HILL-coefficients would be expected to be identical and $S(c_1, c_2) = \frac{1}{1+(\frac{c_1}{M_1}+\frac{c_2}{M_2})^m}$, which for the case $m_i = 1$ reduces to $\frac{1}{S(c_1,c_2)} = \sum_{i=1}^{2} \frac{1}{S(c_i)} - (n - 1)$. If the HILL-coefficients are not identical, then iteration of the following equation can be used to determine the isoeffective doses: $C_1\left[1 - \left(\frac{M_1}{C_1}\right)^{m_a/m_b} \times \frac{c_b}{M_b}\right] - c_1 = 0$. Obviously, the reason for the differences in the HILL-coefficients would need to be investigated.

Mutually non-exclusive agents, which will always give rise to synergism between the agents. For the case that the HILL-coefficients are identical, we get $\left(\frac{E(c_1,c_2)}{S(c_1,c_2)}\right)^{1/m} = \frac{c_a}{M_a} + \frac{c_b}{M_b} + \frac{c_a c_b}{M_a M_b}$, which for $m = 1$ reduces to $S(c_a, c_b) = S(c_a) * S(c_b)$.

A useful equation for combined action of two agents on a target is

$$1 = \frac{c_1}{M_1\left(\frac{E(c_1,c_2)}{1-E(c_1,c_2)}\right)^{1/m_1}} + \frac{c_2}{M_2\left(\frac{E(c_1,c_2)}{1-E(c_1,c_2)}\right)^{1/m_2}} + \frac{\alpha c_1 c_2}{M_1 M_2\left(\frac{E(c_1,c_2)}{1-E(c_1,c_2)}\right)^{(1/m_1+1/m_2)}} \tag{36.12}$$

α is the interaction strength between the agents. Unfortunately, the effect can not be brought to the right side and has to be calculated by iteration [127].

The derivation of these results is not repeated here, the interested reader is referred to [24].

Intuitively, one would suggest the use of strongly synergistic drug combinations to treat bacterial or viral infections and cancer. However, it has been shown [151] that such combinations increase the rate of evolution of **drug resistance**. In such a combination growing resistance to drug 1 would relieve the agent not only of the damage done by that drug, but also of its enhancement on the effect of drug 2. In case of an antagonistic interaction relieving the target of the effect of drug 1 would increase the effect of drug 2, so much so that in some situations viability of the target is actually reduced. Thus if synergistic drug combinations are used it is even more important than with single drugs to use a high enough dose for a long enough time to kill all agents and prevent the evolution of drug resistance.

Drug resistance is a particular concern with *Mycobacterium tuberculosis*, the causative agent of Tuberculosis: 1/3 of world population is infected with this bacterium, of those each year 8×10^6 develop tuberculosis. Each patient with tuberculosis produces on average 10–50 new infections, before they are either cured or die. Standard treatment is a regime of Isoniazid, Ethambutol, Rifampicin and Pyrazinamide developed in the 1950s, it takes between 6 and 9 months to complete and has considerable side effects. For these reasons many patients do not complete their regime, leading to the development of drug resistant bacteria. Currently about 5 % of all tuberculosis patients suffer from multidrug-resistant (MDR) bacteria (usually Isoniazid and Rifampicin), their treatment with second line drugs is 1 400-times more expensive than the standard regime. 2 % of the patients have XDR bacteria, which are also resistant against the second-line drugs. Currently, these bacteria are prevalent mainly in the states of the former Soviet Union, but with modern transcontinental travelling it is only a question of time before these bacteria spread further. Some charitable organisations currently fund tuberculosis research, but industrial efforts are virtually non-existent. Most victims would be unable to pay for newly developed, and hence expensive, drugs. This reminds me of what the great researcher and surgeon R. VIRCHOW said more than a hundred years ago: "When the rich turn away from the needs of the poor, microbes triumph!"

36.4 Stereoselectivity

One of the reasons to use enzymes as catalysts in industry is their stereoselectivity, that is, their ability to either distinguish between stereoisomeric versions of a substrate ((S) or (R) forms), or to produce only one stereoisomer from a prochiral substrate.

Especially for the food and pharmaceutical industry this is an essential ability: In the 1950s, thalidomide (see Fig. 36.5) was introduced mainly as a sedative, but also anti-inflammatory and anti-emetic drug. Because of the anti-emetic properties it was used by pregnant women against morning-sickness. This led to approximately 10 000 children born with severe malformation like phocomelia ("seal-limbs": hands directly attached to the shoulder or feet to the hips) and missing ears or eyes. Since intelligence was not affected, and many of these children have gone into professional careers. It came out later, that only (R)-thalidomide is an anti-emetic, whilst (S)-thalidomide can bind to the major grove of GC-rich segments of DNA. This reduces the expression of certain growth-factors, which control the formation

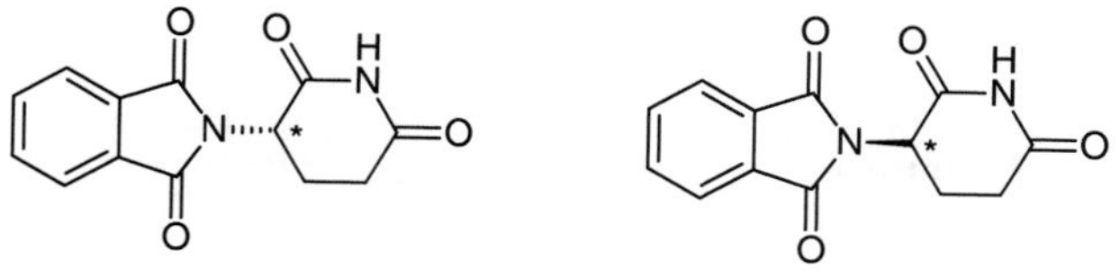

(S)-Thalidomide
(teratogenic)

(R)-Thalidomide
(anti-emetic)

Fig. 36.5 Thalidomide contains a stereoactive carbon and occurs in a (R)- and a (S)-form

of blood vessels in growing limbs [392]. Without oxygen and nutrients, the long bones in arms and legs can not form. Unfortunately, this effect is limited to primates and rabbits, and did not show up during safety testing in rats and mice. However, thalidomide has many beneficial effects in autoimmune-diseases, leprosy, AIDS, and certain forms of cancer. Under strictest supervision it is still in use. Research is ongoing to replace thalidomide with safer, yet even more powerful analogues.

This sad story created great interest in stereoselective synthesis, nowadays it would be next to impossible to bring a racemic drug onto the market. The NOBEL-Prize for Chemistry in 2001 was given to WILLIAM S. KNOWLES, RYOJI NOYORI AND K. BARRY SHARPLESS for their development of chemical methods for stereoselective synthesis. Ironically however, using the pure (R)-thalidomide would not have prevented the catastrophe, as thalidomide undergoes spontaneous racemisation at physiological pH.

The stereoselectivity of an enzyme can be defined either thermodynamically or kinetically:

$$E_{\mathrm{k}} = \frac{v_{\mathrm{S}}}{v_{\mathrm{R}}} \tag{36.13}$$

$$E_{\mathrm{t}} = \frac{(k_{\mathrm{cat}}/K_{\mathrm{m}})_{\mathrm{S}}}{(k_{\mathrm{cat}}/K_{\mathrm{m}})_{\mathrm{R}}}. \tag{36.14}$$

Sometimes, the selectivity is determined by starting with a racemic mixture as substrate, and then measuring the excess of one isomer in either the product or the remaining substrate:

$$e_{\mathrm{s}} = \frac{[S_{\mathrm{S}}] - [S_{\mathrm{R}}]}{[S_{\mathrm{S}}]_0 + [S_{\mathrm{R}}]_0}, \tag{36.15}$$

$$e_{\mathrm{p}} = \frac{[P_{\mathrm{S}}] - [P_{\mathrm{R}}]}{[P_{\mathrm{S}}] + [P_{\mathrm{R}}]} \tag{36.16}$$

as a function of the extend of the reaction

$$c = 1 - \frac{[S_{\mathrm{S}}] + [S_{\mathrm{R}}]}{[S_{\mathrm{S}}]_0 + [S_{\mathrm{R}}]_0}. \tag{36.17}$$

Then E becomes

$$E = \frac{\ln[1 - c * (1 + e_\mathrm{p})]}{\ln[1 - c * (1 - e_\mathrm{p})]} = \frac{\ln[1 - c * (1 - e_\mathrm{s})]}{\ln[1 - c * (1 + e_\mathrm{p})]}. \tag{36.18}$$

These methods yield the same E only under initial rate conditions, the third has the disadvantage of giving a high standard deviation for E. E-values can be quite different for different enzymes and substrates, for example the penicillin amidase of *E. coli* hydrolyses phenylglycine-*O*-methyl ester with $E = 0.5$, but N-acetylphenylglycine-*O*-methyl ester with $E = 13$ [39]. Stereoselectivity may also change with environmental conditions like pH, this can be used to gain information on enzyme mechanism. In particular, the selectivity depends on temperature:

$$\ln(E_\mathrm{t}) = -\frac{(\Delta G^{\ddagger}_\mathrm{S} - \Delta G^{\ddagger}_\mathrm{R}) - (\Delta G^{s}_\mathrm{R} - \Delta G^{s}_\mathrm{S})}{RT} \tag{36.19}$$

where ΔG^s is the change of free energy upon substrate binding and $\Delta G^{\ddagger}$ the change in free energy of converting the enzyme–substrate complex into the transition state $\mathrm{ES}^{\ddagger}$.

Chapter 37
Leaving the Steady State: Analysis of Progress Curves

The HENRI–MICHAELIS–MENTEN-law

$$v = \frac{d[P]}{dt} = -\frac{d[S]}{dt} = \frac{V_{max} \times [S]}{K_m + [S]} \tag{37.1}$$

can be integrated (assuming $[P]_0 = 0$ at $t = 0$), leading to:

$$\frac{[P]_t}{t} = V_{max} + K_m \times \frac{\ln\left(1 - \frac{[P]_t}{[S]_0}\right)}{t} \tag{37.2}$$

or, if the decrease in substrate is measured

$$\frac{[S]_0 - [S]_t}{t} = V_{max} - K_m \times \frac{\ln\left(\frac{[S]_0}{[S]_t}\right)}{t} \tag{37.3}$$

(HENRI-equation). Thus K_m and V_{max} can be determined from a single reaction, plotting $\frac{[P]_t}{t}$ as a function of $\frac{\ln\left(1-\frac{[P]_t}{[S]_0}\right)}{t}$ or, alternatively, $\frac{[S]_0-[S]_t}{t}$ as a function of $\frac{\ln\left(\frac{[S]_0}{[S]_t}\right)}{t}$. This would obviously result in great savings in time and material. For enzymes that use several substrates, all substrates except that being studied have to be held at near saturating concentration (say, $10 \times K_m$).

However, the precision for the parameters V_{max} and K_m determined this way is much lower than by the conventional variation of $[S]$. Also, since all measurements are obtained from a single experiment, they are highly correlated and normal least-square fitting techniques give artificially low error estimates for the parameters. Therefore, this method is rarely used. However, [83, 304] give details on it, if the need arises.

E. Buxbaum, *Biophysical Chemistry of Proteins: An Introduction to Laboratory Methods*, DOI 10.1007/978-1-4419-7251-4_37,

Chapter 38
Reaction Velocities

K_d and K_m values are relatively easy to determine, but the underlying rate constants often require special equipment to measure fast and very fast reactions. This is usually available only in specialised departments.

These techniques require the rapid mixing of reactants. Mixing times of about 1 ms can be achieved by tangential flow mixers, a block with three bores which meet but are slightly offset against each other. Injecting samples through two of these holes results in a highly turbulent flow. The sample jets are turned into thin layers from which the reactants can diffuse toward each other in a short time. The resulting mixture leaves the chamber through the third hole. Mixing dead time is determined by the volume of the apparatus and the efficiency of mixing, about 1 ms is achievable. Even shorter reaction times are accessible by the use of caged compounds and jump-methods.

Recently it has become possible to compare the bulk behaviour of an enzyme with that of single enzyme molecules [437]: A solutions of lactate dehydrogenase (10^{-17}M) turned over substrate in a 20 µm capillary. Each enzyme molecule was surrounded by a zone of NADH (product), the fluorescence of which was excited with an Ar-laser and followed over time under a microscope. Thus the distribution of activities could be determined. In [299] association and dissociation rates of myosin and fluorescently labelled ATP- and ADP-analogues were measured by stopped flow fluorescence in bulk and by TIRF-microscopy on single molecules. The results appear to agree reasonably well. Following single molecules in optical traps has been discussed in Sect. 2.1 on p. 23.

38.1 Near Equilibrium Higher Order Reactions can be Treated as First Order

For the simple second order enzymatic reaction

$$E + S \underset{k_{-1}}{\overset{k_1}{\rightleftharpoons}} X, \tag{38.1}$$

E. Buxbaum, *Biophysical Chemistry of Proteins: An Introduction to Laboratory Methods*, DOI 10.1007/978-1-4419-7251-4_38,

we get the following rate equation:

$$\frac{-\mathrm{d}[E]}{\mathrm{d}t} = k_1[E][S] - k_{-1}X \tag{38.2}$$

which is subject to the following restrictions from mass conservation:

$$[E]_0 = [E] + X, \tag{38.3}$$
$$[S]_0 = [S] + X. \tag{38.4}$$

Near equilibrium (that is, within about 10 % of equilibrium concentration) we can write

$$[E] = [E]_\mathrm{e} + \Delta[E], \tag{38.5}$$
$$[S] = [S]_\mathrm{e} + \Delta[S], \tag{38.6}$$
$$[X] = [X]_\mathrm{e} + \Delta[X] \tag{38.7}$$
$$-\Delta[X] = \Delta[E] + \Delta[S] \tag{38.8}$$

with $[E]_\mathrm{e}$ the equilibrium concentration and $\Delta[E]$ the deviation from equilibrium (and analogous for $[S]$ and $[X]$). If we combine these equations we get

$$\frac{-\mathrm{d}\Delta E}{\mathrm{d}t} = k_1([E]_\mathrm{e} + [S]_\mathrm{e}) + k_{-2}\Delta[E] + k_1[E]_\mathrm{e}[S]_\mathrm{e} - k_{-1}[X]_\mathrm{e} + k_1[E]^2 \tag{38.9}$$

Since $[E]^2$ is very small and $k_1[E]_\mathrm{e}[S]_\mathrm{e} = k_{-1}[X]_\mathrm{e}$ by definition we can simplify to

$$\frac{-\mathrm{d}\Delta E}{\mathrm{d}t} = \frac{\Delta[E]}{\tau} \tag{38.10}$$
$$\Delta[E] = \Delta[E]_0 * \mathrm{e}^{-(t/\tau)} \tag{38.11}$$

with τ the **relaxation time** and $\frac{1}{\tau} = k_1[E]_e + [S]_\mathrm{e} + k_{-1}$. If you determine τ as a function of the equilibrium concentrations and plot $\frac{1}{\tau}$ vs. $([E]_\mathrm{e} + [S]_\mathrm{e})$, the resulting line will have an intercept of k_{-1}, the slope is k_1. This procedure can be used even if the exact equilibrium is unknown, use an assumed K to get approximate values for k_1 and k_{-1}, from which better values for K can be calculated iteratively.

In a similar way, reactions of even higher order can be converted into first order reactions near equilibrium, making them amendable to study.

38.2 Continuous Flow

This was the first method used to describe fast processes developed by HARTRIDGE AND ROUGHTON in 1923 [147]. They wanted to measure the speed of oxygen binding to haemoglobin and used 2 vessels of 401 (sic) each with deoxygenated haemoglobin and oxygen equilibrated buffer, respectively. The solutions were driven by nitrogen pressure through the mixer into a glass tube of 1 cm diameter. At a flow rate of 0.5 l/s the flow was highly turbulent, ensuring that the sample moved as block through the tube. The detection of complex formation was by spectroscopy, the detector could be moved along the aging tube. A minimal distance of 5 mm was possible, equivalent to 1 ms of reaction time. From these truly heroic beginnings rapid kinetics moved by miniaturisation of equipment, by 1936 MILLICAN had managed to construct a continuous flow apparatus that used only 20 ml of sample volume [276]. Mass spectrometry allows particular sensitive detection of the reactants [430]. For a historic perspective on these techniques, see [54].

38.3 Quenched Flow

In quenched flow (see Fig. 38.1), the sample moves from the mixer into a long tube and from there into a test tube with a quenching solution, which stops the reaction immediately. The reaction time is equal to the time the sample needs to flow through

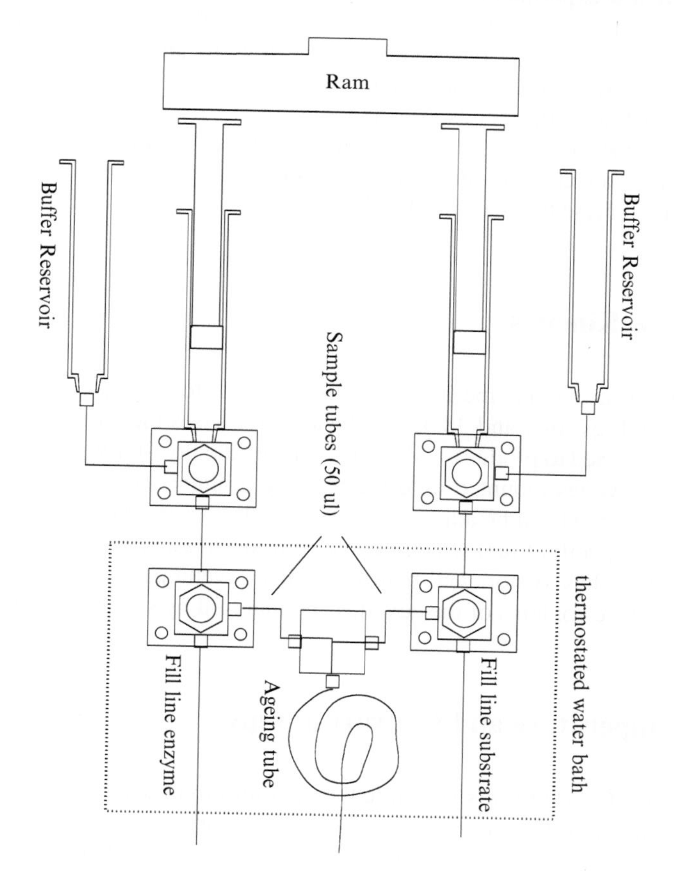

Fig. 38.1 Schematic diagram of a quenched-flow apparatus

the aging tube, which depends on linear flow rate and the length of the tube. The experiment is repeated several times, and each time a little bit of the tube is cut off, so that the reaction time becomes shorter. Compared to continuous flow, sample volumes are much smaller (several 10 μl in modern equipment).

One example for the utility of this method was the elucidation of the CALVIN-cycle, the dark reaction of photosynthesis [17]. An algal suspension was mixed with a solution of radio-labelled bicarbonate and pushed through the ageing tube. The samples were collected in boiling ethanol as quenching solution. The radioactive compounds formed were then separated by 2-D thin layer chromatography. As reaction times became shorter, fewer radioactive spots appeared on the chromatogram, until only one compound remained (3-Phosphoglycerate).

38.4 Stopped Flow

In a stopped flow experiment, the sample is injected after mixing into a flow-through cuvette, and from there into a stop-syringe (see Figs. 38.2 and 38.3). Filling of that syringe pushes its plunger out, switching the flow off and starting measurements (usually by fluorescence, but absorbtion or CD may also be used). Thus measurement starts a few milliseconds after the cuvette has been filled with the reaction mixture, and measurements are made every millisecond or so (highly advanced spectrometers can measure an entire spectrum every ms!). Here too ESI-MS has been used to analyse the reactants [205].

38.5 Flow Kinetics

In flow kinetics, the enzyme is immobilised on a filter at the exit of the mixing chamber (see Figs. 38.4 and 38.5). As the sample passes the enzyme, the substrate is bound, converted to product, and finally appears the flow-through which is caught in a line of cuvettes mounted onto the rotating turntable of a vinyl record player (their rotation speed can be adjusted very precisely). Each cuvette collects the flow through for a specific time, which depends on the rotation speed of the turntable. The liquid flow has to stop well before the first cuvette passes the outlet a second time. Finally, the contents of the cuvettes are analysed by suitable methods.

38.6 Temperature and Pressure Jumps

Following LE CHATELIER's principle the equilibrium of a system will change following application of an outside force in such a way as to counteract that force. For example, if the system is heated, the equilibrium of a chemical reaction will

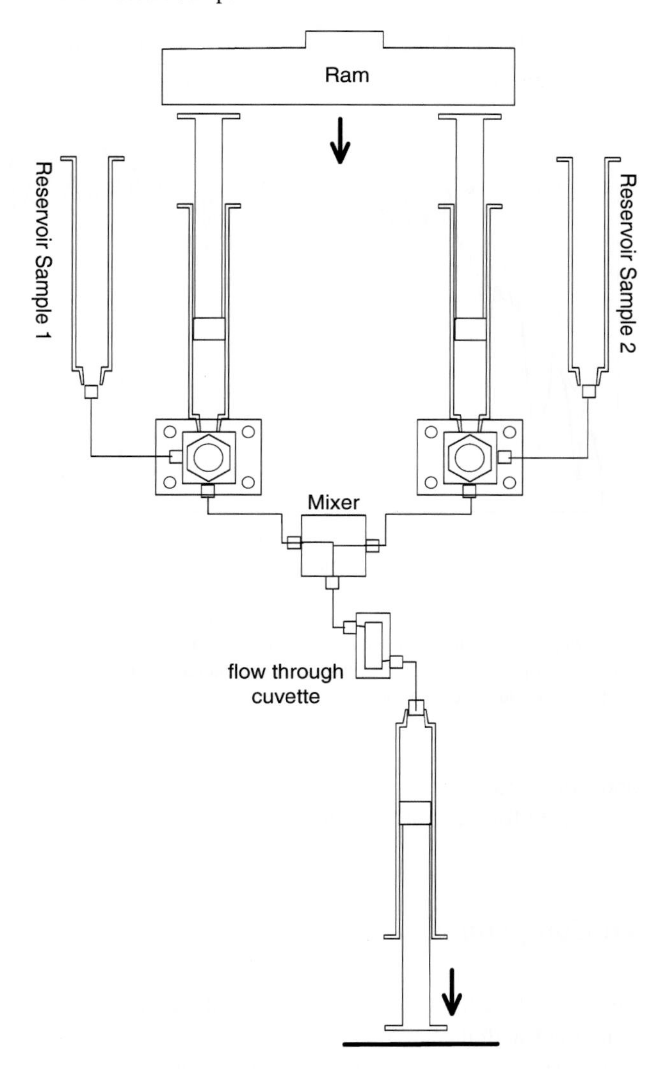

Fig. 38.2 Schematic diagram of a stopped-flow apparatus

change in the endergonic direction. Rapid pressure changes can change the equilibrium in reactions accompanied by a change in volume. The advantage of such jump-methods is that they do not require mixing, thus their dead-time is extremely short (in the order of microseconds).

Temperature increases, for example, can be achieved by discharging a high-voltage capacitor through the sample. The energy added to the system is $E = 1/2\ C\ U^2$ with C the capacity of the capacitor and U the voltage used. The time constant ($\approx$5–10 µs) is $\tau = \frac{RC}{2}$ with R the resistance of the solution, which can

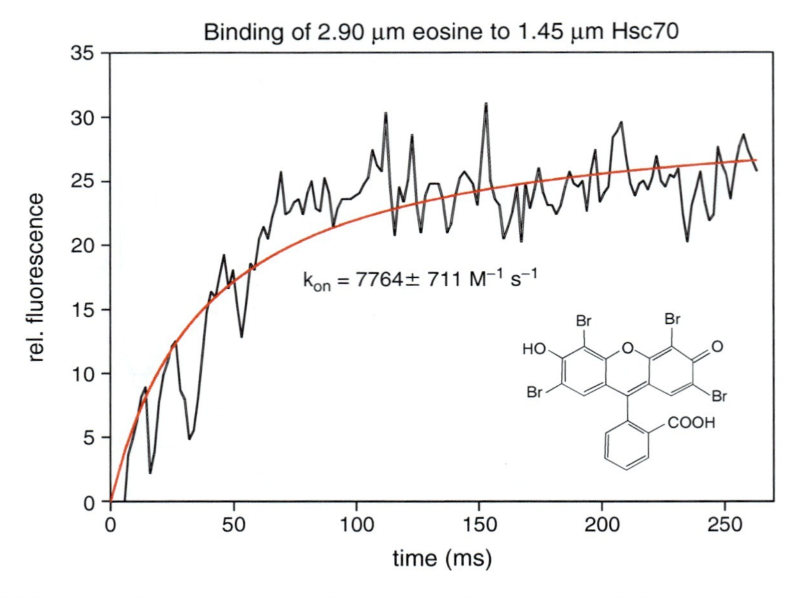

Fig. 38.3 Stopped flow measurements of the binding of eosine to Hsc70. Eosine is a fluorescent dye, which binds to the ATP-binding site of many ATPases. The change to a more hydrophobic environment increases fluorescence intensity. Note that the data are fairly noisy, which is a typical problem with stopped-flow fluorescence measurements

adjusted by addition of an inert electrolyte (typically $\approx 100\,\Omega$). With that type of experiment monitoring for enzyme denaturation is even more important than usual.

38.7 Caged Compounds

If a substrate is bound to a light-sensitive compound (see Fig. 38.6), the derivative may no longer interact with its enzyme. If the mixture of this "caged" substrate with the enzyme is exposed to a sudden light-flash (say, a laser pulse), the substrate is liberated by photolysis and can react with the enzyme. This allows the concentration of the substrate to be raised suddenly (laser pulses can be as short as a few ps). The substrate concentration can be controlled by the intensity of the laser flash.

38.8 Surface Plasmon Resonance

Surface plasmon resonance (SPR) is a quantum mechanical effect which is used to accurately measure the refractive index of a medium close to a surface [216, 358, 359]. Binding of ligands to a receptor on that surface increases the refractive

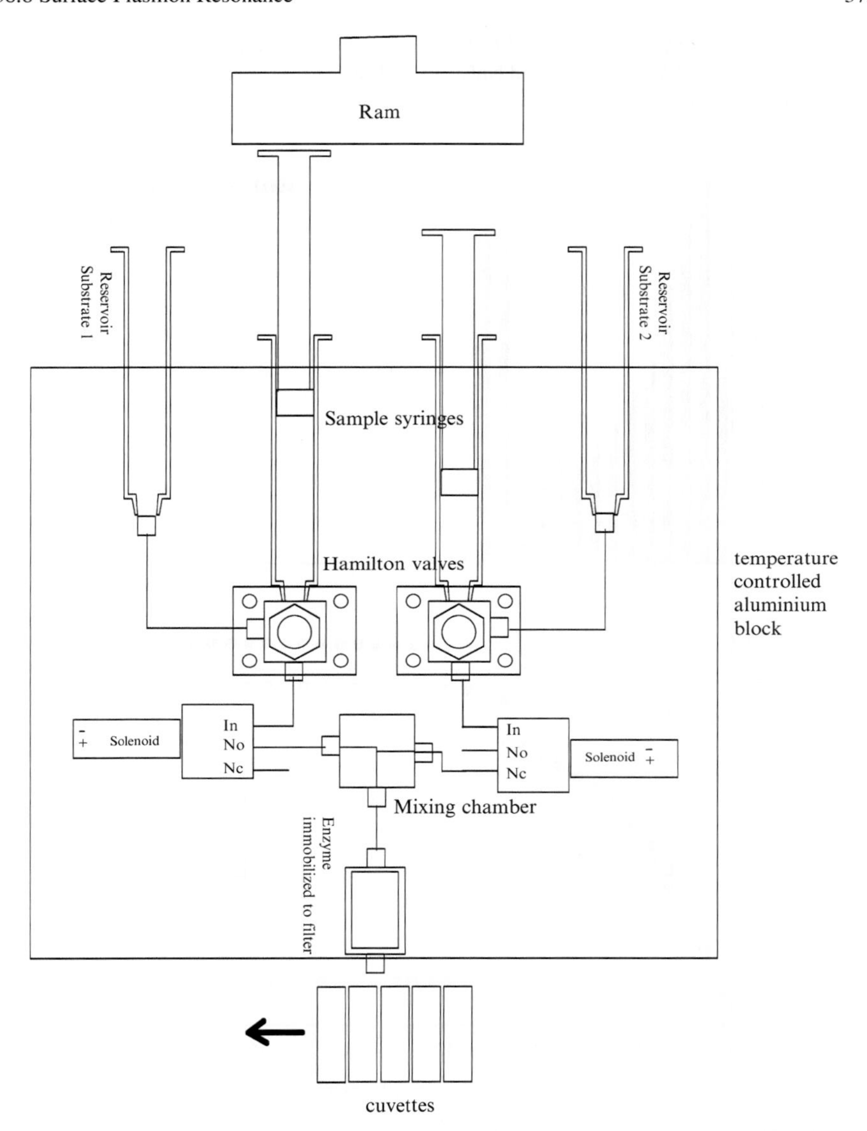

Fig. 38.4 Schematic diagram of a flow-kinetics apparatus. Note that the pistons of the sample syringe are at different heights, the composition of the sample changes during the experiment. When the ram reaches the second syringe, a micro-switch is pressed to either reduce the speed of the ram (in which case the enzyme is exposed to a mixture of the contents of both syringes) or to open a solenoid valve to discard the solution coming from the first syringe (exposing the enzyme to the content of syringe 2 only)

index, dissociation decreases it. Thus SPR can be used to measure k_{on} and k_{off}; from the ratio of k_{on} and k_{off} we get K_d. SPR can also be used to excite fluorescence, fluorescence intensity is then proportional to the distance between the fluorophore and the surface [408, and references therein].

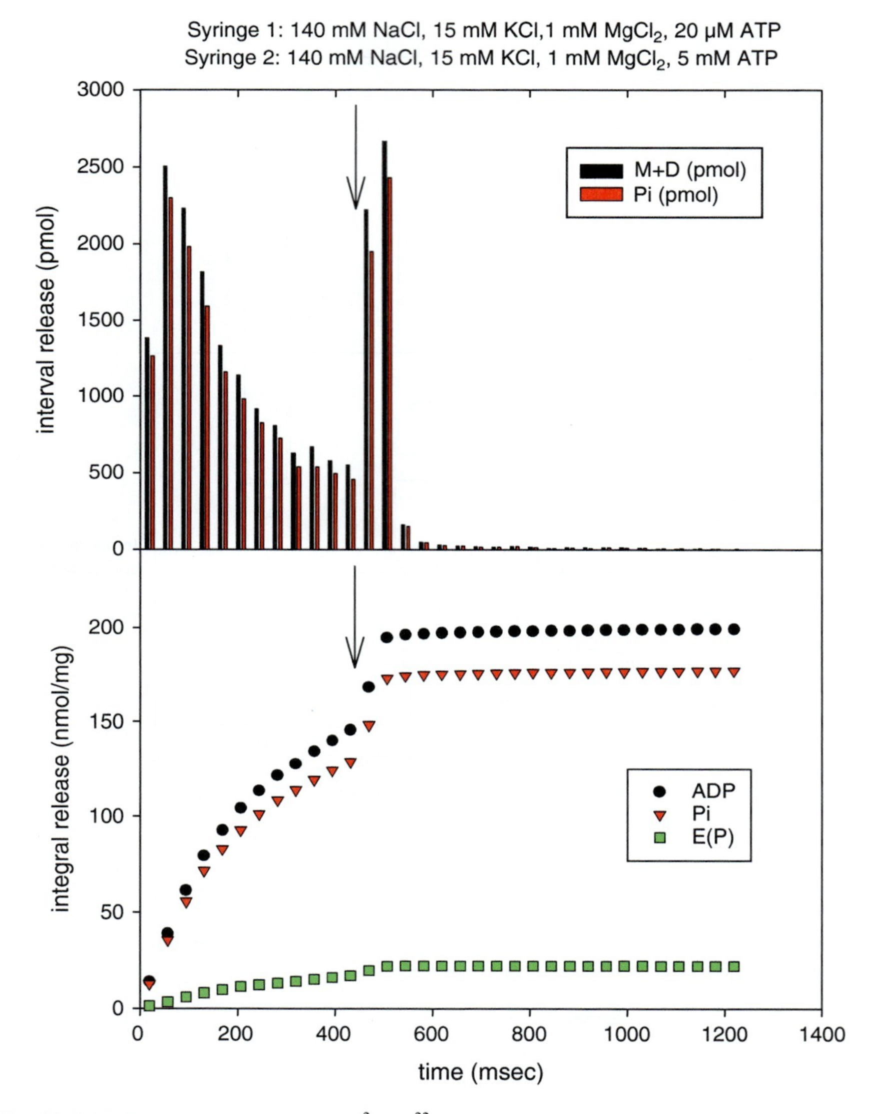

Fig. 38.5 Na/K-ATPase reaction with [^{3}H, γ^{32}P]ATP. The top diagram shows the concentration of ADP and P_i in the cuvettes, the lower diagram the integrated release (sum up to that time point). Note that more ADP is released than P_i, the remaining phosphate is bound to the enzyme (dubbed E(P) since we can not distinguish high energy $E_1{-}P$ and low energy $E_2 \cdot P$ in this experiment). The experiment starts with a low ATP concentration (only E_1-site occupied). The second syringe (*arrow*) contains a high concentration of unlabelled ATP, which also saturates the low-affinity E_2-site. This results in accelerated splitting of labelled ATP bound to the E_1-site, further evidence for coexistence and cooperation of E_1- and E_2-sites in P-type ATPases

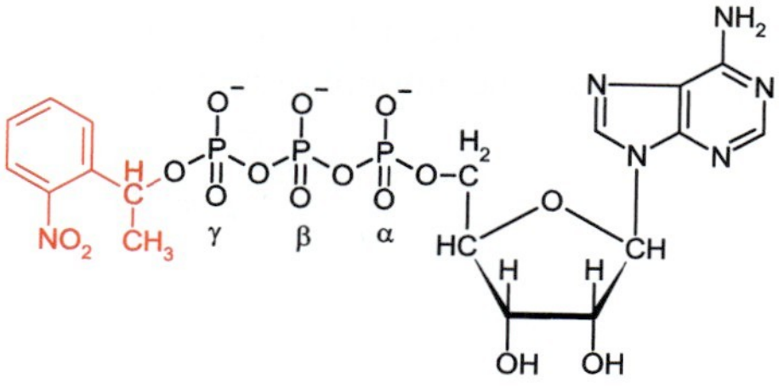

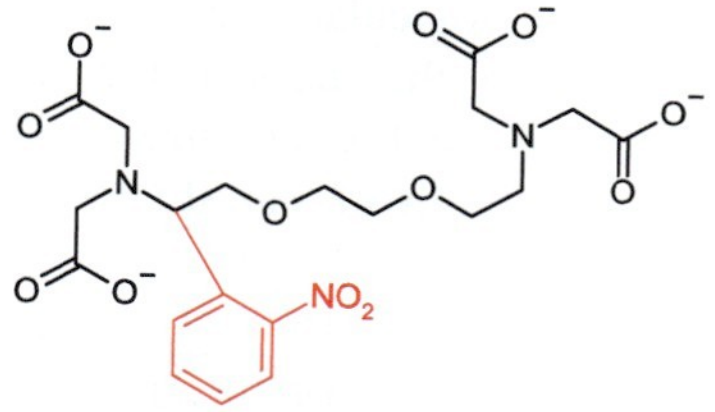

Fig. 38.6 Caged ATP does not bind to most ATP-utilising enzymes. Upon exposure to a laser flash, the nitrophenylethyl-group is photolysed and free ATP becomes available. 2-Nitrophenyl-EGTA binds Ca-ions which are released when the organic molecule is destroyed by photolysis (caged Ca^{2+})

P3-[1-(2-Nitrophenyl)ethyl]-ATP

2-Nitrophenyl-EGTA

Since the receptor is not destroyed during measurement, many ligands can be successively tested once the system has been set up, each measurement takes a couple of minutes. Thus SPR has become a favourite method for large scale screening in the pharmaceutical industry.

The only limitation to this technique is that the molecular mass of the ligand must be large enough to ensure that binding increases the refractive index enough to give a useful signal ($>$1 000Da).

38.8.1 Theory of SPR

The conducting electrons in metals are freely mobile, like molecules in a gas. At the surface of metals, density fluctuations of this plasma can be excited, these are called **plasmons**. In practice, there are two methods for exciting plasmons:

Electrons passing through a thin foil are used in material science and electrochemistry, but have few biological applications.

Light passing in a prism coated with a metal film (usually gold, sometimes silver), whose thickness is $<\lambda$. At a particular angle of incidence the wave vector of the evanescent field (see p. 29) matches the wave vector of the plasmons and energy is transmitted from the photons to the plasmons. At this angle a sharp minimum of transmittance is observed. In effect the electrons of the metal try to move in the electric field of the light beam, but cannot quite keep up because of the high frequency (about 1×10^{15} Hz). Thus the metal becomes transparent.

There is a direct relationship between the refractive index n and the mass density ρ, hence the amount of a pure compound deposited can be calculated:

$$\rho = M_r * N = \frac{M_r}{A\frac{(n^2-1)}{(n^2+2)}} \tag{38.12}$$

$$m = 0.1\frac{M_r}{A\frac{(n^2-1)}{(n^2+2)}} \tag{38.13}$$

With M_r = molecular mass, N = number of moles per unit volume, A = molar refractivity, d thickness of the layer (in nm), m = absorbed mass (μg/cm) and n = refractive index. For mixtures of compounds, the change in refractive index is given by the LORENTZ–LORENZ-equation:

$$\frac{(n^2-1)}{(n^2+2)} = A_1N_1 + A_2N_2 + \cdots + A_nN_n \tag{38.14}$$

For two-component systems (buffer/protein, buffer/lipid) such absolute calculations may still be feasible (try corninfo.chem.wisc.edu for a program to calculate reflectivity curves), but in real systems relative changes of refractive index are often used for characterisation.

The change of the angle at which reflectance is minimal, θ_{SP} is usually measured in relative units RU, 1 RU is defined as a change of θ_{SP} of 0.0001°.

Absorbance by the material deposited near the glas/metal-interface will change the SPR-spectrum, allowing substance identification on top of concentration determination.

38.8.2 Practical Aspects

Usually SPR is measured in the KRETSCHMANN-configuration (see Fig. 38.7), where the metal film is deposited directly onto the glass prism (or to a slide optically coupled to the prism by an index matching fluid). In the OTTO-configuration, which is rarely used in biochemistry, there is an air gap between prism and metal film, which has a thickness of about one wavelength.

For the metal film, gold or silver are the most commonly used materials. Silver has a narrower SPR-spectrum, resulting in higher sensitivity. However, because of its inertness, gold is more frequently used. It is also relatively easy to cross-link proteins to a gold surface, using sulphur groups in the cross-linker.

SPR-imaging is used to perform SPR-measurements in an array format, that is with many samples at the same time. Sample throughput is increased accordingly. The reflected light from such an array is collected on a CCD-camera and then processed by image analysis software.

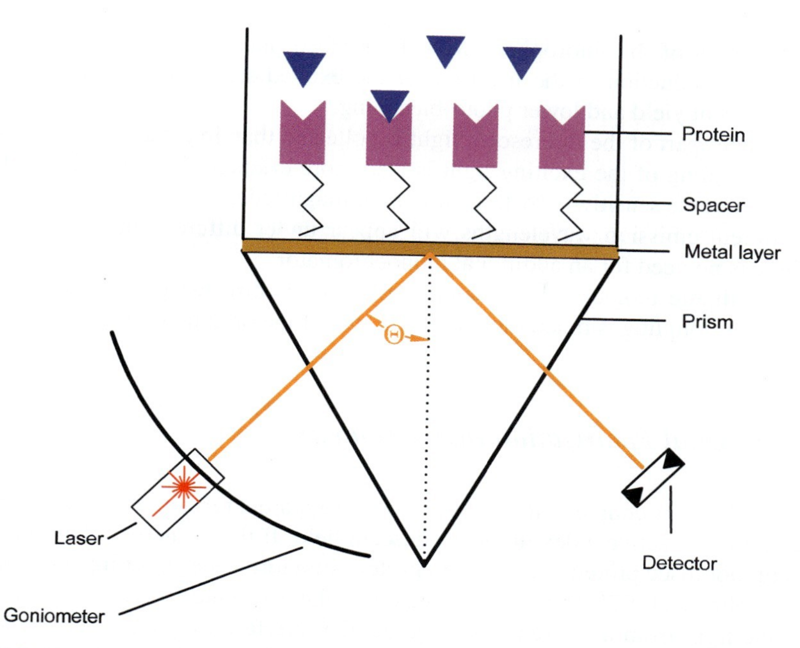

Fig. 38.7 KRETSCHMANN setup for surface plasmon resonance experiments. SPR is excited in a thin metal film that covers a glass prism. On the surface of the metal film the protein of interest is bound, for example, via chemical cross-linkers (as shown here) or embedded into a lipid membrane. The total volume of the sample compartment is only a few microliters. Light source is a laser ($\approx$800 nm, the intensity of reflected light is recorded by a detector. The angle of incidence of the laser light (α) is changed by moving the laser on a goniometer. Alternatively, a diode array detector may be used to measure the reflected light from a focussed laser beam, in this case no moving parts are required. The SPR-spectrum is the plot of reflectance as a function of either wavelength at constant incidence angle or incidence angle at constant wavelength. Because tunable lasers are expensive, the latter method is used more frequently

38.8.3 Surface Plasmon Coupled Fluorescence

If the incidence beam can excite surface plasmon resonance and an evanescent field, then an excited fluorophore interacting with a surface plasmon should in turn create a beam of light leaving the prism in a hollow cone with an opening angle of $2 * \theta_{SP}$ for the emitted wavelength [13, 235]. Note that it is irrelevant whether the fluorophore was excited by an evanescent wave (KRETSCHMANN-configuration) or by light from the opposite side of the prism (reverse KRETSCHMANN-configuration). In the former case only molecules near the surface will be excited, in both the latter and the former only emissions from molecules near the surface are detected. Either way, fluorescence is detected only from fluorophores near the surface of the KRETSCHMANN-prism (20–200 nm), autofluorescence from the remainder of the sample does not interfere. Measuring fluorescence in this way has a number of additional advantages:

- Interaction of the fluorophore with the surface plasmon on the metal surface causes a reduction in the life time of the excited state. This results in higher fluorescent yield and lower photo-bleaching.
- A greater part of the fluorescent light is collected than in a standard fluorimeter, and coupling of the exciting light is also more efficient. Together these effects may increase sensitivity by three orders of magnitude.
- Different emission wavelengths will appear under different angles θ_{SP}, hence there is no need for an additional monochromator.
- If the dipole moment of the sample molecule is directed perpendicular to the surface, coupling is much more efficient than if the orientation is parallel.

38.8.4 Dual Polarisation Interferometry

If a laser beam is coupled into a waveguide the apparent refractive index depends in part on the refractive index in the evanescent field. If the evanescent field interacts with immobilised proteins, changes in protein structure – *e.g.*, after ligand binding – can be followed [397]. The light coming out of the waveguide is allowed to interfere with the light from a reference waveguide, the interference pattern is detected by an electronic camera. Fourier-analysis of the pattern is performed in a computer. A ferroelectric liquid crystal is used to switch the polarisation of the laser light from parallel to the face of the waveguide to vertical. Since the evanescent field profile depends on polarisation, both density and size of the protein can be determined.

38.9 Quartz Crystal Microbalance

The detector in a quartz crystal microbalance (QCM) is a quartz crystal cut in "AT"-direction. Such crystals are **piezoelectric**, *e.g.*, the application of a voltage across the y-direction leads to a shear wave (also in y-direction) and a displacement in x-direction and *vice versa*. The quartz crystal is a round disk with electrodes at both surfaces. The electrodes are usually gold (because it is inert) "glued" to the crystal by a thin chromium layer.

Application of an AC-voltage leads to vibration in the crystal which is strongest at its resonance frequency. Since the shear force at both ends of the crystal must be zero, at the fundamental resonance frequency the standing wave has one node and a wavelength λ of twice the thickness of the crystal D:

$$\lambda = 2D, \tag{38.15}$$

$$\tau = \frac{2D}{v} \tag{38.16}$$

with τ the period and v the velocity of the wave in the crystal. Resonance will also occur at harmonics of the fundamental frequency with n nodes and $n + 1/2$ wavelengths across the y-direction of the crystal:

$$\tau_n = \frac{2D}{(n+1)v}. \tag{38.17}$$

If the thickness is increased by a stiff (=thin) layer of deposited material the resonance frequency is decreased by

$$\frac{\Delta\tau}{\tau} = \frac{\rho_{\mathrm{m}}\Delta D}{\rho_{\mathrm{q}} D} \tag{38.18}$$

with ρ_{m} and ρ_{q} the densities of the deposited material and quartz, respectively. This equation describes the system well up to a frequency change of 10 %. In his paper describing the QCM SAUERBREY [365] used a different equation to describe the decrease in resonance frequency F as a function of deposited mass m:

$$\Delta m = -\frac{c\Delta\nu}{n} \tag{38.19}$$

c is a constant for the given quartz, about 17.7 ng Hz^{-1} cm^2 for a 5 MHz quartz. This SAUERBREY-equation is still used despite being less accurate than (Eqn. 38.18).

As the film becomes thicker, coupling of the deposited material to the crystal becomes less direct and the film becomes viscoelastic. Such a film is characterised by its dissipation ζ:

$$\zeta = \frac{E_{\mathrm{lost}}}{2\pi E_{\mathrm{stored}}} \tag{38.20}$$

If the crystal is vibrated at its resonance frequency and the AC switched off, the amplitude of the free oscillation will decrease with an e-function. Measuring the speed of decrease at different harmonics allows both the thickness and the viscosity of the surface layer to be calculated.

Chapter 39
Isotope Effects

For a brief history of the field and more detailed discussions of the method see [65], [62–64, 68]. Isotope effects can be used to investigate reaction mechanism and transition state structure.

Enzymes are substrate specific, so much so that they can differentiate between different isotopes in a substrate molecule. Measurement of the strength of the isotope effect can tell us something about the transition state structure of the enzyme–substrate complex.

Labelling a substrate with heavy isotopes can change the rate constant (**kinetic isotope effect**) and/or the equilibrium constant (**equilibrium isotope effect**). We define

$$^{x}k = \frac{k_{\mathrm{n}}}{k_{\mathrm{l}}} \tag{39.1}$$

$$^{x}V = \frac{V_{\mathrm{n}}}{V_{\mathrm{l}}} \tag{39.2}$$

$$^{x}K = \frac{K_{\mathrm{n}}}{K_{\mathrm{l}}} \tag{39.3}$$

$$^{x}\left(\frac{V}{K}\right) = \left(\frac{V}{K}\right)_{\mathrm{n}} \Big/ \left(\frac{V}{K}\right)_{\mathrm{l}} = \frac{V_{\mathrm{n}}K_{\mathrm{l}}}{V_{\mathrm{l}}K_{\mathrm{r}}} \tag{39.4}$$

with x = D, T, 13, 14, 15, 17 or 18 for ^{2}H, ^{3}H, ^{14}C, ^{15}N, ^{17}O and ^{18}O. k_{n} is the rate constant for non-labelled, k_{l} for the labelled substrate, similar for the reaction velocity, the equilibrium constant, and the catalytic constant.

The change in the equilibrium constant can be larger or less than unity, depending on whether the isotope is bound more strongly in the reactant or product.

A **primary isotope effect** occurs when the bond to the isotope is created or broken during the reaction, a **secondary isotope effect** occurs when the strength of the bond to the isotope changes in the reaction. The isotope effect is strongest when it occurs in the rate limiting step of a reaction, if other steps are partially rate limiting the isotope effect is masked.

E. Buxbaum, *Biophysical Chemistry of Proteins: An Introduction to Laboratory Methods*, DOI 10.1007/978-1-4419-7251-4_39,

The isotope effect is proportional to the change in reduced mass

$$\mu = m_a m_b/(m_a + m_b) \tag{39.5}$$

with m_a and m_b the mass of the atoms involved in a bond.

The reason for the higher speed of reactions with lower mass isotopes is that the vibration frequency of the bond depends on the mass of the atoms connected; it becomes lower as atoms become heavier:

$$\nu = \frac{1}{2\pi}\sqrt[2]{\frac{k}{\mu}} \tag{39.6}$$

for a di-atomic molecule. ν is the fundamental vibrational frequency of a harmonic oscillator, k its spring constant. In quantum mechanics the energy E_n of the nth harmonic is given by:

$$E_n = h\nu\left(n + \frac{1}{2}\right) \tag{39.7}$$

In other words, the **zero-point energy** (energy of the ground state, $n = 0$) of the bond decreases as we replace an atom by a heavier isotope, and more activation energy is required to break the bond. According to the ARRHENIUS equation that results in lower reaction velocity (see p. 409).

By going from a C–H to a C–D or C–T bond the spring constant k stays the same, but the reduced mass goes from 0.92 Da to 1.71 Da or 2.40 Da, respectively.

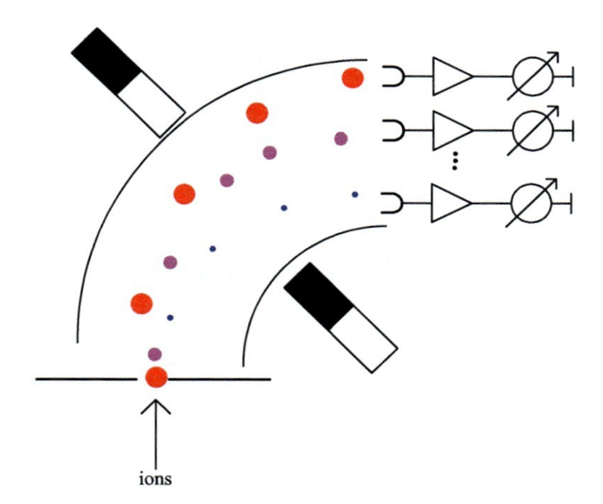

Fig. 39.1 The isotope ratio mass spectrometer allows the determination if isotope ratios with high precision. The sample is combusted or pyrolysed, the gas to be determined (*e.g.*, H_2O, N_2, CO_2, SO_2) purified and ionised. The ioniser has a double inlet, so that sample gas and a reference gas with known isotope ratio can be analysed in rapid succession. The ions are separated in a magnetic sector analyser, the various beams hit a row of FARADAY-cups. The current through these cups is amplified and measured

Thus the zero-point energy changes by $\sqrt{1.71/0.92} = 1.36$ for deuterium and 1.62 for tritium. By comparison changing a C$-$^{12}C bond to C$-$^{13}C changes the reduced mass from 6 Da to 6.24 Da and the zero-point energy by a factor of 1.02.

In addition, tunnelling can significantly increase the kinetic isotope effect, especially with hydrogen.

Several methods are used to measure isotope effects:

- Direct comparison of the kinetics of labelled and unlabelled substrate.
- Equilibrium perturbation by mixing labelled substrate with unlabelled product of a freely reversible reaction.
- Internal competition by mixing the unlabelled substrate with a radioactive trace label and measuring the isotope ratio in the product as a function of reaction time. This is done with a specialised instrument, the **isotope ratio mass spectrometer** (see Fig. 39.1). The reaction must be (made) irreversible.

Chapter 40
Isotope Exchange

Some reactions proceed via an intermediate that is stable enough to be detected with sensitive isotope techniques.

One example is the transfer of the γ-phosphate group to aspartic acid in P-type ATPases, forming an acyl-phosphate intermediate, from which ADP is released followed by dephosphorylation:

$$E \underset{}{\overset{ATP}{\rightleftharpoons}} E^{P}_{ADP} \overset{ADP}{\rightleftharpoons} E^{P} \overset{H_2O}{\rightleftharpoons} E \cdot P_i \overset{P_i}{\rightleftharpoons} E$$

This reaction can be studied by several types of isotope exchange experiments:

40.1 ADP/ATP Exchange

$$E \overset{ATP}{\rightleftharpoons} E^{P}_{ADP} \overset{ADP}{\rightleftharpoons} E^{P} \overset{[^3H]ADP}{\rightleftharpoons} E^{P}_{[^3H]ADP} \overset{[^3H]ATP}{\rightleftharpoons} E$$

The radioactive ATP formed is determined by scintillation counting after chromatographic separation from ADP. Evaluation is complicated by the fact that ATP is constantly hydrolysed to ADP, which can be controlled for by measuring the rate of hydrolysis of [γ-^{32}P]ATP to $^{32}P_i$ [52].

40.2 ^{18}O-Exchange

$$E \underset{H_2O}{\overset{P_i}{\rightleftharpoons}} E-P \overset{H_2{}^{18}O}{\rightleftharpoons} E \cdot [^{18}O]P_i \overset{[^{18}O]P_i}{\rightleftharpoons} E$$

The amount of ^{18}O-labelled phosphate is determined by mass spectrometry or by ^{31}P-NMR.

E. Buxbaum, *Biophysical Chemistry of Proteins: An Introduction to Laboratory Methods*, DOI 10.1007/978-1-4419-7251-4_40,

40.3 Positional Isotope Exchange

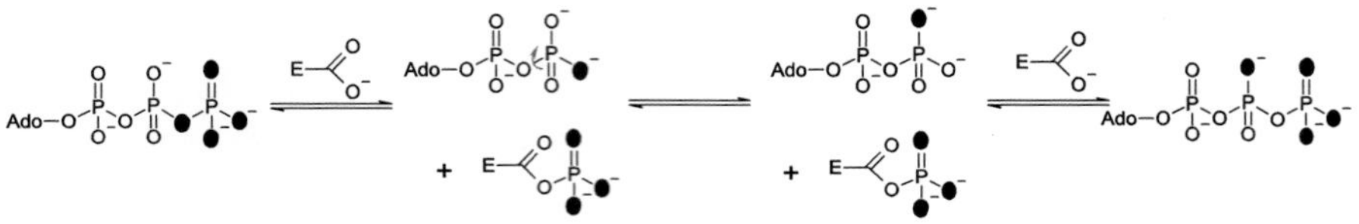

Rotation of the β-phosphate group during reversible acyl-phosphate formation leads to ATP-molecules with a ^{18}O (filled oxygens) in a different position. Detection is by ^{31}P-NMR [345].

Investigations on **partial reactions** like these led to the ALBERS-POST-model for the reaction mechanism of P-type ATPases [259].

Part VII
Protein–Ligand Interactions

Assuming the diameter of an animal cell to be about 10 μm, the minimal concentration of a protein (when this protein is present only in a single copy) would be about 3 pM. For an interaction with such a protein to be physiologically relevant, the K_d would have to be of the same magnitude. On the other hand, some proteins can make up several percent of the total protein of a cell, achieving concentrations of about 1 mM. For such abundant proteins low affinity interactions with K_d in the mM range would still be physiologically relevant. Thus we need methods that can measure K_d over nine orders of magnitude!

Interaction of proteins is key to their function, the **interactome** of cells is a hot research topic [112], a recent review about possible medical applications may be found in [123]. In a connectivity network proteins are presented as nodes, interactions as edges. Such diagrams can reveal signalling pathways, as proteins of a pathway have to interact with each other. It can also reveal central proteins (those with a particularly large number of interacting partners), deficiencies in such hub proteins is often more critical than those in less-connected ones (**centrality–lethality rule**, although there are many exceptions).

40.3.1 Structural Aspects of Protein–Protein Interactions

Hub proteins have a flexible structure which allows interactions with many partners. Such flexible regions are often **intrinsically disordered**, that is, they have relatively little secondary structure when isolated, but can adapt to the shape of the binding site of their partner, with the binding energy driving the conformational change. This represents a technical challenge to protein biochemists, as intrinsically disordered proteins are difficult to isolate (easily digested by proteases), difficult to crystallise and, once crystallised, do not give good diffraction patterns unless bound to a partner. Intrinsically disordered proteins are also involved in the formation of amyloid, precipitates of misfolded proteins involved in about 40 debilitating human diseases, including ALZHEIMER's dementia, PARKINSON's disease, and type II diabetes. In all cases, misfolded proteins auto-catalytically nucleate the conversion of their correctly folded peers into β-helices (see Fig. 34.7 on p. 318), that aggregate and precipitate. The inclusion bodies formed in genetically engineered bacteria overexpressing proteins have a similar structure.

About one-third of all proteins have to form **homo-oligomers** to function. Oligomerisation gives additional stability against denaturation and proteolysis, and additional options for regulation (*e.g.*, co-operativity). However, **dominant negative inheritance**, where one malformed partner renders the entire oligomer non-functional, is a significant disadvantage of oligomerisation.

Binding sites for protein–protein interactions are on average about 1 500 $Å^2$, but may be half or twice as big. Stable interactions are often driven by hydrophobic bonds, while in temporary interactions the binding sites are more polar and the binding energy results from ionic or hydrogen bonding. After all, proteins that undergo temporary interactions have to be stable and soluble without their partners. Only a few amino acids in the binding site are actually involved in bond formation, these **hot spots** are often Trp, Arg and Tyr.

We have discussed already the use of surface plasmon resonance, fluorescence, and absorbtion spectroscopy to monitor binding. However, many ligands bind without changes in absorbtion or fluorescence, and in surface plasmon resonance one of the partners has to be immobilised which may change its properties. In this chapter we take a look at some important techniques to measure molecular interactions.

Chapter 41
General Conditions for Interpretable Results

In principle, any technique that separates bound from unbound ligand, or occupied from unoccupied protein, can be used to monitor binding. As a general rule, highest accuracy is achieved when the concentration of the protein in binding studies is approximately equal to K_d (within about one order of magnitude either way), this makes it easier to determine the concentration difference between bound and total ligand. Since the supply of protein is usually a limiting factor in research, binding studies are expensive and the sample size must be reduced as much as possible. Considerable progress has been made in this direction over the years.

A sufficient number of samples (at least 12) evenly distributed over a sufficiently wide concentration range (0.1–5 K_d) should be collected [354]. Otherwise, deviations from a hyperbolic behaviour (co-operativity) can not be detected reliably. If co-operativity is detected, the concentration range (and hence the number of samples) needs to be increased further, in that case it can be useful to space the data points logarithmically (*e.g.*, 1.0, 1.2, 1.4, 2.5, 4.0, 6.3, 10, 12...). Note that data covering such wide ranges can not be curve-fitted by the least-squares criterion, which would virtually ignore data with small y-value. χ^2 is an appropriate criterium for fitting such data, this cannot be done with the MARQUARDT–LEVENBERG, but for example with the `Simplex`-algorithm of NELDER AND MEAD. See [337] for details on these methods.

Sometimes this leads to ligand concentrations much lower than the concentration in the cell (many transport ATPases show interesting behaviour down to 10 nM ATP, while the physiological concentration is 4 mM). When presenting the results you may get asked why you measured such "unphysiological" concentrations. Enzyme kinetics tries to unravel the reaction mechanism of an enzyme, not the physiology of the ligand. Low ligand concentrations can be very relevant indeed.

Great care should be taken to ensure that the data collected have as small a standard deviation as possible (≤ 2 %), only then can small deviations from hyperbolic kinetics be detected.

Stability of both enzyme and ligand in the respective stock solutions over measurement time must be ensured. For many ligands, stability is pH-dependent, for example CoA and $NAD(P)^+$ are most stable at pH 4 (acetate buffer), while NAD(P)H is more stable in slightly alkaline solutions (bicarbonate pH 8). ATP is most stable around pH 9. Buffer concentration should be low so that addition

E. Buxbaum, *Biophysical Chemistry of Proteins: An Introduction to Laboratory Methods*, DOI 10.1007/978-1-4419-7251-4_41,

of substrate does not change the pH of the reaction mix. Check the enzyme activity both at the beginning and end of a measurement series. Additionally, if you measure the effect of ligand concentration on activity, first use a series of increasing concentrations, then repeat the measurements with decreasing concentrations. An increasing gap between corresponding values indicates inactivation.

Chapter 42
Binding Equations

42.1 The Langmuir-Isotherm: A Single Substrate Binding to a Single Binding Site

We are looking at the following reaction: $\mathsf{R} + \mathsf{L} \rightleftharpoons \mathsf{RL}$. Since for binding experiments the concentration of protein needs to be much higher than for the determination of enzyme turnover to obtain a measurable signal, the free substrate concentration ($[F]$) can no longer be assumed to be equal to the total substrate concentration ($[T]$). Thus the binding equation needs to be modified:

$$
\begin{aligned}
K_\mathrm{d} &= \frac{[R][L]}{[RL]} = \frac{([R]_\mathrm{t} - [RL])([L]_\mathrm{t} - [RL])}{[RL]} \\
&= \frac{[R]_\mathrm{t}[L]_\mathrm{t} - [R]_\mathrm{t}[RL] - [L]_\mathrm{t}[RL] + [RL]^2}{[RL]} \\
0 &= -[RL]^2 + [RL](K_\mathrm{d} + [R]_\mathrm{t} + [L]_\mathrm{t}) - [R]_\mathrm{t}[L]_\mathrm{t} \\
[RL] &= \frac{([R]_\mathrm{t} + [L]_\mathrm{t} + K_\mathrm{d}) - \sqrt[2]{([R]_\mathrm{t} + [L]_\mathrm{t} + K_\mathrm{d})^2 - 4[R]_\mathrm{t}[L]_\mathrm{t}}}{2}
\end{aligned} \tag{42.1}
$$

Note that of the two solutions of the quadratic equation only one is physically meaningful, as there are no negative concentrations. This equation is called the Langmuir-isotherm, because Langmuir derived a similar equation for the binding of gases to a solid matrix if the temperature was kept constant. If the experiment was repeated at different temperatures, a set of binding curves (=isotherms, from Gr.: equal temperature) was obtained.

The other alternative is to calculate $[L]$ as $[L]_\mathrm{t}$-$[RL]$ and to plot $[RL]$ as a function of $[L]$ rather than $[L]_\mathrm{t}$. However, this leads to a mixing of dependent and independent variables, which has unpleasant statistical implications. Thus a plot of the observed variable $[RL]$ as a function of the controlled variable $[L]_\mathrm{t}$ is the better option.

Binding isotherms for more complicated cases may be developed along similar lines. The maths can be daunting, but computer algebra systems can be used to aid manipulations of equations. Under www.wolframalpha.com such a system is available online.

E. Buxbaum, *Biophysical Chemistry of Proteins: An Introduction to Laboratory Methods*, DOI 10.1007/978-1-4419-7251-4_42,

42.2 Binding in the Presence of Inhibitors

Inhibition is usually discussed in the context of enzymatic turnover (see for example the chapter on enzyme kinetics in [44]). However, the inhibition of enzyme turnover may be preceded by an inhibition of ligand binding (see Fig. 42.1). Inhibitor binding may prevent the binding of ligand (competitive inhibition) or reduce the apparent affinity for the ligand (some cases of non-competitive and partial inhibition). It may even increase the apparent affinity (uncompetitive and some cases of non-competitive and partial inhibition). Since we look only at binding and not at substrate turnover, non-competitive (only ES can form product) and partial inhibition (both ES and EIS can form product, albeit with different rates) cannot be distinguished.

42.2.1 Competitive Inhibition

The dissociation constants are defined as

$$K_{\mathrm{D}} = \frac{[R][L]}{[RL]} \tag{42.2}$$

$$K_{\mathrm{i}} = \frac{[R][I]}{[IR]} \tag{42.3}$$

and the conservation equations are

$$[R]_0 = [R] + [IR] + [RL] = B_{\max} \tag{42.4}$$

$$[I]_0 = [I] + [IR] \tag{42.5}$$

$$[L]_0 = [L] + [RL] \tag{42.6}$$

$$B = [RL] \tag{42.7}$$

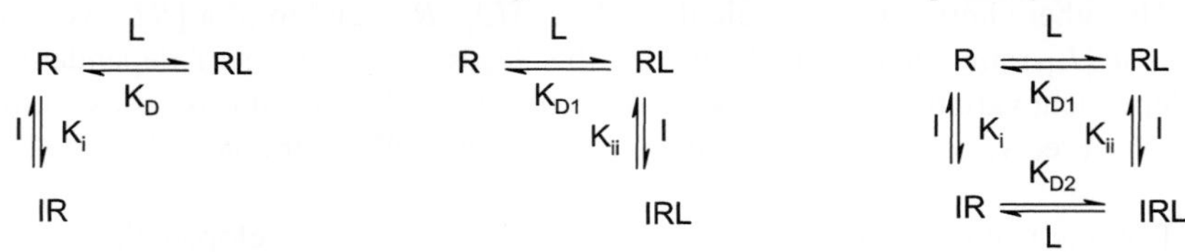

Fig. 42.1 Inhibition of ligand binding to a receptor. Note that for the binding reaction there is no difference between partially and non-competitive inhibition

Then the binding is described by

$$\frac{B}{L} = \frac{[R]_0 - B}{K_D} * \frac{K_i}{K_i + [I]} \tag{42.8}$$

$$[I] = \frac{B - [R]_0 + [I]_0 - K_i}{2} + \sqrt{\frac{(K_i + [R]_0 - B - [I]_0)^2}{4} + [I]_0 * K_i} \tag{42.9}$$

One frequent method to evaluate such binding is to keep $[L]_0$ and $[R]_0$ constant and vary [I] to find the point of 50 % inhibition (IC_{50}). Then the CHENG–PRUSOFF equation states:

$$K_i = \frac{K_D * IC_{50}}{[L]_0 + K_D}. \tag{42.10}$$

However, this method can be used only if a competitive mechanism has already been established, and even then it is less precise than a variation of both $[I]_0$ and $[L]_0$. It thus is a "quick and dirty" technique.

42.2.2 Non-competitive Inhibition

If $K_{ii} > K_i$ then I inhibits ligand binding and negative co-operativity results. $K_{ii} < K_i$ results in positive co-operativity, binding of ligand is increased by the inhibitor (even though the enzymatic reaction rate is decreased). The dissociation constants are:

$$K_{D1} = \frac{[R][L]}{[RL]} \qquad K_{D2} = \frac{[IR][L]}{[IRL]}, \tag{42.11}$$

$$K_i = \frac{[R][I]}{[IR]} \qquad K_{ii} = \frac{[LR][I]}{[IRL]} \tag{42.12}$$

and the conservation equations are

$$[R]_0 = [R] + [IR] + [RL] + [IRL] = B_{max}, \tag{42.13}$$

$$[I]_0 = [I] + [IR] + [IRL], \tag{42.14}$$

$$[L]_0 = [L] + [RL] + [IRL], \tag{42.15}$$

$$B = [RL] + [IRL]. \tag{42.16}$$

Then the binding equation becomes

$$\frac{B}{[L]} = \frac{[R]_0 - B}{K_{D1}} * \frac{1 + \frac{[I]}{K_{ii}}}{1 + \frac{[I]}{K_i}}. \tag{42.17}$$

42.3 Affinity Labelling

Affinity labelling proceeds in a similar way as product formation in an enzymatic reaction – enzyme and inactivator form a transition state $E \cdot I$ in a reversible reaction, from this transition state the stable $E - I$ complex is produced:

$$E + I \underset{k_{-1}}{\overset{k_1}{\rightleftharpoons}} E \cdot I \overset{k_2}{\longrightarrow} E - I. \tag{42.18}$$

The reaction can either stabilise the enzyme in the transition state (Ser-proteases at low *p*H, no deacylation), or stabilises the substrate (SCHIFF-base reduction with $NaBH_4$). Reagents can be designed as substrate analogues (K_s-reagent), transition state analogue ($K_s^{\ddagger}$-reagent, these can have very high affinity to the enzyme) or as suicide (k_{cat})-reagents (pseudo-substrates, that are converted into the inactivating species by the enzyme, they are highly specific) [435]. The same can happen for the backward reaction where the product turns into the substrate:

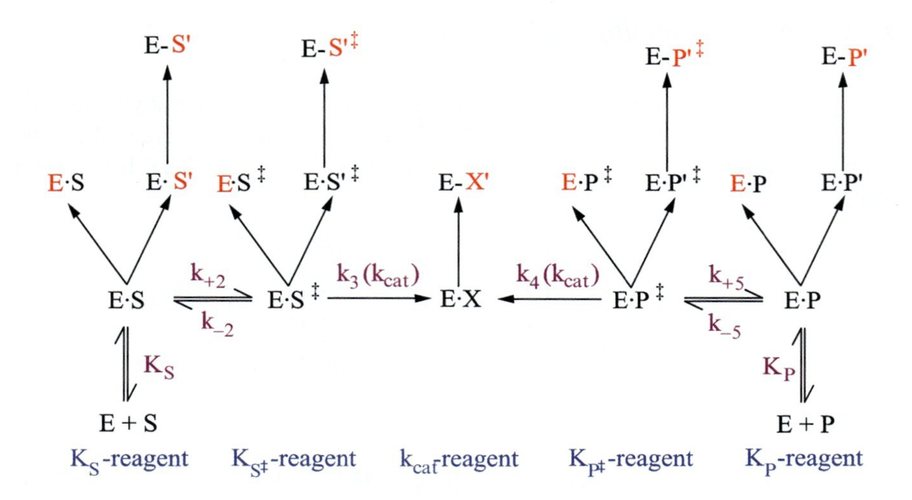

In this diagram the chemically modified species is written in red. For a review on applications of affinity labelling see [330]. The constants are:

$$K_i = \frac{[E][I]}{[E \cdot I]}, \tag{42.19}$$

$$k_{obs} = \frac{k_2[I]}{K_i + [I]}, \tag{42.20}$$

which is a reaction of first order at fixed $[I]$ (that is, $[I] \gg [E]_0$):

$$v_{\text{inact}} = \frac{V_{\text{inact}}}{1 + K_{\text{inact}}(1 + [S]/K_s)/[I]} \tag{42.21}$$

with $K_{\text{inact}} = (k_2 + k_{-1})k_{+1}$, $v_{\text{inact}} = k_2[P \cdot I]$ and V_{inact} the maximum velocity. v_{inact} vs $[I]$ is hyperbola, with competition by $[S]$. A plot of the half-life period $\tau = \ln(2)/k_2$ vs $1/[I]$ is linear with

$$\tau = \frac{1}{[I]} \times \tau_{\text{min}} \times \left(K_{\text{inact}} + \frac{K_{\text{inact}}[S]}{K_s}\right) + \tau_{\text{min}} \tag{42.22}$$

$$= \frac{1}{[I]} \times [S] \times \tau_{\text{min}} \frac{K_{\text{inact}}}{K_s} + \left(\tau_{\text{min}} + \frac{\tau_{\text{min}} K_{\text{inact}}}{[I]}\right) \tag{42.23}$$

42.3.1 Differential Labelling

The enzyme is incubated first with the inactivator in the presence of its natural ligand. Then the ligand is removed and the incubation repeated with labelled (usually radioactive) inactivator. Under this regime all unspecific binding sites for the inactivator are occupied by unlabelled molecules (possibly even without loss of biological function!), the labelled version is bound specifically to ligand-protectable sites. Proteolytic digestion and sequencing of labelled fragments leads to their identification [435].

Chapter 43
Methods to Measure Binding Equilibria

43.1 Dialysis

43.1.1 Equilibrium Dialysis

In equilibrium dialysis, the protein and its ligand are separated by a semi-permeable membrane which allows passage of unbound ligand, but not the protein (see Fig. 43.1). Thus the ligand will diffuse across the membrane until the concentration of free ligand at both sides of the membrane will be identical. Since the RL-complex can not diffuse across the membrane, it will increase the total ligand concentration on the protein side of the membrane. Often, radioactively labelled ligands are used.

This method is technically quite simple, but requires that

- The equilibrium between both sides has been achieved. This can be accelerated by stirring, but still requires incubation over several hours (usually over night) which may inactivate the protein, especially at higher temperatures.
- DONNAN-effects[1] are minimised by working either at elevated salt concentrations (100 mM NaCl or KCl) or near the isoelectric point of the protein.
- Changes in osmotic pressure and hence volume of the protein solution are reduced by including the ligand on both sides of the membrane at the beginning of the experiment. Such changes can be controlled for by measuring the protein concentration after the experiment.

[1] The DONNAN-effect was originally predicted by J.W. GIBBS and then investigated by F.G. DONNAN. It occurs whenever one of the charged components in a dialysis experiment is a macromolecule, which can not pass the membrane. Their charge has an influence on the distribution of the charged low molecular mass components, as the total activity of cations and anions in both chambers needs to be equal $\left(\frac{[K_i]}{[K_o]} = \frac{[A_i]}{[A_o]}\right)$. This results in an electrical potential across the membrane.

E. Buxbaum, *Biophysical Chemistry of Proteins: An Introduction to Laboratory Methods*, DOI 10.1007/978-1-4419-7251-4_43,

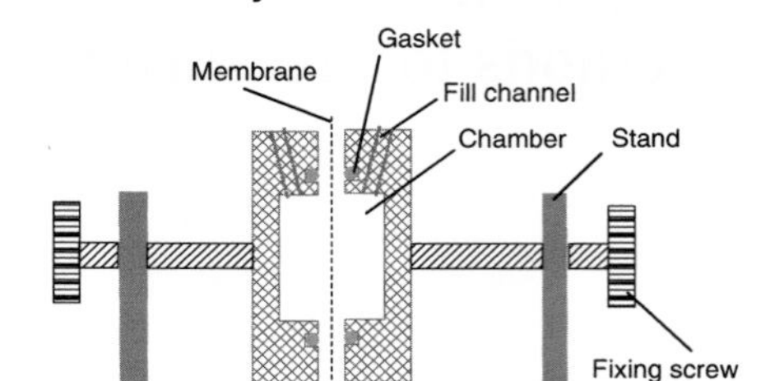

Fig. 43.1 Equilibrium dialysis setup. The sample (30–100 μl) is in a chamber inside a perspex block. Several such blocks may be held in the same stand to measure different concentrations in parallel

- Proteases or bacteria must be absent or inhibited, as they can seriously affect the results.
- No enzymatic turnover of the ligand must occur.

43.1.2 Continuous Dialysis

The relatively long time required for equilibrium dialysis makes this method unsuitable for unstable proteins or ligands, especially if experiments have to be performed at higher temperatures (*e.g.*, 37 °C). In such cases, continuous dialysis (see Fig. 43.2) can be used with advantage. The sample (protein and labelled ligand L*) are dialysed against a buffer. Unbound L* will move from the sample compartment into the buffer, after a short time the transport-rate of ligand into the buffer will become constant. Then the experiment is continued with buffer with a small concentration of unlabelled ligand (L), which will compete for some of the bound L* in the sample compartment. This will increase the rate of transport of L* into the buffer. Once this rate has become constant again, the concentration of L in the buffer compartment is increased by a small amount. This procedure is repeated until the entire concentration range of L (0.1–5 times K_d) has been tested. The condition for this method is that the total concentration of L* inside the sample cell does not change significantly during the course of the experiment. Thus the total time of the experiment needs to be kept as short as possible.

43.2 Ultrafiltration

The mixture of protein and ligand is filtered across a membrane that allows passage of the small ligand, but not the large protein molecules. In those cases where the dissociation of ligand from protein is slow, one can filter the entire solution and allow the protein to stick to the filter. Protein binding membranes made from nitrocellulose or polyvinylidene fluoride (PVDF) may be used. The latter is preferable, as it has a

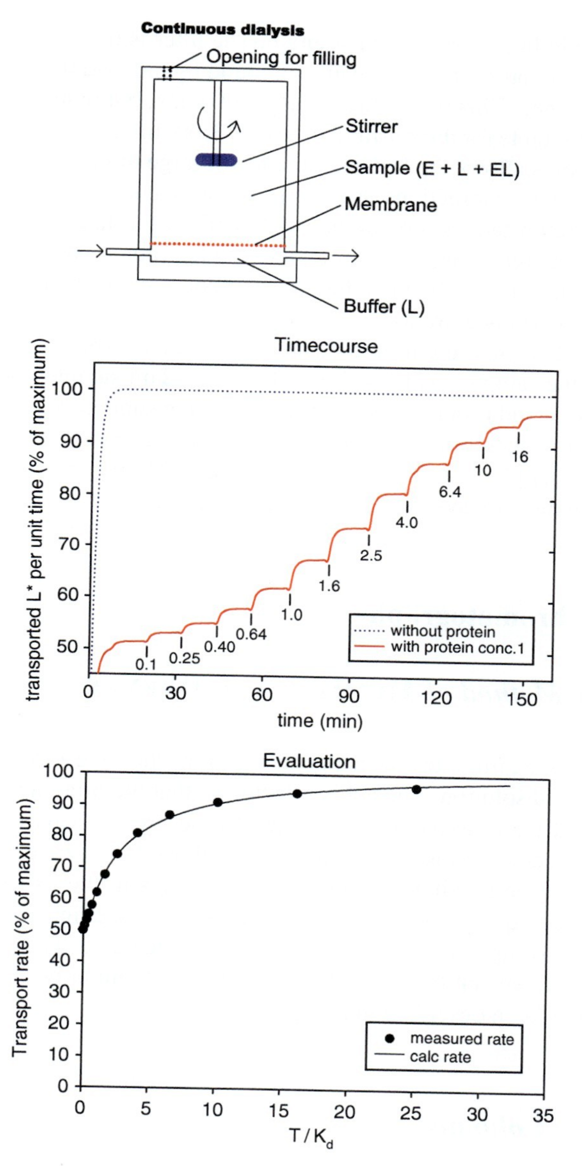

Fig. 43.2 In continuous dialysis the sample (protein plus labelled ligand) is dialysed against a buffer with increasing concentrations of unlabelled ligand. The buffer is constantly replaced, the rate of appearance of labelled ligand in the buffer is measured, resulting in the step-curve *top right*. If the rate at equilibrium is plotted against the ligand concentration, a binding curve with background $\left([B] = [B]_0 + \frac{([B]_{\max} - [B]_0)*[F]}{K_d + [F]}\right)$ is obtained. For the simulation of such an experiment the protein concentration and the K_d were assumed to be 1, the concentration of labelled ligand 0.01

higher protein binding capacity and strength. The filter is then washed two or three times with binding buffer to remove any unbound ligand, and the amount of bound ligand is determined. This procedure is easy and quick, but requires ligand–protein interaction to be stable for the washing time (20–30 s). In this case the use of 96-well filtration manifolds allows the fast examination of large numbers of samples.

If ligand–protein interactions are not stable enough, washing is not possible. Under these circumstances the amount of ligand on the filter is the sum of both bound ligand and free ligand in the trapped sample volume. Sometimes it is possible to measure this volume by including an inert compound labelled with a different isotope at a known concentration.

Filtration with a non protein-binding membrane can also be continued until most, but not all, of the sample has passed the filter (*e.g.*, 2 ml sample volume, 50–60 μl retentate). Free ligand is determined in the filtrate, the sum of free and bound ligand and the protein concentration are determined in the retentate. Then $[EL]/[L_f] = [E]_{tot}/(K_d + [L]_f)$ [271]. This type of experiment is often performed in centrifugal concentrators which are available from several suppliers.

43.3 Gel Chromatography

43.3.1 The Method of Hummel and Dreyer

A gel column is equilibrated with ligand solution, then the protein is added and eluted with ligand solution. The gel is selected so that the protein is eluted with the void, the ligand with the included volume. As the protein leaves the column, the total ligand concentration is increased, this is followed by a zone where the ligand was depleted by the passing protein [173]. The areas in the excess-peak and the trough need to be equal. The trough is deeper the lower the ligand concentration is, thus the amount of ligand to fill the trough can be calculated. The chromatographic separation of antigen, antibody and their complex by quantitative high performance gel filtration has been described in [362].

43.3.2 Spin Columns

In those cases where the dissociation of ligand from protein is slow, spin columns may be used for measurement. Those are 1 ml syringes half filled with a gel filtration medium equilibrated with binding buffer without ligand. The sample is added on top of the column and spun through the medium. Unbound ligand will stay on the column, while the protein with bound ligand is eluted with the void volume.

43.4 Ultracentrifugation

There are several methods for the determination of binding by ultracentrifugation.

43.4.1 *The Method of* DRAPER AND V. HIPPEL

This method [81] is analogous to the chromatographic technique of HUMMEL AND DREYER ([173], *vide supra*) and the electrophoretic method of TAKEDO ([402], see p. 73). The protein is spun through a sucrose gradient containing the ligand (effectively a separation by molecular mass as described by MARTIN AND AMES, see p. 242). In those fractions containing the protein the ligand concentration is higher; those fractions immediately on top lower than in the rest of the gradient.

43.4.1.1 The Method of CHANUTIN et al.

This method [55] uses ultracentrifugation of a solution of ligand and protein to establish self-forming gradients of the protein according to (25.18), while the much smaller ligand is not affected. The gradient is then fractionated and protein and ligand concentrations are determined in the fractions [55]. The total ligand concentration in each of the fractions is plotted against the protein concentration, the data points should form a straight line which intersects the y-axis at $[F]$.

The introduction of desktop-ultracentrifuges, where small sample volumes (0.2 ml) can be spun at high forces (850 000 g) has made this method quite interesting. However, it is important to establish that under the conditions used the ligand does not form a noticeable gradient when spun in the absence of protein.

43.4.2 *The Method of* STEINBACH AND SCHACHMAN

Alternatively, if the ligand absorbs light at a wavelength different from the protein, analytical ultracentrifuges can be used to determine both protein and ligand concentrations at different distances from the meniscus at different centrifugation times [389]. Such measurements may be performed either by sedimentation equilibrium or by sedimentation velocity. Because of the low centrifugal fields involved in sedimentation equilibrium, there is no significant gradient of the free ligand which simplifies the evaluation.

43.5 Patch-Clamping

Patch-clamping is used to measure the activity of single transmembrane proteins or small populations of them [139]. ERWIN NEHER and BERT SAKMANN received the NOBEL Prize in Physiology or Medicine in 1991 for the development of this technique. A fire-polished, smooth capillary is used to form a tight seal ("Gigaohm seal") with a cell membrane rather than penetrate it (see Fig. 43.3). The diameter of this electrode is about 1 μm, so that the patch of membrane enclosed by it often contains only one or at least a few of the proteins to be characterised. In this "**cell attached**" configuration the cell is alive and well. The electrode can then be pulled off the cell, leaving a small patch of membrane on it with any proteins in **inside-out** orientation. For these experiments, the bath-buffer corresponds to the milieu inside the cell, the pipette solution to the extracellular. Alternatively, it is possible to rupture the membrane by stronger suction on the capillary. Then currents across the cell membrane of the **whole cell** are recorded, rather than across a small patch. Cytosol and electrode solution will slowly mix by diffusion through the narrow capillary opening, this "dialysis" takes about 10 min. Mixing is slower if the membrane

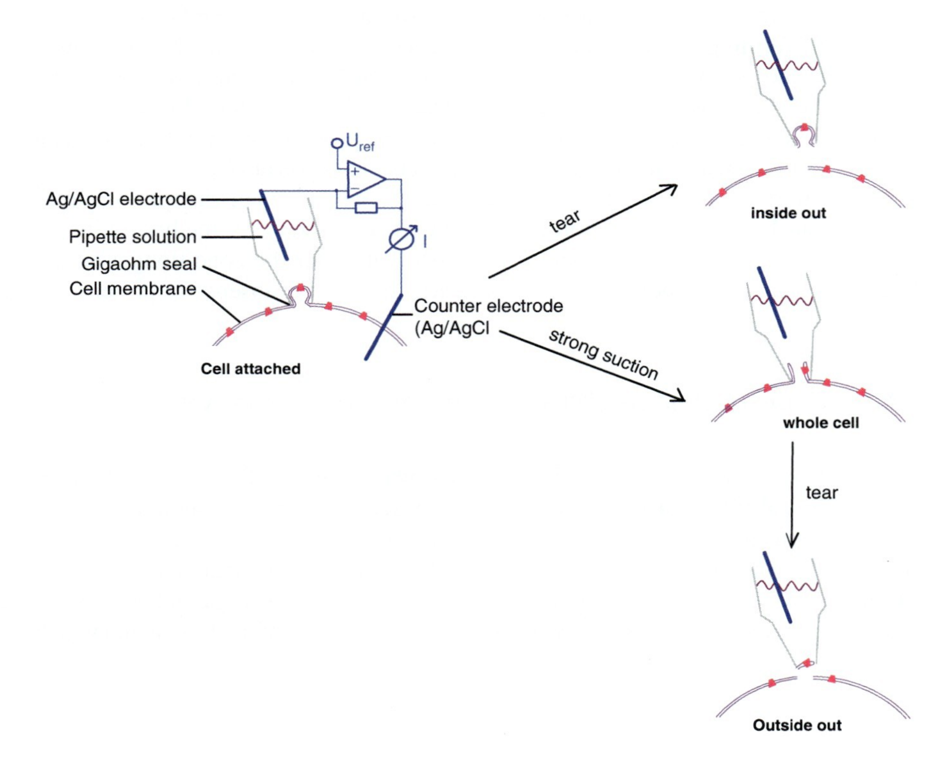

Fig. 43.3 Patch-clamping is used to measure activities of membrane proteins in supravital tissue. For details see text

is not physically ruptured, but made permeable to ions with pore-forming antibiotics like **amphotericin B**. Withdrawing the electrode from a whole cell preparation results in a membrane bleb, which, when the electrode is withdrawn far enough, will cover the electrode opening, forming an **outside out** preparation. Here the electrode solution is in contact with the cytosolic side of the membrane, the buffer bath with the extracellular.

The signal of the microelectrode is fed into an **operational amplifier** (the triangle in Fig. 43.3). This circuit element has the property of keeping the difference between its inverting and non-inverting input at 0 V, at the same time its input resistance is extremely high, so that it does not load the electrodes. A reference voltage U_{ref} is applied to the non-inverting input of the amplifier (the one marked "+"), the amplifier will then send a current I through the reference electrode that is sufficient to bring its inverting input (marked "–") to the same voltage. Dividing the current by the reference voltage gives the conductivity γ of the membrane patch as a function of the reference voltage and the composition of the bath-buffer, to which hormones, pharmaceuticals, or ions may be added. Alternatively, it is also possible to keep the current through the membrane constant, and to measure the voltage required (**current clamp**).

Patch-clamping is performed under a microscope, the electrodes are handled with micro-manipulators, a process that requires some dexterity and experience. The entire setup is placed on a heavy base (*e.g.*, a marble plate) to dampen vibrations. It is also screened against external electric fields by a grounded FARADAY-cage of copper mesh.

Because of the stochastic opening and closing of protein channels in the patch, the conductivity is not constant but fluctuates, representing the number of open channels at any given time. One should remember that the 70 mV or so potential acts across a membrane that is about 5 nm thick, resulting in field strengths of 140 kV/cm (compare this with 200 V over 6 cm = 33 V/cm in a typical SDS–PAGE setup!). Charged amino acid residues in a protein can move in this field, resulting in conformational changes of a protein, opening or closing it for ion traffic (**voltage gating**). In addition, the presence of transmitters also influences the opening probability of a channel. The current typically measured is in the order of a few pA over a few milliseconds. 1 pA is 1×10^{-12} C/s / 96 484 C/mol $\times$ 6.022×10^{23} charges/mol = 6240 charges/ms.

43.6 Mass Spectrometry

Many protein–ligand complexes are stable enough to survive electrospray ionisation, under these conditions complex formation can be followed by mass spectrometry [355]. A continuous-flow apparatus for measuring rapid kinetics has been described in [430], stopped-flow in [205].

43.7 Determination of the Number of Binding Sites: The Job-Plot

Before one can determine the dissociation constant(s) of binding site(s), the number of binding sites (n) needs to be determined. For this purpose, JOB-plots are commonly used [176] (see Fig. 43.4).

To construct a JOB-plot, prepare solutions of protein (P) and ligand (L) of equal molarity (for best results the concentration should be at least five times K_d). Then mix different volumes of both solutions in such a way that the total concentration

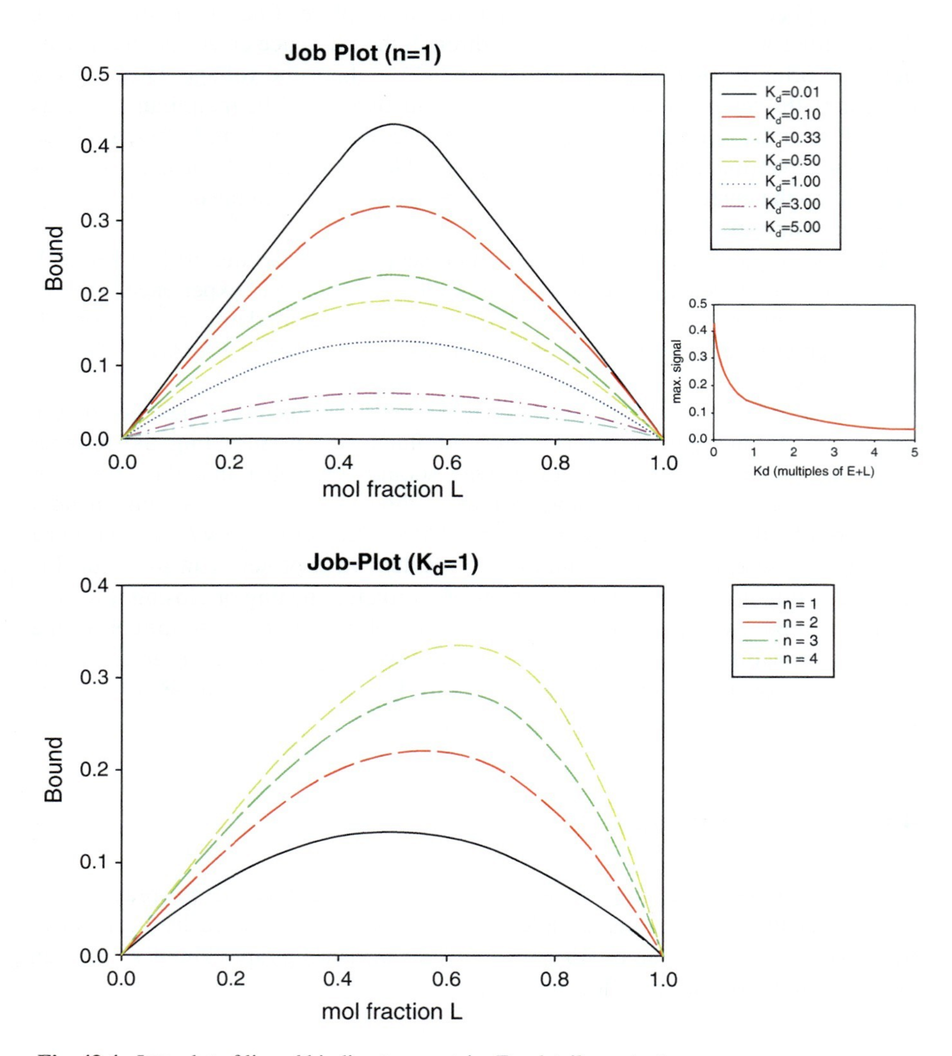

Fig. 43.4 JOB-plot of ligand binding to a protein. For details see text

($c_0 = [P]_0 + [L]_0$) stays constant (*e.g.*, $0 + 20$, $1 + 19$, $2 + 18, \ldots, 19 + 1$, $20 + 0$) and determine for each sample the bound ligand. JOB-plots may also be used for stimulation of enzymatic activity, provided the ligand concentration stays constant during the assay (co-factors, effectors or even substrate, if a cycling assay is used).

Now assume that X is the mol fraction of protein and Y is the mol fraction of ligand (because $c_0 = \text{const} \Rightarrow X + Y = 1$). Plot the signal obtained as a function of X, this will be a maximum curve. Draw in tangents at $X = 0$ (no protein) and $X = 1$ (no ligand). Obtain X_i and Y_i from the intersection of the tangents, then

$$\frac{X_i}{Y_i} = \frac{K_\mathrm{d} + n c_0}{K_\mathrm{d} + c_0} \tag{43.1}$$

which reduces to $X_i/Y_i = n$ if $c_0 \gg K_\mathrm{d}$. Plot the obtained n against c_0 from several experiments to make sure that your c_0 is high enough, you should see a horizontal line.

The precise shape of the curve depends on the mechanism of binding, for a detailed mathematical treatment see [170].

Chapter 44
Temperature Effects on Binding Equilibrium and Reaction Rate

This type of investigation is quite important for the characterisation of an enzyme:

$$\Delta G'^{0} = -R * T * \ln(1/K_{\mathrm{d}}) \qquad \text{at } 25°\mathrm{C} \tag{44.1}$$

$$\ln(K_{\mathrm{d}}) = -\frac{\Delta H'^{0}}{R} * \frac{1}{T} + A \tag{44.2}$$

$$\Delta S'^{0} = \frac{\Delta H'^{0} - \Delta G'^{0}}{T} \tag{44.3}$$

Such experiments can be performed on a FORBES-bar (see Fig. 44.1), that allows measurements at different temperatures in parallel. The standard GIBBS free energy of ligand binding[1] $\Delta G'^{0}$ can be determined from the dissociation constant at standard temperature (25 °C).

The heat of binding $\Delta H'^{0}$ can be obtained from a plot of the logarithm of the K_{d} versus the reciprocal temperature (VAN'T HOFF-plot, see Fig. 44.2). The data fall on a straight line with slope $\Delta H'^{0}/R$.

The binding entropy $\Delta S'^{0}$ is then available from the difference of $\Delta H'^{0}$ and $\Delta G'^{0}$.

44.1 Activation Energy

Assume that colliding molecules can react only if their energy exceeds a certain value, called the activation energy E_{a}. The fraction of molecules exceeding that barrier n/n_0 can be calculated from the BOLTZMANN-distribution :

$$n = n_0 \exp\left(-\frac{E_{\mathrm{a}}}{RT}\right) \tag{44.4}$$

[1] Thermodynamic parameters like ΔG, ΔH or ΔS may be given for a particular set of conditions, but often more useful are data for standard conditions (25 °C, 101 325 Pa, *p*H 0, all reactants at 1 M concentration. In this case, the superscript 0 is added to the abbreviation. Biologists use *p*H 7 instead of 0, indicated by a prime ('^{0}). In the older literature you may find ^{0}' instead, IUPAC has changed the nomenclature a few years ago.

E. Buxbaum, *Biophysical Chemistry of Proteins: An Introduction to Laboratory Methods*, DOI 10.1007/978-1-4419-7251-4_44,

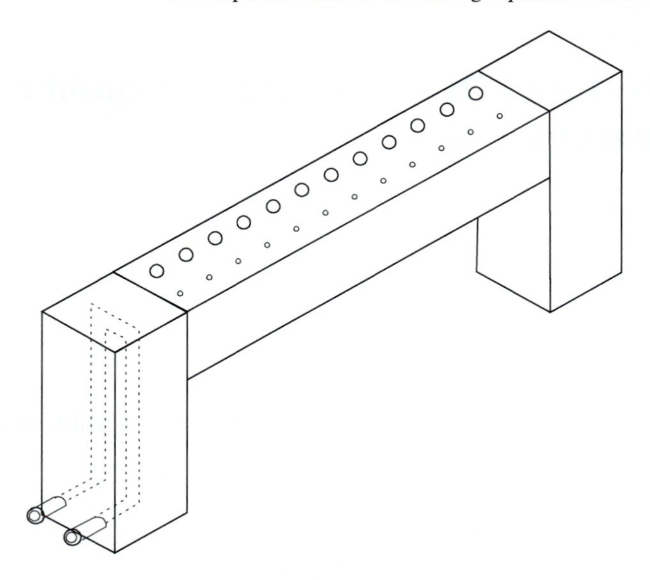

Fig. 44.1 A FORBES-bar [178] allows a reaction to be measured at several different temperatures concurrently. It consists of a long horizontal and two vertical metal blocks. The vertical bars have internal loops which are connected to temperature-controlled water baths, which keep these blocks at defined temperatures. This results in a linear temperature gradient across the horizontal bar. In the larger holes, test tubes with the reaction mixture are kept at different temperatures, the small holes are for a thermometer. All holes are filled with a little distilled water to ensure good thermal contact between sample vials and metal. The bar is about 50 cm long

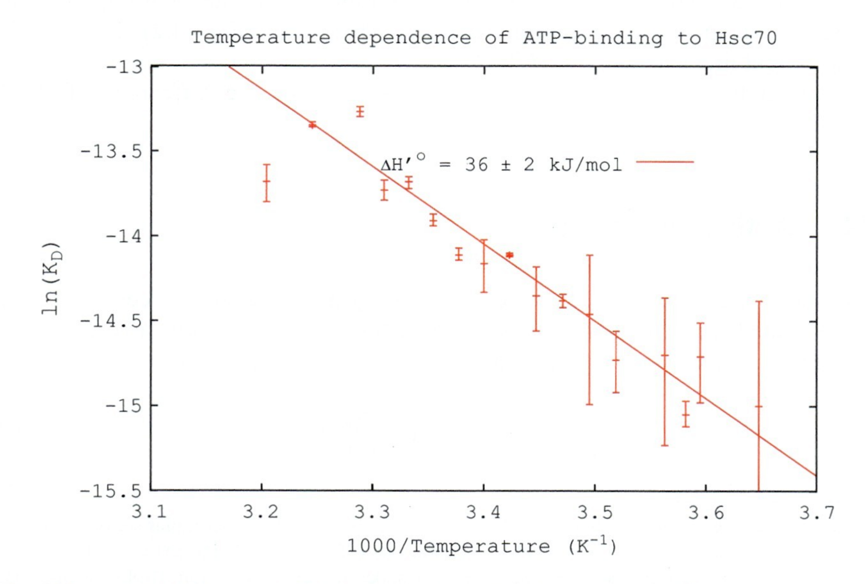

Fig. 44.2 VAN'T HOFF-plot of ATP-binding to the molecular chaperon Hsc70

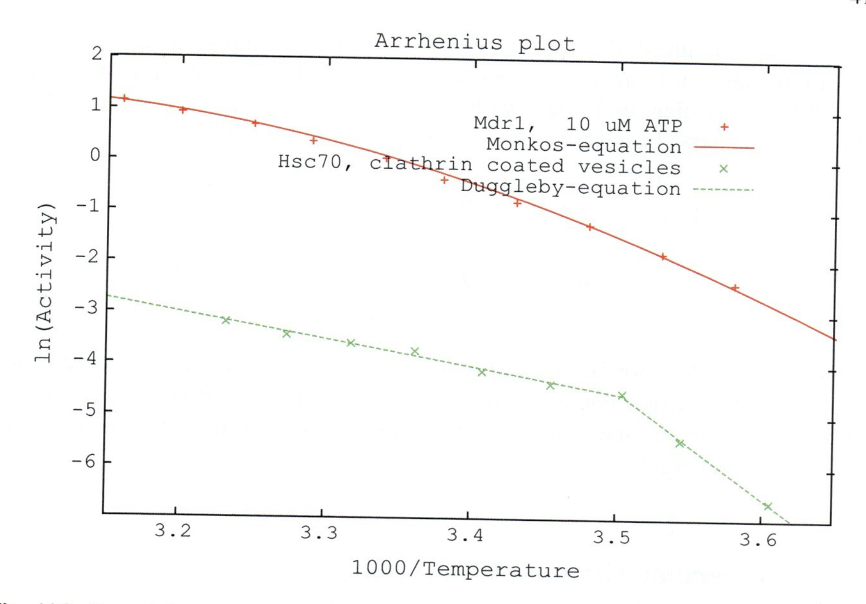

Fig. 44.3 Curved ARRHENIUS-diagrams occur when the pre-exponential term of the ARRHENIUS-equation is temperature dependent, as was observed with the multiple drug resistance transporter Mdr1. Changes in membrane fluidity can change the activation energy of enzymes at the "freezing" point of phospholipids, resulting in ARRHENIUS-diagrams with a kink (here: clathrin coated vesicle stimulated ATPase activity of the molecular chaperon Hsc70, the activation energy changes from 37 to 178 kJ/mol)

Thus from the temperature dependence of the reaction rate one can calculate the activation energy of the reaction by the ARRHENIUS-equation [10]:

$$v = A * \exp\left(-\frac{\Delta E^{\mathrm{a}}}{RT}\right). \tag{44.5}$$

The data points should form straight lines with slope $\Delta E^{\mathrm{a}}/R$ if $\ln(v)$ is plotted against $1/T$ (see Fig. 44.3).

Sometimes the data form curves in ARRHENIUS-coordinates. MONKOS [281] has shown that such curvature arises if the pre-exponential term A is dependent on T. He has also shown that $A(T)$ is an exponential function, yielding:

$$\begin{aligned} v &= \exp(b - d * T) * \exp\left(\frac{-\Delta E^{\mathrm{a}}}{R * T}\right) \\ &= \exp\left(b - d * T - \frac{\Delta E^{\mathrm{a}}}{R * T}\right). \end{aligned} \tag{44.6}$$

Another form of curved ARRHENIUS-diagrams is often encountered in membrane enzymes. At physiological temperature, membrane lipids are quite fluid. If the temperature is reduced, the lipids "freeze", resulting in a fairly sudden increase in activation energy for enzymatic reactions. In mammalian membranes, this kink in the ARRHENIUS-diagram is usually observed around 10–12 °C. According to DUGGLEBY the data can be described by a general hyperbola [82]:

$$\ln(v) = \ln(v)_t - \frac{(\Delta E_l + \Delta E_r) * (1/T - 1/T_t)}{2R} + \frac{(\Delta E_l - \Delta E_r) * \sqrt{(1/T - 1/T_t)^2 * c^2}}{2R} \tag{44.7}$$

where ΔE_r and ΔE_l are the activation energies to the right and left of the transition point, v_t is the reaction rate at the transition temperature T_t, R is the gas constant and c a curvature parameter that describes how sudden the transition between the two activation energies is.

44.2 Isothermal Titration Calorimetry

The heat of reaction can be determined directly in a calorimeter (see also the section on differential scanning calorimetry (DSC) on p. 350). This consists of two well insulated sample chambers, which are kept at identical temperatures by electrical heating elements (see Fig. 44.4). The ligand is added to one of the samples, to the other a buffer blank. If the ligand reacts with the protein in the sample, heat is developed (or used) and the control vial (or sample vial) has to be heated to keep the temperature difference between the vials zero. The current through the heating element is measured, from that ΔH can be calculated.

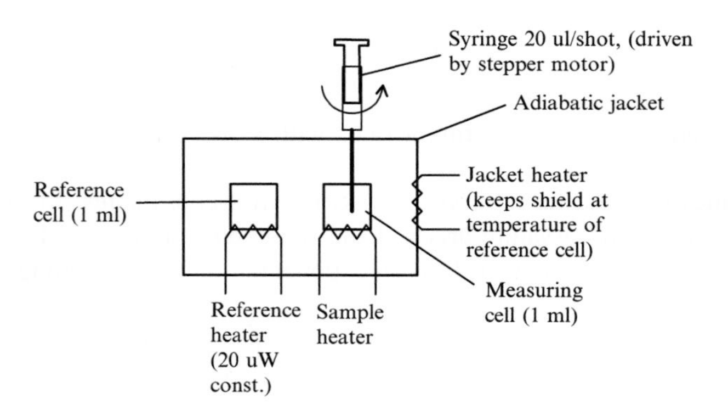

Fig. 44.4 The isothermal titration calorimeter. Two vials, one with sample and one with blank, are heated electrically so that they are both always at the same temperature (to within 0.001 K). The currents required are measured. The ligand is injected step-wise, the heat produced (or absorbed) is measured

Despite the name MICROcalorimetry, large (milligrams) amounts of protein are required for each study (physical chemists and biochemists have different opinions about what a small sample is and the large molecular mass of proteins does not help), and the instrumentation is very expensive. Both problems have limited the application of this elegant method.

44.2.1 Photoacoustic Calorimetry

Besides ITC and DSC, there is a third important calorimetric method, PAC, used to measure the enthalpy required to break bonds (homolytic bond dissociation enthalpy). This method makes use of the photoacoustic effect discovered by A.G. BELL in 1880 and was developed by K. PETERS in 1983 [324]. Photon energy from a brief (ps) laser pulse is used to break chemical bonds. The part of the photon's energy not required for this is converted to heat, causing a steep local increase in temperature. This in turn causes a sound wave that can be detected by a piezoelectric transducer. The reaction enthalpy ΔH is the part of the photons energy used in the reaction, divided by the quantum yield of the reaction. The temporal profile of the photoacoustic signal is used to measure reaction rates in the microsecond range. For further information, see also cqb.fc.ul.pt/menergetics/photoacoustic_calorimetry.htm

Part VIII
Industrial Enzymology

In 2000, 2×10^9€ worth of enzymes were produced for industrial purposes (see Fig. 1), twice what it was in 1995. 90 % of all enzymes used industrially are recombinant and genetically engineered, with substrate specificity, heat stability, and other parameters changed to suit the application. They often contain tags for affinity purification, and the expression rate has been increased to reduce production costs [58]. As a rule of thumb, enzyme costs need to be less than 10 % of the product value to make a process commercially viable. For more information on industrial enzyme applications see [39].

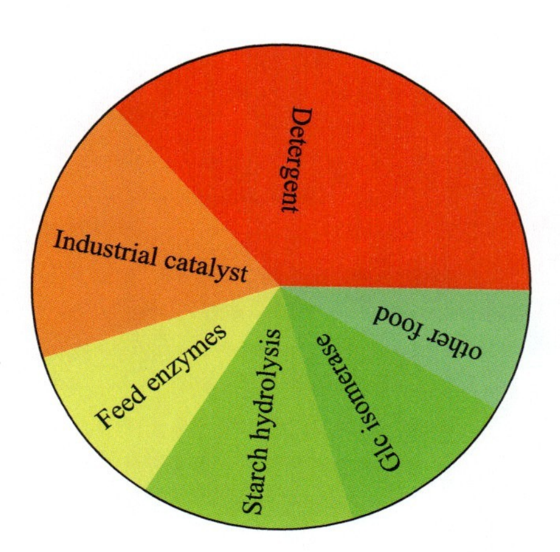

Fig. 1 Estimated market size for industrial enzymes in 2000 [39]. About 2×10^9€ worth of enzymes were produced. 35–40 % went into manufacture of detergents (so called "biological" detergents containing lipases or proteases), 16–21 % were used as industrial catalysts (in production of paper, pharmaceuticals, fine chemicals...). 30–35 % were used by the food industry (11–15 % were starch hydrolysing enzymes, 12 % glucose isomerase for the production of high fructose corn syrup, the rest in cheese production and other food processing). 11 % went into animal production to make feeds more digestible by breaking down anti-nutrients like non-starch carbohydrates or phytate

Chapter 45
Industrial Enzyme Use

Compared to conventional chemical methods, enzymatic processes have the following advantages:

- Milder reaction conditions (temperature, pressure, *p*H) making the process cheaper and safer
- Avoid environmentally unfriendly reactants
- Fewer reaction steps
- Fewer side products leading to easier purification
- Stereoselectivity

The oldest method in biotechnology is fermentation, where the enzymes are provided by whole organisms. This method is still used where several enzymes are required to make a product, or to regenerate cofactors. However, for simple processes (≤ 3 enzymes) the use of purified enzymes

- Provides higher space/time yields
- Minimises the risk that the product is catabolised
- Reduces the cost for product purification
- Allows continuous rather than batch processes

The space/time yield STY is defined as

$$\mathrm{STY} = \frac{[S]_0 - [S]_t}{tV} \tag{45.1}$$

with $[S]_0$ and $[S]_t$ the substrate concentrations at times 0 and t, respectively, and V the reactor volume. A minimum STY is often specified by upstream processes, which feed substrate into the reactor at a certain rate which the reactor then has to handle.

Reaction yields may be equilibrium controlled, however, in some cases the product is destroyed by further reactions (say, hydrolysis), with the product concentration first increasing, then decreasing over time. Such kinetically controlled processes must be stopped once the optimum product concentration has been obtained.

If the reaction is performed as a batch process, the equations of initial rate kinetics do not apply, we have to take into account the depletion of substrate and accumulation of product over time. For a first order reaction:

E. Buxbaum, *Biophysical Chemistry of Proteins: An Introduction to Laboratory Methods*, DOI 10.1007/978-1-4419-7251-4_45,

$$\frac{d[S]}{dt} = \frac{V_{max}e^{-kt}}{f([S][P])} \tag{45.2}$$

If the reverse reaction can be neglected and no substrate inhibition occurs, then

$$f([S][P]) = \frac{K_m + [S]}{[S]} \tag{45.3}$$

However, industrial processes are often performed at high substrate and product concentrations to ease downstream processing. This increases the probability for substrate or product inhibition. For non-competitive substrate inhibition $f([S][P])$ becomes

$$f([S][P]) = \frac{K_m + (1 + K_m/K_i)[S] + [S]^2/K_i}{[S]} \tag{45.4}$$

and for competitive product inhibition

$$f([S][P]) = \frac{K_m + (1 + [S]_0/K_i) + K_m/K_i[S]}{[S]} \tag{45.5}$$

In addition, for many industrially important processes the back-reaction cannot be neglected, and in equilibrium the reaction mixture contains significant amount of substrate (*e.g.*, 50 % for the isomerisation of glucose). In addition, the enzyme may be inactivated during the reaction.

If the reaction produces or consumes protons, the *p*H is usually kept constant by the addition of acid or base. The amount of acid/base required gives a convenient reporter on reaction progress. Choice of *p*H may be made to shift the equilibrium from substrate to product, even if that leads to more rapid enzyme inactivation. In such cases the enzyme may be engineered to improve stability at the process *p*H.

If the product or the substrate have low solubility in water, the reaction may be performed in suspension. The activity of the compound is then determined by its solubility. Removal of product by precipitation can move the equilibrium toward the right. The reaction rate is determined not only by the activity of the enzyme, but also the rate at which the substrate can go into solution.

If the solubility of the substrate is very low, the reaction may be performed using organic solvents (see for example Fig. 45.1), either in a mono- or biphasic system. Several options exist:

Water miscible organic solvent at a concentration of solvent where the substrate is soluble, but the enzyme is not inactivated. Solvent may be one of the substrates (*e.g.*, alcohols).

Monophasic organic Enzyme is dissolved in the organic solvent, with just enough bound water to maintain its activity.

Biphasic aqueous/organic substrate or product dissolved in organic solvent, enzyme in aqueous phase. Activity of substrate and product is determined by their

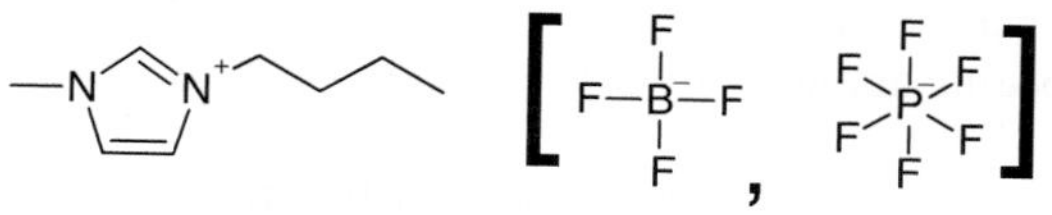

1-butyl-3-methylimidazolium perfluoroborate or perfluorophosphate

Fig. 45.1 The ionic liquids (salts with low melting points) BMIM-BF_4 (water miscible) or BMIM-PF_6 (water immiscible) can be used as solvents when either substrate or product are too insoluble in water. However, data on toxicity and environmental effects are currently limited, they are also very expensive. Supercritical CO_2 (critical temperature 31.1 °C, critical pressure 7.39 MPa) may also be used, which has solvent properties similar to acetone and the advantage that it can be removed easily and without leaving residues

partition ratios. The reaction equilibrium may be shifted to the right if the product is more hydrophobic than the substrate. *p*H-control in aqueous phase may be difficult.

Immobilised enzyme suspended in organic solvent again with enough bound water to ensure maximum enzymatic activity.

Organic solvent may bind to hydrophobic parts of the enzyme, changing its structure. This can be investigated by methods sensitive to structure like fluorescence, CD- or IR-spectroscopy. Effects on K_m, k_{cat} and E depend on the enzyme/solvent pair and can not be predicted. Since the water activity in all these systems is reduced, hydrolysis of the product is less of a problem than in aqueous solutions. This may turn a kinetically controlled process into an equilibrium controlled one.

45.1 Enzyme Denaturation

The following factors can lead to enzyme denaturation and thereby increase enzyme costs in a process:

Temperature may lead to heat or cold denaturation and dissociation of subunits. The enzyme should be used at a temperature where it is stable (rather than at the temperature optimum).

***p*H** may lead to unfolding of protein when ionic bonds are disrupted.

O_2 and oxidising chemicals (*e.g.*, disinfectants) cause disulphide bond formation and Met oxidation.

High $[S]$ may lead to reduced water activity, but also to chemical modification of the enzyme (*e.g.*, SCHIFF-base formation from glucose).

Organic solvents may lead to unfolding or precipitation of enzyme.

Proteases can be introduced with the enzyme or with the substrate. There is a balance between purification costs and residual protease content. Some enzymes are subject to intramolecular auto-proteolysis.

Heavy metals either from the chemicals used, or leaching from the apparatus. They interact with specific amino acids (Cys, His, *etc.*) leading to inhibition of enzyme activity.

Foam formation can be avoided by proper construction of the reactor, and by addition of anti-foaming agents.

Stability may be increased by intramolecular cross-linking or immobilisation. One may also search for or construct a more stable enzyme. Sometimes one has to make a compromise between enzyme stability and product yield.

45.2 Calculation of the Required Amount of Enzyme

Once the conversion factor $([S]_0 - [S]_t)/[S]_0$ and the maximum process time t are fixed, the required V_{max} can be calculated from the integrated rate equations. If there is neither inhibition nor inactivation, this becomes:

$$V_{\mathrm{max},0} = \frac{([S]_0 - [S]_t) + K_\mathrm{m} \ln([S]_0/[S]_t)}{t} \tag{45.6}$$

For competitive product inhibition the integration of (45.5) yields

$$V_{\mathrm{max},0} = \frac{([S]_0 - [S]_t)\left(1 - \frac{K_\mathrm{m}}{K_i}\right) + \left(K_\mathrm{m} + \frac{K_\mathrm{m}}{K_i}[S]_0\right)\ln([S]_0/[S]_t)}{t} \tag{45.7}$$

and for non-competitive substrate inhibition from (45.4) we get

$$V_{\mathrm{max},0} = \frac{([S]_0 - [S]_t)\left(1 + \frac{K_\mathrm{m}}{K_i}\right) + K_\mathrm{m}\ln([S]_0/[S]_t) + \frac{[S]_0^2 - [S]_t^2}{2K_i}}{t} \tag{45.8}$$

If there is enzyme inactivation by a first order process with rate k_i, the denominator t in above equations is replaced by

$$\frac{1 - \exp(-k_i t)}{k_i} \tag{45.9}$$

In either case, the enzyme productivity becomes $([S]_0 - [S]_t)/V_{max}$.

Chapter 46
Immobilised Enzymes

To ease removal of enzymes from the product and their recycling, enzymes are often immobilised. Immobilisation may also be required to convert batch into continuous processes. Immobilisation may be achieved by

- Inclusion of the enzyme in the reactor with semipermeable membranes. Co-factors can be kept in the reactor by linking them to large molecules like polyethylene glycol.
- Binding the protein (or entire cells) to a carrier. This in turn can be achieved by
  - Adsorption
  - Binding via ionic or metal-complexing side-chains
  - Covalent bonds (see p. 191).

The process of immobilisation may lead to inactivation of enzymes due to formation of unwanted protein/substrate contacts, and therefore protein unfolding. This effect can be reduced by using an appropriate, hydrophilic, matrix. On the other hand, immobilisation reduces access of proteases to the enzyme and intramolecular motion in the enzyme, improving stability. The theory behind all these processes is not well understood.

The ideal carrier is stable against chemical and microbial attack, does not swell or shrink, has a narrow particle size distribution and a large inner surface area. It can resist pressure and compression, but is elastic enough to survive contact with a stirrer. It should be safe, cheap and available in food grade. The pore size must be sufficient for proteins to pass through (protein Ø $\approx$ 50 Å, pores must be about 3 times that size for unhindered motion). Materials used include

Porous glass or silica bind via $-\mathrm{OH}$-groups, these can be functionalised. High pressure, but low abrasion resistance. Diatomaceous earth may also be used.

Polysaccharide-based like agarose, dextran, cellulose. Bind via $-\mathrm{OH}$-groups, those can be functionalised.

Synthetic polymers of acrylate, acrylamide, styrene or others, sometimes as copolymer. Different functional groups can be built in for reaction with protein. Molecule size and cross-linking can be controlled.

Tungsten carbide very high density for fluidised bed reactors. Surface covered with other material that provides anchor sites for proteins.

E. Buxbaum, *Biophysical Chemistry of Proteins: An Introduction to Laboratory Methods*, DOI 10.1007/978-1-4419-7251-4_46,

46.1 Kinetic Properties of Immobilised Enzymes

46.1.1 Factors Affecting the Activity of an Immobilised Enzyme

Assume an enzyme molecule immobilised in the core of a porous particle. It is surrounded by other enzyme molecules bound to the same particle. It will see a lower substrate and higher product concentration than a molecule in solution would due to the activity of the surrounding enzyme molecules. Also, diffusion of substrate into and product out of the particle is driven by a concentration gradient between the particle and the surrounding solution. Thus reaction velocity will be lower for immobilised enzyme than for a free enzyme.

The immobilisation matrix will take up part of the volume of the reactor, which is no longer available to the reactants.

Charged groups on the surface of the particle will be balanced by ions of opposite charge from the medium, this creates an electrical double layer in the vicinity of the immobilised enzyme, where *p*H and ionic strength differ significantly from that of the bulk medium. This affects the binding efficiency of the (charged) protein during immobilisation, but also its catalytic properties. If the substrate or product molecules are charged, the situation becomes even more complicated. Charge density on the matrix can exceed the charge density of the enzyme, this can affect reaction mechanism, k_{cat} and the stereoselectivity E.

If protons are produced or consumed during the reaction, a *p*H-gradient may develop between the particle and the bulk medium to which the enzyme is exposed. High buffering capacity ($c \geq 50$ mM with $pK_a \approx pH$) of the medium can counteract that effect. This however increases costs and may lead to problems with downstream processing. Sequences of several smaller reactors with intermittent *p*H-adjustment may also be used in lieu of one large reactor; that also increases process costs.

The surface of the particle sterically hinders approach of substrate molecules to the enzyme, leading to an increase in apparent K_m.

46.1.2 The Effectiveness Factor

For all these reasons, the turnover rate of the entire reactor will be less than that of a similar reactor with freely dissolved enzyme by a **stationary effectiveness factor** $\eta = v_{imm}/v_{free} \leq 1$. Some authors also use the **operational effectiveness factor** which uses progress curves to compare the time the reactors need to convert a certain amount of substrate: $\eta_o = t_{free}/t_{imm}$. Effectiveness will drop as the particle size increases. η is determined by measuring the reaction rate first with the intact particles, and a second time after the particles have been disintegrated with ultrasound.

46.1.3 Maximal Effective Enzyme Loading

As the enzyme concentration in a reactor is increased, V_{max} is also increased. However, if the enzyme is immobilised on particles, increasing the enzyme concentration inside the particles will increase the V_{max} only until diffusion becomes rate limiting. Increasing the enzyme concentration inside the particle further increases the costs for enzyme without benefit. This optimal concentration V'_{max} however depends on the stirring speed which limits the gradient between bulk solution and particle:

$$V'_{max} = \frac{1.5}{r^2} D \text{ Sh } \alpha \; c_s(\infty) \qquad (46.1)$$

where $c_s(\infty)$ is the bulk substrate concentration, α the concentration change in the diffusion layer (≈ 0.1), Sh the **SHERWOOD-number** (ratio of particle radius to thickness of the unstirred diffusion layer, two for spherical particles in unstirred systems, about ten in a well-stirred reactor), and D the diffusion coefficient of the substrate ($\approx 6 \times 10^{-6}$ cm^2 s^{-1} for a typical organic compound). r is the radius of the particle, between 10 and 500 μm in typical applications. The lower limit is determined by filterability, the upper by mass transfer rate.

46.1.4 Decline of V'_{max} Over Time

In addition to destruction of enzyme molecules, as discussed above, immobilised enzymes suffer from the abrasion of the matrix in the reactor due to shear forces and collisions with other particles, the stirrer or the reactor wall. The bond between particles and enzymes can break, and prosthetic groups may be lost. However, under storage conditions, immobilised enzymes tend to be more stable than free.

Loss of enzyme function in a reactor is modelled by a first-order process from the loss of STY over time. The half life period for this process may reach several hundred days. Catalyst is usually replaced when STY is between 20 and 50 % of its original value. The integral of STY over time is the **enzyme productivity**, measured in kg of converted substrate per kg of enzyme. For mass products like fructose, this factor should approach 10 000.

Part IX
Special Statistics

Chapter 47
Quality Control

Quality control and management is an important function for the head of a laboratory, even in places not regulated by GLP. Several cases in the last years where respected labs had to withdraw publications because of measurement error or data forgery by subordinates, demonstrate this need. Textbooks on quality management should be consulted [298, 313, 343]. Several international standards are pertinent, in particular ANSI/ISO 9001-2000 and 17025:1999. Quality of results depend on

Personnel continuously trained in the methods used, dedicated to quality.
Equipment robust, qualified, calibrated.
Method well characterised and suitable for the task, documented in an standard operating protocol (SOP) .
Materials of known and sufficient quality
Environment with known influence on method results (temperature, radiation, vibration, *etc.*)
Management Informs personnel about all relevant issues, supports personnel and gives sufficient time and resources for a task. Possible conflicts are resolved early and in a reasonable manner.

One of the easiest methods of quality assurance is the inclusion of standards in the samples. The results for these standard samples are plotted as a function of the time when the analysis was done into a **control chart**, introduced into the laboratory by LEVEY AND JENNINGS based on earlier work by SHEWHART [232, 373]. Included in this plot are also warning and action limits, for example 2σ and 3σ. SOPs describe what has to happen if these limits are exceeded. For example, if the warning level is exceeded, one might increase the frequency of standard samples. If the action limit is exceeded, one might discard all results produced since the last correctly read standard, check for possible sources of error, and then redo the discarded tests once the process is under control again. From statistical considerations, one would expect that 1 out of $\approx$ 370 samples is outside the $\pm 3\sigma$ threshold. The most useful property of this plot is that one can detect slow changes in the average over time (**drift**). Special events can also be marked in the chart, *e.g.*, a calibration of instruments or a change to a different lot of reagents. Action may be required if those events markedly influence the results.

E. Buxbaum, *Biophysical Chemistry of Proteins: An Introduction to Laboratory Methods*, DOI 10.1007/978-1-4419-7251-4_47,

47.1 Validation

In the development of new assays, considerable effort is required for validation. Validation means the objective proof (by experiment, test, facts and observations) that a procedure can reliably achieve a given purpose within required specifications. In other words, the purpose of validation is to produce conclusions, not merely data. Methods and results of the validation procedure, and the conclusions on suitability must be documented in a **validation report**. Note that validation is always done with a particular purpose in mind, and hence there are specific requirements on cost, risk, and specifications. Several methods may be used for validation of an assay, for example:

Calibration with reference material The content of the standard is determined several (usually six) times, average and standard deviation are calculated and compared to the "correct" value using a *t*-test.

Results of different methods The results obtained with the new method is plotted against the result with the old method for identical samples. The result should be a straight line through the origin with slope 1. This can be confirmed by regression analysis and *t*-test.

Results of different laboratories (ring experiment) The importance of that method for validation is decreasing, however, for quality assurance it is one of best available.

Theoretical understanding of the procedure The uncertainty of each step is determined and the uncertainty of the whole procedure is calculated using the laws of error propagation. If the error contributions of all steps are well understood, the expected uncertainty of a new method can be estimated quickly and cheaply.

If the assay is to be marketed as a product, guidelines by regulatory authorities must be followed. In particular the guidelines of the US Federal Drug Administration (FDA, www.fda.gov/cder/guidance/index.htm) are very detailed, following them ensures that the product can be licensed in most countries of the world without costly additional tests. Even where licensing is not an issue following these guidelines increases the trust placed into the results obtained with the assay.

Formally, standardised methods need not be validated. However, it may be prudent to do so anyway if one is interested in truth rather than in just fulfilling legal requirements.

To **characterise** an assay, the following points need to be addressed:

Accuracy describes how close the arithmetic mean of a measurement series is to the "true" value (see Fig. 47.1).

Precision describes how close the results of repeat measurements are to each other. We distinguish intra-batch, intra-day and inter-batch precision. Precision is usually expressed as **standard deviation (SD)** or **coefficient of variation (CV)**, sometimes also as confidence interval.

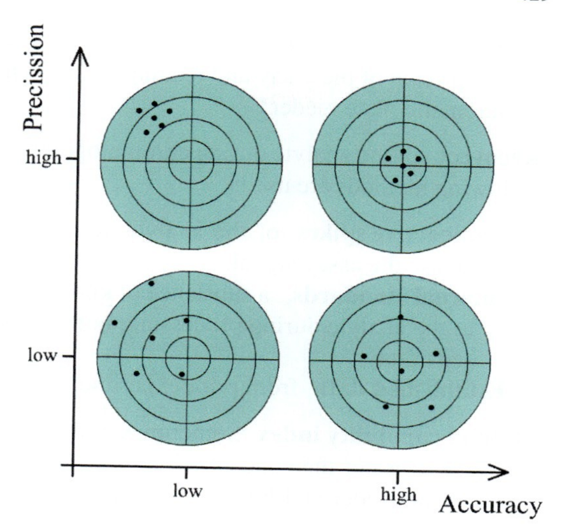

Fig. 47.1 Accuracy describes how close the mean of a series of measurements is to the "true" value. Precision is inversely proportional to the standard deviation

Specificity against matrix effects. Compare the results obtained with known concentrations dissolved in buffer and the biological matrix in question (serum, liquor, urine, *etc.*), at least six independent sources of the matrix should be checked.

Selectivity ability of an assay to determine all analytes of interest in parallel without cross-interference.

Robustness stability of the assay results against changes in condition (temperature, humidity, *p*H, new column, *etc.*).

Sensitivity is usually determined from a standard curve. If the standard curve is linear, 6–8 concentrations (plus blank) covering the expected range are sufficient, for non-linear relationships more data points may be required. The appropriateness of the function used for the standard curve should be determined by tests for goodness of fit (see Chap. 48 on p. 435). The **lower limit of quantitation (LLOQ)** is given by an analyte concentration that gives at least five times the blank signal, a CV of less than 20 % and an accuracy of 80–120 %. The **upper limit of quantitation (ULOQ)** is given by the highest concentration of analyte that can be determined with the same accuracy and precision. Is it possible to determine concentrations higher than ULOQ by diluting the sample?

Reproducibility is the precision of a method when determined in different laboratories or at different times.

Stability of the analyte under a given collection and storage conditions. Stability should be determined for

Long term storage several weeks or months depending on the intended duration of the study.

Short term storage under ambient condition to account for possible destruction of analyte during performance of the assay. Several samples over 4–24 h should be taken.

Freeze/thaw cycles at least three cycles should be measured.

Also important is the signal stability when a sample needs to be stored between completion of the assay and reading of the signal (for example, in the stack of an automatic plate reader).

Recovery of the analyte during chromatography, extraction or processing. The following methods are used:

Radioactive spikes of the analyte at concentrations that do not markedly increase the assay signal.
Internal standards, a substance of known concentration which behaves similar to the analyte during processing, but whose concentration can be determined separately.
Established data from prior experiments.

Method capability index Sometimes the purpose of an analytical method is to test whether or not the concentration of an analyte is within certain **limits of specification** (upper and lower: ULS and LLS). Say, the concentration of an active ingredient of a pharmaceutical preparation is supposed to be between 89 and 91 %. Then a standard deviation s for the analytical method of, say, 1 % would be too high to determine with reasonable certainty whether a given sample is or is not within specification. The method capability index $c_m = \frac{\text{ULS}-\text{LLS}}{6 \times s} = \frac{91-89}{6 \times 1} = 1/3 \leq 1$, hence the method is not capable. The factor 6 comes from the fact that in normal distributed measurements 99.7 % of all outcomes are within ± 3 s. If it were possible to reduce the standard deviation to 0.3 % the method would become capable. It may be necessary to relax the specification, the manufacturing process may deliver better quality, but we are unable to prove it analytically!

If a procedure is well characterised, efforts for validation for a particular application can be significantly reduced.

We distinguish the following levels:

Full validation is used for new or substantially changed method.
Partial validation is used to confirm that an assay works as intended after minor changes in the assay procedure. Partial validation is also used when a different batch of certain reagents is used which may show batch to batch variability (*e.g.*, antibodies) or if the intended application changes (*e.g.*, the analyte is to be determined in a different matrix).
Cross validation is used to compare the results of different assays or different laboratories.

From validation we have to distinguish **qualification**: proof that a certain instrument fulfills its technical specifications. This should be done jointly by manufacturer and customer. Formally, we can distinguish the following phases of qualification:

Design qualification instrument design and manufacturer are chosen according to needs. The specifications of the instrument are part of the purchase contract, often in the form of a **specification sheet**.
Installation qualification The instrument was delivered completely including spare parts, accessories and manual(s). It was installed properly at the intended place, any self-diagnosis shows the expected results.

Operational qualification Correct operation is verified with test samples. This step needs to be repeated after repairs or a change of location. OQ should be performed at the limits of operational range (worst case conditions). The instrument is adjusted and calibrated.

Performance qualification The instrument achieves the required sensitivity and precision with user samples.

Note that there is a very fundamental difference between qualification and validation: The former test technical specifications that are independent of the intended use, the latter test suitability for a given purpose. The sensitivity of a detector may be within its technical specifications, but too low for the purpose at hand!

The following terms have specific meaning:

Measure determine the special value of a physical quantity as multiple of a unit or reference value

Test compare the result of a measurement to some specification

Adjust (permanently) change the settings of an instrument so that the measurement error is minimised

Calibrate determine and document the remaining measurement error

Gauge calibration by a gauging office, which issues a gauging certificate

47.2 Assessing the Quality of Measurements

Assume an assay, where μ_n is the average signal for samples without activity (sham operation, reagent blank), and μ_p the signal obtained for a sample with maximal activity. σ_n and σ_p are the respective standard deviations. Then the

Signal to background is defined as $S/B = \mu_s/\mu_n$. It does, however, help us very little: the means of signal and background may be very different and still the standard deviations overlapping. S/B is therefore rarely quoted.

Dynamic range is the difference between the maximum signal and background: $\mu_p - \mu_n$. Again, without knowledge of the standard deviations, information content of this value is limited.

Signal to noise ratio was originally derived to measure the strength of radio signals and is defined as $S/N = (\mu_s - \mu_n)/\sigma_n$. It thus answers the question: How confident can we be that there is a real signal and not just noise? However, since the standard deviation of the signal is not included, S/N tells us little about the quality of the signal.

Screening window coefficient is calculated from both the means and standard deviations. $|\mu_p - \mu_n| - (3\sigma_p + 3\sigma_n)$ ([446], see Fig. 47.2) is the separation band between the ranges for positive and negative controls. Note that the use of the absolute difference allows for assays that either reduce or increase the signal. Then

$$Z' = \frac{|\mu_p - \mu_n| - (3\sigma_p + 3\sigma_n)}{|\mu_p - \mu_n|} = 1 - \frac{3\sigma_p + 3\sigma_n}{|\mu_p - \mu_n|}. \tag{47.1}$$

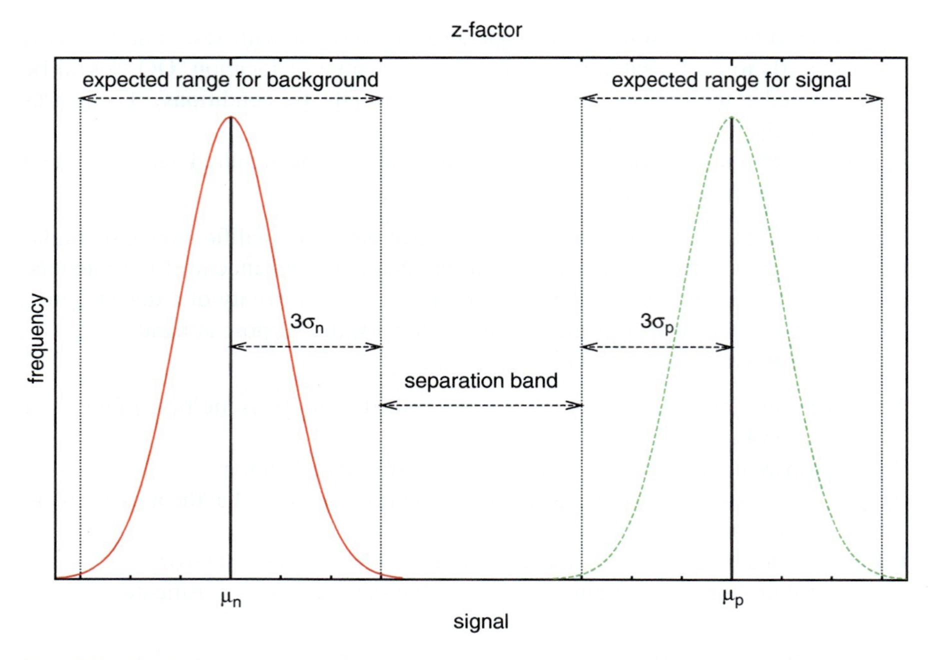

Fig. 47.2 Distribution of signals for negative and positive samples. With a probability of about 99.7 % the value for either will be within ±3 standard deviations σ_{n} and σ_{p} around their respective means μ_{n} and μ_{p}. The bigger the separation band between positive and negative values, the more reliable is the assay. Figure modified from [446]

Ideally, $0.5 \leq Z' < 1.0$ for good separation. If $Z' = 0$ the ranges for positive and negative signal touch (yes/no assays may still be possible), if $Z' < 0$ they overlap, making the assay useless. The quality of individual data point can be calculated in a similar way, replacing $\mu_{\text{p}}, \sigma_{\text{p}}$ with its mean and standard deviation, resulting in Z'. Thus one may compare the quality of results obtained by an assay over its concentration range. This could provide a rationale to set LLOQ and ULOQ.

47.3 Analytical Results Need Careful Interpretation

Regardless of the method used to obtain analytical results, they need careful interpretation. For example, in a study on streptozotozinised rats – an animal model for type I diabetes mellitus – no ketone bodies could be detected in their urine with the usual urine test sticks (BM7, Boehringer Mannheim), although the animals were clearly losing body fat and protein (unpublished observation). In that metabolic situation, mammals would be expected to produce ketone bodies, and some of them should appear in their urine, so the analytical result was unexpected. Studying the manufacturer's documentation for the sticks, it was found that the test field for ketone bodies contained sodium pentacyanonitrosylferrate(III) (nitroprusside sodium), which gives a cherry-red complex with all compounds containing a keto-group. From the three ketone bodies produced in mammals, acetoacetate, acetone and β-hydroxybutyrate, the latter does not contain a keto-group and would therefore not be expected to react with nitroprusside sodium.

A coupled spectrophotometric test for β-hydroxybutyrate, using β-HB dehydrogenase (Sigma), was used to test this hypothesis. It was found that rat urine does contain β-HB, but that the concentration in diabetic animals was slightly, but not significantly, *lower* than that of control animals (1.6 ± 0.5 mg/dl, $n = 9$ vs. 1.8 ± 0.8 mg/dl, $n = 5$), the same also was true for serum. However, if one took into account the much larger urine production in diabetics (101 ± 28 ml/d vs 14 ± 5 ml/d), the excretion of β-HB by diabetic animals is six times higher than by controls (160 ± 65 mg/d vs. 26 ± 16 mg/d).

Thus diabetic rats do get ketosis, but they excrete mostly β-HB, unlike humans who excrete also acetoacetate. It would have been easy to arrive at wrong conclusions during this study, which is presented here as a warning.

47.4 False Positives in Large-Scale Screening

Assume we use a screening test for a rare, but serious disease. Let be A the presence of the disease in a particular test subject, then the prior probability $P(A)$ is the prevalence of the disease in the population, say, 0.001. Further, let B be the event of a positive test result. Then the probability of getting a positive test result for an infected person $P(B|A)$ (read: probability of B, given A) is the sensitivity of the test, which the manufacturer proudly specifies with 99 %. The probability of a positive test for a non-infected person is $P(B|A^c)$ (read: probability of B, given the complement of A), which is 1 - Specificity. If the specificity is 99 %, then $P(B|A^c) = 0.01$. Calculating the number of false-positive and false-negative results can be done by BAYESian statistics, but for our purposes a tree-diagram is more intuitive:

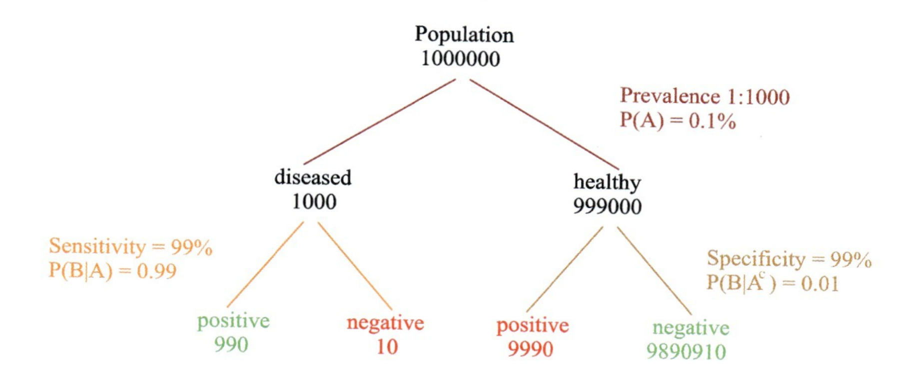

- Out of 1 000 000 randomly selected persons from a population with a prevalence of 1:1 000 we expect 1 000 to be infected and 999 000 to be non-infected.
- From the 1 000 infected persons a test with a sensitivity of 99 % will correctly identify 990, while 10 will be false-negatives.
- From the 999 000 non-infected persons a test with a specificity of 99 % will correctly identify 989 010, while 9 990 will be incorrectly identified as infected (false-positives).
- Of the $9\,990 + 990 = 10\,980$ persons with positive tests, 9 990 will in fact not be infected, that is, for a person with positive test the probability to be non-infected is $P(A^c|B) = 91\,\%$.

In other words, population screening for rare diseases requires extremely high specificities to be useful. Such tests are often difficult and expensive. Hence in many cases cheap screening tests with low specificity are used to identify those cases for which a more expensive confirmation test is warranted. In the above example, only 10 980 out of the 1 000 000 persons in the population (1.1 %) would have to take a confirmation test. There is an ethical dilemma here, however: Informing a person that their screening test for a (potentially fatal) disease has come up positive, and that further tests are required, is not nice when there is only a small chance of this person actually having the disease. One (sometimes) possible solution is to take samples for both tests at the same time, and release positive results only after both search and confirmation tests have been performed. The HIV-antibody test is an example, an ELISA is used for searching, and a Western-blot for confirmation.

Chapter 48
Testing Whether or Not a Model Fits the Data

Usually we use mathematical models to describe our data. Regression methods (linear or non-linear) are then used to determine the parameters of such models that result in the best fit, for example by the least-squares criterium. There is one problem with this approach, however: If the model is inappropriate then the fitting procedure will spit out meaningless parameters without any warning. A classical example – often seen in student's lab reports – is the fitting of a straight line through data that form a curve.

How then can we ensure that the model is appropriate? *Nota bene*: we can not prove by statistical methods that the physicochemical process underlying the model is actually what happens in the experiment. The best we can do is to ensure that the data do not *falsify* the model. Also, the question is not whether or not a more complex model fits the data better. That is usually the case, as equations with more parameters tend to give more flexible curves. However, does this added flexibility allow the curve to better adapt to random noise contained in the data, or can we show that the added parameter(s) is/are needed to explain the results of the experiment, even in the absence of noise?

There is another use of such tests: We may be unsure which of several models is appropriate. For example, is a binding curve hyperbolic or do we have several co-operating binding sites? In such a case one would fit a hyperbolic model and check whether or not the fit is adequate. If not, we test models with an increasing number of binding sites until a reasonable fit is obtained. In a study on enzymatic activity of the Mdr1 multiple drug resistance transporter on ATP-concentration [41] it was found that at least four ATP-binding sites must co-operate to describe the data, models with 1, 2 or 3 binding sites could be rejected. Note that there may well have been more than four binding sites, and models with higher number of binding sites (more flexible curves) may fit the data better. However, a model with four sites explains the data in the statistical sense.

For a parametric t-test, see textbooks of statistics, for example [212, chapter 17]. Unfortunately, this test is quite laborious, and to the best of my knowledge no common data handling software supports this method. The *Null-hypothesis* for this test is that the deviation of the data points from the fitted model can be explained by their standard deviation, that is, the model describes the data adequately. The *alternative*

E. Buxbaum, *Biophysical Chemistry of Proteins: An Introduction to Laboratory Methods*, DOI 10.1007/978-1-4419-7251-4_48,

hypothesis is that the deviation from the model is larger than what would be expected given the standard deviation of the data points, that is, the model is unsuitable.

We can prove smaller differences as significant if the data are of higher quality. As a general rule the standard deviation in enzymological experiments should be less than 2 %, that is achievable with common laboratory equipment and a little practice. In most cases, the result will be obvious by “eyeballing” the data, the test will merely state that the error probability for your conclusion is less than some ridiculously small value. However, during publication reviewers often insist on such tests even where they are superfluous.

48.1 The Runs-Test

For quick-and-dirty tests, a non-parametric method is available, the WALD–WOLFOWITZ **runs test** (also known as STEVENS’ test of iteration frequency). Note that non-parametric tests are not as discriminatory as parametric ones, you may accept the Null-hypothesis in a situation where a parametric test may have rejected it. If the data deviated from the fitted model only by chance fluctuations, each data point would have an equal chance to be either above or below the curve. If, however, the model is inappropriate there will be clusters of data above and below the fitted curve. Our *Null-hypothesis* is that the signs of the residuals $|y - \hat{y}|$ are independent of the controlled variable x. The *alternative hypothesis* is that they are non-randomly distributed.

Figure 48.1 shows a computer-simulation, where data were calculated for a co-operative enzyme with 2 % random noise added. Then curves were fit for both a hyperbolic and a sigmoidal model. Both curves fit the data quite well, but the question is whether the data support the more complex sigmoidal (HILL) model or whether we would have to conclude that the simpler hyperbolic (HENRI–MICHAELIS–MENTEN) model adequately describes the data.

If we sort our data by increasing independent variable and then determine whether or not the dependent variable is larger (+) or smaller (−) than the calculated value (above or below the regression curve), we get strings like + + + + + + + + + + + + + + + + − + − − − − − + + + − − − for the hyperbolic model of Fig. 48.1 and − − − − + − − + + + − + + − − + − + + − − − − − + + − − − for the sigmoidal. For the test, we determine the number of data points N, the number of runs R (blocks of either (+) or (−)), and the number of (+) (n_+) and (−) (n_-) . Both n_+ and n_- must be $\geq$8 for the test to be valid.

The expectation value for R is

$$\mu = \frac{2n_+n_-}{N} + 1 \tag{48.1}$$

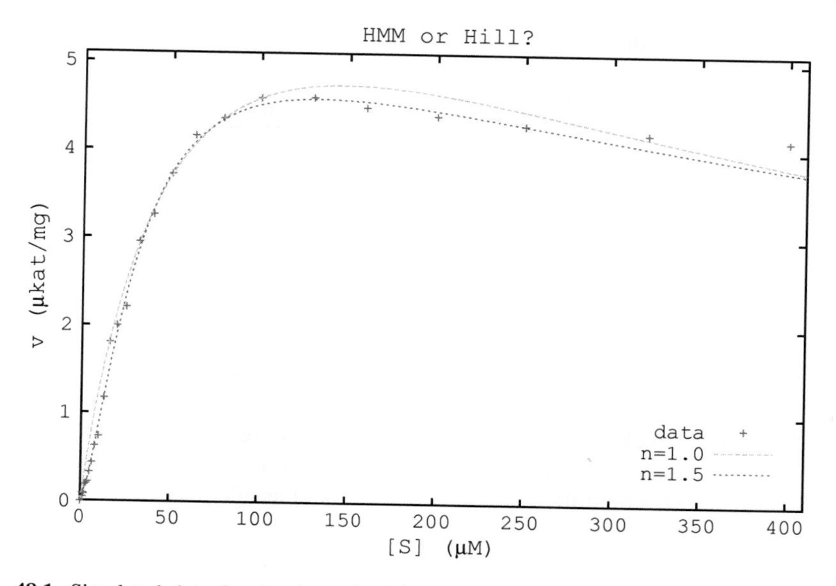

Fig. 48.1 Simulated data for the determination of enzymatic reaction velocity as function of substrate concentration. Data were calculated for a co-operative model with substrate inhibition, and 2 % random GAUSSIAN noise was added. Curve fitting was performed for both the HENRI–MICHAELIS–MENTEN and the HILL-equation. By eyeballing, both curves fit the data reasonably well. Do the data support the more complex co-operative model?

with the standard deviation

$$\sigma = \sqrt{\frac{2n_+n_-(2n_+n_- - N)}{N^2(N-1)}}. \tag{48.2}$$

We can then calculate a test value

$$W = \frac{R - \mu}{\sigma}, \tag{48.3}$$

which is the number of standard deviations away from the mean of a normal distribution. A negative value for W means that we have fewer runs than expected, which might indicate clustering of data. Larger values means we have more runs than expected, this might indicate oscillation. The cut-off value of $|W|$ are

P_0 (%)	$\|W\|$
5.0	1.960
2.0	2.326
1.0	2.576
0.5	2.800
0.2	3.100
0.1	3.300

Returning to the data in Fig. 48.1 we get:

	Hyperbolic	Sigmoidal
N	28	28
n_+	20	11
n_-	8	17
R	6	13
μ	12.4	14.3
σ	2.1	2.5
W	−3.06	−0.55
	$0.2\,\% < P_0 < 0.5\,\%$	$P_0 > 5\,\%$

Thus we would conclude that we have good evidence to reject the hyperbolic model. However, the residuals of the sigmoidal model do not deviate significantly from random noise, we would therefore accept the model as adequate description of the data.

Part X
Appendix

Appendix A
List of Symbols

Table A.1 Often the same symbol is used with different meaning in different fields. In a book that spans many fields of physics, chemistry, and biology, this unfortunately leads to symbol clashes. I have abstained from resolving these clashes by inventing new symbols to maintain consistency with accepted standards (the IUPAC "Green Book" [277] where possible). Thus you will sometimes have to deduce the meaning of a symbol from the context

Symbol	Meaning
$\vec{a}$	Acceleration ($\mathrm{m\,s^{-2}}$)
a	Activity (M)
A	Area ($\mathrm{m^2}$)
	Fluorescence anisotropy (number in 0..1)
B	Magnetic flux density (T)
	Temperature factor in structure models ($\mathrm{\mathring{A}^2}$)
B_2	Second virial coefficient ($\mathrm{cm^3\,g^{-2}}$)
c	Concentration (mol/l)
	Speed of light (2.99793×10^8 m/s in vacuum)
C	Cross-linker concentration in PAGE (%)
C_p	Heat capacity at constant pressure (J/K)
cmc	Critical micellar concentration (M)
d	Penetration depth of evanescent wave (Å)
	Thickness (m)
	Dose (g/d)
D	Diffusion coefficient ($\mathrm{cm^2\,s^{-1}}$)
	Radiation dose (Gy = J/kg)
e	Elementary charge (charge of protons, 1.6022×10^{-19} C/mol)
ΔE^a	Activation energy (J/mol)
$\mathcal{E}$	Absorbance (pure number)
E	Potential difference (V)
	Stereoselectivity of enzymes (pure number)
	Energy (J)
$\vec{\mathfrak{E}}$	Electrical field strength (V/m)
$\vec{\mathfrak{F}}$	Force (N)
F	FARADAY-constant (96 484.61 C/mol)
f	Friction coefficient (kg/s)

(continued)

E. Buxbaum, *Biophysical Chemistry of Proteins: An Introduction to Laboratory Methods*, DOI 10.1007/978-1-4419-7251-4_49,

Table A.1 (continued)

$\vec{g}$	Gravitational acceleration ($g = \mid \vec{g} \mid = 9.8067\,\mathrm{m\,s^{-2}}$)
$\mathbf{g}$	LANDE's **g**-factor, $2\frac{\mu_e}{\mu_0}$
$\Delta G'^0$	GIBBS free energy (under standard biological conditions, J/mol)
$\Delta H'^0$	Heat of reaction (under standard biological conditions, J/mol)
h	PLANCKS quantum (6.6262×10^{-34} J/s)
$\hbar$	$h/2\pi = 1.0546 \times 10^{-34}$ J/s
I	Radiant intensity ($\mathrm{W\,m^{-2}}$)
	Moment of inertia ($\mathrm{kg\,m^2}$)
	Electrical current (A)
I_c	Molar ionic strength (mol/l)
j	Spin quantum number (pure number)
J_{AB}	Coupling constant (Hz)
j	Current density ($\mathrm{A/m^2}$
k	Boltzmann constant (1.3807×10^{-23} J/K)
	Reaction velocity constant (unit depends on order of reaction)
	Pelleting efficiency of rotor (pure number)
k_{cat}	Turnover number ($\mathrm{s^{-1}}$)
K_a	Association constant ($\mathrm{M^{-1}}$)
K_d	Dissociation constant (M)
K_m	MICHAELIS-constant (M)
K_p	Partition ratio (partition coefficient, pure number)
l	Length (m)
m	Mass (kg)
	Modulation depth (pure number)
$\bar{m}$	Aggregation number of a detergent
m_e	Resting mass of electron (9.1091×10^{-31} kg)
M_r	Relative molecular mass (pure number, but Da is often used)
N	Theoretical plates (pure number)
	Number of neutrons in a nucleus
N_A	AVOGADRO's number ($6.022 \times 10^{23}\,\mathrm{mol^{-1}}$)
N_a	Numerical aperture (pure number)
n	HILL-coefficient (pure number)
	number
	Refractive index (pure number)
$\hat{n}$	Complex refractive index
P	Degree of fluorescence polarisation (use anisotropy A instead)
	Probability (pure number between 0 and 1)
$\vec{p}$	Momentum (kg m/s)
pH	Hydrogen ion tension (pure number)
pI	Isoelectric point (pure number)
$p\mathrm{K_a}$	Strength of an acid (pure number)
Q	Electrical charge (C)
r	Radius (m)
r_G	Radius of gyration (Å)
$\bar{r}$	Radius of a molecule (Å)

(continued)

Table A.1 (continued)

R	Universal gas constant ($8.3143\,\mathrm{J\,mol^{-1}\,K^{-1}}$)
	Reflectance (pure number)
	Quality of a molecular model (pure number between 0 and 1)
R_θ	Rayleigh-ratio for angle θ (pure number)
R_f	Relative chromatographic or electrophoretic mobility (pure number)
S	Sedimentation constant (1×10^{-13} s)
$\Delta S'^0$	Entropy of reaction (under standard biological conditions, J/mol)
s	Solubility (mol/l)
t	Time (s)
T	Absolute (thermodynamic) temperature (K)
	Depth of focus (μm)
	Total concentration of a gel (acrylamide + bisacrylamide, %)
U	Voltage (V)
v	Reaction velocity (mol/s)
$\vec{\mathfrak{v}}$	Velocity (m/s)
V_max	Maximal reaction velocity (kat = mol/s)
V	Volume (l)
z	Number of elementary charges transferred in a reaction
Z	Atomic number (number of protons in a nucleus)
Z'	Screening window coefficient (pure number $-\infty < Z' \leq 1$)
α	Polarisability of a molecule ($\mathrm{A\,s\,m^2\,V^{-1}}$)
	Fractional dissociation (pure number)
γ	Electrical conductivity ($\Omega^{-1}\,\mathrm{m^{-1}}$)
	Surface tension ($\mathrm{J\,m^{-2}}$)
	Activity coefficient (pure number)
δ	Hydration of a molecule (g/g)
	Chemical shift (ppm)
ϵ	Molar absorption coefficient ($\mathrm{l\,mol^{-1}\,cm^{-1}}$)
ϵ_0	Permitivity of vacuum (8.854×10^{-12} F/m)
ϵ_r	Permitivity relative to vacuum (pure number)
η	Effectiveness factor (pure number)
	Viscosity ($\mathrm{kg\,m^{-1}\,s^{-1}}$)
$[\eta]$	Intrinsic viscosity (pure number)
θ	Angle of entry (°)
	Molar fraction of enzyme with bound substrate (pure number)
θ_λ	Molar elipticity ($^\circ\,\mathrm{l\,mol^{-1}\,cm^{-1}}$)
λ	Wavelength (nm)
μ	Magnetic moment (V s m)
	Electrophoretic mobility ($\mathrm{cm^2\,V^{-1}\,s^{-1}}$)
	Mean of population
μ_B	BOHR magneton ($\frac{e\hbar}{mc} = 9.2741 \times 10^{-24}$ J/T)
μ_N	Nuclear magneton (5.0508×10^{-27} J/T)
μ_e	Electron magneton (9.2848×10^{-24} J/T)
ν	Frequency (Hz)
$\bar{\nu}$	Partial specific volume (ml/g)
$\tilde{\nu}$	Wavenumber ($\mathrm{cm^{-1}}$)
π	Surface pressure (N/m)

(continued)

Table A.1 (continued)

Π	Osmotic pressure (Pa)
ρ	Mass density (g/ml)
τ	Relaxation time (s)
ϕ	Dihedral angle around the N-C^{α}-bond (°)
	Phase delay (°)
φ_F	Fluorescence yield (number in 0.0–1.0
Φ	Light flux (s^{-1} m^{-2})
ψ	Dihedral angle around the C^{α}-C'-bond (°)
ω_L	LARMOR-frequency (MHz)
$\vec{\omega}$	Angular velocity (radians/s)
σ	Standard deviation
ζ	Dissipation (pure number)

Appendix B
Greek Alphabets

Alpha	α	A
Beta	β	B
Gamma	γ	Γ
Delta	δ	Δ
Epsilon	ϵ, ε	E
Zeta	ζ	Z
Eta	η	H
Theta	θ, ϑ	Θ
Iota	ι	I
Kappa	κ	K
Lambda	λ	Λ
Mu	μ	M
Nu	ν	N
Xi	ξ	Ξ
O	o	O
Pi	π, ϖ	Π
Rho	ρ, ϱ	R
Sigma	σ, ς	Σ
Tau	τ	T
Upsilon	υ	Υ
Phi	ϕ, φ	Φ
Chi	χ	X
Psi	ψ	Ψ
Omega	ω	Ω

E. Buxbaum, *Biophysical Chemistry of Proteins: An Introduction to Laboratory Methods*, DOI 10.1007/978-1-4419-7251-4_50,

Appendix C
Properties of Electrophoretic Buffers

E. Buxbaum, *Biophysical Chemistry of Proteins: An Introduction to Laboratory Methods*, DOI 10.1007/978-1-4419-7251-4_51,

Table C.1 Properties of electrophoretically important buffers and ions. Mobilities are in $10^9\ m^2s^{-1}V^{-1}$

Protolyte	Structure	M_r	pK_a (0 °C)	pK_a (25 °C)	dpK_a/dT	μ_∞ (0 °C)	μ_∞ (25 °C)
ACES		182.21	7.38	6.84	−0.020	−14.24	−27.86
Acetic acid		60.05	4.78	4.76	0.0002	−22.18	−42.39
Alanine (1)		90.09	2.51	2.43		17.46	32.67
Alanine (2)	$H_3C{-}CH(NH_3^+){-}COO^- \overset{H^+}{\rightleftharpoons} H_3C{-}CH(NH_2){-}COO^-$	89.09		9.69			
β-Alanine (1)	$H_3N^+{-}CH_2{-}CH_2{-}COOH \overset{H^+}{\rightleftharpoons} H_3N^+{-}CH_2{-}CH_2{-}COO^-$	90.09	3.65	3.55		17.22	29.30
β-Alanine (2)	$H_3N^+{-}CH_2{-}CH_2{-}COO^- \overset{H^+}{\rightleftharpoons} H_2N{-}CH_2{-}CH_2{-}COO^-$	89.09		10.19			
Ammediolinium	$HO{-}CH_2{-}C(NH_3^+)(CH_3){-}CH_2{-}OH \overset{H^+}{\rightleftharpoons} HO{-}CH_2{-}C(NH_2)(CH_3){-}CH_2{-}OH$	106.14	9.56	8.83		13.37	27.86
Ammonium	$NH_4^+ \overset{H^+}{\rightleftharpoons} NH_3$	18.03	10.12	9.24	−0.031	40.32	76.23
Asparagine (1)	$H_3N^+{-}C(=O){-}CH_2{-}CH(NH_3^+){-}COOH \overset{H^+}{\rightleftharpoons} H_3N^+{-}C(=O){-}CH_2{-}CH(NH_3^+){-}COO^-$	151.12		2.02			
Asparagine (2)	$H_2N{-}C(=O){-}CH_2{-}CH(NH_3^+){-}COO^- \overset{H^+}{\rightleftharpoons} H_2N{-}C(=O){-}CH_2{-}CH(NH_2){-}COO^-$	150.12	9.04	8.86		−14.98	−29.78
Bicine	$(HO{-}CH_2{-}CH_2)_2NH^+{-}CH_2{-}COO^- \overset{H^+}{\rightleftharpoons} (HO{-}CH_2{-}CH_2)_2N{-}CH_2{-}COO^-$	163.17	8.78	8.34	−0.018	−19.70	−36.04

(continued)

Table C.1 (continued)

BisTris	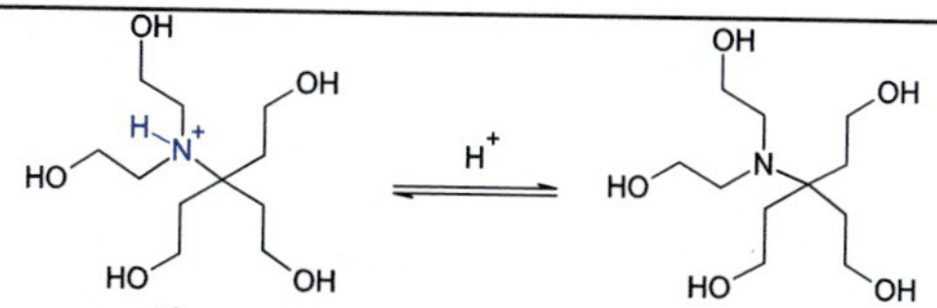 $(HOCH_2CH_2)_2NH^+C(CH_2OH)_3 \overset{H^+}{\rightleftharpoons} (HOCH_2CH_2)_2NC(CH_2OH)_3$	210.24	6.88	6.50	−0.017	12.00	24.01
Boric acid	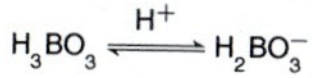 $H_3BO_3 \overset{H^+}{\rightleftharpoons} H_2BO_3^-$	61.83	9.50	9.24	−0.008	−18.21	−33.63
Cacodylic acid	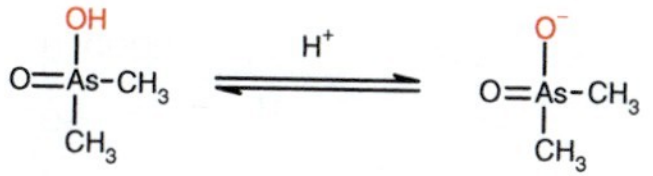 $O{=}As(OH)(CH_3)CH_3 \overset{H^+}{\rightleftharpoons} O{=}As(O^-)(CH_3)CH_3$	137.99	6.20	6.31		−14.52	−25.34
Ethanolaminium	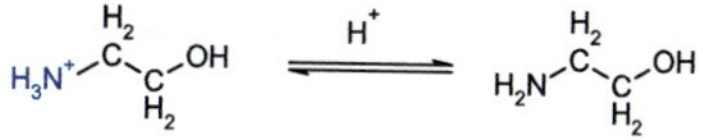 $H_3N^+{-}CH_2{-}CH_2{-}OH \overset{H^+}{\rightleftharpoons} H_2N{-}CH_2{-}CH_2{-}OH$	62.08	10.35	9.54		23.92	47.59
Formic acid	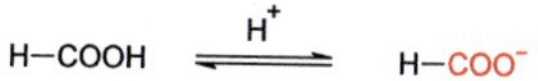 $H{-}COOH \overset{H^+}{\rightleftharpoons} H{-}COO^-$	46.03	3.78	3.75	0.0	−30.88	−56.60
GABA (1)	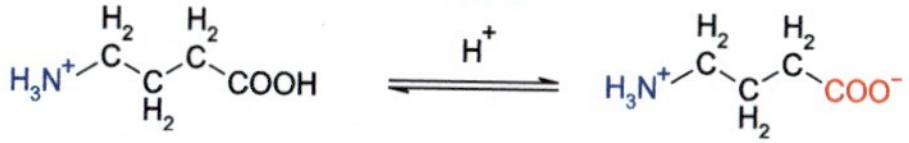 $H_3N^+{-}CH_2{-}CH_2{-}CH_2{-}COOH \overset{H^+}{\rightleftharpoons} H_3N^+{-}CH_2{-}CH_2{-}CH_2{-}COO^-$	104.12	4.09	4.03		18.46	33.63
GABA (2)	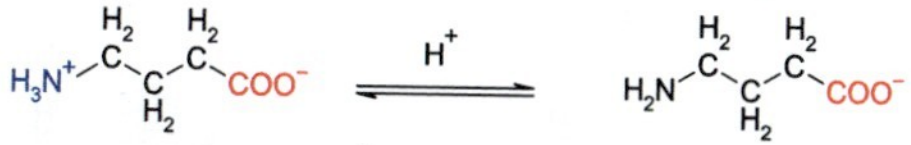 $H_3N^+{-}CH_2{-}CH_2{-}CH_2{-}COO^- \overset{H^+}{\rightleftharpoons} H_2N{-}CH_2{-}CH_2{-}CH_2{-}COO^-$	103.12	11.73	10.56		−16.22	−32.19
Glycine (1)	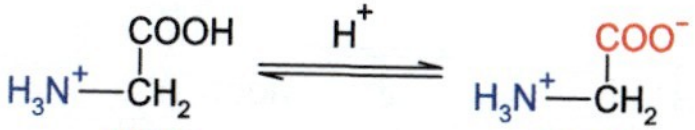 $H_3N^+{-}CH_2{-}COOH \overset{H^+}{\rightleftharpoons} H_3N^+{-}CH_2{-}COO^-$	76.07	2.44	2.35		15.73	36.52
Glycine (2)	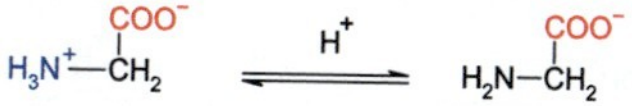 $H_3N^+{-}CH_2{-}COO^- \overset{H^+}{\rightleftharpoons} H_2N{-}CH_2{-}COO^-$	75.07	10.50	9.78	−0.025	−20.94	−38.45
Glycylglycine (1)	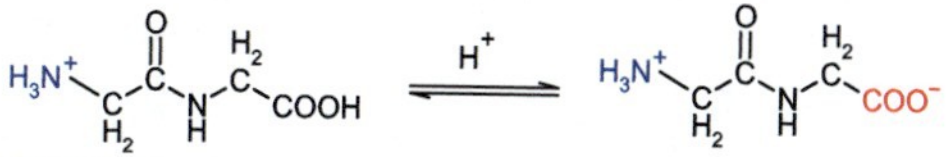 $H_3N^+{-}CH_2{-}C({=}O){-}NH{-}CH_2{-}COOH \overset{H^+}{\rightleftharpoons} H_3N^+{-}CH_2{-}C({=}O){-}NH{-}CH_2{-}COO^-$	133.12		3.12			

(continued)

Table C.1 (continued)

Glycylglycine (2)	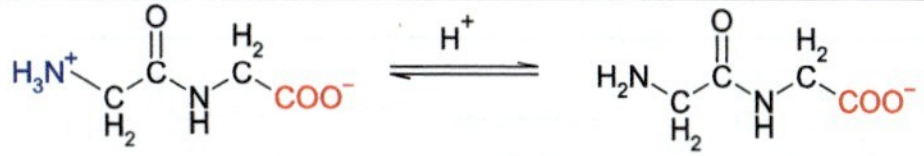 $H_3N^+CH_2CONHCH_2COO^- \overset{H^+}{\rightleftharpoons} H_2NCH_2CONHCH_2COO^-$	132.12	8.98	8.25	−0.025	−16.97	−30.26
HEPES	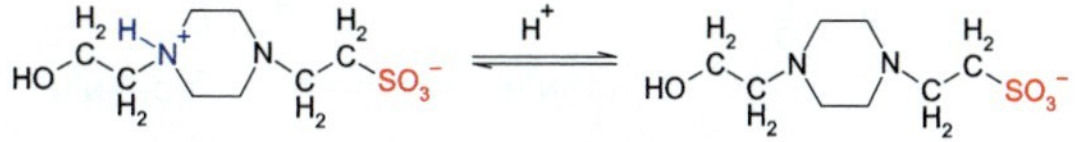 $HOCH_2CH_2N(C_4H_8)N^+HCH_2CH_2SO_3^- \overset{H^+}{\rightleftharpoons} HOCH_2CH_2N(C_4H_8)NCH_2CH_2SO_3^-$	238.30	7.89	7.52	−0.014	−9.52	−17.27
Histidine (1)	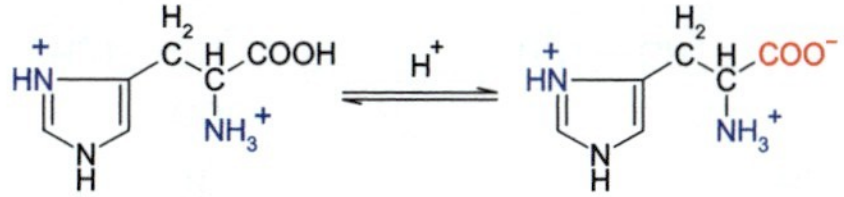 $(HN^+\text{-imidazole})CH_2CH(NH_3^+)COOH \overset{H^+}{\rightleftharpoons} (HN^+\text{-imidazole})CH_2CH(NH_3^+)COO^-$	157.16		1.78			
Histidine (2)		156.16	6.35	6.04		14.73	24.01
Histidine (3)	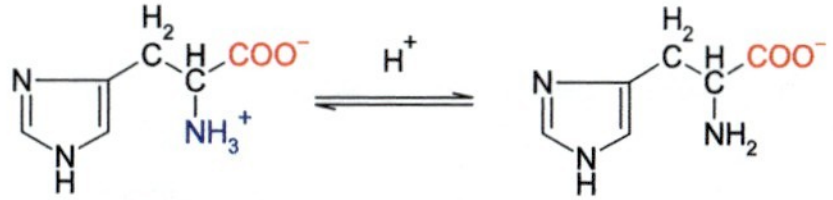 $(\text{imidazole})CH_2CH(NH_3^+)COO^- \overset{H^+}{\rightleftharpoons} (\text{imidazole})CH_2CH(NH_2)COO^-$	155.16		8.97			
Hydrochloric acid	$HCl \overset{H^+}{\rightleftharpoons} Cl^-$	36.46				−42.95	−79.12
Imidazole	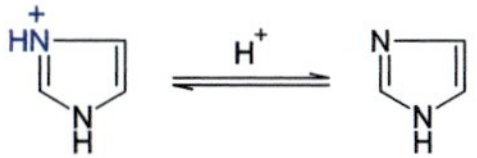 $(HN^+\text{-imidazole}) \overset{H^+}{\rightleftharpoons} (\text{imidazole})$	69.08	7.46	6.99	−0.020	26.65	43.74
Lactic acid	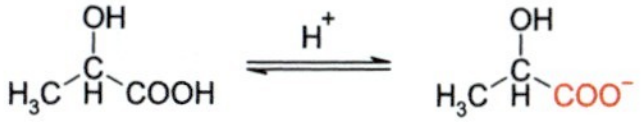 $H_3CCH(OH)COOH \overset{H^+}{\rightleftharpoons} H_3CCH(OH)COO^-$	90.08	3.88	3.84		−18.95	−36.50
Lutidinium	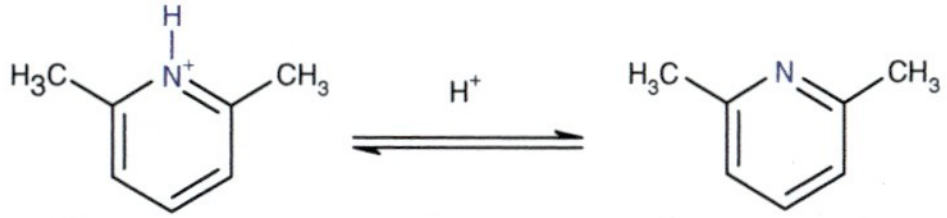 $(H_3C)_2C_5H_3N^+H \overset{H^+}{\rightleftharpoons} (H_3C)_2C_5H_3N$	107.15	7.00	6.79		10.96	29.08
MES	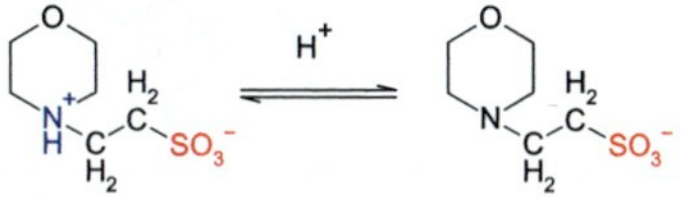 $O(C_4H_8)N^+HCH_2CH_2SO_3^- \overset{H^+}{\rightleftharpoons} O(C_4H_8)NCH_2CH_2SO_3^-$	195.20	6.45	6.16	−0.011	−13.94	−25.45

(continued)

Table C.1 (continued)

Morpholine	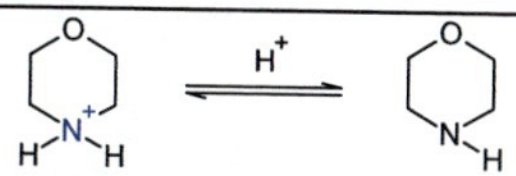	88.12	8.85	8.60		20.69	40.37
Phosphate (1)	$H_3PO_4 \overset{H^+}{\rightleftharpoons} H_2PO_4^-$	98.00	2.06	2.15	0.0044	−17.46	−32.19
Phosphate (2)	$H_2PO_4^- \overset{H^+}{\rightleftharpoons} HPO_4^{2-}$	97.00	7.32	7.20	−0.0028	−28.33	−59.08
Phosphate (3)	$HPO_4^{2-} \overset{H^+}{\rightleftharpoons} PO_4^{3-}$	96.00		12.33	−0.026		
Picolinium	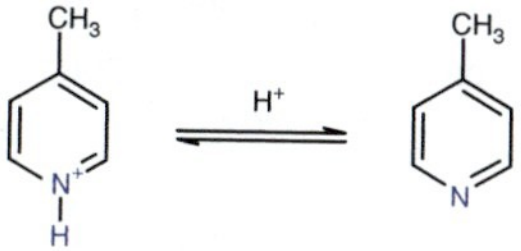	93.13	6.21	5.98		19.45	37.90
Potassium	K^+	39.10	–	–	–	39.57	76.18
Propionic acid	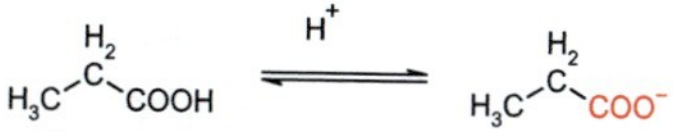 $H_3C-CH_2-COOH \overset{H^+}{\rightleftharpoons} H_3C-CH_2-COO^-$	74.08	4.90	4.87		−18.21	−37.48
Pyridinium	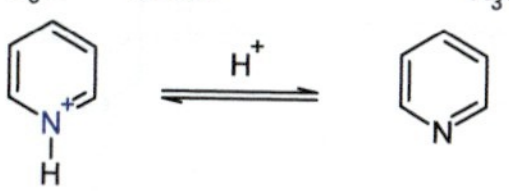	80.10	5.50	5.21	−0.014	23.92	42.30
Proline	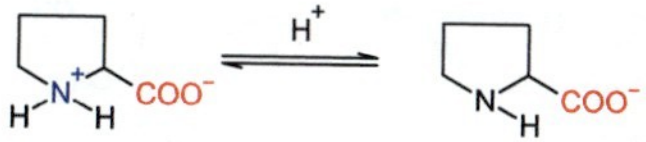	115.13	11.33	10.64		−17.22	−39.89
Sodium	Na^+	22.99	–	–		27.40	51.92
Sulphate (1)	$H_2SO_4 \overset{H^+}{\rightleftharpoons} HSO_4^-$	98.08				−25.40	−52.00

(continued)

Table C.1 (continued)

Sulphate (2)	$HSO_4^- \overset{H^+}{\rightleftharpoons} SO_4^{2-}$	97.08				−42.49	−82.94
Taurine	$H_3N^+\text{-}CH_2\text{-}CH_2\text{-}SO_3^- \overset{H^+}{\rightleftharpoons} H_2N\text{-}CH_2\text{-}CH_2\text{-}SO_3^-$	125.14	9.74	9.06	−0.022	−17.71	−34.11
TES	$(HOCH_2)_3C\text{-}N^+H_2\text{-}CH_2CH_2\text{-}SO_3^- \overset{H^+}{\rightleftharpoons} (HOCH_2)_3C\text{-}NH\text{-}CH_2CH_2\text{-}SO_3^-$	229.24	8.00	7.48	−0.020	−12.75	−26.90
Tricine	$(HOCH_2)_3C\text{-}N^+H_2\text{-}CH_2\text{-}COO^- \overset{H^+}{\rightleftharpoons} (HOCH_2)_3C\text{-}NH\text{-}CH_2\text{-}COO^-$	179.17	8.64	8.09	−0.021	−10.51	−24.01
Triethanolamonium	$(HOCH_2CH_2)_3NH^+ \overset{H^+}{\rightleftharpoons} (HOCH_2CH_2)_3N$	150.19	8.35	7.80	−0.020	16.22	30.26
Tris	$(HOCH_2)_3C\text{-}NH_3^+ \overset{H^+}{\rightleftharpoons} (HOCH_2)_3C\text{-}NH_2$	122.14	8.84	8.07	−0.028	12.75	27.86
Veronal	$O{=}C(NH\text{-}CH(OH))_2C(CH_2\text{-}CH_3)_2 \overset{H^+}{\rightleftharpoons} O{=}C(NH\text{-}CH(O^-))(NH\text{-}CH(OH))C(CH_2\text{-}CH_3)_2$	184.19	8.40	7.98		−13.99	−24.97
Water (1)	$H_3O^+ \overset{H^+}{\rightleftharpoons} H_2O$	19.02		−1.74		248.74	362.56
Water (2)	$H_2O \overset{H^+}{\rightleftharpoons} OH^-$	18.02	16.63	15.74		−108.83	−205.52

Appendix D
Bond Properties

Table D.1 VAN DER WAALS-radii of atoms

Element	Radius Å
H	1.2
C	2.0
N	1.5
O	1.4
S	1.85
P	1.9

Table D.2 Properties of bonds

Type	Length Å	Energy kJ/mol
C−C	1.54	350
C=C	1.33	610
C≡C		
N−N	1.45	161
N=N	1.24	161
C−N	1.47	290
C=N	1.26	610
C−O	1.43	350
C=O	1.14	720
C−S	1.82	260
O−H	0.96	463
N−H	1.01	390
S−H	1.35	340
C−H	1.09	410
H−H	0.74	430
$-\mathrm{OH}\cdots\mathrm{O} <$	2.76	18.8
$-\mathrm{OH}\cdots\mathrm{O{=}C} <$	2.80	34.3
$-\mathrm{OH}\cdots\mathrm{N} <$	2.78	29
$> \mathrm{NH}\cdots\mathrm{O{=}C} <$	2.93	8
$> \mathrm{NH}\cdots\mathrm{N} <$	3.07	5.4
$> \mathrm{NH}\cdots\mathrm{S} <$	3.40	
$\mathrm{CH}\cdots\mathrm{O} <$	3.23	

E. Buxbaum, *Biophysical Chemistry of Proteins: An Introduction to Laboratory Methods*, DOI 10.1007/978-1-4419-7251-4_52,

Appendix E
Acronyms

AC	alternating current
ACF	autocorrelation function, statistical method to determine diffusion coefficients by light scattering or fluorescence
ADC	analog/digital converter, electronic circuit that allows data from analog sensors to be handled by a digital computer
ADP	adenosine diphosphate
AFA	alcohol/formalin/acetic acid, a fixative
AFM	atomic force microscopy, technique that reconstructs the shape of an object by scanning it with a fine tip, allowing molecular resolution
AGTC	N-(4-azidobenzoylglycyl)-*S*-(2-thiopyridyl)-cysteine, photoactivatable and cleavable cross-linker
AIDS	acquired immune deficiency syndrome, human disease caused by infection with HIV
Akt	AKR mouse directly transforming retrovirus associated oncogene, also known as protein kinase B
ALARA	as low as reasonably achievable, principle used in radiation protection
ALEX	alternating laser excitation, method in multi-colour fluorescence spectroscopy
ANS	aminonaphtalene sulphonic acid, WEBERS reagent, dye that fluorescs only in a hydrophobic environment
APS	ammonium peroxodisulphate, initiator used for the polymerisation of acrylamide
ARF	ADP-ribosylating factor, small G-protein in ER vesicle transport
ATCC	American type culture collection, source of cultured cells and microorganisms
ATP	adenosine triphosphate, energy carrier in cellular metabolism
ATPase	adenosine triphosphatase, enzymes that hydrolyse ATP to ADP and P_i
BCA	bicinchonic acid, used for protein quantitation
BCIP	5-Bromo-4-chloro-3-indolylphosphate, reagent for phosphatase activity
BLM	black lipid membrane, synthetic double membrane in a small orifice

E. Buxbaum, *Biophysical Chemistry of Proteins: An Introduction to Laboratory Methods*, DOI 10.1007/978-1-4419-7251-4_53,

βME β-mercaptoethanol, reducing agent to protect protein -SH groups
BRET bioluminescence resonance energy transfer, like FRET, but with luciferase as light source
BSA bovine serum albumin, relatively cheap protein often used as carrier in experiments
CBB-G250 Coomassie Brilliant blue G250, dye used in the Bradford protein assay
CBB-R250 Coomassie Brilliant blue R250, dye used for staining proteins after electrophoresis
CCD charge coupled device, a type of electronic camera
CD circular dichroism, measures differences in the absorbtion coefficient for right and left circular polarised light
cDNA complementary DNA, obtained from mRNA with reverse transcriptase
CHAPS 3-[(3-Cholamidopropyl)dimethylammonio]propanesulphonic acid, a detergent with both negative and positive charges, sometimes also used as buffer
cmc critical micellar concentration, concentration at which detergent molecules in solution aggregate into micelles
COSY correlated spectroscopy, technique in protein NMR-spectroscopy
cpm counts per minute, commonly used unit of radioactivity
CROX chromium oxalate, used as a quencher in ESR-spectroscopy
CTAB cetyltrimethylammonium bromide, a positively charged detergent
CV coefficient of variation, statistical measure of the precision of an assay
DAPI 4',6-diamidino-2-phenylindole, DNA intercalating fluorescent stain
DC direct current
DCCD dicyclohexylcarbodiimide, used to activate carboxylic acid groups for coupling reactions
DEPC diethylpyrocarbonate, reagent to modify His residues in proteins
DESI Desorption electrospray ionization, method to produce ions for MS from molecules on a surface, *e.g.*, skin
DESY Deutscher Elektronensynchrotron, nuclear research facility in Hamburg, Germany
DIGE differential gel electrophoresis, method where two or more labeled samples are subjected to 2-D gel electrophoresis on the same gel. Intensity of the various labels in each spot is later compared.
DIN Deutsches Institut für Normung e.V., German organization for standardization
DISC discontinous electrophoresis, isotachyphoresis in a collecting gel resulting in narrow bands followed by zone electrophoresis in a separation gel.
DMSO dimethyl sulphoxide, organic solvent, water miscible but able to dissolve many hydrophobic compounds. Warning: skin permeable!
DNA desoxyribonucleic acid, carrier of genetic information

DOC	desoxycholate, detergent
DPH	1,6-diphenyl-1,3,5-hexatriene, compound with high fluorescence in hydrophobic, but low fluorescence in hydrophilic environment
DPI	dual polarisation interferometry, optical method to measure changes in protein structure
DRM	detergent resitant membrane, the part of the plasma membrane that is insoluble in mild detergents like Triton-X100 (equivalent to rafts?)
DSC	differential scanning calorimetry, method used to measure heat capacity as a function of temperature
DTSP	3,3'-dithiobis(succimidylpropionate), cleavable homobifunctional cross-linker (with sulphonic acid group: Sulpho-DTSP for increased solubility)
DTT	dithiotreitol, reducing agent with two -SH groups
EC	Enzyme Commission, organisation within the IUBMB responsible for enzyme nomenclature
ECACC	European collection of cell cultures, source of culture cells and microorganisms
EDAC	1-ethyl-3-(3-dimethylaminopropyl)carbodiimide, water soluble carbodiimide for labeling COOH-groups
ECD	electron capture dissociation, protein sequencing technique in MS, that also characterises post-translational modification
EDTA	ethylene diamine tetraacetic acid, complex former
EGTA	ethyleneglycol tetraacetic acid, complex former
EI	electron ionisation, ionisation method in mass spectrometry
ELISA	enzyme linked immunosorbent assay, very sensitive method for detecting antigens or antibodies
EM	electron microscopy, using electrons instead of light for microscopy results in higher resolution
EMBL	European Molecular Biology Organisation
EMIT	enzyme multiplied immunoassay technique, immunological method for the rapid detection of small molecules like drugs
ENDOR	electron nuclear double resonance, technique in ESR-spectroscopy
EN	Euronorm, standard in the EU
EPR	electron paramagnetic resonance, synonym for ESR
Erk	extracellular signal regulated kinase, protein in the **MAP!**-kinase pathway
ESI	electrospray ionisation, method for ionisation in mass spectrometry
ESR	electron spin resonance, method for measuring the state of single electrons
EU	European Union, organization of (at the time of writing) 27 European countries
excimer	exited dimer, complex of two dye molecules with different fluorescent properties than the single molecules
FAB	fast atom bombardment, method for ionisation in mass spectrometry

FACS	fluorescence activated cell sorter, used to separate particles by fluorescent properties
FAD	flavine adenine dinucleotide, oxidised
$FADH_2$	flavine adenine dinucleotide, reduced
FBS	fetal bovine serum, a synonym for FCS
FCS	fetal calf serum, used in cell culture to improve cell growth rate
FDA	Food and Drug Administration, regulatory authority in the **US!**
FFT	fast Fourier transform, computer method to transform data from the intensity versus time to the intensity versus frequency domain
FID	flame ionisation detector, used in gas chromatographs
FITC	fluorescein isothiocyanate, activated form of fluorescein, reacts with primary amino-groups
FMOC-	9-fluorenyl-methoxycarbonyl-, used as protection group during peptide synthesis
FRAP	fluorescence recovery after photo-bleaching, technique to measure diffusion rates of in living cells
FRET	FÖRSTER resonance energy transfer, method to determine the distance of two labels
FTIR	Fourier transformed infrared, modern way of obtaining IR-spectra fast and with high resolution
GC	gas chromatography
GFP	green fluorescent protein, 27 kDa protein that autocatalytically forms a fluorophore and that can be grafted onto a protein of interest to follow its fate in a living cell by fluorescence microscopy
GLP	good laboratory practice, method to ensure that results are obtained and documented in a reliable manner (also expanded: a great lot of paperwork)
GPI	glycosylphosphatidylinositol, anchors membrane-associated proteins to the membrane
HAC	hydroxylapatite chromatography, hydroxylapatite is a crystal form of calcium phosphate
HEPES	*N*-(2-hydroxyethyl)piperazine-*N*'-(2-ethanesulphonic acid), buffer substance
HIC	hydrophobic interaction chromatography, separating proteins by hydrophobicity
HIV	human immunodeficiency virus, causative agent of acquired immune deficiency syndrome (AIDS)
HLB	hydrophilic lipophilic balance number, physico-chemical property of a detergent
HPLC	high performance liquid chromatography (**not** high pressure!)
HRP	horseradish peroxidase, marker enzyme linked to antibodies in immunological detection procedures
Hsc70	70 kDa heat shock cognate, constitutively expressed cytosolic isoform of the 70 kDa heat shock protein
IEF	isoelectric focussing, separates proteins by isoelectric point

IC	internal conversion, pathway for the return of electrons from an exited to the ground state
IEC	ion exchange chromatography, separates proteins or small molecules by charge
IgG	immunoglobulin G
IgY	immunoglobulin Y, obtained from chick eggs
IMAC	immobilised metal affinity chromatography, separation method used mainly for genetically engineered proteins with "His-tag"
IR	infrared, light with wavelength 1 mm > λ > 780 nm
ISC	intersystem crossing, pathway for the return of an excited electron to the ground state
ISO	International Organization of Standardization
ITC	isothermal titration calorimetry, method to measure a heat of reaction
IUBMB	International Union for Biochemistry and Molecular Biology
IUPAC	International Union for Pure and Applied Chemistry
KLH	keyhole limpet haemocyanine, protein of high molecular mass used as carrier for haptens in immunisation experiments
LC	liquid chromatography
LDAO	lauryl dimethyl amine oxide, detergent used for membrane solubilisation
LILBID	laser induced liquid beam ion desorption, ionisation technique for mass spectrometry of proteins
LLS	laser light scattering, method to determine the size of large molecules and small particles in suspension
LLOQ	lower limit of quantitation, minimal concentration of an analyte that allows quantitation at reasonable accuracy and precision
mAb	monoclonal antibody, obtained from fusing antibody-producing B-cells with immortal hybridoma cells
MAD	multiple anomalous dispersion, method of obtaining the phase information from X-ray scattering experiments, using X-rays of different wavelengths
MALDI	matrix assisted laser desorbtion/ionisation, method for the gentle ionisation of large molecules in mass spectrometry
MA-MEF	microwave-accelerated metal-enhanced fluorescence, rapid detection method for biomolecules
MAS	magic angle spinning, method to investigate membrane proteins, protein crystals and other solid samples by NMR
Mdr1	multiple drug resistance transporter 1, ABC-type transmembrane ATPase, product of the ABCB1 gene
MECC	micellar electrokinetic capillary electrophoresis, separation method for neutral molecules, which partition between the solution (stationary phase) and SDS-micelles (mobile phase)

MIR multiple isomorphous replacement, method of obtaining the phase information from X-ray scattering experiments by binding heavy metals to specific sites of the protein in a crystal
MPD 2-methylpentan-2,4-diol, precipitant used for protein crystallization
mRNA messenger RNA, working copy of genetic information
MS mass spectrometry, analytical method for separating molecules by molecular mass
MSP amphipatic membrane scaffold proteins, proteins that can stabilise phospholipids in small disks of 90 Å diameter
MT-MEC Microwave-triggered metal-enhanced chemiluminescence, fast and sensitive detection method for biomolecules
MudPIT multi-dimensional protein identification technology, method for proteomics
NAD^+ nicotine adenine dinucleotide, oxidised, acceptor of activated hydrogen, usually in catabolic reactions
NADH nicotine adenine dinucleotide, reduced, carrier of activated hydrogen
$NADP^+$ nicotine adenine dinucleotide phosphate, oxidised, acceptor of activated hydrogen
NADPH nicotine adenine dinucleotide phosphate, reduced, donor of activated hydrogen, usually in anabolic reactions
NAD(P) either NAD^+ or $NADP^+$
NAD(P)H either NADH or NADPH
NBT nitroblue tetrazolium, reagent for phosphatase detection
NC nitrocellulose, material for blotting membranes
NEPHGE non-equilibrium *p*H-gradient electrophoresis, IEF-like technique for basic proteins
NHSS N-hydroxysulphosuccinimide, used to prepare sulphosuccimidyl esters as reactive probes
NIST National Institute of Standards and Technology, federal agency in the USA
NMR nuclear magnetic resonance, method for determining the distances of nucleï in a molecule
NOE nuclear OVERHAUSER effect, used in multi-dimensional NMR-spectroscopy
NOESY nuclear OVERHAUSER effect spectroscopy
NTA nitrilotriacetic acid, complexon, often used as stationary phase for IMAC
ODMR optically detected magnetic resonance, detection of triplet states in ESR by phosphorescence
OG octyl glucoside, non-ionic detergent
OPA *ortho*-phtaldialdehyde, fluorescent reagent for primary amines
ORD optical rotary dispersion, detecting different refractive indices for left and right circular polarised light
PAC photoacoustic calorimetry, method to determine the enthalpy required to break a bond

PAGE	polyacrylamide gel electrophoresis, electrical separation of charged polymers
PAPDIP	Methyl-3-(p-azidophenyl)dithiopropioimidate, reversible cross-linker
PAS	periodic acid/SCHIFF, reaction that gives a bright red colour with sugars
PBS	phosphate buffered saline, solution that resembles the interstitial fluid in its ionic composition
PCR	polymerase chain reaction, method to amplify specific DNA or RNA sequences
PDB	protein data base, data bank with protein structures available at http://www.rcsb.org/pdb/home/home.do
PEG	polyethylene glycol
PET	photoinduced electron transfer, process where fluorescent molecules either donate the excited electron to an acceptor (becoming oxidised) or accept an additional electron from a donor (becoming reduced)
PGP	permeability glycoprotein. syn. for Mdr1
P_i	inorganic phosphate
PIE	pulsed interleaved excitation, method in multi-colour fluorescence spectroscopy
PITC	phenyl isothiocyanate, compound used in Edman-degradation of proteins
PMSF	phenyl methyl sulphonyl fluoride, inhibitor of Ser-dependent hydrolases (proteases, esterases)
PMT	photomultiplier tube, sensitive device to measure light down to single photons
POPOP	1,4-Bis-2-(5-phenyl-2-oxazolyl)-benzene, compound used in scintillation fluids for counting β-radiation
PPO	2,5-Diphenyloxazole, compound used in scintillation fluids for counting β-radiation
PTFE	polytetrafluorethylene, temperature-resistant, inert and unwettable plastic
PVDF	polyvinylidene fluoride, $-CH_2-CF_2-$, in the literature often incorrectly expanded -difluoride, plastic membrane with high protein binding capacity
QCM	quartz crystal microbalance, sensitive device to measure the deposition of substances on a surface
QIP	quench-identification parameter, used to detect quenching in β-scintillation counters
RIA	radioimmunoassay, method for detecting antigens or antibodies
RIPA	radioimmunoprecipitation assay
RMS	root mean square, or quadratic mean: $x_{\mathrm{rms}} = \sqrt{\frac{1}{n}\sum x_i^2} = \bar{x}^2 + \sigma_x^2$
RNA	ribonucleic acid

RPC	reversed phase chromatography, chromatographic separation on hydrophobic stationary phases with polar solvent
rpm	revolutions per minute, common unit for centrifugal speed
Rubisco	ribulose bisphosphate carboxylase, enzyme that fixes carbon in C3-plants
RF	radio frequency
S	singlet state, excited state of electrons without spin-reversal
SAXS	small angle X-ray scattering, method to determine the size of molecules
SANS	small angle neutron scattering
SCR	sample channel ratio, parameter used to identify quenching in β-scintillation counters
SD	standard deviation, statistical measure of precision of a measurement
SDS	sodium dodecyl sulphate, negatively charged detergent
SEC	size exclusion chromatography, syn. gel filtration
SELDI	surface enhanced laser desorption ionisation, MALDI-like technique for MS, sample is captured on a protein chip first
SELEX	systematic evolution of ligands by exponential enrichment, method to generate aptamers
SEM	scanning electron microscopy, method for obtaining 3-dimensional pictures of small objects
SNL	signal-to-noise levels, value that characterises the quality of experimental results
SNOM	scanning near-field optical microscope, fluorescent microscopy where the illumination is supplied by a fine scanning tip, the diameter of which determines the resolution of the microscope
SOP	standard operating protocol, detailed description how a certain method is to be performed
SPIM	single plane imaging microscopy, technique to increase contrast in fluorescence microscopy by eliminating out-of-focus light
SPR	surface plasmon resonance, method to measure the refractive index of a medium in atomic distance from a surface
STM	scanning tunnelling microscope, scans a surface by keeping the tunnelling current between a tip and the sample constant
SVD	singular value decomposition, computer method to separate noise and signal in measured data
T	triplet state, excited state of an electron with spin reversal
Tat	twin arginine translocon, system that transports folded proteins through the plasma membrane of bacteria into their periplasmic space
TBS	Tris buffered saline, 10 mM Tris-HCl, 150 mM NaCl pH 8.8 used instead of PBS in cases were phosphate interferes with a measurement
TCA	trichloroacetic acid, used to denature and precipitate proteins

TCEP tris-(2-carboxyethyl)-phosphine, reagent to reduce cystine to cysteine

TEA triethylammonium, used as buffer-substance that can be removed by lyophilisation

TEM transmission electron microscopy, microscopy with electrons instead of light, may achieve atomic resolution

TEMED *N,N,N',N'*-Tetramethylethylenediamine, catalyst used for the polymerisation of acrylamide

TEMPO 2,2,6,6-tetramethylpiperidin-1-oxide, stable spin label used in ESR-spectroscopy

TIR total internal reflection, method to measure the optical properties of a surface-bound sample by evanescent waves

TIRF total internal reflection fluorescence (microscopy), method to observe reactions on single molecules

TLC thin layer chromatography

TLCK *N*-Tosyl-L-lysyl-chloromethane, inactivator of trypsin

TNBS 2,4,6-trinitrobenzene sulphonic acid, substance used to determine free amino groups in proteins

TOF time of flight, detector in mass spectrometry (MS)

TPA tripropylamine, redox reagent in electrochemiluminescence

TPCK *N*-Tosyl-L-phenylalanyl chloromethane, inactivator of chymotrypsin

Tris tris-hydroxymethyl aminomethane, buffer substance

ULOQ upper limit of quantitation, maximum concentration of an analyte that allows quantitation with reasonable accuracy and precision

UPLC ultrahigh performance liquid chromatography, column chromatography with bead sizes of less than 2 μm

UV ultraviolet, light with wavelength $380\,\text{nm} > \lambda > 1\,\text{nm}$

UVIS visible and ultraviolet

WHO World Health Organisation, charged by the United Nations with improving human health

YAG yttrium aluminium garnet, doped with rare earth ions (*e.g.*, Nd) used for IR-lasers

References

1. E. ABBE: Beiträge zur theorie des mikroskops und der mikroskopischen wahrnehmung. *Arch. Mikr. Anat.* **9** (1873) 413–420. doi: 10.1007/BF02956173
2. G. ACKERS: Analytical gel chromatography of proteins. In: *Mechanisms and pathways of heterotrimeric G protein signaling*, C. ANFINSEN, J. EDSALL *et al.*, eds., vol. 24 of *Adv. Protein Chem.* chap. 5, pp. 343–441. (Academic, New York), 1970. ISBN 978-0120342242
3. G. ADAIR: A critical study of the direct method of measuring the osmotic pressure of hœmoglobin. *Proc. R. Soc. Lond. A* **108** (1925)(748) 627–637
4. G. ADAIR: The osmotic pressure of hœmoglobin in the absence of salts. *Proc. R. Soc. Lond. A* **108** (1925) 292–300
5. C. AISENBREY, P. BERTANI *et al.*: Structure, dynamics and topology of membrane polypeptides by oriented ^{2}H solid-state NMR spectroscopy. *Eur. Biophys. J.* **36** (2007) 451–460. doi: 10.1007/s00249-006-0122-2
6. K. ALTLAND: IPGMaker: A program for IBM compatible personal computers to create and test recipes for immobilized pH gradients. *Electrophoresis* **11** (1990) 140–147. doi: 10.1002/elps.1150110207
7. H. AN, J. FROEHLICH *et al.*: Determination of glycosylation sites and site-specific heterogeneity in glycoproteins. *Curr. Opin. Chem. Biol.* **13** (2009)(4) 421–426. doi: 10.1016/j.cbpa.2009.07.022
8. K. ANDERSON, A. POTTER *et al.*: Protein expression changes in spinal muscular atrophy revealed with a novel antibody array technology. *Brain* **126** (2003)(9) 2052–2064. doi: 10.1093/brain/awg208. URL http://brain.oxfordjournals.org/cgi/reprint/126/9/2052.pdf
9. L. ANDERSSON, H. JÖRNVALL *et al.*: Separation of isozymes of horse liver alcohol dehydrogenase and purification of the enzyme by affinity chromatography on an immobilized AMP-analogue. *Biochim. Biophys. Acta* **364** (1974) 1–8. doi: 10.1016/0005-2744(74)90126-0
10. S. ARRHENIUS: Über die reaktionsgeschwindigkeit bei der inversion von rohrzucker durch säuren. *Z. phys. Chem.* **4** (1889) 226–248
11. C. ARTOM: Labeled compounds in the study of phospholipid metabolism. *Methods Enzymol.* **4** (1957) 809–840. doi: 10.1016/0076-6879(57)04081-1
12. A. ASHKIN, J. M. DZIEDZIC *et al.*: Observation of a single-beam gradient force optical trap for dielectric particles. *Opt. Lett.* **11** (1986) 288. URL http://ol.osa.org/abstract.cfm?URI=ol-11-5-288
13. K. ASLAN & C. GEDDES: New tools for rapid clinical and bioagent diagnostics: Microwaves and plasmonic nanostructures. *Analyst* **133** (2008) 1469–1480. doi: 10.1039/b808292h
14. S. AUDIC, F. LOPEZ *et al.*: SAmBa: An interactive software for optimizing the design of biological macromolecules crystallization experiments. *Proteins* **29** (1997) 252–257. doi: 10.1002/(SICI)1097-0134(199710)29:2<252::AID-PROT12>3.0.CO;2-N
15. A. AYED, A. KRUTCHINSKY *et al.*: Quantitative evaluation of protein-protein and ligand-protein equilibria of a large allosteric enzyme by electrospray ionization time-of-flight mass spectrometry. *Rapid Commun. Mass Spectrom.* **12** (1998) 339–344. doi: 10.1002/(SICI)1097-0231(19980415)12:7<339::AID-RCM163>3.0.CO;2-6

E. Buxbaum, *Biophysical Chemistry of Proteins: An Introduction to Laboratory Methods*, DOI 10.1007/978-1-4419-7251-4,

16. P. Barth, T. Alber *et al.*: Accurate, conformation-dependent predictions of solvent effects on protein ionization constants. *Proc. Natl. Acad. Sci. USA* **104** (2007) 4898–4903. doi: 10.1073/pnas.0700188104
17. J. Bassham, A. Benson *et al.*: Isotope studies in photosynthesis. *J. Chem. Educ.* **30** (1953) 274–283. doi: 10.1021/ed030p274
18. A. Batista, W. Vetter *et al.*: Use of focused open vessel microwave-assisted extraction as prelude for the determination of the fatty acid profile of fish - a comparison with results obtained after liquid-liquid extraction according to Bligh and Dyer. *Eur. Food Res. Technol.* **212** (2001) 377–384. doi: 10.1007/s002170000240
19. T. Bayburt & S. Sligar: Self-assembly of single integral membrane proteins into soluble nanoscale phospholipid bilayers. *Protein Sci.* **12** (2003) 2476–2481. doi: 10.1110/ps.03267503
20. H.-J. Böckenhauer & D. Bongartz: *Algorithmische grundlagen der bioinformatik.* 1st edn.. (Teubner, Wiesbaden), 2003. ISBN 978-3-519-00398-8
21. J. Bechhoefer & S. Wilson: Faster, cheaper, safer optical tweezers for the undergraduate laboratory. *Am. J. Phys.* **70** (2002) 393–400. doi: 10.1119/1.1445403. URL http://link.aip.org/link/?AJP/70/393/1
22. M. Belov, M. Gorshkov *et al.*: Zeptomole-sensitivity electrospray ionization - Fourier transform ion cyclotron resonance mass spectrometry of proteins. *Anal. Chem.* **72** (2000) 2271–2279. doi: 10.1021/ac991360b
23. A. Bensadoun & D. Weinstein: Assay of proteins in the presence of interfering materials. *Anal Biochem.* **70** (1976) 241–250. doi: 10.1016/S0003-2697(76)80064-4
24. M. Berenbaum: What is synergy? *Pharmacol. Rev.* **41** (1989) 93–141. URL http://pharmrev.aspetjournals.org/cgi/reprint/41/2/93
25. M. Berezovski, M. Lechmann *et al.*: Aptamer-facilitated biomarker discovery (AptaBiD). *J. Am. Chem. Soc.* **130** (2008) 9137–9143. doi: 10.1021/ja801951p
26. J. Bernal & D. Crowford: X-ray photographs of crystalline pepsin. *Nature* **133** (1934) 794–795. doi: 10.1038/133794b0
27. H. Bisswanger: *Enzyme kinetics, theories and methods.* 3rd edn.. (Wiley-VCH, Weinheim), 2008. ISBN 978-3527319572
28. B. Bjellqvist, K. Ek *et al.*: Isoelectric focusing in immobilized pH gradients: principle, methodology and some applications. *J. Biochem. Biophys. Methods* **6** (1982) 317–339. doi: 10.1016/0165-022X(82)90013-6
29. G. Blackburn, H. Shah *et al.*: Electrochemiluminescence detection for development of immunoassays and DNA probe assays for clinical diagnostics. *Clin. Chem.* **37** (1991) 1534. URL http://www.clinchem.org/cgi/reprint/37/9/1534.pdf
30. B. Blagoev, I. Kratchmarova *et al.*: A proteomics strategy to elucidate functional protein-protein interactions. *Nat. Biotechnol.* **21** (2003) 315–318. doi: 10.1038/nbt790
31. E. Bligh & W. Dyer: A rapid method for total lipid extraction and purification. *Can. J. Biochem. Physiol.* **37** (1959)(8) 911–917. doi: 10.1139/y59-099
32. J. Blinks, W. Wier *et al.*: Measurement of Ca^{2+} concentrations in living cells. *Prog. Biophys. Mol. Biol.* **40** (1982) 1–114. doi: 10.1016/0079-6107(82)90011-6
33. D. Bolen: Effects of naturally occuring osmolytes on protein stability and solubility: Issues important in protein crystallization. *Methods* **34** (2004) 312–322. doi: 10.1016/j.ymeth.2004.03.022
34. D. Bolen & I. Baskakov: The osmophobic effect: Natural selection of a thermodynamic force in protein folding. *J. Mol. Biol.* **310** (2001) 955–963. doi: 10.1006/jmbi.2001.4819
35. P. Booth & P. Curnow: Folding scene investigation: Membrane proteins. *Curr. Opin. Struct. Biol.* **19** (2009) 8–13. doi: 10.1016/j.sbi.2008.12.005
36. M. Bradford: A rapid and sensitive method for the quantitation of microgram quantities of protein utilizing the principles of protein-dye binding. *Anal. Biochem.* **72** (1976) 248–254. doi: 10.1016/0003-2697(76)90527-3
37. S. Brahms & J. Brahms: Determination of protein secondary structure in solution by vacuum ultraviolet circular dichroism. *J. Mol. Biol.* **138** (1980) 149–178. doi: 10.1016/0022-2836(80)90282-X

38. G. U. BUBLITZ & S. G. BOXER: STARK spectroscopy: Applications in chemistry, biology, and materials science. *Annu. Rev. Phys. Chem.* **48** (1997) 213–242. doi: 10.1146/annurev.physchem.48.1.213
39. K. BUCHHOLZ, V. KASCHE *et al.*: *Biocatalysts and enzyme technology*. 2nd edn.. (Wiley-VCH, Weinheim), 2005. ISBN 9-783527-304974
40. D. BUNKA & P. STOCKLEY: Aptamers come of age – at last. *Nat. Rev. Microbiol.* **4** (2006) 588–596. doi: 10.1038/nrmicro1458
41. E. BUXBAUM: Co-operating ATP sites in the multiple drug resistance transporter Mdr1. *Eur. J. Biochem.* **265** (1999) 54–63. doi: 10.1046/j.1432-1327.1999.00643.x
42. E. BUXBAUM: Co-operative binding sites for transported substrates in the multiple drug resistance transporter Mdr1. *Eur. J. Biochem.* **265** (1999) 64–70. doi: 10.1046/j.1432-1327.1999.00644.x
43. E. BUXBAUM: Cationic electrophoresis and electrotransfer of membrane glycoproteins. *Anal. Biochem.* **314** (2003) 70–76. doi: 10.1016/S0003-2697(02)00639-5
44. E. BUXBAUM: *Introduction to protein structure and function*. (Springer, New York), 2007. ISBN 978-0387-26352-6
45. E. BUXBAUM: Cationic electrophoresis and eastern blotting. In: *Protein blotting and detection – Methods and protocols*, B. KURIEN & R. SCOFIELD, eds., vol. 536 of *Methods Mol. Biol.* (Humana, Dordrecht). ISBN 978-1-934115-73-2, 2009 pp. 115–128
46. E. BUXBAUM & W. SCHONER: Investigation of subunit interaction by radiation inactivation: The case of Na^+/K^+-ATPase. *J. Theor. Biol.* **155** (1992) 21–23. doi: 10.1016/S0022-5193(05)80546-6
47. E. BUXBAUM & P. WOODMAN: Binding of ATP and nucleotide analogues to Hsc70. *Biochem. J.* **318** (1996) 923–929. URL http://www.ncbi.nlm.nih.gov/pmc/articles/PMC1217706/pdf/8836139.pdf
48. M. CAFFREY: A lipids eye view of membrane protein crystallization in mesophases. *Curr. Opin. Struct. Biol.* **10** (2000) 486–497. URL http://www.ul.ie/~ces/PDF%20files/Lipid.pdf
49. C. CANTOR & P. SCHIMMEL: *Biophysical chemistry vol 1-3*. (W.H. Freeman, San Francisco), 1980
50. C. CARTER & R. SWEET (eds.): *Macromolecular crystallography vol. A*, vol. 276 of *Methods Enzymol.* (Academic, New York, London), 1997. ISBN 978-0-12-182177-7
51. N. CATSIMPOOLAS & J. KENNEY: Analytical isotachophoresis of human serum proteins with ampholine spacers. *Biochim. Biophys. Acta - Protein Structure* **285** (1972) 287–292. doi: 10.1016/0005-2795(72)90312-1
52. J. CAVIERES: Fast reversal of the initial reaction steps of the plasma membrane ($Ca^{2+} + Mg^{2+}$)-ATPase. *Biochim. Biophys. Acta* **899** (1987) 83–92. doi: 10.1016/0005-2736(87)90242-2
53. R. CELIKEL, E. PETERSON *et al.*: Crystal structure of a therapeutic single chain antibody in complex with two drugs of abuse – methamphetamine and 3,4-methylenedioxymethamphetamine. *Protein Sci.* **18** (2009) 2336–2345. doi: 10.1002/pro.244
54. B. CHANCE: The stopped-flow method and chemical intermediates in enzyme reactions – a personal essay. *Photosynth. Res.* **80** (2004) 387–400. doi: 10.1023/B:PRES.0000030601.50634.0c
55. A. CHANUTIN, S. LUDEWIG *et al.*: Studies on the calcium-protein relationship with the aid of the ultracentrifuge. *J. Biol. Chem.* **143** (1942) 737–751. URL http://www.jbc.org/content/143/3/763.full.pdf
56. L. CHARBONNIÈRE & N. HILDEBRANDT: Lanthanide complexes and quantum dots: A bright wedding for resonance energy transfer. *Eur. J. Inorg. Chem.* **2008** (2008) 3241–3251. doi: 10.1002/ejic.200800332
57. A. CHATTOPADHYAY & E. LONDON: Fluorimetric determination of critical micelle concentration avoiding interference from detergent charge. *Anal. Biochem.* **139** (1984) 408–412. doi: 10.1016/0003-2697(84)90026-5
58. J. CHERRY & A. FIDANTSEF: Directed evolution of industrial enzymes: an update. *Curr. Opin. Biotechnol.* **14** (2003) 438–443. doi: 10.1016/S0958-1669(03)00099-5

59. P. CHONG & R. HODGES: A new heterobifunctional cross-linking reagent for the study of biological interactions between proteins. I. Design, synthesis, and characterization. *J. Biol. Chem.* **256** (1981) 5064–5070. URL http://www.jbc.org/content/256/10/5064.full.pdf
60. A. CHRAMBACH: *The practice of quantitative gel electrophoresis.* (VCH, Weinheim), 1985. ISBN 978-3527260393
61. S. CHRISTENSEN, E. H. MOELLER *et al.*: Preliminary studies of the physical stability of a glucagon-like peptide-1 derivate in the presence of metal ions. *Eur. J. Pharm. Biopharm.* **66** (2006) 366–371. doi: 10.1016/j.ejpb.2006.11.019
62. W. CLELAND: Determination of equilibrium isotope effects by the equilibrium perturbation method. *Methods Enzymol.* **87** (1982) 641–646. doi: 10.1016/S0076-6879(82)87033-X
63. W. CLELAND: The use of isotope effects to determine transition-state structure for enzymatic reactions. *Methods Enzymol.* **87** (1982) 625–641. doi: 10.1016/S0076-6879(82)87033-X
64. W. CLELAND: Use of isotope effects to elucidate enzyme mechanism. *CRC Crit. Rev. Biochem.* **13** (1982) 385–428. doi: 10.3109/10409238209108715
65. W. CLELAND: The use of isotope effects to determine enzyme mechanism. *J. Biol. Chem.* **278** (2003) 51 975–51 984. doi: 10.1074/jbc.X300005200
66. M. COMISAROW & A. MARSHALL: Fourier transform ion cyclotron resonance spectroscopy. *Chem. Phys. Lett.* **25** (1974)(2) 282–283. doi: 10.1016/0009-2614(74)89137-2
67. I. H. G. S. CONSORTIUM: Initial sequencing and analysis of the human genome. *Nature* **409** (2001) 860–921. doi: 10.1038/35057062
68. P. COOK & W. CLELAND: *Enzyme kinetics and mechanism.* (Garland, New York), 2007. ISBN 978-0-8153-4140-6
69. R. COPELAND: *Enzymes. A practical Introduction to structure, mechanism and data analysis.* 2nd edn.. (Wiley-VCH, New York), 2000. ISBN 978-0471359296
70. L. CRAIG: Identification of small amounts of organic compounds by distribution studies. *J. Biol. Chem.* **155** (1944) 519–534. URL http://www.jbc.org/content/155/2/519.full.pdf
71. T. CREIGHTON: Electrophoretic analysis of the unfolding of proteins by urea. *J. Mol. Biol.* **129** (1979) 235–264. doi: 10.1016/0022-2836(79)90279-1
72. L. CUMMINGS, M. SNYDER *et al.*: Protein chromatography on hydroxyapatite columns. *Methods Enzymol.* **463** (2009) 387–404. doi: 10.1016/S0076-6879(09)63024-X
73. J. DABAN, S. BARTOLOMÉ *et al.*: Use of the hydrophobic probe nile red for the fluorescent staining of protein bands in sodium dodecyl sulfate polyacrylamide gels. *Anal. Biochem.* **199** (1991) 169–174. doi: 10.1016/0003-2697(91)90085-8
74. M. DAS, T. MIYAKAWA *et al.*: Specific radiolabeling of a cell surface receptor for epidermal growth factor. *Proc. Natl. Acad. Sci. USA* **74,** (1977) 2790–2794. URL http://www.pnas.org/content/74/7/2790.full.pdf
75. B. DAVIS: Disc electrophoresis – II Method and application to human serum proteins. *Ann. N.Y. Acad. Sci.* **121** (1964) 404–427. doi: 10.1111/j.1749-6632.1964.tb14213.x. URL http://www.pipeline.com/~lenornst/DiscEle2.pdf
76. A. DERFUS, W. CHAN *et al.*: Probing the cytotoxicity of semiconductor quantum dots. *Nano Lett.* **4** (2004)(1) 11–18. doi: 10.1021/nl0347334. URL http://homepages.wmich.edu/~sobare/courses/Nano%20Lett%202004-4-11.pdf
77. P. DEVINE & J. WARREN: Glycoprotein detection on immobilon PVDF transfer membrane using the periodic acid / SCHIFF reagent. *Biotechniques* **8** (1990) 492–495
78. R. DISCIPIO: Preparation of colloidal gold particles of various sizes using sodium borohydride and sodium cyanoborohydride. *Anal. Biochem.* **236** (1996) 168–170. doi: 10.1006/abio.1996.0146
79. E. DOLK, M. VAN DER VAART *et al.*: Isolation of llama antibody fragments for prevention of dandruff by phage display in shampoo. *Appl. Environ. Microbiol.* **71** (2005) 442–450. doi: 10.1128/AEM.71.1.442–450.2005
80. A. DOUNCE, R. WITTER *et al.*: A method for isolating intact mitochondria and nuclei from the same homogenate, and the influence of mitochondrial destruction on the properties of cell nuclei. *J. Biophys. Biochem. Cytol.* **1** (1955) 139–153. doi: 10.1083/jcb.1.2.139
81. D. DRAPER & P. VON HIPPEL: Measurements of macromolecule binding constants by a sucrose gradient band. *Biochemistry* **18** (1979) 753–760. doi: 10.1021/bi00572a003

82. R. DUGGLEBY: Regression analysis of nonlinear ARRHENIUS plots: An empirical model and a computer program. *Comput. Biol. Med.* **14** (1984) 447–455. doi: 10.1016/0010-4825(84)90045-3
83. R. DUGGLEBY: Quantitative analysis of the time courses of enzyme-catalyzed reactions. *Methods* **24** (2001) 168–174. doi: 10.1006/meth.2001.1177
84. S. DUNN: Effects of the modification of transfer buffer composition and the renaturation of proteins in gels on the recognition of proteins on western blots by monoclonal antibodies. *Anal. Biochem.* **157** (1986) 144–153. doi: 10.1016/0003-2697(86)90207-1
85. J.-L. EISELÉ & J. ROSENBUSCH: Crystallization of porin using short chain phospholipids. *J. Mol. Biol.* **206** (1989) 209–212. doi: 10.1016/0022-2836(89)90533-0
86. E. ELSON & D. MAGDE: Fluorescence correlation spectroscopy. I. Conceptual basis and theory. *Biopolymers* **13** (1974) 1–27. doi: 10.1002/bip.1974.360130102
87. H. ENGELHARDT & J. PLITZKO: 3D-Structuren prokaryotischer organellen. *BIOspektrum* **12** (2006) 477–478. URL http://www.biospektrum.de/blatt/d_bs_pdf&_id=932616
88. J. F. ENGELHART: *Commentatio de vera materiae sanguini purpureum colorem impertientis natura.* (Dieterich, Göttingen), 1825
89. E. ENGVALL & P. PERLMAN: Enzyme-linked immunosorbent assay of immunoglobulin G. *Immunochemistry* **8** (1971) 871–874. doi: 10.1016/0019-2791(71)90454-X
90. L. ERIKS, J. MAYOR *et al.*: A strategy for identification and quantification of detergents frequently used in the purification of membrane proteins. *Anal. Biochem.* **323** (2003) 234–241. doi: doi:10.1016/j.ab.2003.09.002
91. S. ERIKSSON, I. CARAS *et al.*: Direct photoaffinity labeling of an allosteric site on subunit protein MI of mouse ribonucleotide reductase by dTTP. *Proc. Natl. Acad. Sci. USA* **79** (1982) 81–85. URL http://www.pnas.org/content/79/1/81.full.pdf+html
92. L. ESTELA & T. HEINRICHS: Evaluation of the counterimmunoelectrophoretic (CIE) procedure in a clinical laboratory setting. *Am. J. Clin. Pathol.* **70** (1978) 329–343
93. S. FAHAM & J. BOWIE: Bicelle crystallization: a new method for crystallizing membrane proteins yields a monomeric bacteriorhodopsin structure. *J. Mol. Biol.* **316** (2002) 1–6. doi: 10.1006/jmbi.2001.5295
94. G. FAIRBANKS, T. STECK *et al.*: Electrophoretic analysis of the major polypeptides of the human erythrocyte membrane. *Biochemistry* **10** (1971) 2606–2617. doi: 10.1021/bi00789a030
95. A. FELINGER: *Data analysis and signal processing in chromatography.* (Elsevier, Amsterdam), 1998. ISBN 978-0-444-82066-2
96. H. FENTON: Oxidation of tartaric acid in presence of iron. *J. Chem. Soc. Trans.* **65** (1894) 899 – 910. doi: 10.1039/CT8946500899
97. K. FERGUSON: Starch gel electrophoresis: application to the classification of pituitary proteins and peptides. *Metabolism* **13** (1964) 985–1002. URL http://www.his.com/~djt/FergusonKA-Metabolism-October-1964.pdf
98. M. FERRERAS, J. GAVILANES *et al.*: A permanent Zn^{2+} reverse staining method for the detection and quantification of proteins in polyacrylamide gels. *Anal. Biochem.* **213** (1993) 206–212. doi: 10.1006/abio.1993.1410
99. A. FERSHT, A. MATOUSCHEK *et al.*: Physical-organic molecular biology: Pathway and stability of protein folding. *Pure Appl. Chem.* **63** (1991) 187–194. URL http://www.iupac.org/publications/pac/1991/pdf/6302x0187.pdf
100. S. FINET, D. VIVARÈS *et al.*: Controlling biomolecular crystallization by understanding the distinct effects of PEGs and salts on solubility. *Methods Enzymol.* **368** (2003) 105–129. doi: 10.1016/S0076-6879(03)68007-9
101. H. FISCHER, I. POLIKARPOV *et al.*: Average protein density is a molecular-weight-dependent function. *Protein Sci.* **13** (2004) 2825–2828. doi: 10.1110/ps.04688204
102. N. FITZKEE, P. FLEMMING *et al.*: Are proteins made from a limited parts list? *Trends Biochem. Sci.* **30** (2005) 73–80. doi: 10.1016/j.tibs.2004.12.005
103. S. FLEISCHER & M. KERVINA: Subcellular fractionation of rat liver. *Methods Enzymol.* **31** (1974) 6–41. doi: 10.1016/j.tibs.2004.12.005

104. J. FLOSSDORF & U. SÜSSENBACH: Bestimmung des partiellen spezifischen volumens von gelöst-stoffen in lösungen unbekannter konzentration. *Die Makromolekulare Chemie* **179** (1978) 1061–1068. doi: 10.1002/macp.1978.021790420
105. J. FOLCH, M. LEES *et al.*: A simple method for the isolation and purification of total lipids from animal tissues. *J. Biol. Chem.* **226** (1957) 497–509. http://www.jbc.org/content/226/1/497.full.pdf+html?sid=1a04415e-c1d6-406a-ab02-70674cab2d2f
106. P. FREDERIX, T. AKIYAMA *et al.*: Atomic force bioanalytics. *Curr. Opin. Chem. Biol.* **7** (2003) 641–647. doi: 10.1016/j.cbpa.2003.08.010
107. T. FÖRSTER: Zwischenmolekulare energiewanderung und fluoreszenz. *Ann. Phys.* **437** (1948) 55–75. doi: 10.1002/andp.19484370105
108. Y. FUJIKI, A. HUBBARD *et al.*: Isolation of intracellular membranes by means of sodium carbonate treatment: Application to endoplasmic reticulum. *J. Cell Biol.* **93** (1982) 97–102. URL http://jcb.rupress.org/cgi/reprint/93/1/97
109. S. C. GAD (ed.): *Handbook of Pharmaceutical Biotechnology*. (Wiley Interscience, Hoboken, NJ), 2007. ISBN 978-0471213864
110. Z. GANIM, H. CHUNG *et al.*: Amide I two-dimensional infrared spectroscopy of proteins. *Acc. Chem. Res.* **41** (2008) 432–441. doi: 10.1021/ar700188n
111. P. GAST, P. HEMELRIJK *et al.*: Determination of the number of detergent molecules associated with the reaction center protein isolated from the photosynthetic bacterium *Rhodopseudomonas viridis*. *FEBS Lett.* **337** (1994) 39–42. doi: 10.1016/0014-5793(94)80625-X
112. A.-C. GAVIN, M. BÖSCHE *et al.*: Functional organisation of the yeast proteome by systematic analysis of protein complexes. *Nature* **415** (2002) 141–147. doi: 10.1038/415141a
113. A. GEORGE & W. WILSON: Predicting protein crystallization from dilute solution property. *Acta Crystallogr. D* **50** (1994) 361–365. doi: 10.1107/S0907444994001216
114. R. GILBERT, P. FUCINI *et al.*: Three-dimensional structures of translating ribosomes by cryo-EM. *Mol. Cell* **14** (2004) 57–66. doi: 10.1016/S1097-2765(04)00163-7
115. S. GILL & P. VON HIPPEL: Calculation of protein extinction coefficients from amino acid sequence data. *Anal. Biochem.* **182** (1989) 319–326. doi: 10.1016/0003-2697(89)90602-7
116. R. GIVENS, G. TIMBERLAKE *et al.*: A photoactivated diazopyruvoyl cross-linking agent for bonding tissue containing type-I collagen. *Photochem. Photobiol.* **78** (2003) 23–29
117. R. GLASER: *Biophysics: An Introduction*. 5th edn.. (Springer, Berlin), 2000. ISBN 978-3540670889
118. M. GOLDBERG & A. CHAFFOTTE: Undistorted structural analysis of soluble proteins by attenuated total reflectance infrared spectroscopy. *Protein Sci.* **14** (2005) 2781–2792. doi: 10.1110/ps.051678205
119. L. GOLDSCHMIDT, D.R. COOPER *et al.*: Toward rational protein crystallization: A Web server for the design of crystallizable protein variants. *Protein Sci.* **16** (2007) 1569–1576. doi: 10.1110/ps.072914007
120. D. GOOD & J. COON: Advancing proteomics with ion/ion chemistry. *Biotechniques* **40** (2006) 783–789. doi: 10.2144/000112194
121. V. GOODFELLOW, M. SETTINERI *et al.*: p-Nitrophenyl 3-diazopyruvate and diazopyruvamides, a new family of photoactivatable cross-linking bioprobes. *Biochemistry* **28** (1989) 6346–6360. doi: 10.1021/bi00441a030
122. B. GOODSON: Nuclear magnetic resonance of laser-polarised noble gases in molecules, materils and organisms. *J. Magn. Reson.* **155** (2002) 157–216. doi: 10.1006/jmre.2001.2341
123. S. GORDO & E. GIRALT: Knitting and untying the protein network: Modulation of protein ensembles as a therapeutic strategy. *Protein Sci.* **18** (2009) 481–493. doi: 10.1002/pro.43
124. P. GRABAR & C. WILLIAMS: Méthode permettant l'étude conjuguée des propriétés électrophorétiques et immunochimiques d'un mélange de protéines. Application au sérum sanguin. *Biochim. Biophys. Acta* **10** (1953) 193–194. doi: 10.1016/0006-3002(53)90233-9
125. K. GRAHAM & J. SHIVELY: Improved initial yields in C-terminal sequence analysis by thiohydantoin chemistry using purified diphenylphosphoryl isothiocyanate: NMR evidence for a reaction intermediate in the coupling reaction. *Anal. Biochem.* **307** (2002) 202–211. doi: 10.1016/0003-2697(89)90602-7

126. M. GREASER & C. WARREN: Efficient electroblotting of very large proteins using a vertical agarose electrophoresis system. In: *Protein blotting and detection – Methods and protocols*, B. KURIEN & R. SCOFIELD, eds., vol. 536 of *Methods Mol. Biol.* (Humana, Dordrecht). ISBN 978-1-934115-73-2, 2009 pp. 221–227
127. W. GRECO, G. BRAVO *et al.*: The search for synergy: A critical review from a response surface perspective. *Pharmacol. Rev.* **47** (1995) 331–385. URL http://pharmrev.aspetjournals.org/cgi/reprint/47/2/331
128. D. GREENBAUM, C. COLANGELO *et al.*: Comparing protein abundance and mRNA expression levels on a genomic scale. *Genome Biol.* **4** (2003) 117. URL http://genomebiology.com/2003/4/9/117
129. W. GRIFFIN: Classification of surface-active agents by "HLB". *J. Soc. Cosmet. Chem.* **1** (1949) 311
130. W. GRIFFIN: Calculation of HLB values of non-ionic surfactants. *J. Soc. Cosmet. Chem.* **5** (1954) 249
131. B. GRIFFITH & A. DOYLE: *Cell culture: Essential techniques.* (J. Wiley & Sons, New York), 1997. doi: 978-0471970576
132. G. GRIMSLEY, J. SCHOLTZ *et al.*: A summary of the measured *p*K values of the ionizable groups in folded proteins. *Protein Sci.* **18** (2009) 247–251. doi: 10.1002/pro.19
133. E. GROSS & B. WITKOP: Selective cleavage of the methionyl peptide bonds in ribonuclease with cyanogen bromide. *J. Am. Chem. Soc.* **83** (1961) 1510–1511. doi: 10.1021/ja01467a052
134. D. GUTMANN, E. MIZOHATA *et al.*: A high-throughput method for protein solubility screening: The ultracentrifugation dispersity sedimentation assay. *Protein Sci.* **16** (2007) 1422–1428. doi: 10.1110/ps.072759907
135. S. GYGI, B. RIST *et al.*: Quantitative analysis of complex protein mixtures using isotope-coded affinity tags. *Nat. Biotechnol.* **17** (1999) 994–999. doi: 10.1038/13690
136. S. GYGI, Y. ROCHON *et al.*: Correlation between protein and mRNA abundance in yeast. *Mol. Cell Biol.* **19** (1999) 1720–1730. URL http://www.ncbi.nlm.nih.gov/pmc/articles/PMC83965/
137. J. HAJDU: Single molecule X-ray diffraction. *Curr. Opin. Struct. Biol.* **10** (2000) 569. doi: 10.1016/S0959-440X(00)00133-0
138. A. VON HALL, M. HAMACHER *et al.*: 5 Jahre human brain proteom project. *BIOspektrum* **12** (2006) 727–730. URL http://www.biospektrum.de/blatt/d_bs_pdf&_id=932513
139. O. P. HAMILL, A. MARTY *et al.*: Improved patch-clamp techniques for high-resolution current recording from cells and cell-free membrane patches. *Pflügers Arch.* **1981** (391) 85–100. doi: 10.1007/BF00656997
140. K. HANCOCK & V. TSANG: India ink staining of proteins on nitrocellulose paper. *Anal. Biochem.* **133** (1983) 157–162. doi: 10.1016/0003-2697(83)90237-3
141. Q. HANLEY, K. LIDKE *et al.*: Fluorescence lifetime imaging in an optically sectioning programmable array microscope (PAM). *Cytometry* **67A** (2005) 112–118. doi: 10.1002/cyto.a.20177
142. Q. HANLEY, P. MURRAY *et al.*: Microspectroscopic fluorescence analysis with prism-based imaging spectrometers: Review and current studies. *Cytometry* **69A** (2006) 759–766. doi: 10.1002/cyto.a.20265
143. K. HANNIG: New aspects in preparative and analytical continuous free-flow cell electrophoresis. *Electrophoresis* **3** (1982) 235–243. doi: 10.1002/elps.1150030502
144. E. HARDY & L. CASTELLANOS-SERRA: "Reverse-staining" of biomolecules in electrophoresis gels: Analytical and micropreparative applications. *Anal. Biochem.* **328** (2004) 1–13. doi: 10.1016/j.ab.2004.02.017
145. P. HARE, D. VON ENDT *et al.*: Protein and amino acid diagenesis dating. In: *Chronometric dating in archaeology*, R. TAILOR & M. AITKEN, eds., vol. 2 of *Adv. archaeology mus. sci.* (Springer, New York). ISBN 978-0-306-45715-9, 1997 pp. 261–296
146. E. HARLOW & D. LANE: *Antibodies – A Laboratory Manual.* 1 edn.. (Cold Spring Harbor Laboratory, Cold Spring Harbor), 1988. ISBN 978-0879694145
147. H. HARTRIDGE & F. ROUGHTON: A method of measuring the velocity of very rapid chemical reactions. *Proc. R. Soc. Lond. A* **104** (1923) 376–394. doi: 10.1098/rspa.1923.0116

148. L. HAYFLICK & P. S. MOORHEAD: The serial cultivation of human diploid cell strains. *Exp. Cell Res.* **25** (1961) 585–621. doi: 10.1016/0014-4827(61)90192-6
149. J. HEDRICK & A. SMITH: Size and charge isomer separation and estimation of molecular weights of protein by disk gel electrophoresis. *Arch. Biochem. Biophys.* **126** (1968) 155–64. doi: 10.1016/0003-9861(68)90569-9
150. H. HEERKLOTZ & J. SEELIG: Correlation of membrane/water partition coefficients of detergents with the cmc. *Biophys. J.* **78** (2000) 2435–2440. doi: 10.1016/S0006-3495(00)76787-7
151. M. HEGRENESS, N. SHORESH *et al.*: Accelerated evolution of resistance in multidrug environments. *Proc. Natl. Acad. Sci. USA* **105** (2008) 13 977–13 981. doi: 10.1073/pnas.0805965105
152. A. HELENIUS & K. SIMONS: Solubilization of membranes by detergents. *Biochim. Biophys. Acta* **415** (1975) 29–79. doi: 10.1016/0304-4157(75)90016-7
153. P. HELFMAN & J. BADA: Aspartic acid racemization in tooth enamel from living humans. *Proc. Natl. Acad. Sci. USA* **72** (1975) 2891–2894. URL http://www.pnas.org/content/72/8/2891.full.pdf
154. S. HELL & J. WICHMANN: Breaking the diffraction resolution limit by stimulated emission: Stimulated-emission-depletion fluorescence microscopy. *Opt. Lett.* **19** (1994) 780–782. doi: 10.1364/OL.19.000780
155. R. HENDERSON, J. BALDWIN *et al.*: Model for the structure of bacteriorhodopsin based on high-resolution electron cryo-microscopy. *J. Mol. Biol.* **213** (1990) 899–929. doi: 10.1016/S0022-2836(05)80271-2
156. P. HENGEN: Chemoluminescent detection methods. *Trends Biochem. Sci.* **22** (1997) 313–314. doi: 10.1016/S0968-0004(97)01095-5
157. A. HENKEL & S. BIEGER: Quantification of proteins dissolved in an electrophoresis sample buffer. *Anal. Biochem.* **223** (1995) 329–331. doi: 10.1006/abio.1994.1595
158. S. HENNING, J. PETER-KATALINIC *et al.*: Structure analysis of N-glycoproteins. In: *Mass Spectrometry of Proteins and Peptides*, L. PAŠA-TOLIC & M. LIPTON, eds., vol. 492 of *Methods Mol. Biol.*, 2nd edn., pp. 181–200. (Humana, Dordrecht). ISBN 978-1-934115-48-0, 2009 doi: 10.1007/978-1-59745-493-3
159. G. HERMANSON: *Bioconjugate techniques*. 2nd edn.. (Academic, Amsterdam), 2008. ISBN 978-0-12-370501-3
160. J. HEUKESHOVEN & R. DERNICK: Improved silver staining procedure for fast staining in phastsystem development unit I. Staining of sodium dodecyl sulfate gels. *Electrophoresis* **9** (1988) 28–32. doi: 10.1002/elps.1150090106
161. A. HINCHLIFFE: *Molecular modelling for beginners*. 2nd edn.. (Wiley, Chichester), 2008. ISBN 978-0470513149
162. T. HISABORI, K. INOUE *et al.*: Two-dimensional gel electrophoresis of membrane-bound protein complexes, including photosystem I, of tylacoid membranes in the presence of sodium oligooxyethylene alkyl ether sulfate / dimethyl dodecylamine oxide and sodium dodecyl sulfate. *J. Biochem. Biophys. Methods* **22** (1991) 253–260. doi: 10.1016/0165-022X(91)90073-6
163. L. HJELMELAND & A. CHRAMBACH: Formation of natural *p*H gradients in sequential moving boundary systems with solvent counterions I. Theory. *Electrophoresis* **4** (83) 20–26. doi: 10.1002/elps.1150040104
164. S. HJERTÉN, L. ÖFVERSTEDT *et al.*: Free displacement electrophoresis (isotachophoresis): An absolute determination of the KOHLRAUSCH functions and their use in interaction studies. *J. Chromatogr. A* **194** (1980) 1–10. doi: 10.1016/S0021-9673(00)81044-4
165. H.-D. HÖLTJE, W. SIPPL *et al.*: *Molecular Modeling: Basic Principles and Applications*. 3rd edn.. (Wiley, Weinheim), 2008. ISBN 978-3527315680
166. F. HOFMEISTER: Zur lehre der wirkung von salzen. *Arch. Exp. Pathol. Pharmakol.* **24** (1888) 247–260. doi: 10.1007/BF01838161
167. K. VAN HOLDE, W. JOHNSON *et al.*: *Principles of physical biochemistry*. 2 edn.. (Pearson, Upper Saddle River, NJ), 2006. ISBN 0-13-046427-9
168. P. HOLLOWAY: A simple procedure for the removal of Triton-X100 from protein samples. *Anal. Biochem.* **53** (1973) 304–308. doi: 10.1016/0003-2697(73)90436-3

169. F. HOPPE-SEYLER & K. BUTZ: Peptide aptamers: Powerful new tools for molecular medicine. *J. Mol. Med.* **78** (2000) 426–430. doi: 10.1007/s0010900000140
170. C. HUANG: Determination of the binding stoichiometry by the continuous variation method: The JOB-plot. *Methods Enzymol.* **87** (1982) 509–525. doi: 10.1016/S0076-6879(82)87029-8
171. W. HUEMOELLER & H. TIEN: A simple setup for black lipid membrane experiments. *J. Chem. Educ.* **47** (1970) 469–470. doi: 10.1021/ed047p469
172. P. HUGHES, D. MARSHALL *et al.*: The costs of using unauthenticated, over-passaged cell lines: how much more data do we need? *Biotechniques* **43** (2007) 575–586. doi: 10.2144/000112598
173. J. HUMMEL & W. DREYER: Measurment of protein-binding phenomena by gel filtration. *Biochim. Biophys. Acta* **63** (1962) 530–532. doi: 10.1016/0006-3002(62)90124-5
174. S. JACKSON: How do small single-domain proteins fold? *Fold Des* **3** (1998) R81–R91. doi: doi:10.1016/S1359-0278(98)00033-9
175. N. JAIN & I. ROY: Effect of trehalose on protein structure. *Protein Sci.* **18** (2009) 24–36. doi: 10.1002/pro.3
176. P. JOB: Recherches sur la formation de complexes minéraux en solution, et sur leur stabilité. *Ann. Chim. (Paris)* **9** (1928) 113–203
177. J. JOHNSON, R. YOST *et al.*: Tandem-in-space and tandem-in-time mass spectrometry: triple quadrupoles and quadrupole ion traps. *Anal. Chem.* **62** (1990)(20) 2162–2172. doi: 10.1021/ac00219a003
178. S. JOHNSON & A. BANGHAM: Potassium permeability of single compartment liposomes with and without valinomycin. *Biochim. Biophys. Acta* **193** (1969) 82–91. doi: 10.1016/0005-2736(69)90061-3
179. G. JONES & M. DOLE: The viscosity of aqueous solutions of strong electrolytes with special reference to barium chloride. *J. Am. Chem. Soc.* **51** (1929) 2950–2964. doi: 10.1021/ja01385a012
180. N. JONES & P. PEVZNER: *An introduction to bioinformatics algorithms.* (MIT, Cambridge, London), 2004. ISBN 978-0-262-10106-6
181. J. JØRGENSEN & K. LUKACS: Zone electrophoresis in open-tubular glass capillaries. *Anal. Chem.* **53** (1981) 1298–1302. doi: 10.1021/ac00231a037
182. T. JOVIN: Multiphasic zone electrophoresis. I. Steady-state moving-boundary systems formed by different electrolyte combinations. *Biochemistry* **12** (1973) 871–879. doi: 10.1021/bi00729a014
183. T. JOVIN: Multiphasic zone electrophoresis. II. Design of integrated discontinuous buffer systems for analytical and preparative fractionation. *Biochemistry* **12** (1973) 879–890. doi: 10.1021/bi00729a015
184. T. JOVIN: Multiphasic zone electrophoresis. III. Further analysis and new forms of discontinuous buffer systems. *Biochemistry* **12** (1973) 890–898. doi: 10.1021/bi00729a016
185. T. JOVIN: Multiphasic zone electrophoresis. IV Design and analysis of discontinuous buffer systems with a digital computer. *Ann. NY Acad. Sci.* **209** (1973) 477–496. doi: 10.1111/j.1749-6632.1973.tb47551.x
186. J. KALINICH & D. MCCLAIN: An *in vitro* method for radiolabeling proteins with ^{35}S. *Anal.Biochem.* **205** (1992) 208–212. doi: 10.1016/0003-2697(92)90425-7
187. E. KALTSCHMIDT & H. G. WITTMANN: Ribosomal proteins. VII: Two-dimensional polyacrylamide gel electrophoresis for fingerprinting of ribosomal proteins. *Anal. Biochem.* **36** (1970) 401–412. doi: 10.1016/0003-2697(70)90376-3
188. A. KAPANIDIS, N. LEE *et al.*: Fluorescence-aided molecule sorting: Analysis of structure and interactions by alternating-laser excitation of single molecules. *Proc. Natl. Acad. Sci. USA* **101** (2004) 8936–8941. doi: 10.1073pnas.0401690101
189. K. KARLSSON, B. SAMUELSSON *et al.*: The lipid composition of the salt (rectal) gland of spiny dogfish. *Biochim. Biophys. Acta* **337** (1974) 365–376. doi: 10.1016/0005-2760(74)90111-8
190. V. KASCHE & R. ZÖLLNER: Tris (hydroxymethyl) methylamine is acylated when it reacts with acyl-chymotrypsin. *Hoppe Seylers Z. physiol. Chem.* **363** (1982) 531–534

191. S. VAN KASTERENA, S. CAMPBELLB *et al.*: Glyconanoparticles allow pre-symptomatic *in vivo* imaging of brain disease. *Proc. Natl. Acad. Sci. USA* **106** (2009) 18–23. doi: 10.1073/pnas.0806787106
192. A. KATDARE & M. CHAUBAL (eds.): *Excipient Development for Pharmaceutical, Biotechnology, and Drug Delivery Systems.* (Informa HealthCare), 2006. ISBN 978-0849327063
193. S. KELLY, T. JESS *et al.*: How to study proteins by circular dichroism. *Biochim. Biophys. Acta* **1751** (2005) 119–139. doi: 10.1016/j.bbapap.2005.06.005
194. T. KERPPOLA: Visualization of molecular interactions by fluorescence complementation. *Nature Rev. Mol. Cell Biol.* **7** (2006) 449–456. doi: 10.1038/nrm1929
195. F. KIENBERGER, G. KADA *et al.*: Single molecule studies of antibody-antigen interaction strength versus intra-molecular antigen stability. *J. Mol. Biol.* **347** (2005) 597–606. doi: 10.1016/j.jmb.2005.01.042
196. H. KIM, D. BYRNE *et al.*: Perceiving mitosis in eukaryotic cells. *In Vitro Cell. Dev. Biol. Plant* **24** (1988) 100–107. doi: 10.1007/BF02623886
197. M.-S. KIM, J. SONG *et al.*: Determining protein stability in cell lysates by pulse proteolysis and western blotting. *Protein Sci.* **18** (2009) 1051–1059. doi: 10.1002/pro.115
198. M. KIMBER *et al.*: Data mining crystallization data bases: Knowledge-based approaches to optimize protein crystal screens. *Proteins* **51** (2003) 562–568. URL http://papers.gersteinlab.org/e-print/xtalmine/reprint.pdf
199. M. KINTER & N. SHERMAN: *Protein sequencing and identification using tandem mass spectrometry.* (Wiley Interscience, New York), 2000. ISBN 978-0471322498
200. M. KLAPPER & I. KLOTZ: Acylation with dicarboxylic acid anhydrides. *Methods Enzymol.* **25** (1972) 531–552. doi: 10.1016/S0076-6879(72)25050-9
201. B. KLEINE, G. LÖFFLER *et al.*: Reduced chemical and radioactive liquid waste during electrophoresis using polymerized electrode gels. *Electrophoresis* **13** (1992) 73–75. doi: 10.1002/elps.1150130114
202. J. KLOSE & U. KOBALZ: Two-dimensional electrophoresis of proteins: An updated protocoll and implications for functional analysis of genomes. *Electrophoresis* **16** (1995) 1034–1059. doi: 10.1002/elps.11501601175
203. F. KOHLRAUSCH: Über concentrations-verschiebungen durch electrolyse im inneren von lösungen und lösungsgemischen. *Ann. Phys. Chem.* **62** (1897) 209–239. doi: 10.1002/andp.18972981002
204. J. KOHN: A cellulose acetate supporting medium for zone electrophoresis. *Clin. Chim. Acta* **2** (1957) 297–303. doi: 10.1016/0009-8981(57)90005-0
205. B. KOLAKOWSKI & L. KONERMANN: From small-molecule reactions to protein folding: Studying biochemical kinetics by stopped-flow electrospray mass spectrometry. *Anal. Biochem.* **292** (2001) 107–114. doi: 10.1006/abio.2001.5062
206. A. KOLLER, M. WASHBURN *et al.*: Proteomic survey of metabolic pathways in rice. *Proc. Natl. Acad. Sci. USA* **99** (2002)(18) 11 969–11 974. doi: 10.1073/pnas.172183199. URL http://www.pnas.org/content/99/18/11969.full.pdf+html
207. J. KOST: Blotting from phastgel to membranes by ultrasound. In: *Protein blotting and detection – Methods and protocols*, B. KURIEN & R. SCOFIELD, eds., vol. 536 of *Methods Mol. Biol.* (Humana, Dordrecht). ISBN 978-1-934115-73-2, 2009 pp. 173–179
208. M. KOSZELAK-ROSENBLUM, A. KROL *et al.*: Determination and application of empirically derived detergent phase boundaries to effectively crystallize membrane proteins. *Protein Sci.* **18** (2009) 1828–1839. doi: 10.1002/pro183
209. U. KRAGH-HANSEN, M. LE MAIRE *et al.*: The mechanism of detergent solubilization of liposomes and protein containing membranes. *Biophys. J.* **75** (1998) 2932–2946. doi: 10.1016/S0006-3495(98)77735-5
210. S. KRAJEWSKI, J. ZAPATA *et al.*: Detection of multiple antigens on western blots. *Anal. Biochem.* **236** (1996) 221–228. doi: 10.1006/abio.1996.0160
211. O. KRATKY, H. LEOPOLD *et al.*: The determination of the partial specific volume by the mechanical oscillator technique. *Methods Enzymol.* **27** (1973) 98–110. doi: 10.1016/S0006-3495(98)77735-5

212. E. KREYSZIG: *Statistische Methoden und ihre Anwendungen*. 7th edn.. (Vandenhoeck & Ruprecht, Göttingen), 1979. ISBN 978-3-525-40717-2
213. A. KROGH, B. LARSSON *et al.*: Predicting transmembrane protein topology with a hidden MARKOW model: Application to complete genomes. *J. Mol. Biol.* **305** (2001) 567–580. doi: 10.1006/jmbi.2000.4315
214. J. KUIPER, R. PLUTA *et al.*: A method for site-specific labelling of multiple protein thiols. *Protein Sci.* **18** (2009) 1033–1041. doi: 10.1002/pro.113
215. B. KURIEN & R. SCOFIELD: Non-electrophoretic bi-directional transfer of a single SDS-PAGE gel with multiple antigens to obtain 12 immunoblots. In: *Protein blotting and detection – Methods and protocols*, B. KURIEN & R. SCOFIELD, eds., vol. 536 of *Methods Mol. Biol.* (Humana, Dordrecht). ISBN 978-1-934115-73-2, 2009 pp. 55–65
216. K. KURIHARA & K. SUZUKI: Theoretical undestanding of an absorption-based surface plasmon resonance sensor based on KRETSCHMANNS theory. *Anal. Chem.* **74** (2002) 696–701. doi: 10.1021/ac010820+
217. J. KYTE & R. DOOLITTLE: A simple method for displaying the hydropathic character of a protein. *J. Mol. Biol.* **157** (1982) 105. doi: 10.1016/0022-2836(82)90515-0
218. U. LAEMLI: Cleavage of structural proteins during the assembly of the head of bacteriophage T4. *Nature* **227** (1970) 680–685. doi: 10.1038/227680a0
219. J. LAKOWICZ: *Principles of fluorescence spectroscopy*. 3rd edn.. (Kluwer, New York), 2006. ISBN 978-0387312781
220. A. LAMANDA, A. ZAHN *et al.*: Improved ruthenium(II) tris (bathophenantroline disulfonate) staining and destaining protocol for a better signal-to-background ratio and improved baseline resolution. *Proteomics* **4** (2004) 599–608. doi: 10.1002/pmic.200300587
221. J. LANDRY & S. DELHAYE: A simple and rapid procedure for hydrolyzing minute amounts of proteins with alkali. *Anal. Biochem.* **243** (1996) 191–194. doi: 10.1006/abio.1996.0503
222. T. LAUE, A. HAZARD *et al.*: Direct determination of macromolecular charge by equilibrium electrophoresis. *Anal. Biochem.* **182** (1989) 377–382. doi: 10.1016/0003-2697(89)90611-8
223. R. LAUGHLIN: *The Aqueous Phase Behavior of Surfactants*. 2nd edn.. (Elsevier, Amsterdam), 1996. ISBN 9780124377608
224. C. LAURELL: Quantitative estimation of proteins by electrophoresis in agarose gel containing antibodies. *Anal. Biochem.* **15** (1966) 45–52. doi: 10.1016/0003-2697(66)90246-6
225. M. LAURIERE: A semidry electroblotting system efficiently transfers both high- and low-molecular-weight proteins separated by SDS-PAGE. *Anal. Biochem.* **212** (1993) 206–211. doi: 10.1006/abio.1993.1313
226. J. LAZAR & F. TAUB: A highly sensitive method for detecting peroxidase in situ hybridization or immunohistochemical assays. In: *Non-radioactive labeling and detection of biomolecules*, C. KESSLER, ed., pp. 135–142. (Springer, Berlin), 1992
227. H. LÜDI & W. HASSELBACH: Excimer formation of ATPase from sarcoplasmic reticulum labeled with N-(3-pyrene)maleimide. *Eur. J. Biochem.* **130** (1983) 5–8. doi: 10.1111/j.1432-1033.1983.tb07108.x
228. A. LEACH: *Molecular Modelling: Principles and Applications*. 2nd edn.. (Prentice Hall, New York), 2001. ISBN 978-0582382107
229. J. LEBOWITZ, M. LEWIS *et al.*: Modern analytical ultracentrifugation in protein science: A tutorial review. *Protein Sci.* **11** (2002) 2067–2079. doi: 10.1110/ps.0207702
230. S. LEE, J. STEVENS *et al.*: A liquid gelatin blocking reagent for western blotting with chemiluminescent detection. *Biotechniques* **17** (1994) 60–62
231. P. LEMEY, M. SALEMI *et al.*: *The Phylogenetics Handbook: A Practical Approach to DNA and Protein Phylogeny*. 2nd edn.. (Cambridge University Press, Cambridge), 2009. ISBN 978-0521730716
232. S. LEVEY & E. JENNINGS: The use of control charts in the clinical laboratory. *Am. J. Clin. Pathol.* **20** (1950) 1059–1066. doi: 10.1016/0009-8981(60)90150-9
233. H. LEVY, P. LEBER *et al.*: Inactivation of myosin by 2,4-dinitrophenol and protection by adenosine triphosphate and other phosphate compounds. *J. Biol. Chem.* **238** (1963) 3654–3659. URL http://www.jbc.org/cgi/reprint/238/11/3654.pdf

234. L. Lewis, M. Robson *et al.*: Interference with spectrophotometric analysis of nucleic acids and proteins by leaching of chemicals from plastic tubes. *Biotechniques* **48** (2010) 297–302. doi: 10.2144/000113387
235. T. Liebermann & W. Knoll: Surface-plasmon field-enhanced fluorescence spectroscopy. *Colloids Surf A* **171** (2000) 115–130. doi: 10.1016/S0927-7757(99)00550-6
236. H. Lindley: A new synthetic substrate for trypsin and its application to the determination of the amino-acid sequence of proteins. *Nature* **178** (1956) 647–648. doi: 10.1038/178647a0
237. A. Link, J. Eng *et al.*: Direct analysis of protein complexes using mass spectrometry. *Nat. Biotechnol.* **17** (1999)(7) 676–682. doi: 10.1038/10890. URL http://www.worldinfocus.net/Papers/Proteomics/mudpit.pdf
238. T. Liu, E. Diemann *et al.*: Self-assembly in aqueous solution of wheel-shaped Mo_{154} oxide clusters into vesicles. *Nature* **426** (2003) 59–62. doi: 10.1038/nature02036
239. P. Loll, M. Allaman *et al.*: Assessing the role of detergent-detergent interactions in membrane protein crystallization. *J. Cryst. Growth* **232** (2001) 432–438. doi: 10.1016/S0022-0248(01)01076-4
240. A. Lomant & G. Fairbanks: Chemical probes of extended biological structures: Synthesis and properties of the cleavable protein cross-linking reagent ^{35}S-dithiobis(succinimidyl propionate). *J. Mol. Biol.* **104** (1976) 243–261. doi: 10.1016/0022-2836(76)90011-5
241. J. Loo: Studying non-covalent protein complexes by electrospray ionization mass spectrometry. *Mass Spectrom. Rev.* **16** (1997) 1–23. doi: 10.1002/(SICI)1098-2787(1997)16:1<1::AID-MAS1>3.0.CO;2-L
242. M. Lopez, W. Patton *et al.*: Effect of various detergents on protein migration in the second dimension of two-dimensional gels. *Anal. Biochem.* **199** (1991)(1) 35–44. doi: 10.1016/0003-2697(91)90266-V
243. O. Lowry, N. Rosebrough *et al.*: Protein measurement with the Folin phenol reagent. *J. Biol. Chem.* **193** (1951) 265–275. URL http://www.jbc.org/content/193/1/265.full.pdf+html
244. C. López, M. Fleissner *et al.*: Osmolyte perturbation reveals conformational equilibria in spin-labeled proteins. *Protein Sci.* **18** (2009) 1637–1652. doi: 10.1002/pro180
245. P. Lu, C. Vogel *et al.*: Absolute protein expression profiling estimates the relative contributions of transcriptional and translational regulation. *Nat. Biotechnol.* **25** (2006) 117–124. doi: doi:10.1038/nbt1270
246. G. Lubec, M. Weninger *et al.*: Racemization and oxidation studies of hair protein in the *Homo tirolensis*. *FASEB J.* **8** (1994)(14) 1166–1169. URL http://www.fasebj.org/cgi/reprint/8/14/1166.pdf
247. M. Luckey: *Membrane Structural Biology: With Biochemical and Biophysical Foundations.* (Cambridge University Press, Cambridge; New York), 2008. ISBN 9780521856553
248. R. Lundblad & C. Noyes: *Chemical reagents for protein modification.* 3rd edn.. (CRC, Boca Raton), 2005. ISBN 0-8493-1983-8
249. M. Lunder, T. Bratkovic *et al.*: Ultrasound in phage display: A new approach to nonspecific elution. *Biotechniques* **44** (2008) 893–900. doi: 10.2144/000112759
250. A. Lustig, A. Engel *et al.*: Molecular weight determination of membrane proteins by sedimentation equilibrium at the sucrose or nycodenz-adjusted density of the hydrated detergent micelle. *Biochim. Biophys. Acta* **1464** (2000) 199–206. doi: 10.1016/S0005-2736(99)00254-0
251. G. MacBath: Protein microarrays and proteomics. *Nat. Genet. Suppl.* **32** (2002) 526–532. doi: 10.1038/ng1037
252. D. Magde & E. Elson: Fluorescence correlation spectroscopy. II. An experimental realization. *Biopolymers* **13** (1974) 29–61. doi: 10.1002/bip.1974.360130103
253. M. le Maire, P. Champeil *et al.*: Interaction of membrane proteins and lipids with solubilizing detergents. *Biochim. Biophys. Acta* **1508** (2000) 86–111. doi: 10.1016/S0304-4157(00)00010-1
254. M. le Maire, J. Møller *et al.*: Protein-protein contacts in solubilized membrane proteins, as detected by cross-linking. *Anal. Biochem.* **362** (2007) 168–171. doi: 10.1016/j.ab.2006.11.025

255. K. MANDAL, B. PENTELUTE *et al.*: Racemic crystallography of synthetic protein enantiomeres used to determine the X-ray structure of plectasin by direct methods. *Protein Sci.* **18** (2009) 1146–1154. doi: 10.1002/pro.127

256. M. MANNING: Use of infrared spectroscopy to monitor protein structure and stability. *Expert Rev. Proteomics* **2** (2005)(5) 731–743. doi: 10.1586/14789450.2.5.731

257. J. MARGOLIS & K. KENRICK: Polyacrylamide gel electrophoresis in a continuous molecular sieve gradient. *Anal. Biochem.* **25** (1968) 347–362. doi: 10.1016/0003-2697(68)90109-7

258. R. MARTIN & B. AMES: A method for determining the sedimentation behaviour of enzymes: Applicaton to protein mixtures. *J. Biol. Chem.* **236** (1961) 1372–1379. http://www.jbc.org/content/236/5/1372.full.pdf+html?sid=49a396bd-5a66-45a8-b0fd-cdae82b24c59

259. A. MARTONOSI (ed.): *The enzymes of biological membranes*, vol. 3. (Plenum, New York, London), 1985. ISBN 978-0306414534

260. J. MATHER & P. ROBERTS: *Introduction to cell and tissue cultrure: Theory and techniques.* (Plenum, New York), 1998. ISBN 978-0306458590

261. A. MATOUSCHEK & A. FERSHT: Protein engineering in analysis of protein folding pathways and stability. *Methods Enzymol.* **202** (1991) 82–112. doi: 10.1016/0076-6879(91)02008-W

262. P. MATSUDAIRA: Sequence from picomole quantities of proteins electroblotted onto polyvinylidene difluoride membranes. *J. Biol. Chem.* **262** (1987) 10 035–10 038. URL http://www.jbc.org/cgi/reprint/262/21/10035.pdf

263. K. MATSUO, Y. MATSUSHIMA *et al.*: Vacuum-ultraviolet circular dichroism of amino acids as revealed by synchrotron radiation spectrophotometer. *Chem. Lett.* **31** (2002) 826–827. http://home.hiroshima-u.ac.jp/x070010/HiSOR-HSRC-BL15.files/VUVCD%20amino%20acids.pdf

264. T. MATSUYA, N. HOSHINO *et al.*: Design of lanthanide complex probes for highly sensitive time-resolved fluorometric detection methods and its application to biochemical, environmental and clinical analyses. *Curr. Anal. Chem.* **2** (2006) 397–410. doi: 10.2174/157341106778520526

265. C. MATTHEWS: Effect of point mutations on the folding of globular proteins. *Methods Enzymol.* **154** (1987) 498–511. doi: 10.1016/0076-6879(87)54092-7

266. G. MAYER, B. LUDWIG *et al.*: Studying membrane proteins in detergent solution by analytical ultracentrifugation: Different methods for density matching. *Prog. Colloid Polym. Sci.* **113** (1999) 176–181. doi: 10.1007/3-540-48703-4_25

267. C. MCCUDDEN & V. KRAUS: Biochemistry of amino acid racemization and clinical application to musculoskeletal disease. *Clin. Biochem.* **39** (2006) 1112–1130. doi: 10.1016/j.clinbiochem.2006.07.009

268. L. MCINTOSH, G. HAND *et al.*: The pK_a of the general acid/base carboxyl group of a glycosidase cycles during catalysis: A ^{13}C-NMR study of bacillus circulans xylanase. *Biochemistry* **35** (1996) 9958–9966. doi: 10.1021/bi9613234. URL http://startrek.ccs.yorku.ca/~pjohnson/publications/bio03.pdf

269. A. MCPHERSON: *Crystallization of biological macromolecules.* (Cold Spring Harbor Laboratory, Cold Spring Harbor, NY), 1999. ISBN 978-0879695279

270. W. MELCHIOR & D. FAHRNEY: Ethoxyformylation of proteins. *Biochemistry* **9** (1970) 251–258. doi: 10.1021/bi00804a010

271. T. MENGUY, S. CHENEVOIS *et al.*: Ligand binding to macromolecules or micelles: Use of centrifugal ultrafiltration to measure low-affinity binding. *Anal. Biochem.* **264** (1998) 141–148. doi: 10.1006/abio.1998.2854

272. C. MERRIL, D. GOLDMAN *et al.*: Ultrasensitive stain for proteins in polyacrylamide gels shows regional variations in cerebrospinal fluid proteins. *Science* **211** (1981) 1437–1438. doi: 10.1126/science.6162199

273. M. MESELSON & W. STAHL: The replication of DNA in *Escherischia coli*. *Proc. Natl. Acad. Sci. USA* **44** (1958) 671–682. URL http://www.pnas.org/content/44/7/671.full.pdf+html

274. X. MICHALET, F. PINAUD *et al.*: Quantum dots for live cells, *in vivo* imaging and diagnostics. *Science* **307** (2005) 538–544. doi: 10.1126/science.1104274

275. M. DE MICHELIS & R. SPANSWICK: H^+-pumping driven by the vanadate-sensitive ATPase in membrane vesicles from corn root. *Plant Physiol.* **81** (1986) 542–547. URL http://www.plantphysiol.org/cgi/reprint/81/2/542
276. G. A. MILLICAN: Photoelectric methods of measuring the velocity of rapid reactions. III. A portable micro-apparatus applicable to an extended range of reactions. *Proc. R. Soc. Lond. A* **155** (1936) 277–292. doi: 10.1098/rspa.1936.0099
277. I. MILLS, T. CVITA *et al.*: *Quantities, units and symbols in physical chemistry*. 2nd edn.. (Blackwell, Oxford), 1993. ISBN 0-632-03583-8. http://old.iupac.org/publications/books/gbook/index.html
278. L. MITIC, V. UNGER *et al.*: Expression, solubilization, and biochemical characterization of the tight junction transmembrane protein claudin-4. *Protein Sci.* **12** (2003) 217–218. doi: 10.1110/ps.0233903
279. M. MIYAGI & K. C. S. RAO: Proteolytic ^{18}O-labeling strategies for quantitative proteomics. *Mass Spectrom. Rev.* **26** (2007) 121–136. doi: 10.1002/mas.20116
280. B. MÜLLER, E. ZAYCHIKOV *et al.*: Pulsed interleaved excitation. *Biophys J.* **89** (2005) 3508–3522. doi: 10.1529/biophysj.105.064766
281. K. MONKOS: Concentration and temperature dependence of viscosity in lysozyme aqueous solution. *Biochim. Biophs. Acta* **1339** (1997) 304–310. doi: 10.1016/S0167-4838(97)00013-7
282. M. MONTAL & P. MUELLER: Formation of bimolecular membranes from lipid monolayers and a study of their electrical properties. *Proc Natl. Acad. Sci. USA* **69** (1972) 3561–3566. http://www.pnas.org/content/69/12/3561.full.pdf
283. N. MORGNER, H.-D. BARTH *et al.*: A new way to detect noncovalently bonded complexes of biomolecules from liquid micro-droplets by laser mass spectrometry. *Aust. J. Chem.* **59** (2006) 109–114. doi: 10.1071/CH05285
284. K. MOSBACH: Separation of isozymes. *Methods Enzymol.* **34** (1974) 595–598. doi: 10.1016/S0076-6879(74)34079-7
285. P. MUELLER, D. RUDIN *et al.*: Reconstitution of cell membrane structure *in vitro* and its transformation into an excitable system. *Nature* **194** (1962) 979–980. doi: 10.1038/194979a0
286. M. MULISCH & U. WELSCH (eds.): *Romeis – Mikroskopische technik*. 18th edn.. (Urban und Schwarzenberg, München), 2010. ISBN 978-3827416766. (This is a textbook for students. The 16th edition from 1968, the last one edited by ROMEIS, is an encyclopedic laboratory handbook.)
287. E. MULLER & T. DAVIS: Protein localization by cell imaging. In: *Proteomics for biological discovery*, T. VEENSTRA & J. Y. III, eds.. (Wiley, Hoboken, NJ). ISBN 978-0-471-16005-2, 2006 pp. 137–155
288. S. MUYLDERMANS, C. CAMBILLAU *et al.*: Recognition of antigens by single-domain antibody fragments: The superfluous luxury of paired domains. *Trends Biochem. Sci.* **26** (2001) 230–235. doi: 10.1016/S0968-0004(01)01790-X
289. L. NANNINGA & W. MOMMAERTS: Studies on the formation of an enzyme-substrate complex between myosin and adenosinetriphosphate. *Proc. Natl. Acad. Sci. USA* **46** (1960)(8) 1155. http://www.ncbi.nlm.nih.gov/pmc/articles/PMC223016/pdf/pnas00207-0149.pdf
290. A. VON NEUHOFF: Die kolloidale coomassie-blau-färbung von proteinen und peptiden in polyacrylamid-gelen: Schnell, untergrundfrei und mit nanogramm-empfindlichkeit. *Biol. Chem. Hoppe Seyler* **371** (1990) 10–11
291. S. NEWSTEAD, S. FERRANDON *et al.*: Rationalizing α-helical membrane protein crystallization. *Protein Sci.* **17** (2008) 466–472. doi: 10.1110/ps.073263108
292. N. NIRMALAN, P. HARNDEN *et al.*: Mining the archival formalin-fixed paraffin embedded tisue proteome: Opportuniuties and challenges. *Mol. Biosyst* **4** (2008) 712–720. doi: 10.1039/b800098k
293. T. NISHIBU, K. HIRAYASU *et al.*: Quantitative filtration-blotting of protein in the presence of sodium dodecyl sulfate and its use for protein assay. *Anal. Biochem.* **319** (2003) 88–95. doi: 10.1016/S0003-2697(03)00255-0
294. H. NISHIZAWA, N. KITA *et al.*: Determination of molecular weights of native proteins by polyacrylamide gradient gel electrophoresis. *Electrophoresis* **9** (1989) 803–806. doi: 10.1002/elps.1150091203

295. M. ÜNLÜ, M. MORGAN *et al.*: Difference gel electrophoresis: A single gel method for detecting changes in protein extracts. *Electrophoresis* **18** (1997) 2071–2077. doi: 10.1002/elps.1150181133
296. U. NUBER (ed.): *DNA Microarrays.* (Taylor & Francis, New York), 2005. ISBN 978-0415358668
297. P. O'FARRELL, H. GOODMAN *et al.*: High resolution two-dimensional electrophoresis of basic as well as acidic proteins. *Cell* **12** (1977) 1133–1142. doi: 10.1016/0092-8674(77)90176-3
298. G. OGG: *A Practical Guide to Quality Management in Clinical Trial Research.* (Informa HealthCare, Boca Raton), 2005. ISBN 978-0849397226
299. K. OIWA, J. ECCLESTON *et al.*: Comparative single-molecule and ensemble myosin enzymology: Sulfoindocyanine ATP and ADP derivatives. *Biophys. J.* **78** (2000) 3048–3071. doi: 10.1016/S0006-3495(00)76843-3
300. I. OLSEN & H. WIKER: Diffusion blotting for rapid production of multiple identical imprints from sodium dodecylsulfate polyacrylamide gel electrophoresis on a solid support. In: *Protein blotting and detection – Methods and protocols*, B. KURIEN & R. SCOFIELD, eds., vol. 536 of *Methods Mol. Biol.* (Humana, Dordrecht). ISBN 978-1-934115-73-2, 2009 pp. 35–38
301. S. OPELLA & F. MARASSI: Structure determination of membrane proteins by NMR spectroscopy. *Chem. Rev.* **104** (2004) 3587–3606. doi: 10.1021/cr0304121
302. V. ORLANDO, H. STRUTT *et al.*: Analysis of chromatin structure by *in vivo* formaldehyde cross-linking. *Methods* **11** (1997) 205–214. doi: 10.1006/meth.1996.0407
303. L. ORNSTEIN: Disk electrophoresis: I. Background and theory. *Ann. N.Y. Acad. Sci.* **121** (1962) 321–351. doi: 10.1111/j.1749-6632.1964.tb14207.x
304. B. ORSI & K. TIPTON: Kinetic analysis of progress curves. *Methods Enzymol.* **63** (1979) 159–183. doi: 10.1016/0076-6879(79)63010-0
305. J. ORTH, I. PREUSS *et al.*: *Pasteurella multocida* toxin activation of heterotrimeric G-proteins by deamination. *Proc. Natl. Acad. Sci. USA* **106** (2009) 7179–7184. doi: 10.1073/pnas.0900160106
306. C. OSTERMEIER, A. HARRENGA *et al.*: Structure at 2.7 Å resolution of the *Paracoccus denitrificans* two-subunit cytochrome c oxidase complexed with an antibody F_v fragment. *Proc. Natl. Acad. Sci. USA* **94** (1997) 10547–10553. http://www.pnas.org/content/94/20/10547.full.pdf+html
307. O. OUCHTERLONY: Antigen-antibody reactions in gels. *Acta Pathol. Microbiol. Scand.* **26** (1949) 507. doi: 10.1111/j.1600-0463.2007.apm_678a.x
308. W. OUTLAW, JR., S. SPRINGER *et al.*: Histochemical technique: A general method for quantitative enzyme assays of single cell 'extracts' with a time resolution of seconds and a reading precision of femtomoles. *Plant Physiol.* **77** (1985) 659–666. doi: 10.1104/pp.77.3.659
309. W. OUTLAW, JR. & S. ZHANG: Single-cell dissection and microdroplet chemistry. *J. Exp. Bot.* **52** (2001) 605–614. doi: 10.1093/jexbot/52.356.605
310. C. PACE: Determination and analysis of urea and guanidine hydrochloride denaturation curves. *Methods Enzymol.* **131** (1986) 266–280. doi: 10.1016/0076-6879(86)31045-0
311. D. PANTAZATOS, J. KIM *et al.*: Rapid refinement of crystallographic protein construct, definition employing enhanced hydrogen/deuterium exchange MS. *Proc. Natl. Acad. Sci. USA* **101** (2004) 751–756. http://www.pnas.org/content/101/3/751.full.pdf+html
312. J. PARK, M. MABUCHI *et al.*: Visualization of unstained protein bands on PVDF. In: *Protein blotting and detection – Methods and protocols*, B. KURIEN & R. SCOFIELD, eds., vol. 536 of *Methods Mol. Biol.* (Humana, Dordrecht). ISBN 978-1-934115-73-2, 2009 pp. 527–531
313. M. PARKANY: *Quality Assurance and Total Quality Management for Analytical Laboratories.* (Royal Society of Chemistry, London), 1993. ISBN 978-0851867052
314. M. PASCHKA & W. HÖHNE: A twin-arginine translocation (Tat)-mediated phage display system. *Gene* **350** (2005) 79–88. doi: 10.1016/j.gene.2005.02.005
315. L. PATTHY & E. L. SMITH: Reversible modification of arginine residues. Application to sequence studies by restriction of tryptic hydrolysis to lysine residues. *J. Biol. Chem.* **250** (1975)(2) 557–564. http://www.jbc.org/cgi/reprint/250/2/557.pdf

316. C. Paweletz, L. Charbonneau *et al.*: Reverse phase protein microarrays which capture disease progression show activation of pro-survival pathways at the cancer invasion front. *Oncogene* **20** (2001) 1981–1989. doi: 10.1038/sj.onc.1204265
317. R. Pearlman & Y. Wang: *Formulation, characterization, and stability of protein drugs: Case histories*. (Springer, New York), 1996. ISBN 978-0306453328
318. M. Pedersen, N. Sorensen *et al.*: Effect of different hapten-carrier conjugation ratios and molecular orientations on antibody affinity against a peptide antigen. *J. Immunol. Methods* **311** (2006) 198–206. doi: 10.1016/j.jim.2006.02.008
319. T. Pederson: Turning a PAGE: The overnight sensation of SDS-polycrylamide gel electrophoresis. *FASEB J.* **22** (2008) 949–953. doi: 10.1096/fj.08-0402ufm
320. J. Peng, D. Schwartz *et al.*: A proteomics approach to understanding protein ubiquitination. *Nat. Biotechnol.* **21** (2003) 921–926. http://www.fbmc.fcen.uba.ar/~3-2-2005/SEM6JM2007.pdf
321. D. Perkins, D. Pappin *et al.*: Probability-based protein identification by searching sequence databases using mass spectrometry data. *Electrophoresis* **20** (1999) 3551–3567. doi: 10.1002/(SICI)1522-2683(19991201)20:18<3551::AID-ELPS3551>3.0.CO;2-2
322. S. Perkins: Protein volumes and hydration effects. *J. Biochem.* **157** (1986) 169–180. doi: 10.1111/j.1432-1033.1986.tb09653.x
323. F. Perrin: Polarisation de la lumière de fluorecence. Vie moyenne des molécules dans l´etat excité. *J. Phys.* **Series VI, 7** (1926) 390–401. doi: 10.1051/jphysrad:01926007012039000
324. K. Peters & G. Snyder: Time-resolved photoacoustic calorimetry: probing the energetics and dynamics of fast chemical and biochemical reactions. *Science* **241** (1988) 1053 – 1057. doi: 10.1126/science.3045967
325. M. Petoukhov, N. Eady *et al.*: Addition of missing loops and domains to protein models by X-ray solution scattering. *Biophys. J.* **83** (2002) 3113–3125. doi: 10.1016/S0006-3495(02)75315-0
326. E. Petricoin, A. Ardekani *et al.*: Use of proteomic patterns in serum to identify ovarian cancer. *Lancet* **359** (2002) 572–577. doi: 10.1016/S0140-6736(02)07746-2
327. G. Pfennig, H. Kleve-Hebenius *et al.*: *Karlsruher nuklidkarte*. 7th edn.. (Forschungszentrum Karlsruhe, Karlsruhe), 2006. ISBN 978-3921879184
328. A. Pitre, Y. Pan *et al.*: On the use of ratio standard curves to accurately quantitate relative changes in protein levels by western blot. *Anal. Biochem.* **361** (2007) 305–307. doi: 10.1016/j.ab.2006.11.008
329. J. Pitts: Crystallization by centrifugation. *Nature* **355** (1992) 117. doi: 10.1038/355117a0
330. B. Plapp: Application of affinity labeling for studying structure and function of enzymes. *Methods Enzymol.* **87** (1982) 469–499. doi: 10.1016/S0076-6879(82)87027-4
331. J. Poduslo: Glycoprotein molecular-weight estimation using sodium dodecyl sulfate-pore gradient electrophoresis: Comparison of tris-glycine and tris-borate-EDTA buffer systems. *Anal. Biochem.* **114** (1981) 131–139. doi: 10.1016/0003-2697(81)90463-2
332. A. Pollard & C. Heron: *Archaeological Chemistry*. 2nd edn.. (RSC, Cambridge, UK), 2008. ISBN 978-0-85404-262-3
333. R. Pollet, B. Haase *et al.*: Macromolecular characterization by sedimentation equilibrium in the preparative ultracentrifuge. *J. Biol. Chem.* **254** (1979) 30–33. http://www.jbc.org/content/254/1/30.full.pdf+html?sid=31e44cff-cf18-49d6-93a4-7d116b4ef52e
334. C. Poole: *Electron spin resonance: A comprehensive treatise on experimental techniques*. 2nd edn.. (Wiley, New York), 1983. ISBN 978-0471046783
335. J. Porath & P. Flodin: Gel filtration: a method for desalting and group separation. *Nature* **183** (1959) 1657–1659. doi: 10.1038/1831657a0
336. V. Potter & C. Elvehjem: A modified method for the study of tissue oxidations. *J. Biol. Chem.* **114** (1936) 495–504. http://www.jbc.org/content/114/2/495.full.pdf+html?sid=49ae7180-3734-425c-8084-8372de3a0269
337. W. Press, B. Flannery *et al.*: *Numerical recipes in Pascal: The art of scientific computing*. (Cambridge University Press, Cambridge), 1989. ISBN 978-0521375160

338. T. RABILLOUD, V. GIRARDOT *et al.*: One- and two-dimensional histone separation in acidic gels: Usefulness of methylene blue-driven photopolymerization. *Electrophoresis* **17** (1996) 67–73. doi: 10.1002/elps.1150170112
339. T. RABILLOUD, J.-M. STRUB *et al.*: A comparison between sypro ruby and ruthenium(II) tris (bathophenanthroline disulfonate) as fluorescent stains for protein detection in gels. *Proteomics* **1** (2001) 699–704. http://www.uhkt.cz/files/proteomika/sypro_versus_ruthenium.pdf
340. M. RADIN: Extraction of tissue lipids with a solvent of low toxicity. *Methods Enymol.* **72** (1981) 5–7. doi: 10.1016/S0076-6879(81)72003-2
341. R. RAGONE, F. FACCHIANO *et al.*: Flexibility plot of proteins. *Protein Eng.* **2** (1989)(7) 497–504. doi: 10.1093/protein/2.7.497
342. K. RAINES, S. SALHA *et al.*: Three-dimensional structure determination from a single view. *Nature* **463** (2010) 214–217. doi: 10.1038/nature08705
343. T. RATLIFF: *The laboratory quality assurance system: A manual of quality procedures and forms.* 3rd edn.. (Wiley-Interscience, Hoboken, NJ), 2003. ISBN 978-0471269182
344. H. RATTLE: *An NMR Primer for life scientists.* (Partnership Press, Fareham (GB)), 1995. ISBN 978-0951643631
345. F. RAUSHEL & J. VILLAFRANCA: Positional isotope exchange. *Crit. Rev. Biochem. Mol. Biol.* **23** (1988) 1–26. doi: 10.3109/10409238809103118
346. S. RAYMOND & L. WEINTRAUB: Acrylamide gel as a supporting medium for zone electrophoresis. *Science* **130** (1959) 711. doi: 10.1126/science.130.3377.711
347. J. REIG & D. KLEIN: Submicron quantities of unstained proteins are visualized on polyvinylidene difluoride membranes by transillumination. *Appl. Theor. Electrophor.* **1** (1988) 59–60
348. C. RETAMAL & J. BABUL: Determination of the molecular weight of proteins by electrophoresis in slab gels with a transverse pore gradient of cross-linked polyacrylamide in the absence of denaturing agents. *Anal. Biochem.* **175** (1988) 544–547. doi: 10.1016/0003-2697(88)90581-7
349. J. REYNOLDS & C. TANFORD: Determination of molecular weight of the protein moiety in protein-detergent complexes without direct knowledge of detergent binding. *Proc. Natl. Acad. Sci. USA* **73** (1976) 4467–4470. http://www.pnas.org/content/73/12/4467.full.pdf+html
350. G. RHODES: *Crystallography made crystal clear.* 3rd edn.. (Academic, Amsterdam), 2006. ISBN 978-0-12-587073-3
351. J. RIGAUD: Membrane proteins: Functional and structural studies using reconstituted proteoliposomes and 2-D crystals. *Braz. J. Med. Biol. Res.* **35** (2002) 753–766. doi: 10.1590/S0100-879X2002000700001. http://www.scielo.br/pdf/bjmbr/v35n7/4512.pdf
352. J. RIGAUD, M. CHAMI *et al.*: Use of detergents in two-dimensional crystallization of membrane proteins. *Biochim. Biophys. Acta* **1508** (2000) 112–128. doi: 10.1016/S0005-2736(00)00307-2
353. J.-L. RIGAUD, D. LEVY *et al.*: Detergent removal by non-polar polystyrene beads. *Eur. Biophys. J.* **27** (1998) 305–319. doi: 10.1007/s002490050138
354. R. RITCHIE & T. PRVAN: A simulation studie on designing experiments to measure the K_m of MICHAELIS-MENTEN kinetics curves. *J. Theor. Biol.* **178** (1996) 239–254. doi: 10.1006/jtbi.1996.0023
355. H. ROGNIAUX, S. SANGLIER *et al.*: Mass spectrometry as a novel approach to probe cooperativity in multimeric enzymatic systems. *Anal. Biochem.* **291** (2001) 48–51. doi: 10.1006/abio.2000.4975
356. G. ROUSER & S. FLEISCHER: Isolation, characterisation and determination of polar lipids of mitochondria. *Methods Enzymol.* **10** (1967) 385–406. doi: 10.1016/0076-6879(67)10072-4
357. A. SADRA, T. CINEK *et al.*: Multiple probing of an immunoblot membrane using a non-block technique: Advantages in speed and sensitivity. *Anal. Biochem.* **278** (2000) 235–237. doi: 10.1006/abio.1999.4453
358. Z. SALAMON, H. MACLEOD *et al.*: Surface plasmon resonance spectroscopy as a tool for investigating the biochemical and biophysical properties of membrane protein systems. I: Theoretical principles. *Biochim. Biophys. Acta* **1331** (1997) 117–129. doi: 10.1016/S0304-4157(97)00004-X

359. Z. SALAMON, H. MACLEOD *et al.*: Surface plasmon resonance spectroscopy as a tool for investigating the biochemical and biophysical properties of membrane protein systems. II: Application to biological systems. *Biochim. Biophys. Acta* **1331** (1997) 131–152. doi: 10.1016/S0304-4157(97)00003-8
360. J. SAMBROOK & D. RUSSEL: *Molecular cloning: A laboratory manual*. 3rd edn.. (Cold Spring Harbor Laboratory, Cold Spring Harbor, N.Y.), 2001. ISBN 978-0879695774
361. C. SANDERS & R. PROSSER: Bicelles: a model membrane system for all seasons. *Structure* **6** (1998) 1227–1234. doi: 10.1016/S0969-2126(98)00123-3
362. C. SANNY & J. PRICE: Analysis of antibody-antigen interactions using size exclusion high-performance (pressure) chromatography. *Anal. Biochem.* **246** (1997) 7–14. doi: 10.1006/abio.1996.9995
363. T. SANO, C. SMITH *et al.*: Immuno-PCR: very sensitive antigen detection by means of specific antibody-DNA conjugates. *Science* **258** (1992) 120–122. doi: 10.1126/science.1439758
364. M. SARGENT: Fiftyfold amplification of the LOWRY protein assay. *Anal. Biochem.* **163** (1987) 476–481. doi: 10.1016/0003-2697(87)90251-X
365. G. SAUERBREY: Verwendung von schwingquarzen zur wägung dünner schichten und zur mikrowägung. *Z. Phys. A Hadrons Nucl.* **155** (1959) 206. doi: 10.1007/BF01337937
366. H. SCHÄGGER: Tricine-SDS-PAGE. *Nat. Protoc.* **1** (2006) 16–22. doi: 10.1038/nprot.2006.4
367. H. SCHÄGGER, W. CRAMER *et al.*: Analysis of molecular masses and oligomeric states of protein complexes by blue native electrophoresis and isolation of membrane protein complexes by two-dimensional native electrophoresis. *Anal. Biochem.* **217** (1994) 220–230. doi: 10.1006/abio.1994.1112
368. H. SCHÄGGER & G. VON JAGOW: Tricine-sodium dodecyl sulfate polyacrylamide gel electrophoresis for the separation of proteins in the range from 1-100 kDalton. *Anal. Biochem.* **166** (1987) 368–379. doi: 10.1016/0003-2697(87)90587-2
369. S. SCHRIER, A. ZACHOWSKI *et al.*: Mechanisms of amphipath-induced stomatocytosis in human erythrocytes. *Blood* **79** (1992) 782–786. http://bloodjournal.hematologylibrary.org/cgi/reprint/79/3/782.pdf
370. P. SCHUCK: Size-distribution analysis of macromolecules by sedimentation velocity ultracentrifugation and LAMM equation modeling. *Biophys. J.* **78** (2000) 1606–1619. http://www.pubmedcentral.nih.gov/picrender.fcgi?artid=1300758&blobtype=pdf
371. E. SCRIVEN (ed.): *Azides and nitrenes*. (Academic, New York), 1984. ISBN 978-0126334807
372. I. SEGEL: *Enzyme Kinetics*. (Wiley, New York), 1975, reprinted 1993. ISBN 978-0471303091
373. W. SHEWHART: *Economic Control of Quality of Manufactured Product*. (D. van Nostrand, New York), 1931. ISBN 0-87389-076-0
374. A. SHIOI, M. HARADA *et al.*: Protein extraction in a tailored reversed micellar system containing nonionic surfactants. *Langmuir* **13** (1997) 609–616. doi: 10.1021/la960250m
375. K. SHIRAHAMA, K. TSUJII *et al.*: Free boundary electrophoresis of sodium dodecyl sulphate-protein polypeptide complexes with special reference to SDS-polyacrylamide gel electrophoresis. *J. Biochem.* **75** (1974) 309–319
376. G. SIEGEL & T. DESMOND: Effects of tetradecyl sulfate on electrophoretic resolution of kidney Na, K-ATPase catalytic subunit isoforms. *J. Biol. Chem.* **264** (1989) 4751–4754. http://www.jbc.org/content/264/9/4751.full.pdf
377. K. SIMONS & E. IKONEN: Functional rafts in cell membranes. *Nature* **387** (1997) 569–572. doi: 10.1038/42408
378. S. SINGER & G. NICOLSON: The fluid mosaic model of the structure of cell membranes. *Science* **175** (1972) 720–731. doi: 10.1126/science.175.4023.720
379. A. SIVASHANMUGAM, V. MURRAY *et al.*: Practical protocols for production of very high yields of recombinant proteins using *Escherichia coli*. *Protein Sci.* **18** (2009) 936–948. doi: 10.1002/pro.102
380. G. SMITH: Filamentous fusion phage: Novel expression vectors that display cloned antigens on the virion surface. *Science* **228** (1985) 1315–1317. doi: 10.1126/science.4001944
381. I. SMITH, R. CROMIE *et al.*: Seeing gel wells well. *Anal. Biochem.* **169** (1988) 370–371. doi: 10.1016/0003-2697(88)90297-7

382. P. SMITH, R. KROHN *et al.*: Measurement of protein using bicinchonic acid. *Anal. Biochem.* **150** (1985) 76–85. doi: 10.1016/0003-2697(85)90442-7
383. P. SNEATH & R. SOKAL: *The principles and practice of numerical taxonomy*. (Freeman, San Francisco), 1973. ISBN 9780716706977
384. A. SOLOVYOVA, P. SCHUCK *et al.*: Non-ideality by sedimentation velocity of halophilic malate dehydrogenase in complex solvents. *Biophys. J.* **81** (2001) 1868–1880. doi: 10.1016/S0006-3495(01)75838-9
385. L. SONG & J. GOUAUX: Membrane protein crystallization: Application of sparse matrices to the α-hemolysin heptamer. *Methods Enzymol.* **276** (1997) 60–74. doi: 10.1016/S0076-6879(97)76050-6
386. T. SPANDE, B. WITKOP *et al.*: Selective cleavage and modification of peptides and proteins. *Adv. Protein Chem.* **24** (1970) 97–260
387. V. SPASSOV & L. YAN: A fast and accurate computatioal approach for protein ionization. *Protein Sci.* **17** (2008) 1955–1970. doi: 10.1110/ps.036335.108
388. B. SPENGLER, M. KARAS *et al.*: Excimer laser desorption mass spectrometry of biomolecules at 248 and 193 nm. *J. Phys. Chem.* **91** (1987)(26) 6502–6506. doi: 10.1021/j100310a016
389. I. STEINBERG & H. SCHACHMAN: Ultracentrifugation with absoption optics. Analysis of interacting systems involving macromolecules and small molecules. *Biochemistry* **5** (1966) 3728–3747. doi: 10.1021/bi00876a003
390. N. STELLWAGEN: Apparent pore size of polyacrylamide gels. *Electrophoresis* **19** (1998) 1542–1547. doi: 10.1002/elps.1150191004
391. A. STENSBALLE, O. JENSEN *et al.*: Electron capture dissociation of singly and multiply phosphorylated peptides. *Rapid Commun. Mass Spectrom.* **14** (2000)(19) 1793–1800. http://www.bmms.uu.se/pubs/article003.pdf
392. T. STEPHENS, C. BUNDE *et al.*: Mechanism of action in thalidomide teratogenesis. *Biochem. Pharmacol.* **59** (2000) 1489–1499. doi: 10.1016/S0006-2952(99)00388-3
393. M. SUCHANEK, A. RADZIKOWSKA *et al.*: Photo-leucine and photo-methionine allow identification of protein-protein interactions in living cells. *Nat. Methods* **2** (2005) 261–267. doi: 10.1038/NMETH752
394. I. SUYDAM, C. SNOW *et al.*: Electric fields at the active site of an enzyme: Direct comparison of experiment with theory. *Science* **313** (2006) 200–204. doi: 10.1126/science.1127159
395. T. SVEDBERG & K. PEDERSEN: *The ultracentrifuge*. (Oxford University Press, Oxford), 1940. ISBN 978-0384588905
396. D. SVERGUN: Restoring low resolution structure of biological macromolecules from solution scattering using simulated annealing. *Biophys. J.* **76** (1999) 2879–2886. doi: 10.1016/S0006-3495(99)77443-6
397. M. SWANN, L. PEEL *et al.*: Dual-polarization interferometry: An analytical technique to measure changes in protein structure in real time, to determine the stoichiometry of binding events, and to differentiate between specific and nonspecific interactions. *Anal. Biochem.* **329** (2004) 190–198. doi: 10.1016/j.ab.2004.02.019
398. G. TAGUCHI: *Introduction to quality engineering*. (Asian Productivity Organisation UNIPUB, Tokyo), 1986. ISBN 978-9283310846
399. G. TAGUCHI: *System of experimental design*. (Asian Productivity Organisation UNIPUB / Kraus Intern. Publications, New York), 1987. ISBN 978-0527916213
400. K. TAKAHASHI: The Reaction of Phenylglyoxal with Arginine Residues in Proteins. *J. Biol. Chem.* **243** (1968)(23) 6171–6179. http://www.jbc.org/cgi/reprint/243/23/6171.pdf
401. K. TAKAHASHI: Specific Modification of Arginine Residues in Proteins with Ninhydrin. *J. Biochem. (Tokyo)* **80** (1976)(5) 1173–1176. http://jb.oxfordjournals.org/cgi/content/abstract/80/5/1173
402. K. TAKEO: Affinity electrophoresis: Principles and application. *Electrophoresis* **5** (1984) 187–195. doi: 10.1002/elps.1150050402
403. T. TAKI, T. GONZALEZ *et al.*: TLC blot (far-eastern blot) and its applications. In: *Protein blotting and detection – Methods and protocols*, B. KURIEN & R. SCOFIELD, eds., vol. 536 of *Methods Mol. Biol.* (Humana, Dordrecht). ISBN 978-1-934115-73-2, 2009 pp. 545–556

404. Z. TAKÁTS, J. WISEMAN *et al.*: Mass spectrometry sampling under ambient conditions with desorption electrospray ionization. *Science* **306** (2004) 471–473. doi: 10.1126/science.1104404
405. H. TAN, T. NG *et al.*: Effects of light spectrum in flatbed scanner densitometry of stained polyacrylamide gels. *Biotechniques* **42** (2007) 474–478. doi: 10.2144/000112402
406. K. TANAKA, H. WAKI *et al.*: Protein and polymer analyses up to m/z 100,000 by laser ionization time-of-flight mass spectrometry. *Rapid Commun. Mass Spectrom.* **2** (1988)(8) 151–153. http://masspec.scripps.edu/mshistory/timeline/time_pdf/1988_Tanaka2.pdf
407. H. TAO, W. LIU *et al.*: Purifying natively folded proteins from inclusion bodies using sarkosyl, Triton X-100, and CHAPS. *Biotechniques* **48** (2010) 61–64. doi: 10.2144/000113304
408. K. TAWA & K. MORIGAKI: Substrate-supported phospholipid membranes studied by surface plasmon resonance and surface plasmon fluorescence spectroscopy. *Biophys. J.* **89** (2005) 2750–2758. doi: 10.1529/biophysj.105.065482
409. C. TAYLOR & H. LIPSON: *Optical transforms. Their preparation and application to X-ray diffraction problems.* (G. Bell and Sons, London), 1964
410. P. TESSIER, A. LENHOFF *et al.*: Rapid measurement of protein osmotic second virial coefficients by self-interaction chromatography. *Biophys.J.* **82** (2002) 1620–1631. doi: 10.1016/S0006-3495(02)75513-6
411. B. THOMSON: Radio frequency quadrupole ion guides in modern mass spectrometry. *Can. J. Chem.* **76** (1998) 499–505. doi: 10.1139/cjc-76-5-499
412. A. TISELIUS: A new apparatus for analysis of colloidal mixtures. *Trans. Faraday Soc.* **33** (1937) 524–531. doi: 10.1039/TF9373300524
413. H. TOWBIN, T. STAEHELIN *et al.*: Electrophoretic transfer of proteins from polyacrylamide gels to nitrocellulose sheets: Procedure and some applications. *Proc. Natl. Acad. Sci. USA* **76** (1979) 4350–4354. http://www.pnas.org/content/76/9/4350.full.pdf+html
414. D. TRONRUD & B. MATTHEWS: Sorting the chaff from the weat at the PDB. *Protein Sci.* **18** (2009) 2–5. doi: 10.1002/pro.13
415. R. TSIEN: The green fluorescent protein. *Annu. Rev. Biochem.* **67** (1998) 509–544. doi: 10.1146/annurev.biochem.67.1.509
416. C. TSOU: Relation between modification of functional groups of proteins and their biological activity. I. A graphical method for the determination of the number and type of essential groups. *Sci. Sin.* **11** (1962) 1535–1558. http://en.cnki.com.cn/Article_en/CJFDTOTAL-SHWL196203006.htm
417. E. VALERO & F. GARCÍA-CARMONA: Optimizing enzymatic cycling assays: Spectrophotometric determination of low levels of pyruvate and L-lactate. *Anal. Biochem.* **239** (1996) 47–52. doi: 10.1006/abio.1996.0289
418. J. VENTER *et al.*: The sequence of the human genome. *Science* **291** (2001) 1304–1351. doi: 10.1126/science.1058040. http://www.stanford.edu/class/cs273a/papers.spr07/09/celera.pdf
419. H. VESTERBERG & H. SCENSSON: Isoelectric fractionation, analysis, and characterization of ampholytes in natural pH-gradients. *Acta Chim. Scand.* **20** (1966) 820–834
420. T. WAGENKNECHT, R. GRASSUCCI *et al.*: Electron microscopy and computer image averaging of ice-embedded large ribosomal subunits from *Escherichia coli*. *J. Mol. Biol.* **1988** (199) 137–47. doi: 10.1016/0022-2836(88)90384-1
421. B. WALLACE & R. JANES: Synchrotron radiation circular dichroism spectroscopy of proteins secondary structure, fold recognition and structural genomics. *Curr. Opin. Chem. Biol.* **5** (2001) 567–571. doi: 10.1016/S1367-5931(00)00243-X
422. O. WARBURG & W. CHRISTIAN: Isolierung und kristallisation des gärungsferments enolase. *Biochem. Z.* **310** (1942) 384–421
423. V. WASINGER, S. CORDWELL *et al.*: Progress with gene-product mapping of the Mollicutes: *Mycoplasma genitalium*. *Electrophoresis* **16** (1995) 1090–1094. doi: 10.1002/elps.11501601185
424. K. WEBER & M. OSBORN: The reliability of molecular weight determinations by dodecyl sulfate-polyacrylamide gel electrophoresis. *J. Biol. Chem.* **244** (1969) 4406–4412. http://www.jbc.org/content/244/16/4406.full.pdf+html?sid=146d9763-7bb6-4c46-a775-d43c3ac79d26

425. X. Wei & L. Li: Comparative glycoproteomics: approaches and applications. *Brief. Funct. Genomic Proteomic* **8** (2009) 104–113. doi: 10.1093/bfgp/eln053
426. J. Weiland, J. Anderson *et al.*: Inexpensive chemifluorescent detection of antibody-alkaline phosphatase conjugates on western blots using 4-methylumbelliferyl phosphate. *Anal. Biochem.* **361** (2007) 140–142. doi: 10.1016/j.ab.2006.04.012
427. D. Wessel & U. Flügge: A method for the quantitative recovery of protein in the presence of detergents and lipids. *Anal. Biochem.* **138** (1984) 141–143. doi: 10.1016/0003-2697(84)90782-6
428. R. Westermeier: *Electrophoresis in practice.* 4 edn.. (Wiley-VCH, Weinheim), 2005. ISBN 3-527-31181-5
429. C. Whitehouse, R. Dreyer *et al.*: Electrospray ionization for mass-spectrometry of large biomolecules. *Science* **246** (1989)(4926) 64–71. http://www.uni-ulm.de/aok/bernhardt/Teaching/ClusterWS06-07/uebungen/Cluster06-07-Seminar02.pdf
430. D. Wilson & L. Konermann: A capillary mixer with adjustable reaction chamber volume for millisecond time-resolved studies by electrospray mass spectrometry. *Anal. Chem.* **75** (2003) 6408–6414. doi: 10.1021/ac0346757
431. D. Winzor: Determination of the net charge (valence) of a protein: A fundamental but elusive parameter. *Anal. Biochem.* **325** (2004) 1–20. doi: 10.1016/j.ab.2003.09.035
432. G. Wisdom: Conjugation of antibodies to horseradish peroxidase. *Methods Mol. Biol.* **295** (2005) 127–30. doi: 10.1385/1592598730
433. A. Wollrab: *Gaschromatographie.* (Diesterweg Moritz, Braunschweig), 1983. ISBN 978-342-505455-1
434. V. Woods, Jr. & Y. Hamuro: High resolution, high throughput amide deuterium exchange mass spectrometry (DXMS) determination of protein binding sites structure and dynamics. *J. Cell. Biochem.* **37** (2001) 89–98. doi: 10.1002/jcb.10069
435. F. World: Affinity labelling - an overview. *Methods Enzymol.* **46** (1977) 3–14. doi: 10.1016/S0076-6879(77)46005-1
436. K. Wuthrich: *NMR of proteins and nucleic acids.* 2nd edn.. (J. Wiley, New York), 1986. ISBN 978-0471828938
437. Q. Xue & E. Yeung: Differences in the chemical reactivity of individual molecules of an enzyme. *Nature* **373** (1995) 681–683. doi: 10.1038/373681a0
438. J. Yago, F. G. Sevilla *et al.*: A Windows program for the derivation of steady-state equations in enzyme systems. *Appl. Math. Comput.* **181** (2006) 837–852. doi: 10.1016/j.amc.2006.02.016
439. R. Yalow & S. Berson: Assay of plasma insulin in human subjects by immunological methods. *Nature* **184** (1959) 1648–1649. doi: 10.1038/1841648b0
440. J. Yankeelov, Jr., C. Mitchell *et al.*: A simple trimerization of 2,3-butanedione yielding a selective reagent for the modification of arginine in proteins. *J. Am. Chem. Soc.* **90** (1968) 1664–1666. doi: 10.1021/ja01008a056
441. K. Yasui, M. Uegaki *et al.*: Enhanced solubilization of membrane proteins by alkylamines and polyamines. *Protein Sci.* **19** (2010) 486–493. doi: 10.1002/pro.326
442. H. Yom & R. Bremel: Xerographic paper as a transfer medium for western blots: Quantification of bovine alpha s1-casein by western blot. *Anal. Biochem.* **200** (1992) 249–253. doi: 10.1016/0003-2697(92)90461-F
443. V. Yue & P. Schimmel: Direct and specific photochemical cross-linking of adenosine-5'-triphosphate to an aminoacyl-tRNA-synthase. *Biochemistry* **16** (1977) 4678–4684. doi: 10.1021/bi00640a023
444. A. Zamyatnin: Amino acid, peptide, and protein volume in solution. *Annu. Rev. Biophys. Bioeng.* **13** (1984) 145–165. doi: 10.1146/annurev.bb.13.060184.001045
445. A. Zarrine-Afsar & A. Davidson: The analysis of protein folding kinetic data produced in protein engineering experiments. *Methods* **34** (2004) 41–50. doi: 10.1016/j.ymeth.2004.03.013
446. J.-H. Zhang, T. Chung *et al.*: A simple statistical parameter for use in evaluation and validation of high throughput screening assays. *J. Biomol. Screen.* **4** (1999) 67–73. http://genecruiser.broadinstitute.org/bbbc/linked_files/Zhang_JBiomolScr_1999.pdf

447. W. ZHUANG, D. ABRAMAVICIUS *et al.*: Two-dimensional vibrational optical probes for peptide fast folding investigation. *Proc. Natl. Acad. Sci. USA* **103** (2006) 18 934–18 938. doi: 10.1073/pnas.0606912103
448. A. ZIEGLER & G. KÖHLER: Analytical isoelectric focusing in polymerizable thin layers containing sephadex. *FEBS Lett.* **64** (1976) 48–51. doi: 10.1016/0014-5793(76)80245-1
449. A. ZIEGLER & G. KÖHLER: Resolving power of isotachophoresis and isoelectric focussing for immunoglobulins. *FEBS Lett.* **71** (1976) 142–146. doi: 10.1016/0014-5793(76)80917-9
450. X. ZUO & D. SPEICHER: A method for global analysis of complex proteomes using sample prefractionation by solution isoelectrofocusing prior to two-dimensional electrophoresis. *Anal. Biochem.* **284** (2000) 266–278. doi: 10.1006/abio.2000.4714

Index